MYCORRHIZAS

A MOLECULAR ANALYSIS

MYCORRHIZAS

A MOLECULAR ANALYSIS

K.R. Krishna

Science Publishers, Inc.

Enfield (NH), USA Plymouth, UK

CIP data will be provided on request

SCIENCE PUBLISHERS, INC.
Post Office Box 699
Enfield, New Hampshire 03748
United States of America

Internet site: *http://www.scipub.net*

sales@scipub.net (marketing department)
editor@scipub.net (editorial department)
info@scipub.net (for all other enquiries)

ISBN 1-57808-362-1

Published by Science Publishers, Inc., NH, USA
Printed in India.

This book is dedicated to Global Agricultural Fraternity

Preface

Mycorrhiza is a worldwide symbiotic phenomenon involving most plant species and certain species of fungi. Mycorrhizas have great relevance to soil ecosystematic functions, especially nutrient dynamics, microbial processes, plant ecology and agriculture. A review of the early history of discoveries pertaining to mycorrhizas indicates that certain important techniques created spurts in research activity, resulting in rapid progress in our knowledge on this phenomenon. Microscopic and wet sieving techniques, classification charts, production of single spore isolates, axenic cultures, laboratory media, glass house, nursery and field techniques induced rapid progress in mycorrhizal research in the 1980s. Mycorrhizas have once again become a sought after biological phenomenon for researchers, attracting the attention a large number molecular biologists, as also agricultural and environmental specialists. During the past 8 years, the advent of molecular techniques has enthused rapid progress in cellular and molecular aspects of mycorrhizas. Considerably vast knowledge has been accrued in a short span of time. In fact, the present trend is to analyse mycorrhizas in greater detail, using advanced molecular tools and build up capability to manipulate the symbionts to our advantage.

Mycorrhizas: A Molecular Analysis belongs to current wave of interest in the molecular biology of mycorrhizas. It begins with a chapter on molecular evolution and phylogeny of mycorrhizas, which is in lieu of the usual introductory chapter. Lucid discussions on cellular physiology, molecular genetics, and molecular regulation of nutrient exchange phenomenon in mycorrhizas form the core of this book. A comparative analysis of the molecular aspects of symbiosis and pathogenesis has been presented in Chapter 5. It also includes certain agriculturally useful aspects of disease control via mycorrhizas. Discussions on recent developments in molecular ecology of mycorrhizas, including most recently enunciated concepts such as 'Nurse Functions', 'Mycoheterotrophy' are available in Chapter 6. Transformation, Transgenics and Genetic Engineering of Mycorrhizas is a unique and futuristic chapter. Applications of genetic engineering of mycorrhizas, as well as recently developed techniques of genetic transformation and production of viable transgenic mycorhizal fungi have been delineated in Chapter 7.

It is my fervent hope that this book will prove useful to researchers/students in the fields of microbiology, molecular biology, plant biology, agriculture and environment sciences.

Dated: 2005

Dr K.R. Krishna, Ph.D.

Acknowledgment

During the past as well as while preparing this volume on mycorrhiza, several researchers have offered their research material, publications and unpublished reports. Some have interacted continuously on various aspects of mycorrhizas. A few others have been a source of inspiration, either directly or indirectly. Following is a list of researchers and their institutions that I wish to thank.

Drs *Anne Ashford*, Department of Biological Sciences, University of New South Wales, Sydney, Australia; *Linnette Abbott*, Department of Soil Science, University of Western Australia, Nedlands, Australia; *Mark Brundrett* and others, Center for Mediterranean Agricultural Research, CSIRO, Wembley, Western Australia; Australia; *Peter Dart*, Department of Agriculture, University of Queensland, Brisbane, USA; and *Sally Smith*, Department of Soil Science, University of Adelaide, Adelaide, South Australia, USA.

Vivienne Gianinazzi-Pearson and *Silvio Gianinazzi*, INRA, Dijon Cedex, France; *Phillip Franken*, Max Plank Institute for Tropical Ecology and Laboratory for Microbiology, Phillips University, Marburg, 35043, Germany; *Manuella Giovanetti*, Departimento di Chimica e Biotechnologie Agrarie, Universita di Pisa, via del Borghetto 80, 56124 Pisa, Italy; and *Paola Bonfonte*, Centro di Studio sulla Micologia del Terreno (CNR), Dipartimento di Biologia Vegetale, Universita di Torina, Vialle Mattioli 25, Torino, Italy.

Viktor Tsyganov and *Igor Tikanovich*, Institute for Agricultural Microbiology, St Petersburg, Russia *Pal Axel Olsson*, Department of Microbial Ecology, Ecology Building, Lund University, SE-23362, Lund, Sweden; *Roger Finlay*, SVU, Swedish University of Agriculture, Uppsala, Sweden.

Dirk Redecker, *Verena Weimken, Thomas Boller*, Institute of Botany, University of Basel, Hebelstrasse 1, CH-4056, Switzerland.

Martin Parniske, Sainsbury, Norwich, United Kingdom; Martin Schultz, University of Yorkshire, Yorkshire, United Kingdom; *Peter Young*, Department of Biology, University of York, York, United Kingdom; *Mycorrhiza Specialists* from Department of Biological Sciences, University of Dundee, Dundee, Scotland and IACR, Rothamsted Experimental Station, Harpenden, England, United Kingdom.

David Andrews, Department of Agronomy, University of Nebraska, Lincoln, Nebraska, USA; *Robert Auge*, College of Agriculture, University of Tennessee at

Knoxville, Knoxville, USA; *John Bever*, Department of Botany, Duke University, Durham, North Carolina 27708, USA; *Tom Bruns*, Department of Plant Microbiology, University of California at Berkeley, Berkeley, USA; *Jim Graham*, Citrus Center, Lake Alfred, University of Florida, Florida, USA; *Mary Harrison*, Department of Plant Biology, Samuel Roberts Nobel Foundation, Ardmore, Oklahoma, USA; *Shawn Kaeppler*, Department of Crop Sciences, University of Wisconsin, Madison, Wisconsin, USA; *Roger Koide*, Department of Plant Pathology, University of Pennsylvania, Pennsylvania, USA; *Bradley Kropp*, Department of Ecology, Utah State University, Logan, Utah, USA; *David Sylvia*, Department of Soil Science, University of Florida, Gainesville, USA; *Ann Pringle*, Department of Plant Microbiology, University of California at Berkeley, Berkeley, California, USA. Special thanks to *Robert Auge* and staff maintaining International Mycorrhiza Directory and related websites—www.mycorrhizas.ag.utk.edu. Colleagues and friends at University of Agricultural Sciences, GKVK Kendra, Bangalore; G.B. Pant University of Agriculture and Technology, Nainital Dt, Uttaranchal; Indian Agricultural Research Institute, Pusa, New Delhi; Administration and Scientists at the International Crops Research Institute for the Semi-arid Tropics (ICRISAT), Hyderabad, India; Faculty and Staff at the Soil and Water Science Department, University of Florida, Gainesville, Florida, USA.

I wish to express appreciation to my wife *Dr Uma Krishna* and son *Sharath Kowligi.*

Cover Photo credit: ICRISAT Mycorrhiza Collection, International Crops Research Institute for the Semi-Arid Tropics, Patancheru, Andhra Pradesh, India.

Contents

1

EVOLUTION AND PHYLOGENY OF MYCORRHIZAS: CLASSICAL AND MOLECULAR APPROACHES

A background about evolutionary aspects of both the symbionts, mycorrhizal fungi ana plants, can provide a better perspective while discussing their phylogeny. The current trend is to assess evolutionary and phylogenic aspects of fungus and/or plant host using the molecular method, then, compare it with previous inferences made using morphological and functional traits. In tune with this strategy, this chapter firstly highlights the classical aspects followed by detailed discussions on molecular approaches employed to study the evolution and phylogeny of mycorrhizas.

1. A. Evolutionary Aspects of Mycorrhizal Symbionts

Fossil and molecular evidences suggest that Paleozoic fungi diverged from other living organisms in the Proterozoic period. They colonized the land surface long before plants. Fossil study indicates that Paleozoic fungi resembled oomycetes and zygomycetes (chytrids), but were not related to the mycorrhizal lineages (Taylor and Taylor, 1997). Parasitic associations with plants are easily discernible among the Paleozoic fungi. Evidences for saprophytism or lichen type associations during the Paleozoic period are rare. Molecular methods suggest that bryophyte-like land plants were associated with VAM-like endophytes, early during the Devonian period (400 million years ago). It is believed that limiting energy and nutrients drove the evolution of endophytes towards more balanced interdependent symbiotic associations involving intricate nutrient exchange patterns.

The present-day AM fungal genera *Glomus*, *Acaulospora*, *Entrophospora*, *Scutellospora*, *Gigaspora*, *Paraglomus* and *Archaeospora* are considered to be primitive because of their simple asexual spores and lack of sexual reproduction. Molecular diversity noticed within ribosomal gene is in consonance with the absence of a sexual cycle among these fungi (Pringle et al. 2000). Glomeromycota are considered to be one of the oldest true fungi. Based on the phylogenic assessment of DNA sequence data, Simon et al. (1993) opined that the age of Glomeromycota and land plants is similar. Other types of mycorrhizal fungi are evolutionarily younger to Glomales (Tehler, 2000). Fungi resembling the genera of Glomales were recorded on the roots of Triassic plants, thus confirming the existence of mycorrhizal association during that period. The molecular analysis of a small subunit (SSU) rDNA (18S) sequence data for *Geosiphon* indicates that it is a primitive glomeromycetous fungus. The *Geosiphon* associations involve swollen hyphae similar to arbuscules in AM symbiosis. Therefore, it is believed that characteristics such as arbuscules required for effective endomycorrhizal symbiosis might have evolved with Cyanobacteria during the early Devonian period. Brundrett (2002) suggests that Glomales descended from endophytes of algal precursors, mainly because their fossils do not show any close resemblance to parasitic fungi (e.g. Oomycetes and Chytrids) found on early plants. Glomales are a unique monophyletic mycorrhizal fungal lineage that has co-evolved with land plants, whereas, other types of mycorrhizal fungi are polyphyletic. These species represent either parallel or convergent evolution. Actually, two main branches of AM symbiotic fungi have been hypothesized that must have evolved from a common arbuscular ancestor. One branch consisted of *Gigaspora* and *Scutellospora*, both characterized by the extraradical auxiliary cells and spores formed within a thin wall of sporogenous cell. *Scutellospora* evolved from *Gigaspora* with the development of a membranous inner wall, separable wall groups and germination shield. The second branch consisted of *Glomus*, *Sclerocystis*, *Acaulospora* and *Entrophospora*. *Glomus* and *Sclerocystis* were closely related by a unique transformation series from spores formed radially in compact sporocarps. *Acaulospora* and *Entropshospora* were sister monophyletic groups with a common ancestor that possessed stalked spores. Species under these genera are distinguished by chlamydospores formed laterally from or within a hypha, terminating in sporiferous saccule.

Over 6000 species of ECM fungi belonging mainly to Basidiomycetes, some Ascomycetes and Zygomycetes have been recognized based on their morphological and/or molecular assessments (Molina et al. 1992). Many ECM fungi form epigeous fruiting bodies, but not all. A few of them produce hypogeous sequestrate (truffle-like) or resupinate (crusting) fruiting bodies; some others such as *Cenococcum geophilum* fruit infrequently, but propagate via sterile mycelia (Lobuglio et al. 1996; Shinohara et al. 1999). Phylogenetic analyses of DNA sequence data suggest that mushroom-forming agarics and others such as Boletales or Russulales are derived from wood-rotting fungi. The ability to form ECM is an important evolutionary trait among agarics. Several taxa of basidiomycetous ECM fungi such as Cortinariaceae, Boletaceae, Amanitaceae and Russulaceae probably occurred in the early Cretaceous period. Their evolutionary diversity was guided predominantly by host and habitat specificity. It is believed that ECM fungi evolved from saprophytic fungi, which is indicated by their ability to produce cell wall-digesting enzymes. The ECM fungi are polyphyletic have several saprophytic relatives. According to Brundrett (2002), ascomycetous ECM fungi are polyphyletic with separate origins. They possess considerable functional diversity. For example, a few ascomycetous ECM fungi utilize inorganic N, but most use organic N sources. The host range of ECM could be narrow or broad. Their preference to particular hosts is indicative of co-evolution with host tree species.

1. B. Evolutionary Aspects of Host Plants

Bryophytes were the earliest of land plants, possessing a horizontally spreading thallus, exhibiting separate saprophytic and gametophytic stages (see Table 1.1; Renzeglia et al. 2000). Mosses are common bryophytes, but they do not form mycorrhizal symbiosis. However, they do harbor endophytic hyphae. Liverworts and

Table 1.1 Evolutionary relationships between plant lineages and mycorrhizas derived based on fossil records and molecular evidences

Period	*Plant*	*Mycorrhizas*
Ordovician to Silurian (476-432 M yr)	The first Bryophyte-like land plants Liverworts and Hornworts	Limited fossils without roots, mycorrhizas unknown VAM-like with arbuscules in thallus
Silurian (415-425 M yr) to the present	Mosses (the largest living group of Bryophytes)	No roots, NM or endophytic Glomean fungi
Early Devonian (400 M yr)	Aglaophyton major: An early land plant of uncertain affinities	VAM-like arbuscules in specialized rhizome meristem
Devonian (395 M yr) to the present	Lycopods: *Lycopodium, Selaginella* etc.	VAM in saprophyte, underground gametophyte may have exploited association
Devonian (395 M yr) to the present	Spenophytes: *Equisetum*	*Equisetum* facultative VAM or NM
Mid-Devonian (385 M yr) to the present	Sphenophytes, Lycopods, Pteridophytes, Ferns	First plants with roots resembling those of modern VAM, some facultative with fine roots and hairs
Mid-Devonian to the present	Cycads	VAM
Triassic (215-235 M yr)	Cycad from Antarctica (*Antarcticycas* sp.)	Earliest fossil VAM association in roots
Permian (265 M yr) to the present	Ginkgoales: Ginko	Tree with VAM
Triassic (235 M yr) to the present	Southern Hemisphere Conifers: Araucariaceae, Podocarpaceae	Trees with VAM
Early Jurassic (190 M yr) to the present	Northern Hemisphere conifers (except Pinaceae): Cupraceae, Taxodiaceae, Taxales	Trees with VAM
Early Jurassic (190 M yr) to the present	Gnetales: *Ephedra, Gnetum, Welwitschia*	*Welwitschia*-VAM, *Gnetum*-ECM
Early Cretaceous ? (120 M yr) to the present	Conifers in Pinaceae: *Larix, Picea,Tsuga*	ECM trees with heterorhizic roots
Early Cretaceous (120 M yr) to the present	Angiosperms	Several plant species
Cretaceous (100 M yr) to the present	Fagales: Betulaceae, Casuarinaceae, Fagaceae, Juglandaceae, Myricaceae, Nothofagaceae	Single lineage of ECM trees or shrubs with heterorhizic roots (some VAM also)
Cretaceous (100 M yr) to the present	NM families: Protaceae, Cyperaceae, Restionaceae,	Oldest known fossils of plants likely to have NM roots
Late Cretaceous, Oligocene or Eocene to the present (–30 M yr)	Caesalpiniaceae, Fabaceae, Mimosaceae, Myrtaceae, Salicaceae, Tiliaceae	Several separate ECM lineages
Cretaceous (100 M yr) to the present	Orchidaeae	Several endophyte species
Late Cretaceous (80 M yr)	Ericalean plants	Oldest known fossils likely to have ericoid Mycorrhizas

Source: Brundrett et al., 1996

hornworts establish AM-like associations with glomeromycetous fungi. Arbuscules can be easily noticed in their thalli. Bryophytes, in fact, frequently support fine endophytes and other *Glomus* species (Schubler, 2000). These fine endophytes proliferate well in the narrow rhizoids and other free spaces within bryophyte. The liverwort rhizoids support the growth of ericoid mycorrhizas, but exact benefits from such an association is unclear.

Details on mycorrhizas within rhizomes of many primitive plants are available (see Taylor et al. 1995; Phipps and Taylor, 1996). Hyphae, arbuscules and vesicles similar to AM fungi have been detected in their rhizomes. As an example, an early Devonian plant *Aglaophyton major* contained arbuscules that were restricted to cortical zones of rhizomes. Spenophytes, Lycopodophytes and Pteredophytes were the earliest plants with the roots. They occurred in the Mid-Devonian period. *Lycopodium*, *Selaginella* and Isoetes are some of the surviving descendents. They exhibit a separate gametophyte phase without roots. However, the saprophyte with roots and leaves forms AM-like associations (Schmidt and Oberwinkler, 1993). *Lycopodium* species *harbor* AM-like fungi, which do not produce arbuscules; instead, a fine hyphal coil occurs with in host cells. Isoetes are associated with AM fungi commonly, even when growing under submerged conditions in water. Equisetum is a fern that does not support AM association in its gametophytic state, but its saprophyte phase is often mycorrhizal (Pryer et al. 2001; Brundrett, 2002).

Gymnosperms were prominent flora on the earth during Jurassic and Cretaceous periods, some examples being *Araucaria*, *Podocarpus* and *Phylloclades*. Gymnosperms other than members of Pinaceae form AM associations. Pinaceaea forms ectomycorrhizal associations. Gnetales are assemblages of gymnosperms. They are the only known non-Pinaceae to form ectomycorrhizal association (Table 1.1). Fossils and preserved imprints of roots indicate that true ECM occurred during middle Eocene period. Some lower Cretaceous plants are known to possess swollen lateral roots called 'mycorrhizal nodular roots'.

Ferns (Pteredophyta) formed the dominant flora during the Silurian to Paleozoic period. In fact, even now they flourish as a major component of certain ecosystems. Ferns have roots that associate with AM fungi. Fine roots and long root hairs perhaps support AM associations rather inconsistently. Facultative mycorrhizal associations are common among advanced ferns such as Filiales. On the other hand, certain primitive ferns (e.g. Ophioglomus) possess thick roots and are consistently mycorrhizal (Table 1.1). Brundrett (2002) opines that epiphytic and epilithic ferns are less likely to become mycorrhizal than terrestrial ferns that flourish in the soil.

Angiosperms became common on earth by the early-Cretaceous period (Taylor and Taylor, 1993; Brundrett, 2002; see Table 1.1). Amborellaceae, Austrobaileyceae, Nymphaceae, Iliaceae and Schisandraceae are some examples of primitive angiosperms. Trappe (1987) reported that 67% of over 6500 angiosperms examined form AM symbiosis and nearly 18% were non-mycorrhizal in natural conditions. Brundrett (2002) remarks that commonly, it is believed that 90% of land plants are mycorrhizal. However, actual proportions seem much lower, at 82 to 83%. The relative cover of mycorrhizal plants in an ecosystem may range from 100% (96% AM, 4% ECM and <1% NM) to around 52% in the Canadian Temperate Zone (35% AM, 17% ECM and 45% NM). With regard to angiosperm phylogeny, Soltis et al. (2000) state that multiple gene sequence data has allowed better assessment of mycorrhizal lineages. They have also provided better resolution of major and minor clades of angiosperms that consistently support mycorrhizal fungal colonization. Based on inferences from both morphological and molecular methods, Brundrett (2002) suggests that evolution of ECM coincided with the origin of Pinaceae and Fagaceae around the Cretaceous period. The ECM plants are frequently encountered in the rosid branch of endicots. These are woody and accustomed to cool climates. Dipterocarpaceae and Cistacaceae seem to share a common ECM-forming ancestor. Greater details on mycorrhizal lineages and the angiosperm family tree are available (see Soltis et al. 2000; Brundrett, 2002).

According to Brundrett (2002), evolutionary aspects of Ericales are complex, beginning from a VAM forming ancestor, progress to ECM then to arbutoid ECM, and finally into Ericoid mycorrhizas. Their evolution has also produced exploitative ECM or myco-heterotropic plants like *Monotropa*. Soltis et al. (2000) state that while dealing with the evolution of ericoid mycorrhizas, VAM should be deemed as the basal state of Ericales. Ericoid mycorrhizas are mainly seen on Ericaceae and Epacridaceae plants, which are at least 80 million-year-old in terms of evolution. Phylogenetically, plants forming arbutoid ECM are related to ericoid mycorrhizal lineages because of the monophyletic descent (Cullings, 1996). Proteraceae and Restionaceae that existed 100 million years ago are supposed to contain earliest of the non-mycorrhizal (NM) plant species. Among angiosperms, around 10 lineages may totally harbor NM species. Phylogenetically, certain orders and families are prone to contain NM species. These are mostly herbaceous, but shrubs and trees could also be non-mycorrhizal in nature. Families such as Amaranthaceae, Brassicaceae, Caryophyllaceae, Chenopodiaceae, Cyperaceae, Juncaceae and Polygonaceae are some examples of NM families. Many epiphytes are also known to be non-mycorrhizal. Frequently, plant species accustomed to harsh habitats such as extreme salinity, acidity, temperature, drought or aridity may show up non-mycorrhizal species.

Within the context of this chapter, above discussions on evolution of the host plant are pertinent. However, even more important is perhaps the knowledge of evolution of a prime organ of host plant—the roots. Roots must have provided the most crucial environment for co-evolution of symbiotic interaction through the ages. Brundrett (2002) states that mycorrhizal fungal evolution has been strongly influenced by both internal and external environment generated by the host roots. Root-like extracted fossils belonged to the lower Devonian era, but fossils of plants with true roots belong to the Mid-Devonian period (Taylor and Taylor, 1993; Gensel et al. 2001). During this period, diversification of plants was rapid. Roots evolved from subterranean rhizomes. Actually, structural and functional differentiation of underground rhizomes seems to have initiated the evolution of plant roots. The evolutionary series for roots, as suggested by Brundrett (2002), includes firstly a coarse dichotomously branched roots (e.g. *Selaginella*), progressing to roots with apical cell and organized branching (e.g. ferns), then continuing to gymnosperm roots with distinct cell layers, finally, culminating with the present-day angiosperm roots with highly organized cells, tissue differentiation and branching. Within the roots, cortex is a crucial portion vis-à-vis the co-evolution of symbionts. Cortex is the largest section of roots. During evolution, mycorrhizal interactions must have taken place actively and evolved within the cortical tissue. We may note that barring a few exceptions, physically, mycorrhizal fungi confine to cortical tissue during their interactions with the host plant. The cortex colonized with mycorrhizas has highly vacuolated cells, but the volume of cytoplasm increases if they are mycorrhizal. Obviously, a good deal of selection forces affecting evolution and phylogenetic differentiation of both the symbionts must have operated within the cortical area of roots.

2. Classical Taxonomy of Mycorrhizal Fungi: A Background

2. A. Arbuscular Mycorrhizas—Classical Approaches

The emphasis within this chapter is on molecular phylogeny, which is comparatively a recent trend among mycorrhiza researchers. Chronologically, the classical approaches precede the use of molecular phylogenetic methods. In accordance with this fact, a summary of historical development, and a highly abridged version of classical taxonomy have been provided. The classical taxonomy of arbuscular mycorrhizas began with Thaxter's (1922) revision of fungi belonging to Endogoneceae. At that time, involvement of many species belonging to this group of fungi in symbiosis was yet unclear. Around the same period, Peyronel (1923, 1924) grouped these fungi under the order Endogonales. Detailed investigation and characterization of individual AM fungal species became a trifle easier after Gerdemann and Nicholson (1963) standardized the wet sieving and decanting technique. This technique became frequent with mycorrhizasts, and lead to the realization that soil-borne AM fungi are worldwide in distribution and common with most plant species. Gerdemann and Trappe (1974) then delineated Endogonales into 7 genera and described in all 31 zygomycetous mycorrhizal species in their monograph. Trappe (1982) states that a spurt in the interest to classify and study AM fungi led to the identification of additional 70 plus AM species in 1970s and 1980s. Later, Berch and Trappe (1987), in their revision on the classification of AM fungi, stated that 107 species existed. In fact, considering the zeal and rapidity of reports on new AM fungal species (e.g. Schenk and Smith, 1982; Walker and Trappe, 1980; Walker et al. 1984; Smith and Schenk, 1985; Berch and Trappe, 1985; Koske, 1985; Warcup, 1985), it was forecasted that the list may reach over 200 species by 1990 (Trappe, 1982; 1987). However, at present, slightly over 180 AM fungal species have been identified and described (Walker and Trappe, 1993; Morton et al. 1993, 1995; Morton and Bentivenga, 1994; Trappe and Schenck, 1982, see Table 1.2).

The classical taxonomy of Glomales depends predominantly on the morphology of soil-borne resting spores. Traits related to AM fungal developmental physiology have also been utilized, but sparingly so during taxonomic confirmation (Morton, 1988; 1993; Walker, 1992). As yet, production of pure in vitro culture of AM fungi is not possible. However, accurate taxonomic identification and description of AM fungi, which at the least requires pure cultures maintained on a host, is possible. Using such cultures, relevant developmental stages and spore morphology can be accurately deciphered. Morton (1993) cautions that spores collected from field soils may obscure proper taxonomic analysis. Unknown age of spore samples, environmental influences, microbial activity, and insufficient information about developmental aspects introduces variability to diagnostic features. Field samples are generally compared with original descriptions (Schenk and Perez, 1990). As a caution, we should note that changes or incompleteness of any previous description might affect the taxonomic positions of the collected samples (Morton, 1993). A complete list of Glomelean fungi is provided in Table 1.2. Now, let us consider the important characteristics of glomelean fungal spores, which are utilized frequently by taxonomists.

Table 1.2. The alphabetical list of Arbuscular Mycorrhizal Fungal species

Family: GLOMACEAE
Genus: ***Glomus***
Species: *aggregatum; albidum; ambisporum; antarcticum; arborens; arenarium; atrouva; aurim; australe; avelingiae; boreale; botryoides; brohulti; caesarius; caledonium; callosum; canadense; canum; cerebriforme; chimenobombusae; citricola; claroidium; clarum; clavisporum; constrictum; convolutum; coremioides; coronatum; corymbiforme; cuneatum; delhiens; deserticola; diaphanum; dimorphicum; dolichosporum; dominikii; eburneum; epigaeum; etunicatum; fasciculatum; fecundisporum; flavisporum; formosanum; fragile; fragilistratum; fuegianum; fulvum; geosporum; gerdemanii; gibbosum; globiferum; glomerulatum; halonatum; heterosporum; hoi; infrequens; intraradices; invermaium; kerguelense; laccatum; lacteum; lamellosum; leptotichum; liquidambaris; luteum; macrocarpum; maculosum; magnicaule; manihotis; melanosporum; merredum; microaggregatum; microcarpum; minutum; monosporum; mortonii; mosseae; multicaule; multiformum; multisubstensum; nanolumen; occultum; pallidum; pansihalos; pellucidum; przelewicense; proliferum; pubescens; pulvinatum; pustulatum; radiatum; reticulatum; rubiforme; scintillans; segmentatum; sinuosum; spinosum; spinuliferum; spurcum; sterilum; taiwanese; tenebrosum; tenerum; tenue; tortuosum; trimurales; tubiforme; versiforme; verruculosum; viscosum; vesiculiferum; warcupii*

Genus: ***Sclerocystis***
Species: *clavispora; coccogena; coremeoides; dussii; indica; microcarpus; pachycaulis; pakistanica; rubiformis; sinuosa*

Family: PARAGLOMACEAE
Genus: ***Paraglomus***
Species: *brasillianum; occultum*

Family: ACAULOSPORACEA
Genus: ***Acaulospora***
Species: *appendiculata; bireticulata; capsicula; cavernata; colossica; delicata; denticulata; dilatata; elegans; excavata; foveata; gedanensis; gerdemanii; koskei; lacunosa; laevis; longula; mellae; morrowiae; myriocarpa; nicolsonii; paulinae; polonica; rehmii; rugosa; scrobiculata; spinosa; splendida; sporocarpia; taiwania; thomii; trappei; tuberculata; undulata; wakeri*

Genus: ***Entrophospora***
Species: *baltica; colombiana; infrequens; kentinensis; schenckii*

Family: ARCHAEOSPORACEAE
Genus: ***Archaeopsora***
Species: *gerdemanii; leptoticha; trapei*

Family: GEOSIPHONACEAE
Genus ***Geosiphon***
Species: *pyriformis*

Genus: ***Sclerogone***

Family: GIGASPORACEA
Genus: ***Gigaspora***
Species: *albida; alboaureantiaca; aurigloba; calospora; candida; corolloidea; decipiens; fulgida; giganteaa; gilmorei; lazzarii; margarita; nigra; ramisporophora; rosea; verrucosa*

Genus: ***Scutellospora***
Species: *alborosea; aurenicola; armeniaca; aurigloba; bicolor; biornata; calospora; castanae; cerredensis; corolloidea; crenulata; dipurpurescens; erythropa; fulgida; gilmorei; gregaria; hawaiinsis; heterogama (dipapillosa); minuta; nigra; nodusa; pellucida; persica; projecturata; reticulata; rubra; rubiformis; savannicola; scutata; spinosissima; tricalypta; trirubiginopa; verrucosa; weresubiae*

(*Contd.*)

(*Contd.*)

Family: ENDOGONECEA
Genus: ***Endogone***
Species: *abriolata; acrogena; alba; colombiana; contigua; flamicorana; incrassata; pisiformis; organensis; straton*

Source: List predominately derived from www.tu-darmstadt.de/fb/bio/bot/schussler/amphylo/amphylo_species.html; INVAM websites; Gerdemann and Trappe, 1974; Morton et al. 1993. Note: List includes Endogoneceae. Some members of this family may not be mycorrhizal in nature.

Spore Development: The major genera of glomelean fungi are distinguished on the basis of characteristic spore development patterns. Spores of the genus *Glomus* are normally formed on narrow or flaring hyphae, whereas, *Scutellospora* and *Gigaspora* produce spores on bulbous subtending hyphae. Spores of *Acaulospora* and *Entrophospora* become sessile due to detachment from the sporiferous saccule (Fig. 1.1).

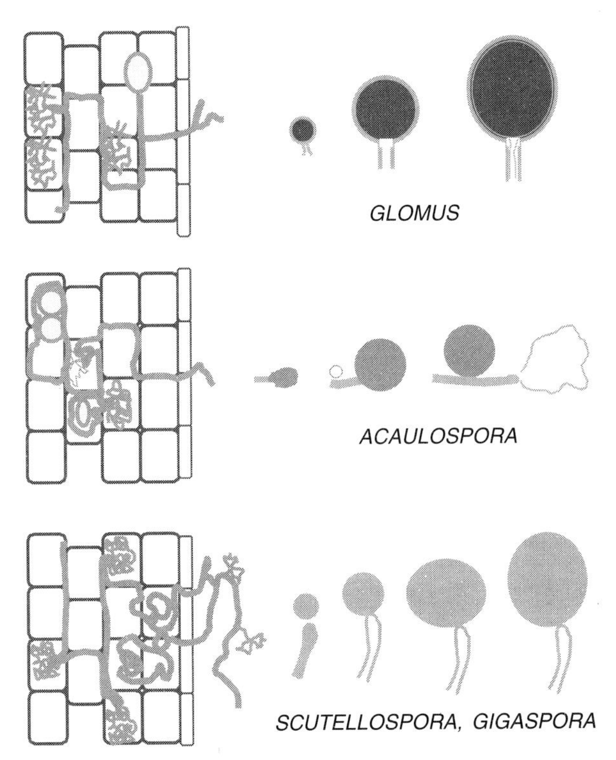

Fig. 1.1 Differences in the development of eq. infection inside host roots and maturation of spores in the three major genera of AM fungi. Note: The genus *Glomus* is easily identifiable by typical profusely branched arbuscules and oval vesicles. Spore maturation is relatively simple, leading to a spore with multi-layered wall structures. The Genus *Acaulospora* produces a sporiferrous saccule that collapses after transferring its contents into developing spore. The genera *Scutellospora* and *Gigaspora* produce spores with a bulbous subtending hyphae. Mature spores contain germinating shield and multi-layered wall. The Extramatrical vesicles are easily identifiable.

Sporocarp: Glomelean fungi produce spores either singly or in aggregates, which are called sporocarps. Brundrett et al. (1996) point out that sporocarps of glomelean fungi are typically simpler in structure compared with those found among mushrooms and truffles that belong to Ascomycetes or Basidiomycetes. Glomelean sporocarps usually contain soil materials, along with hyphae, which is enclosed in a peridium. These sporocarps could be small (<0.5 mm in dia.) or larger. In addition to size, spore surface, its color and nature of peridium are also important. Peridium may be absent or found on a loose tangle of hyphae, or on tightly woven cutis, or even on the pubescence of erect hyphal tips. The arrangement of spores within the sporocarpic structure is also important. For example, the genus *Sclerocystis* is separated from *Glomus* based on the radial arrangement of spores. With regard to tendency of AM fungi to form sporocarps, Brundrett et al. (1996) believe that sporocarp formation requires sufficient energy, which needs to be accumulated by the fungus through several generations of growth. Consequently, AM fungi deemed to produce only single spores may produce sporocarps, if the reserve energy exists within the mature pot-grown cultures.

Spore Dimensions: Spore size may vary widely even within an isolate, and is comparatively a less useful trait for taxonomic purposes. However, distinctions between AM fungal species are possible when differences in spore size are significantly large between two species. Spore length is usually measured along the axis of the point of attachment, and width measured at right angles to it. Glomelean spores range from 200 to 1000 μm in size. Fine endophytes produce spores that are smaller than 5 μm.

Spore Shapes: Spherical (globose) spores are most frequent among glomelean fungi, but some species produce oval or oblong ones. The subtending hypha (attachment) can be cylindrical, conical or swollen.

Spore Color: Spore color is frequently used to characterize the AM fungi, but it is a trait variable, both within and between the isolates of glomelean fungi. Spore colors are usually matched using a color chart. For example, Morton (1992) defines colors as CYM percentages (cyan, yellow and magenta). Color matching is easily accomplished using a Pantone process color system guide or CYMK system (Brundrett et al. 1996).

Spore Ornamentation: Papillae, reticulations, spines, warts, pits and wrinkles are some of the frequently encountered ornamentation. The combinations of such ornamentation, their pattern, height and size are useful during taxonomic diagnosis of AM fungi. Such ornamentation are common within the genera *Scutellospora* and *Acaulospora* (see Brundrett et al. 1996). As a cautionary measure, we should note that physico-chemical weathering and biological factors could obscure the proper diagnosis of ornamentation.

Spore Walls and Murography: Spore walls are an important aspect of AM fungal taxonomy. Normally, spores begin with thin walls, but as they mature, the walls thicken (1-20 μm). Glomelean spores may have one or more layers of walls. Spore walls vary in thickness, structure, appearance and staining reactions. Details of spore walls are generally described using murographs (Walker, 1983; Morton, 1988). Murographs are simpler for the genus *Gigaspora,* whereas, spores of *Glomus* have several wall layers. Outer walls of young spores are delicate, and could be lost as spore ages (Brundrett et al. 1996). Typically, complex wall structure consisting of a thick outer wall and layers of thin walls are common to genera such as *Acaulospora*, *Entrophospora* and *Scutellospora*. Reaction of spore walls to Melzer's stain is a useful trait. Typically, one or more wall layers may turn pink or purple with Melzer's reagent, but it is dependent on spore age, preservatives used and storage conditions. Comparison of developmental characters such as spore wall may also provide a basis for diversity. For example, study of spore development seems to confirm the rationale about placement of *Gigaspora* and *Scutellospora* as sister groups.

Spore Contents: Spore contents and descriptions about their nature can be useful during taxonomic diagnosis. Most spores may contain colorless inclusions, but some species such as *Glomus convulutum* may show up yellow globules. Arrangement of lipid globules that vary in color and size can be helpful during identification.

Mode of Spore Germination: Glomelean fungi can also be distinguished by studying the differences in spore germination mechanisms. For example, *Scutellospora* spores germinate through hyphae that emanate from compartment within germination shields (Brundrett et al. 1996). In case of many other species, germ tubes arise from the point of spore attachment.

Extramatrical Soil Hyphae: Mycelia of glomelean fungi vary widely in case of characteristics such as wall thickness, staining, associated structures, etc. Hyphae of some species are hyaline or lightly pigmented. Isolates of *Scutellospora* may have melanized hyphae that remain brown even after clearing. Hyphae could be coarse 'runner' like or finely branched. Coarse hyphae of *Glomus* may be as thick as 20 μm, whereas, fine hyphae of *Acaulospora* reach just 5 μm thickness, and those of fine endophytes only 2 μm. The hyphae of *Scutellospora* and *Gigaspora*, which are localized in the soil, may produce auxiliary vesicles, that are useful during identification (Fig. 1.1). With regard to comparative anatomy of AM fungal hyphae, Dodd et al. (2000) remind us that lack of deeper insights into the genetics, nuclear cycle, plus the inability to culture AM fungus in vitro has hampered progress towards using hyphal characters during taxonomy. Nevertheless, AM fungi are relatively easy to identify at the genus level using hyphal morphology, especially based on structure, branching, staining pattern with Trypan Blue, Chlorazol Black E or Acid Fuschin. The ink vinegar stain and autoflourascence have also been utilized during taxonomy. Dodd et al. (2000) summarize the hyphal morphology as follows:

Glomus: Straight hyphae that ramify throughout the root cortex. Hyphae stain intensely. Vesicles are oval to elliptical when formed intracellularly. Fine endophytes (e.g. *G. tenue*) have uniquely thin hyphae (< 1.0 μm). They may produce odd fan-like structures or small hyphal swellings as the hyphae ramify in the root cortex.

Acaulospora: Irregular, H-branched hyphae spread into root cortex. Looped or coiled hyphae are frequently encountered. Hyphae are thin walled, contain lipid droplets and stain at variable intensities. Vesicles occur both inter- and intra-cellularly, which may be either obovoid or lobed.

Gigaspora/Scutellospora: They produce thick looping hyphae. Arbuscules possess swollen trunks. Intra-radical hyphae have small projections that stain deeply. Ornamental auxiliary cells are formed on extra radical mycelium. These are pale yellow or brown in *Scutellospora*, but hyaline to pale colored in *Gigaspora*.

Root Colonization: The patterns of root colonization differ widely between glomelean fungi. It can be used as distinguishing characteristics (Brundrett et al. 1996). Hart and Reader (2002) suspect that some gross variations in colonization strategies of the AM fungus could be deciphered right at the taxonomic level, instead of a recourse to genetic analysis at the isolate level. Accordingly, they tested 21 AM fungal isolates belonging to the three major families, i.e. Glomaceae, Acaulosporaceae and Gigasporaceae on different plant hosts. They considered parameters such as root colonization, root fungal biomass, hyphal length, hyphal length in soil and fungal biomass over a period of 12 weeks. Generally, isolates of *Glomus* consistently colonized before those of *Acaulospora* and *Gigaspora*. Fastest colonizers were also most extensive in both root and soil. However, most *Glomus* species were rapid to spread in roots but slow to spread in the soil, and the exactly opposite trend occurred with *Gigaspora* isolates.

2. B. Ectomycorrhizas—Classical Approaches

The ectomycorrhizas (ECM) are formed mainly by Basidiomycetes and Ascomycetes. A few Zygomycetes and Hyphomycetes also form ectomycorrhizas (Table 1.3). Molina et al. (1992) estimated that nearly 5 to 6 thousand fungal species form ectomycorrhizal associations. The classification schemes for ECM fungi may vary. Broadly, all ECM fungi belong to the kingdom Eumycota (fungi), which encompasses Zygomycetes, Ascomycetes and Basidiomycetes. Marguilis and Schwartz (1988) bifurcated the fungal kingdom into phylum Zygomycotina, Ascomycotina and Basdiomycotina, whereas, Hawksworth (1995) classified fungi into divisions Basidiomycota, Ascomycota and Zygomycota. Kendrick (1992) has classified fungi into divisions Zygomycota and Dikaryomycota. The Dikaryomycota comprises subdivisions such as Ascomycotina and Basidiomycotina.

Ascomycetes: Ascomycetes that form ectomycorrhizal association with higher plants belong mainly to Discomycetes. They are mostly grouped under the orders Pezizales, Elaphomycetales and Leotiales (Trappe, 1979; Maia et al. 1996). Pezizales are saprophytic on wood, soil and dung, and they survive both above and below the ground. In nature, members of Pezizales which form conspicuous fruiting bodies known as cups, morels, truffles are both mycorrhizal and saprophytic, e.g. Peziza, Morchella, Tuber, Labyrinthomyces.

Basidiomycetes: Most of the ectomycorrhizal fungi described so far belong to Basidiomycetes. At the same time, not all Basidiomycetes are mycorrhizal. This fungal class includes numerous obligate pathogens. The mycorrhizal species are mainly grouped under Gastromycetes and Hymenomycetes (e.g. Aphyllophorales, Agaricales; see Table 1.4).

Macroscopic morphology of Ectomycorrhizal fungi and taxonomy: Information on macroscopic morphology of ECM is useful during accurate identification and taxonomy. Generally, base and flesh are studied in order to classify epigeous fungi producing mushroom-like and truffle-like fruit bodies, descriptions on cap, hymenium, stipe (stem), partial and universal veil (Burndrett et al. 1996). For hypogeous fungi, traits such as peridium, gleba, locules and columella are analyzed. As a precaution, such macroscopic characteristics are to be recorded on young, expanding and mature fungal specimens because such morphological features change with age. Table 1.5 provides a summary of the macro-morphological details recorded routinely on certain ECM.

Table 1.3 Ectomycorrhiza and Ectendomycorrhiza recorded from different continents

Family	*Genera*	*Remarks*
HYPHOMYCETES	*Cenococcum,*	Conidial
ZYGOMYCETES	*Endogone, Sclerogone*	H
ASCOMYCETES		
Ascobolaceae	*Sphaerosoma*	H
Balsamiaceae	*Balsamia, Picoa*	H
Elaphomycetacea	*Elaphomyces*	H
Geneaceae	*Genea, Geneaba*	H
Geoglossaceae	*Geoglossum, Leotia, Trichoglossum*	E
Helvallaceae	*Gyromitra, Helvalla,*	E
	Barrsia, Fischerulla, Gymnohydnotrya, Hydnotrya, Mycoclelandia,	H
Pezizaceae	*Aleuria, Peziza, Phillipsia, Pulvinia,*	E
	Amylascus, Hydnotryopsis, Muciturbo, Pachyphloeus, Peziza, Ruhlandiella	H
Pyronemataceae	*Geopora, Humaria, Jafnedelphus, Lamprospora, Sphaerosporella, Trichophaeae, Wilcoxina*	E
	Choiromyces, Dingleya, Elderia, Geospora, Hydnobolites, Hydnocystis, Labrynthosmyces, Paurocystis, Reddellomyces, Sphearozine, Stephensia	H
Sarcoscyphaceae	*Plectania, Pseudoplectania, Sarcocypha*	E
Terfaziaceae	*Choiromyces, Terfezia*	H
Tuberaceae	*Mukagomyces, Paradoxa, Tuber*	
BASIDIOMYCETES		
Amanitaceae	*Amanita, Limacella*	E
	Torendia	H
Astraeaceae	*Astraeus*	E
	Pyrenogaster, Radiigera	H
Boletaceae	*Astraboletus, Boletellus, Boletochaeta, Boletus, Buchwaldoboletus*	E
	Chalciporus, Fistulinella, Gyrodon, Gyroporus, Heimella, Leccinum, Phlebopus, Phylloporus, Pulveroboletus, Rubinoboletus, Suillus, Tylopilus, Xanthoconium, Alpova, Boughera, Chamonixia, Gastroboletus, Rhizopogon, Royoungia, Truncocolumella	H
Cantherellaceae	*Cantherellus, Craterellus*	E
Chondrogastraceae	*Chondrogaster*	H
Clavaraiaceae	*Aphelaria, Clavaria, Claviriadelphus, Clavicorana, Clavulina, Clavulinopsis Ramaria, Ramariosis*	E
Corticaceae	*Amphinema, Byssocoticium, Byssosporia, Piloderma*	E
Cortinariaceae	*Astrosporina, Cortinarius, Cuphocybe, Dermocybe, Descolae, Hebeloma, Inocybe, Leucocortinarius, Rozites, Stephanopus, Cortinarius, Cortinomyces, Descomyces, Destuntzia, Hymenogaster, Quadrispora, Setchelliogaster, Thaxterogaster, Timgrovea*	H
Cribbiaceae	*Cribbia, Mycolevis*	H
Entomolataceae	*Clitopilus, Entoloma, Leptonia, Rhodocybe*	E
	Rhodogaster, Richoniella	H
Elasmomycetaceae	*Elasmomyces, Gymnomyces, Mortellia, Zelleromyces*	H
Gelopellidaceae	*Gellopellis*	H
Gomphaceae	*Gomphus*	E
Gomphidiaceae	*Chroogomphus, Cystogomphus, Gomphidius*	E

(*Contd.*)

(*Contd.*)

Hydnaceae	*Bankera, Dentinum, Hydnellum, Hydnum, Phellodon*	E
Hygrophoraceae	*Beratrandia, Camarophyllus, Gliophorus, Humidicutus, Hygrocybe, Hygrophorus*	E
Hysterangiaceae	*Hystrangium, Pseudohysterangium, Trappea*	H
Leucogastraceae	*Leucogaster, Leucophelps*	H
Lycoperdaceae	*Lycoperdon*	P
Melanogasteraceae	*Melanogaster*	H
Mesophelliaceae	*Castoreum, Diplodera, Gummiglobus, Malajczukia, Mesophellia*	H
Octavianinaceae	*Octavianina, Sclerogaster*	E
Paxillaceae	*Paxillus*	P
Pisolithaceae	*Pisolithus*	E
Polyporaceae	*Albatrellus*	E
Russulacea	*Lacterius, Russula, Archangiellela, Cystangium, Macowanities*	H
Sclerodermataceae	*Scleroderma, Horakiella*	P
Sedeculeceae	*Sedecula*	H
Stephanosporaceae	*Stephanospora*	H
Strobilomycetaceae	*Strobilomyces, Austrogautieria, Chmonixia, Gautieria, Wakefieldia*	H,E
Thelophoraceae	*Clitocybe, Cystoderma, Cantherellula, Catathelesma, Laccaria, Lepista, Leucopaxillus, Tricholoma, Tricholomopsis, Gigasperma, Hydnangium, Podohydnangium*	H,E

Source: Mainly from Brundrett et al. 1996. Other sources are Miller, 1982; Kendrik, 1992, Molina et al. 1992; and Bougher, 1995

Table 1.4 Broad classification of mycorrhizal and non-mycorrhizal basidomycetous species

Class Teliomycetes: These fungi do not form large fruit bodies. Most fungi grouped as rusts (Uridinales) and smuts (Ustilaginales) are obligate pathogens in nature.

Class Phragmomycetes: These are identified by basidia devided by septa, and are grouped under the order Tremallales, Auriculariales, Tulasnellales and Septobasidiales. All are non-mycorrhizal.

Class: Holobasidiomycetes: Majority of fungi with the ability to form ECM symbiosis are grouped under the two informal groupings; namely, Gastromycetes and Hymenomycetes.

Gastromycetes: Gastromycetes is an artificial series of orders of Basidiomycetes. They are characterized by sequestered basidiomycetes (no enclosed spore), which are exposed when mature, mainly as spore mass in case of puffballs (e.g. Truffes and Puffballs). These spores are not discharged forcibly, but the dispersal is effected via wind, rain, animals, etc. Gastromycetes may be saprophytic or mycorrhiza. Ectomycorrhizal taxa are abundant such as Mesophelliaceae—Costoreum and Mesophellia, and the Sclerodermatales—Pisolithus and Sclerodermium.

Hymenomycetes: This group includes several series wherein spores are exposed before maturity and discharged forcibly from the fruit body (mushrooms, toadstools, brackets, corals and clubs). Hymenomycetes are divided into orders namely,

Exobasidiales: Parasitic in plants and non-mycorrhizal.

Dacrymycetales: Jelly fungi with tuning fork-like basidia, saprophytic on wood, non-mycorrhizal.

Aphyllophorales: These fungi are called club-like, bracket-shaped, sometimes mushroom-like, they are tough and woody in consistency. Several fungal species grouped under Aphyllophorales possess mycorrhizal ability, while a few others are non-mycorrhizal.

Agaricales: These are mushroom fungi, their above-ground fruit bodies are fleshy and borne on a stalk, whereas truffles produce underground fruit bodies. Again, some species of Agraicales are mycorrhizal, but several others do not form mycorrhizal association, a few others are parasitic on plants. Sometimes mushroom fungi are classified based on color of spore prints (e.g. white—Amanita, Liimacella, Russulaceae: brown—Cortiniracea, Paxillacea; black—Gomphidaceae, pink, etc.

Source: Kendrick, 1992; Brundrett et al. 1996

Table 1.5 Macroscopic traits of fruit bodies of mushrooms and truffles helpful during taxonomic analysis

A. Mushrooms
Cap (pileus)—size, shape, color, surface texture and moisture
Gills (lamellae) —color, attachment, lamellules
Stem (stipes)—Partial veil—presence or absence, form
Universal veil—presence or absence, form
Flesh—color, texture, bruising
Chemical reactions—15% KOH, 10% $FeSO_4$, etc.
B. Truffles
Fruit body—size, color, surface texture and moisture
Skin (peridium)—thickness, layers, texture, bruising
Internal fertile tissue (gleba)—color, texture, size, and orientation of chambers
Sterile tissue (columella)—presence or absence, form
Sterile base—presence or absence, form
Chemical reactions—15% KOH, 10% $FeSO_4$,
Odor and taste

Source: Brundrett et al. 1996

Mushroom Caps: Pileus or caps constitute the prominent and easily studied macro-morphological traits. Their size is ascertained by measuring the width and height (at center) of the smallest and largest of the mature fruiting bodies collected. The margins of the cap vary. Margin is assessed by way of the cross section when cut into half. Edges and the surface view are also noted (Fig. 1.2). Mushroom caps could be moist or dry, sometimes gelatinous, slimy or sticky and viscid. Mushroom caps may change color due to desiccation, which may be either a rapid or slow process in nature. Caps are translucent when moist and healthy, but turn opaque with drying. The texture of mushroom caps varies widely. Firstly, texture caused by superficial structures that are easily removed, and those forming an inherent part of the cap need to be ascertained. Actually, a large number of descriptive terms are utilized to describe the cap texture. They are as follows:

Pits could be scrobiculate (shallow), aveolate (deep) or lacunose (surrounded with ridges). *Splits* could be rimose (radial splits), or areolate (split into polygonal or irregular-shaped blocks). *Wrinkles* may be rivulose (meandering or wavy), rugulose (undulating), or rugose (coarse wrinkles). The *surface powder, hairs and scales* occurring on caps are generally classified into glabrous bald, waxy surface, hispid (covered with stiff hairs), or scabrous (surface roughened due to scales). *Orientation of particles or scales* on caps may be oppressed, recurved or erect (perpendicular to the cap surface). *Fibrils and surface* could be matted (interwoven fibrils), downy (soft downy surface), floccose (coarse cotton-like), or fibrillose (with fibrils). The fibrillar hairs could also be pubescent (very short), velutinous (short, fine, flexible hairs), hirsute (abundant stiff, inflexible hairs), villose (long fine hairs). In many cases, the fibrils combine to form scales. In that case, they could be squamose (scaly surface), squamulose (small scales), squamose (erect scales), punctate (dotted with minute scales), scabrous (roughened surface). *Cap surface and margins* could be smooth and entire, wavy, crenate, appendiculate, eroded, rimose, striate, plicate or zoned concentrically (Fig. 1.2).

Mushroom Gills: Mushroom gills vary with respect to following aspects. Attachment to stipes could be deccurent (extending down the stem), adnate (broadly attached),

Fig. 1.2 Types of Mushroom Caps. Note: Their surface characteristics and margins are frequently utilized during classification of mushrooms.

free (not attached), sinuate (notched near the attachment), receding (attached initially but free later) (Fig. 1.3). Gill spacing could be crowded or distant. Thickness may range from thin to broad or ventricose (swollen in the middle). Margins of gills could be smooth, wavy, serrate, crenate, eroded, fibriate or glistening. Branching and anastomoses of gills could vary from being unbranched, dichotomously branched (bifurcate) or interconnected with anastomoses. Lamellulae, which are short gills, may be absent or abundant. The number of gills is related to spacing and net size of the cap. Usually, they are recorded as the number of gills (L) and lamullelae (l) per half cap and depicted, for example, as 'L = 12-18; l = 32-58'.

Mushroom Stem (Stipes): The mushroom stipes varies with reference to their shape and tissue composition. Together with other traits, it is useful in judging the gross taxonomic position of an ECM fungal sample. Size of the stem is generally measured at the base, middle and apex of the smallest and largest fruit bodies from the collection. The shape in cross section could be circular or compressed, and in longitudinal section, it could be central or lateral. Texture of mushroom stipes range from reticulate (interlocking pattern of ridges or wrinkles), glandular (colored dots or small scales), striated, ridged or veined. Consistency of stem could be robust, fragile, fibrous, woody, chalky, corky and cartilaginous. Stem base could be tapering, clavate, ventricose, and equal. There are also annulus and volva types of mushroom stems, which possess volva that could be powdery, friable, saccate or marginate.

Mushroom Veil: The young mushroom-like fruiting bodies are completely enclosed in an universal veil. This veil ruptures as the mushroom expands, and in mature fruiting bodies, it may be absent or may remain as scales on pilea or volva. Universal

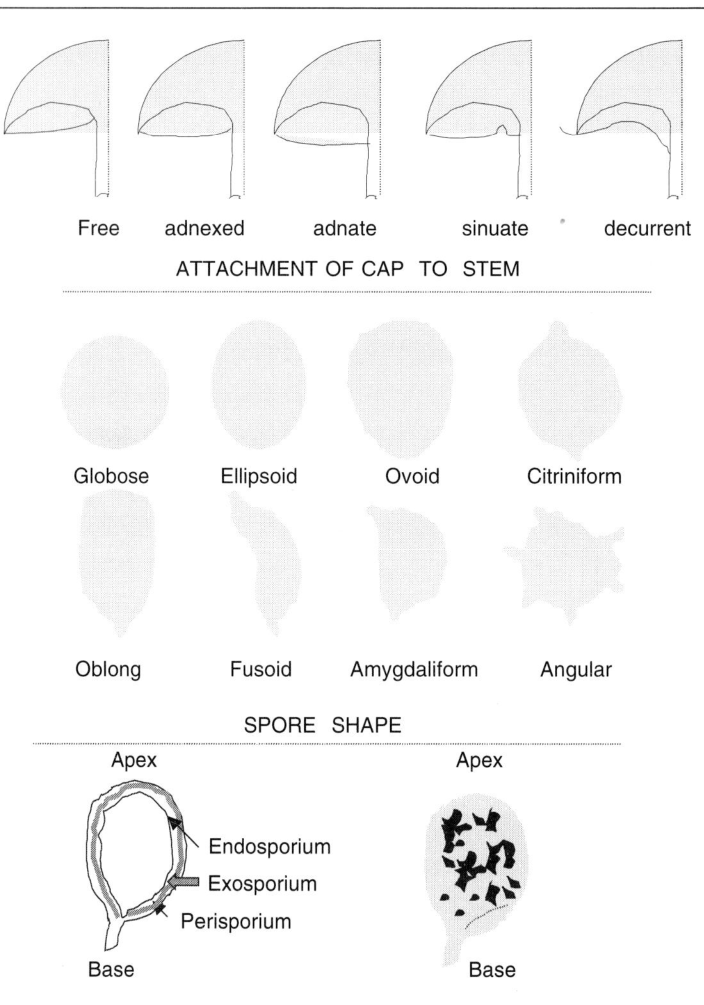

Fig. 1.3 Attachment of mushroom gills, spore shapes and structure encountered during morphotyping of ECM fungi—some examples (*Source:* Redrawn based on Brundrett et al., 1996).

veils also occur on stipes. The volva is actually derived from the veil, and it occurs at the base of the stipe. Volva could be a separate tissue or present only as a rim attached to the bulbous base of the stem. Major types of volva encountered are saccate, membranous, powdery and friable. Sometimes, a partial veil extends from cap margins to the stipe covering the gills. Gills get exposed later on as the mushroom grows and expands.

Chemical Tests: Chemical tests are prepared by placing a drop of test solution directly on any portion of the fruit body, mainly on the cap, stipe, hymenium or flesh. Some of the spot tests utilized during taxonomic characterization of ECM fungi are: (a) 15% (w/v) aqueous KOH; (b) 30% (v/v) aqueous ammonia; and (c) 10% (w/v) aqueous $FeSO_4$.

Other Features: A few other macro-morphological characteristics utilized during the classification of ECM fungi are:

(a) The color and consistency of flesh.
(b) Effect of bruising fruit bodies or exposure to air on its color.
(c) Exudations as droplets or as latex on the cap.
(d) Color and abundance of hyphae at the base of the stem.
(e) Odor of the fruit bodies.
(f) Taste of the mushroom fungi, mainly whether it takes immediately or slow on the tongue.

Microscopic Morphology of ECM Fungi: In addition to macro-morphological analysis, ECM fungal taxonomists bestow appreciable importance to investigations on the microscopic features of these symbiotic fungi. Details on spores and specialized hyphae associated with fruit bodies are essential in order to arrive at accurate taxonomic positions of fungal samples at the genus and species levels. Major features investigated microscopically are spores, basidia, gills, cap, stem and cystidia.

Spore Size and Shape: The ECM fungi belonging to Basidiomycetes (basidiospores) and Ascomycetes (ascospores) produce sexual spores with distinctive microscopic features. ECM fungi vary widely with regard to spore color, shapes, sizes and walls. Often, their unique characteristics need to be recognized and described (Brundrett, 1996; Fig.1.3).

Spore Color: The spore color of ECM fungi is examined after mounting the samples in water and 3% KOH, or 10% ammonia or PVLAG. In addition, reaction with Melzer's reagent allows distinction of spores into amyloid (blue reaction) and dextrinoid (red/brown) types.

Spore Size: Truffle-like ECM fungi generally produce larger spores than mushroom-like fungi. Also, the ascomycetous types have larger spores than basidiomycetous ones. The spore size is generally measured on 20 to 25 specimens. The shortest and longest, as well as the narrowest and broadest positions on spores are measured and averaged. Sometimes, the size measurements may include apiculus and perisporium/ornamentations, if any.

Spore Symmetry: The fruit bodies of basidiomycetous fungi contain spores, which are bilaterally asymmetrical. Basidiospores that are discharged forcibly into the air are called ballistospores. These are also called heterotrophic spores whenever they are borne obliquely to facilitate easy discharge into the air. Truffle-like ECM fungi usually produce bilaterally symmetrical spores called the statismospores. These spores are not discharged forcibly into the air.

Spore Shape: The shapes of spores encountered among mushroom-like and truffle-like fungi vary widely ranging from ellipsoid, ovoid, globose, oblong/cylindrical, fusiform, citriniform, amydaliform, phaseoliform, angular, etc. (see Largent et al. 1977; Fig. 1.3). Spore shapes in between the broad categories stated above also occur, for example, subglobose.

Spore Wall Structure: The spore wall is an important taxonomic feature. A spore wall could be thin, smooth/rough or complex and highly ornamented. For example, certain genera under Amanitaceae possess a smooth spore wall, while others may have rough and ornamented spores. Details on spore wall and other ornamentations are recorded using high magnification (100×) and, in some cases, scanning electron microscopic observation at confirmatory stage. Spore wall thickness varies from 0.5 to 4 or 5 μm. Sometimes, the spore wall thickens at apex to form a callus. For all spores, knowledge about the number of spore wall layers is crucial. Light microscopy reveals only three spore wall layers, namely the perisporium, exosporium and endosporium, but under an electron microscopy, a multitude of spore walls can be recognized. The number and thickness differ on the basis of ECM fungal species. Spores could be utriculate if the spore sac completely encloses the spore, or calyptrate if it partially encloses the spore. The types of ornamentation on ECM spores, their density, size, shape, color are helpful in taxonomy. The six main types of ornamentations encountered are verrucose (warts), nodulose (knobby), ridged (striated, ribbed), punctate (small projections), echinate (spiny), and reticulate (irregular patterns) (Brundrett et al. 1996; Fig. 1.4).

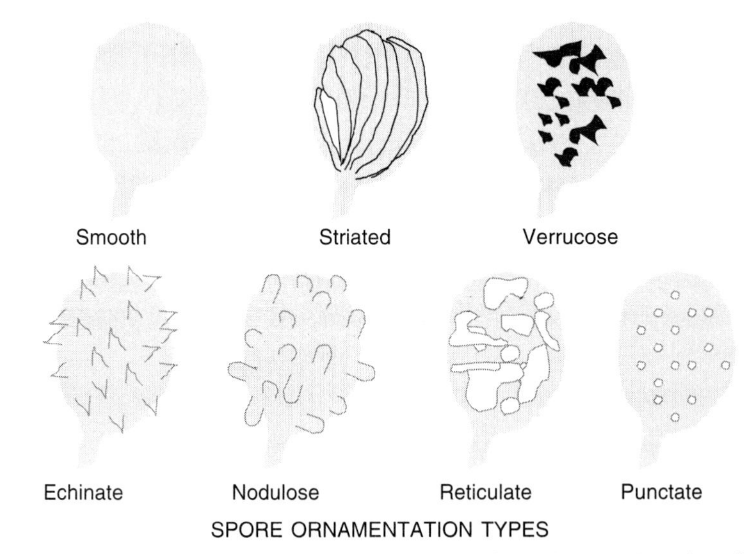

Fig. 1.4 Examples of types of spore ornamentations useful during morphotyping of ectomycorrhizal fungi (*Source:* Redrawn from Brundrett et al., 1996).

3. Molecular Phylogeny

3. A. Methods for Molecular Phylogeny

Traditionally, taxonomists dealing with AM/ECM fungi have depended on spore morphology and ontogenetic traits (Giovannetti et al. 1991; Bentivenga and Morton, 1995; Morton, 1995; Morton et al. 1995; Brundrett et al. 1996; Singer, 1986). However, such a classification system based on morphology and ontogeny may still mask or overlook the relevance of genetic and physiologic differences among AM/EM fungal species or isolates. A total lack of correlation between groupings based on spore morphology, with those based on genetic analysis or physiologic function is a clear possibility. As a consequence, a multidisciplinary approach combining as many relevant methodologies was proposed, so that biodiversity among AM or ECM fungi could be deciphered accurately and understood better (Giovannetti and Gianinazzi-

Pearson, 1994; Dodd et al. 1996). A range of biochemical, serological and molecular techniques have been adopted to attain greater insights into phylogenetic aspects of mycorrhizas. Some of the relevant ones currently in vogue with mycorrhizasts are summarized in the following paragraphs.

Soluble Proteins: Taxonomic identification based on protein or isozyme patterns have been efficiently utilized to discriminate between morpho-species and identify isolates of many organisms, including fungi. With reference to mycorrhizas, one-dimensional SDS-polyacrylamide gel electrophoresis (1D SDS-PAGE) of spore proteins can be helpful in detecting intra- and inter-specific variations among isolates (Giovannetti and Gianinazzi-Pearson 1994; Dodd et al. 1996), and in deciphering the phylogenetic positions (Schellenbaum et al. 1992; Thingstrup et al. 1995). However, obtaining stable reproducible patterns could be a constraint. Accuracy may depend on several factors, including achieving standard cultural conditions and extraction procedures. In a study by Avio and Giovannetti (1998), the consistency of spore protein patterns was assessed using cultures of the same AM species from different geographic locations, those produced on different hosts, and those maintained for different lengths of time (1 to 5 years) on a host plant. Profiles of soluble proteins from spore were consistent, and unaffected by the geographic location, host, or number of generations. For example, the similarity index was 98% for two geographically distinct *G. coronatum* isolates. Similarly, for *G. mosseae*, the similarity index was 100%, although the isolates were derived from four different hosts (Fig. 1.5). The physiological status of spores is an important factor that influences both quantitative and qualitative aspects of protein band patterns. Germinating spores can show a marked variation in protein profiles compared with quiescent stage. For

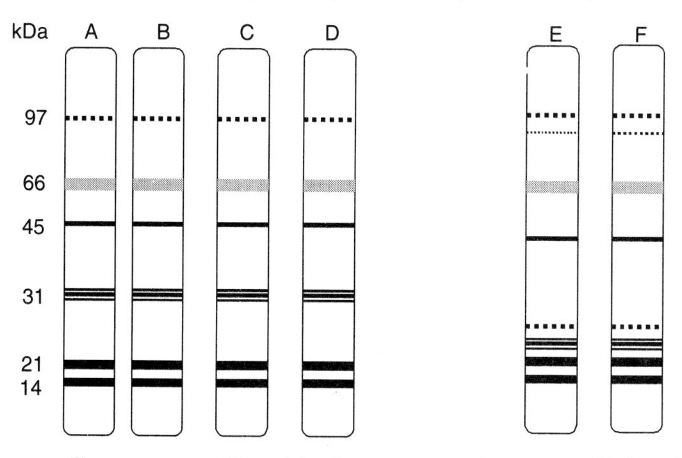

Fig. 1.5 Diagrammatic representation of electrophoretic patterns of soluble proteins extracted from spores of *G. mosseae* grown on four different hosts, namely A. *Allium cepa*; B. *Frageria vesca*; C. *Medicago sativa* and D. *Myrtus communis*. The electrophoretic patterns E and F were obtained from *G. mosseae* isolates collected from two different geographic locations (*Source*: Drawn based on Avio and Giovannetti, 1998). Note: The consistency in band patterns despite large variation in sample source, both in terms of host species and geographic location proves the usefulness of this technique during classification of AM fungi. kDa refers to the size of protein.

example, as germination proceeds, certain polypeptides may disappear, particularly those related with storage proteins (Avio and Giovannetti, 1998). Therefore, spores or hyphae of a comparable metabolic state from different isolates should be selected. Dodd et al. (1996) having analyzed an appreciably large number of isolates of AM fungi, concluded that different AM species (and their isolates) show distinctive SDS polypeptide profiles, allowing taxonomic separation of isolates. In many cases, the phylogenetic discrimination obtained using protein patterns matched those derived based on isozyme patterns and RFLP analysis. Xavier et al. (2000) collected 500 spores each from monoxenic cultures of four different AM fungi, namely, *G. fasciculatum, G. etunicatum, G. clarum* and *G. mosseae.* They subjected the protein aliquots to 1D SDS polyacrylamide gel electrophoresis. The AM fungal species could be easily separated using the electrophoretic pattern of soluble proteins from their spores. Further, each of the fungal species possessed signature protein bands that were reproducible and consistent. The signature protein may be useful as a marker while studying ecological aspects such as persistence, spatial distribution and proliferation in the soil.

Isozymes: These electrophoretic patterns of enzymes have been utilized to differentiate between mycorrhizal fungi. Polyacrylamide gel electrophoresis (PAGE), followed by staining for specific enzymes has been successfully used to identify and even to quantify the mycorrhizal fungi (Kjoller and Rosendahl, 2002). Generally, it is better if the enzyme patterns detected in root extracts of mycorrhizal plants are matched with those obtained using spores. It is easier to run the analysis of a large number of samples once a set of diagnostic enzymes for AM/ECM fungus has been decided upon. In a study on Danish peas by Kjoller and Rosendahl (2002), the electrophoretic patterns suggested that roots were generally dominated by a single species, namely *Glomus geosporum*, and the AM flora did not change significantly during the crop period. Several species of AM fungi could be traced in the soil through trap cultures, but none appeared to be present in root extracts other than *G. geosporum.* Such enzyme profiles have been utilized to study a wide range of AM and ECM fungal species.

A step further, polymorphism in isozymes encountered among the members of an ECM/AM fungus may improve the phylogenetic analysis. For example, Wipf et al. (1996) utilized isozyme profiles to identify and quantify Morels, as well as characterize the intraspecific crosses among monosporal strains of Morchella and improve systematics. Electrophoretic information on enzymes such as glutamine synthetase, NAD-glutamate dehydrogenase, glucose phosphate isomerase, super oxide dismutase, malate dehydrogenase, aspartate aminotransferase and glyceraldehyde phosphate dehydrogenase were helpful in discriminating the mycorrhizal fungi at inter- and intraspecific levels of phylogeny. According to Wipf et al. (1996), enzyme polymorphism-based discrimination of these ECM fungi also helped in analyzing the interactions of monosporal strains of *Morchella esculenta.* Alia and Dodd (1998) utilized the electrophoretic patterns of isozymes of malate dehydrogenase and esterases in order to identify the isolates of *Glomus etunicatum.* Similarly, Giovannetti et al. (2003) utilized polymorphism in malate dehydrogenase (MDH) and esterases (EST) to differentiate between *G. mosseae* and *G. caledonium.* The MDH profiles were also useful in separating *Gigaspora* at the subgeneric level.

Phylogeny of AM/ECM fungi can also be attempted using variations in the nucleotide sequence of genes that code for specific enzymes. Similarity in amino acid sequence of enzymes has also been utilized to infer the phylogenic status of mycorrhizal fungi. Phylogenetic analysis of *Glomus mosseae* conducted by Harrier et al. (1998) involved the study of similarity in a nucleotide sequence within the gene encoding one such enzyme 3-phosphoglycerate kinase (PGK). The nucleotide sequence in PGK gene from *G. mosseae* was similar to other fungi, but not with PGK gene derived from plants. The similarity in amino acid sequence of *G. mosseae* PGK with wheat was 49%, 47% with *Spinaceae oleracea*, and 48% with *Arabidopsis taliana*. In contrast, similarity of amino acid sequence with fungi such as *Aspergillus oryzae* was 79%, and 69% with *Kluveromyces lactis*. More importantly, phylogenic groupings derived using nucleotide sequence of PGK gene and amino acid sequence of the enzyme confirmed a previous classification primarily based on morphological features.

Fatty Acid Profiles: Quantitative and qualitative assessment of fatty acids can be indicative of fungal biomass in soil and sometimes help in detecting specific AM/EM fungi (Olsson et al. 1995; 1997; Muller et al. 1994: Frostgaard, 1996; Jabaji-Hare, 1988). Larson et al. (1998) examined the AM fungus, *G. intraradices* and several other saprophytic fungi such as *Fusarium culmorum, Penicillium hordei, Rhizoctonia solani* and *Trichoderma harzonium* for fatty acid signatures. Mainly, the phospholipids (PLFA) and neutral lipid fatty acids (NLFA) were assessed. Analysis revealed that 18 : 2 omega 6, 9 fatty acid dominated in saprophytic fungi, but was negligible in *G. intraradices*. Also, there were certain unique fatty acids in *G. intraradices*, which were not encountered in *F. culmorum*. It is interesting to note that fatty acid composition of *G. intraradices* was similar to several other *Glomus* species (Olsson et al. 1995) and isolates of *G. intraradices*. Therefore, in this case, fatty acid signature can help us distinguish between the saprophytic and symbiotic fungi. Further, Larsen et al. (1998) suggest that fatty acid signatures seem to be valuable while studying AM fungal mycelia in soil as well as in characterizing them. In a study by Graham et al. (1995) the fatty acid methyl ester (FAME) profiles were used to characterize and classify AM fungi. Among AM fungal isolates tested, they could obtain tight clusters, comparable to phylogeny achieved using the PCR-RAPD analysis. A similar effort by Jansa et al. (1999) compared the lipid profiles of AM fungi, saprophytic fungi and a few unrelated organisms. They concluded that most AM fungi gave characteristic fatty acid profiles which were sharply different from other organisms. They also identified three groupings based on fatty acid profiles, such as a group comparing *G. deserticolus* and *Scutellospora hetrogama*, a second one comprising *Glomus* species and a third consisting of *Endogone pisiformis*. Let us consider a more recent example regarding the utility of fatty acid signature in AM fungal taxonomy and soil ecological studies. The fatty acid methyl esters were analyzed in the spores of different AM fungi (*Glomus coronatum, G. mosseae, Gigaspora margarita* and *Scutellospora calospora*) by Madan et al. (2002) who traced 16:1 omega 9c to be the dominant fatty acid. In addition to it, spores of *Gi margarita* contained large quantities of 18:1 omega 9c and three 20c fatty acids (20:1 omega 9c; 20:1 omega 6c and 22:1 omega 9c), but these were not observable in the other two species studied by them. Such fatty acid-based traits were tested further to identify the occurrence of AM fungal species in the soil. They introduced known quantities of different AM fungal spores, and analyzed the fatty acid signatures of spores extracted from the natural soil. The ratio of fatty acid signatures obtained tallied with the expected ratios. Hence, Madan et al. (2002) concluded that 16:1 omega 5 could be a good fatty acid marker during the identification of at least certain AM fungal species. They also suggest 22:1 omega 9c as a possible useful marker to detect *Gi. margarita*. The fatty acid profiles have also helped in gaining important insights into phylogenetic hypothesis and developing the family tree. Fatty acid profiles could be effectively used to distinguish over 15 fifteen species of *Gigaspora* and *Scutellospora*. Fatty acid methyl ester profiles from AM colonized roots of citrus were effective as a measure of colonization development. It was also indicative of carbon allocation in the intraradical hyphae and extent of lipid storage in the fungus.

Serological Methods: Serological tests such as fluorescent-antibody and enzyme-linked immunosorbent have been applied to identify the differences between species (isolates) of AM/ECM fungi (Aldewell and Hall, 1987). In case of AM fungi, the presence of beta glucans and their variations have been detected using monoclonal and polyclonal antibodies. For example, distribution of beta (1-3) glucans among AM genera suggests that genera in Glominae, genera such as *Glomus* and *Acaulospora* may represent an outlying group in the Zygomycetes, whereas, *Gigaspora* would remain firmly placed in the Zygomycetes. There are several examples pertaining to serological analysis of AM fungi. Sanders et al. (1992) raised polyclonal antibodies against soluble extracts from *A. laevis* and *Gi. margarita*. They reported the presence of cross-reacting antigens related to non-specific glycoconjugates. Aldwell et al. (1983) successfully tested the differences in AM fungi using the polyclonal fluorescent antibody technique. Most of the early attempts utilized spore wall antigens. Gobel et al. (1995) were successful in obtaining monoclonal and polyclonal antibodies against hyphae of AM fungi. The specificity could be easily tested by dot-immunoblot assay. Interestingly, cross reactivity was generally less and not at all detected in case of genera from *Acaulosporaceae*. Enzyme-linked immunosorbent assay have also standardized for both ECM and AM fungi. To suggest an example, three species of the genus *Tuber* were easily distinguished from an unidentified ECM fungus occurring on *Quercus pubescens* using the ELISA technique (Zambonelli et al. 1993).

Small Sub Unit (SSU) Ribosomal Genes and Internal Transcribed Spacer (ITS): The genomic region encoding ribosomal RNA contains genes and spacers that have evolved at different rates, hence, making it easy to design nucleic acid markers at different taxonomic levels. Their rDNA analysis indicates a high degree of conservation in coding regions (18, 25, 5.8 and 5S genes), but considerable sequence differences are discernable in the spacers in ITS and intergeneric spacer (IGS). The rDNA offers a set of target regions with differing levels of genetic resolution from groups like families to populations (Buscot et al. 2000). A diagrammatic representation of rDNA, ITS and IGS region is as follows:

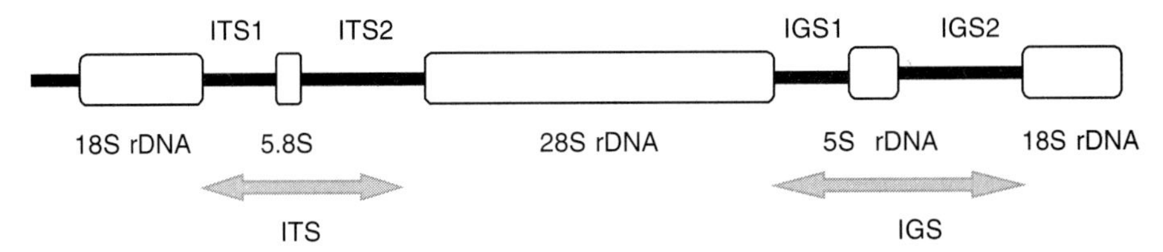

Note: Rectangular regions represent highly conserved regions and dark lines correspond to sequences of the spacer genes that are more variable.

Random Amplified DNA Polymorphism: Rapid Amplified DNA polymorphism (RAPD) is one of the most frequently used techniques to identify AM fungi. This method can be used even when the sample is very small, say, a single spore. The use of PCR methods generally require at least short sequences for the design of oligonucleotides, but in RAPD, short randomly chosen primers are used. A PCR on such primers can then lead to species-specific band patterns from very minute amounts of DNA as template. For example, over 30 to 120 reactions could be conducted on single spores from AM fungi. Such analyzes lead to easy identification and differentiation of not only species and subspecies, but even isolates. One of the major problems initially attached with RAPD was the low reproducibility caused by the high sensitivity of this method. However, it was overcome by designing primers and specific amplification of DNA from spores as well as mycorrhizal roots from the same sample. Through such a technique, Lanfranco et al. (1995) proved that RAPD approach can help in detecting the fungus accurately, overcoming contaminations and interferences, even when the sample size is very small. Later, Edwards et al. (1997) used the same primer set and employed competitive PCR and an internal standard to identify/quantify the degree of colonization by *G. mosseae*. Similarly, specific primers have been designed and utilized to identify as well as quantify several other AM fungi (Franken, 1999).

The amplified fragment length polymorphism (AFLP), originally devised to analyze bacteria and plants, could also be utilized to detect polymorphism in the DNA of a single AM fungal spore (Rosendahl and Taylor, 1997). Generally, DNA is extracted and digested with restriction enzymes to obtain fragments. These fragments are coupled with a linker and amplified with oligo-nucleotides binding to this linker. The PCR products are then further amplified so as to obtain reproducible and characteristic polymorphic patterns.

Polymerase Chain Reaction (PCR) on Ribosomal RNA Genes: The ribosomal genes encountered in all organisms harbor both highly conserved and variable regions. It is possible to construct universal primers and amplify the DNA sequences. Small quantities of DNA suffice to run a PCR for rRNA genes because they usually occur in high copy numbers. The fragments so obtained allow for easy distinction between the AM/ECM fungi. Franken (1999) says that because of these advantages, analysis of rRNA genes is one of the most common techniques utilized during the study of molecular phylogeny. Historically, one of the earliest genes to be amplified and utilized as an indicator of phylogenic status of AM fungus was rRNA from *G. mosseae* by Simon et al. (1992). Studies of cDNA libraries and specific sequences lead to the design of specific primers that helped in taxonomic judgment at different levels, such as at the generic, species or isolate level. Later, to enhance the abundance of specific sequences in root tissues, Clapp et al. (1995) devised a subtractive hybridization step. In other words, selective enrichment of sequences was attained. To decipher species level differences in sequences, a restriction enzyme digest of 18s rRNA may not be sufficient. Hence, single-strand conformation polymorphism (SSCP) or analysis of the complete sequences is needed. Franken (1999) states that in contrast, ITS and the 25s rRNA contain regions that are highly variable. Hence, differences in AM fungi at the species level can be easily judged using a combination of DNA amplification and restriction fragment polymorphism (PCR-RFLP) (Van Tuinen et al. 1994; Sander et al. 1995). Currently, molecular phylogenies of AM/ECM fungi are based on analyses of either complete sequences or restriction enzyme digest patterns developed on PCR products.

Nested PCR-SSCP (Single Stranded Confirmation Polymorphism)**:** Firstly, we should note that phylogenetic delineation of AM fungi based on hyphal morphology and colonization patterns, or species-specific protein patterns or even isozymes may get confounded due to interference by the host (Kjoller and Rosendhal, 2000). Techniques developed later, such as PCR-based sequence analysis, can generally overcome this problem, because specific sequences of AM fungal origin can be targeted. Despite such a phenomenon, a high concentration of plant-host DNA compared with AM fungal DNA in roots, can limit the accuracy of detection (Clapp et al. 1995). According to Kjoller and Rosendhal (2000), such a problem could be solved to a certain extent by precipitating host DNA prior to fungal specific PCR, or by amplifying fungal DNA directly or by using specific primers in a nested PCR (Van Tuinen et al. 1998). However, existence of suitable primers and degree of polymorphism in the amplified target sequence affects the specificity. In the nested PCR methods utilized by Kjoller and Rosendhal (2000), species and even isolates could be separated based on polymorphism found in the gene coding for a large subunit of ribosome. According to them, potentially, SSCP can detect single-base substitutions. The use of mini-gels allows the screening of a large number of samples. The PCR-SSCP technique also allows sensitive detection of both known and unknown AM fungi within roots.

The Repetitive Sequences: All higher eukaryotic organisms possess repetitive DNA sequences, and it can be effectively used to identify and distinguish the individual organism. In case of AM fungi, Zeze et al. (1994) were the earliest to obtain such a repetitive sequence. They constructed a small genomic plasmid library from *Scutellospora castanae* and screened clones from DNA. One of these repetitive sequences that hybridized with clones was analyzed further and used to detect the symbiont (*S. castanae*) in roots and distinguish it from others. A direct approach involves the use of microsatellite sequences as targets. Longato and Bonfonte (1997) initially amplified such repeats from as many as seven different species of glomalean fungi using GTG primers. The variations obtained in banding patterns were used to discriminate between the isolates of a single species. Obviously, here the taxonomic resolution is greater on reaching the isolate level. A sensitive method was used by Gadkar et al. (1997) to detect isolates of *Gigaspora* species. It involved the use of M13 core sequence for PCR. This method can detect intersporal genetic variation. For example, Zeze et al. (1997) have been able to detect intersporal genetic variation in one isolate of *Gi. margarita*. Similarly, Pringle et al. (2000) have discovered intersporal and even intrasporal variations in an *Acualospora* using this technique. Franken (1999) states that such repetitive sequences account for a large part of host genome and are amenable for analysis even with very small sample of fungus in soil or root. Such analysis, based on repetitive sequences, can be applied in situations when the PCR methods are not permissible.

3. B. Molecular Phylogeny of Order Glomales

The order Glomales represents a group of fungi which are obligately biotrophic in habit. They form mycorrhizas in association with higher plant roots. They also occur in a wide range of environments, including Antarctic soils (e.g. *G. antarcticum, G. kerguelense*; see Dalpe et al. 2002). At present, over 150 species of these AM fungi have been described (see Table 1.2), and their taxonomy is largely based on spore morphology and related characters. The framework for such classical taxonomy has been constantly updated during the past three decades (Gerdeman and Trappe, 1974; Berch and Trappe, 1987; Walker, 1992; Schenk and Perez, 1990; Morton and Benny, 1990). Phylogeny of AM fungi based on molecular analysis is a comparatively recent trend (Bruns et al. 1991; Simon; 1996). According to Morton and Benny (1990), the Glomales are grouped as a separate fungal order assuming that the species evolved originating from a common ancestor. Further, the proposed suborders, families and genera that delineate natural groups actually reflect the phylogenetic history of the AM fungi. The phylogenetic analysis using SSU rRNA suggests that Glomales are a distinct clade of fungi. They diverged some time before the emergence of Ascomycetes and Basidiomycetes (Bruns et al. 1992; Willmotte et al. 1993; Simon et al. 1993). These estimates derived using the molecular clock procedures were also confirmed using the classical archeological studies using fossils from Devonian land plants (Pyrozynscki and Dalpe, 1989). Remy et al. (1994) reported that mycorrhizas existed as early as 400 million years ago, which is in confirmation with molecular clock estimates of 462 to 482 million years ago.

The molecular phylogeny of the order Glomales is based on sequencing and analysis of nuclear gene that encodes for SSU rRNA and internal transcribed sequences (ITS) regions. It provides a reliable basis for phylogenetic analysis (Bruns et al. 1991; Bruns and Szaro, 1992; Bruns et al. 1992). Using a combined data set, Simon et al. (1996) conducted parsimony and developed a phylogenetic tree for the order Glomales (Fig. 1.6). The phylogenetic tree developed based on molecular analysis depicts the three families—Glomaceae, Gigasporaceae and Acaulosporaceae. Broadly, such grouping based on SSU analysis is clearly in consonance with the one developed previously, using morphological traits. Further, species under each genus cluster together, thus confirming the utility of the molecular method of identification and classification of AM fungi. Quite often, the resolution of intrageneric divergences might be difficult, if based on SSU rRNA sequence analysis mainly because they evolve into distinctiveness very slowly (Simon, 1996). On the other hand, large divergence observed in 18S rDNA sequences of *Glomus* isolates is an indication, that it could further split into a few more homogenous groups. Some of these aspects are discussed in greater detail under each family of Glomales. In the example provided by Simon (1996), seven SSU sequences obtained that fit the previous morphological groups can be easily identified into two clusters of *Glomus* species under the family *Glomaceae.*

Molecular Phylogeny of Glomaceae: With a view to arrive at accurate taxonomic positions of a species and isolates of Glomaceae, Dodd et al. (1996) suggested developing procedures that utilize a range of tests, including morphological characterization, isozyme and soluble protein profiles (Rosendahl, 1989); lipoprotein (Wright et al. 1987; Thingstrup et al. 1995) and fatty acid signatures (Jabaji-Hare, 1988; Bentivenga and Morton (1994) along with molecular analysis using PCR-RFLP of ribosomal RNA (Sanders et al. 1996); RAPD-PCR (Lanfranco et al. 1995) and ITS sequences (Lloyd-McGilp et al. 1996; Redecker et al. 2003). A combination of these tests can help us in developing more accurate phenograms as also in identifying *Glomus* isolates drawn from geographic locations.

Dodd et al. (1996) applied molecular analysis on the same *G. mosseae* isolate that had been examined earlier for morphological variations. Three approaches were adopted to analyze the DNA polymorphism: (i) PCR with ITS ribosomal primers;

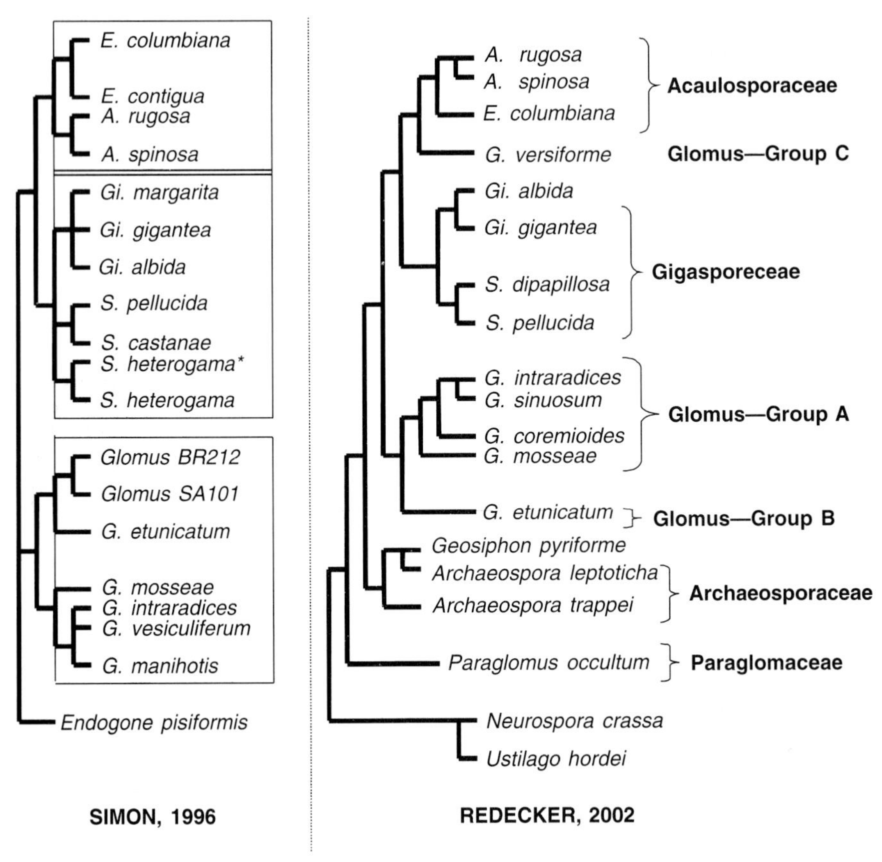

Fig. 1.6 Improvements and changes in the classification of arbuscular mycorhhizal fungi based on molecular methods: An example. Note: Under Simon's classification, boxes from top represent Acaulosporaceae, Gigasporaceae and Glomaceae; in all, only three families were identified. Under Redecker's classification, two more new families have been erected, and several genera are rearranged into different groups. * = *S. heterogama (dipapillosa)*. (*Source*: Simon, 1996; Redecker, 2002).

(ii) PCR with specific primers POM 3; and (iii) PCR using primers of random sequences (RAPDs). The *G. mosseae* specific primers for DNA fingerprints obtained through these procedures provided clear evidence that the *G. mosseae* isolate cluster together. It also confirmed the phylogenetics based on isozymes and soluble protein profiles. In fact, primers PO-M3 designed for *G. mosseae* isolates (Lanfranco et al. 1995) confirmed the specific grouping.

Phylogeny of *Glomus mosseae* based on ITS region of RNA: Lloyd McGilp et al. (1996) assessed *G. mosseae* isolates by analyzing the ITS region of nuclear ribosomal RNA. It was amplified, cloned and sequenced from at least 5 different *G. mosseae* isolates, and one each of *G. fasciculatum, G. dimorphicum* and *G. coronatum.* The phylogenetic analyses were made based on nucleotide substitutions as well as insertion and deletions. The trends were fairly similar with all three types of analyses. Based on nucleotide substitution analysis, the sequence from *G. coronatum* were distinct from all other groups, with 11 to 16% divergence, whereas, all other sequences from *G. mosseae* isolates, *G. fasciculatum* and *G. dimorphicum* were closely related. An independent analysis of the same isolates based on nucleotide insertions and deletions provided a remarkably similar phylogenic pattern. It again confirmed the distinctions of *G. coronatum*, and genetic closeness of *G. fasciculatum* and *G. dimorphicum* to *G. mosseae* isolates.

In the above study, wherein Llyod-McGilp et al. (1996) examined ITS divergence on a wide its range of *G. mosseae* isolates from different continents; the phylogenetic grouping did not match its geographical origins. No worthwhile relationship patterns could be deciphered. In fact, variations among isolates from different continents were no greater than the divergence noticed for isolates from smaller geographical zone. Also, the divergence between isolates from different countries/continents was scarcely greater than divergence noticed within single spores. Overall, inferences drawn based on ITS region divergence indicate that *G. mosseae* isolates worldwide are similar. These comparisons also revealed that although *G. fasciculatum* and *G. dimorphicum* are taxonomically distinct if based on morphological evaluations, they are not that highly distinct if based on molecular analysis. In order to explain this situation, Llyod-McGilp et al. (1996) have remarked that our present knowledge regarding the relationship between genetic divergence and spore morphology is inadequate to arrive at any tangible inferences. Clearly, there is need to understand the biochemical genetics of spore formation and maturation, as well as the molecular basis for variations in spore color, texture size and several other traits, before we arrive at authentic links between spore phenology versus genetic divergence estimates. A recent report by Giovannetti et al. (2003) shows that *G. mosseae* isolates could be characterized using ITS-RFLP analysis. Previous opinions suggested that accuracy of this method was limited to species level differences. Now, ITS-RFLP profiles can be utilized to assess and differentiate between the isolates of AM fungi. For example, in their study, molecular diversity between two European isolates of *G. mosseae*, namely IMA1 and BEG 25, could be confirmed using ITS-RFLP-based technique. In fact, IMA1 was genetically more similar to two American isolates studied than to the other European strain-BEG 25. These assessments were confirmed using protein profiling.

Phylogeny of *Glomus* species based on soluble protein and Isozyme analysis: Classification based on protein and isozyme analyses has been in vogue with a variety of organisms, including AM fungi. In the case of *G. mosseae*, Dodd et al. (1996) utilized the isozyme profiles of malate dehydrogenase (MDH), esterase and even soluble protein profiles, to make a phylogenetic analysis of various isolates of this AM fungus. The isolates examined were derived from different geographic locations, climates, soils and vegetation. The electrophoretic tests proved that *G. mosseae* isolates formed a well-defined group when compared with other species such as *G. coronatum* or *G. caledonium* or *G. geosporum*. They were able to trace only a single MDH locus in the yellow vacuolated *G. mosseae* spores, and its Rf values varied little among isolates. In comparison, *G. coronatum*—the funnel-shaped spore—had two putative loci, a slow-moving monomorphic and a fast-moving polymorphic locus. The esterase pattern showed a slight variation within the YV *G. mosseae* spores. Further, SDS-protein profiles confirmed that *G. mosseae* isolates clustered into one group, exemplifying the close link. In addition, Dodd et al. (1996) found only minor differences among the range of phenograms generated using different clustering techniques. Isozyme profiles derived from both MDH and esterase clearly help characterize and group the YV *G. mosseae* isolates, and differentiate them from the brown funnel-shaped *G. coronatum*. Morphological comparisons may not lead us to such accurate distinctions. For example, using isozyme pattern, *G. dimorphicum* could be easily grouped along with *G. mosseae* isolates. Hence, Dodd et al. (1996) suggested that *G. dimorphicum* could even be re-classified as an isolate of *G. mosseae*.

The genus *Sclerocystis* is characterized by the formation of complex sporocarps. Recently, most of the species under this genus were transferred under the genus *Glomus*. Redecker et al. (2000) remark that despite these transfers, taxonomic and phylogenetic controversy about this genus seems to continue. They analyzed the 18S ribosomal subunit of *G. sinuosum* (=*S. sinuosum*) and compared it with closely related *Sclerocystis coremiodes*. It showed that both these AM fungi were taxonomically close and fall within a monophyletic clade that consists of several other well-characterized *Glomus* species such as *G. mosseae*, *G. intraradices*, *G. vesiculiferum*. In fact, Redecker et al. (2000) have inferred that the formation of a complex sporocarp is a mere character among certain *Glomus* species, but it is not a trait conspicuous or important enough to erect a new genus such as *Sclerocystis*.

Phylogenic identification of *Glomus* species and their isolates can be extended into compatible and incompatible groups based on their ability to form anastomoses and exchange genetic material (Giovannetti et al. 2003). Successful anastomoses between compatible groups was generally characterized by complete hyphal fusion, protoplasmic continuity and occurrence of nuclei in the middle of hyphal bridges, indicating transfer of genetic material. Incompatibility between isolates can be traced easily by ascertaining the negative tropism of hyphal tips, retraction of protoplasm from hyphal tips, lack of anastomes and nuclear transfer. These traits could be effectively used to classify *Glomus* isolates into compatibility/incompatible groups. Usual molecular tests such as protein patterns, and RFLP-PCR profiles have confirmed the differences between compatibility groups.

Molecular Phylogeny of Gigasporaceae

Gigaspora: Bago et al. (1998) made an effort to integrate the phylogeny arrived using molecular methods such as isozyme and SSU with morphological taxonomy of the genus *Gigaspora*. They sequenced SSU of 24 isolates of AM fungi belonging to Gigasporaceae and were able to identify three conspicuous groups within *Gigapsora*. It was actually based on 6 nucleotide long molecular signatures. The groups are: (a) *Gigaspora rosea* group (*Gi. rosea, Gi. albida*); (b) *Gi. margarita* group (*Gi. margarita, Gi. decipens*); and (c) *Gi. gigantea* group (*Gi. gigantea*). Next, the malate dehydrogenase isozyme profiles were studied for 12 isolates of *Gigaspora*. The groupings arrived at using isozyme pattern compared well and it supported the groupings developed using SSU nucleotide sequencing. Bago et al. (1998) suggested that intrageneric molecular signature is an unambiguous and quick method to recognize and arrive at the phylogenic status of *Gigaspora* isolates.

Lanfranco et al. (2000) explored the utility of several PCR-based techniques to study the genetic diversity within the genus *Gigaspora*. The ITS (internal transcribed spacers) of the nuclear ribosomal unit seems to reveal useful information on species-specific DNA polymorphisms. They obtained multispore DNA preparations and amplified with ITS/ITS4 of different isolates of Gigasporaceae. Single bands of about 550 to 600 bp were obtained. They also designed two *Gi. margarita* specific primers that could detect this fungus when in symbiotic phase.

In a later study, Lanfanco et al. (2001) characterized isolates of *Gigaspora* using both morphotyping and nucleotide sequence analysis. They analyzed portions of 18S gene and ITS region amplified by PCR. Later, based on the above data, oligonucleotides that specifically distinguished *Gi. rosea* from *Gi. margarita* and *Gi. gigantea* were prepared. The ITS sequence indicated substantial genetic variability among *Gi. rosea* isolates and those of *G. margarita* derived from different geographic locations.

Scutellospora: Like other AM fungi, molecular phylogeny of *Scutellospora* is also based on the information about its nucleic acid sequences, mainly derived after analyzing the small subunit ribosomal RNA gene. These are high copy number genes existing in tandem arrays and, in most species, least sequence variations occur between copies. The differences and similarities in base sequence of rRNA are generally indicative of the true phylogenetic relationships among the isolates of *Scutellospora*. Clapp et al. (1999a) analyzed spores of *Scutellospora* species and infected blue bell roots by amplifying the SSU rRNA gene, using the SS38/VAGIGA primer combination. They cloned the products and sequenced them. Cloned sequences from both spores and roots could be easily classified and most of them were attributable to *Scutellospora*. The region sequenced in this study represents a well-conserved area of SSU. We may note that sequences from *Gi. margarita* (now *Gi. rosea*) and those of *Scutellospora* isolates cluster together, supporting their groupings under the same family.

In their investigations, Clapp et al. (1995) found that clones from single spores could differ in sequences. Such heterogeneity and sequence polymorphism in the ITS region has also been identified with other AM fungi. Based on their study of primers targeting variable regions of the SSU of rRNA, Clapp et al. (1999b) inferred that amplification of DNA from all homologous genes present in a single spore of *Scutellospora* sp. is unlikely. This aspect may affect accuracy of in situ identification of species from soil, and species composition of the infected roots. Hence, it was suggested that until the frequencies, diversity and significance of different sequences occurring within a single taxa have been assessed, molecular phylogenies based on the SSU of rRNA should be interpreted with caution.

Hosny et al. (1999a) analyzed the genome size using flow cytometry. They reported it be around 0.14 pg to 1.12 pg, depending on the isolate. This is slightly higher than that known for primitive fungi (Zygomycetes and Phycomycetes). The DNA denaturation and re-naturation studies indicated a high proportion—more than 50% of repeat DNA sequences in *S. castanae*. Analyses performed on the ribosomal-RNA specifying genes showed that repeats were 50 to 70 copies, and several families occurred in the bulk DNA of *S. castanae*. Moreover, PCR performed on single spores gave the clear evidence for genetic heterogeneity, and it was also possible to demonstrate that spores were multikaryotic. They suspected that nuclear transfers by interhyphal conjugation might have caused the variation.

Through a re-analysis of rDNA ITS sequences in *S. castanae*, Redecker et al. (1999) have suggested that extremely divergent sequences noticed in nuclei within single spores need to be handled with caution. Such variation in a genome has been attributable to heterokaryotic diversity in these coenocytic organisms (Hijri et al. 1999; Hosny et al. 1999b). However, a recourse to a thorough analysis and comparison with known sequences is worthwhile because it is possible that AM spores are contaminated with remnants of unrelated fungi. In one such analysis, Redecker et al.

(1999) have clearly shown that highly divergent copies of ITS region reported in *S. castanae* actually belong to Ascomycetes. They suggested that the degree of divergence, hetero-karyotic nature and phylogenetic status based on previous studies need to be reassessed. In fact, they concluded that origin of such highly divergent ITS sequences was attributable to Ascomycete contaminants on the spore surface. Incidentally, obtaining completely sterile spores of AM fungi from natural soils is not easy. It could also be due to unduly higher proportion of Ascomycete spores within the *S. castanae* spore samples, or perhaps these excessively divergent sequences belong to the glomalean genome.

Molecular Phylogeny of Acaulosporaceae

The morphological differences between the two genera of Acaulosporaceae, namely *Acualospora* and *Entrophospora* may not be great. Morphologically, the main characteristic that separates *Entrophospora* from *Acaulospora* is the position of the soporiferous saccule. Actually, the absence of conspicuous developmental difference—except for a small positional shift in spore formation with no consequence to subcellular differentiation—suggests that errection of new genus may be artificial. The evolutionary genetics of such a minor shift in location of spore in relation to neck of saccule is unknown. At present, proportion of *Acaulospora* to *Entrophospora* species is 5:1. It is suspected that low number of *Entrophospora* is due to small genetic or regulatory change that converted each *Acaulospora* to *Entrophospora* under rare circumstances. Analysis of rDNA sequences indicated a monophyletic origin. The *Entrophospora* species encountered may be the result of a small genetic or regulatory change that converted *Acaulospora* to *Entrophospora*. The rDNA sequences suggest a monophyletic origin. Molecular differences between *Entrophospora* and *Acaulospora* have been discerned. Rodriguez et al. (2001) investigated the sequence variations within the D2 region of large subunit (LSU) ribosomal RNA gene in *Entrophospora*. The LSU rRNA genes were actually analyzed using a combination PCR-single-strand conformational polymorphism (PCR-SSCP). Pringle et al. (2000) have studied the nuclear rDNA sequences in the ITS region of *A. collossica*. They traced a high degree of ITS sequence variation even among spores of the same fungus. Phylogeny based on 5.8s rDNA showed the occurrence of at least seven distinct types of sequences within a single species.

Developments and Arguments about the Molecular Phylogeny of Glomeromycota: Taxonomy based on spore morphology and other features have, thus far, led us to a three-family structure for the order Glomales. The grouping into three families, Gigasporaceae, Acaulosporaceae and Glomaceae were supported well by the rDNA analysis. However, Redecker (2002) points out that there is large diversity even within the genus *Glomus,* which is not reflected within the classical taxonomy. Hence, basing the phylogeny of Glomales entirely on morphological features is not a tangible suggestion in many cases. There are problems in morphology-based taxonomy. For example, Morton et al. (1997) reported that spores of *Acaulospora gerdemanii* and *Glomus leptotichum* were observed to emanate on the same hyphae. This is a situation not easily explainable using classical taxonomic principles. Since these spores did not represent a sexual stage, they have preferred to term them "synanomorphs". Molecular analysis using 18s rDNA sequence indicated that it is a dimorphic organism, which neither belongs to *Acaulospora* nor *Glomus*; instead it could be an ancestral species, with deeply divergent lineages (Redecker, 2000).

Acaulospora gerdemanii/Glomus leptotichum, because of their large phylogenetic distance to all other known glomalean genera, were renamed under a new genus as *Archaeospora leptoticha* and *Ar. gerdemanii. Archaeospora leptoticha* is a dimorphic AM fungus for which specific PCR primers that aid their identification are available (Kojima et al. 2004). Yet another species in this lineage, which is known to form small, hyaline spores of the acaulospora type, was known as *Acaulospora trapeii.* It is now grouped into *Archaeospora trapeii* (Morton and Redecker, 2001). Two other species of *Glomus*, namely *G. occultum* and *G. brasilianum*, are deeply diverged and have close affliction with *Archaeospora* or *Geosiphum*, but they are grouped under the family Paraglomaceae. The genus *Paraglomus* is, however, distinguishable from *Glomus* based on rDNA sequences, unique fatty acids and certain morphological features (Redecker, 2002).

The genus *Sclerocystis* forms sporocorps, but the spores have greater morphological similarity to *Glomus* spores. Analysis of 18s rDNA indicated that complex sporocarp-forming fungi were related to the *Glomus* group A. Redecker's suggestions (2002) indicate that within the family Gigasporaceae, *Gigaspora* is the advanced genus, and not primitive one. The genus *Gigaspora* forms a narrow clade when compared to a large variation traced within the genus *Scutellospora.* It is possible that *Scutellospora* is polyphyletic. Now, regarding Glomaceae, previously, *Glomus versiforme* was grouped within a weakly supported monophyletic group of *Glomus* species comprising *G. intraradicies, G. mosseae* and *G. etunicatum.* However, recently, it has been placed in a clade of its own, but with a slight affinity to *Acaulospora* (Schwarzott et al. 2001; Redecker, 2002).

Geosiphum pyriforme, shown in the phylogenetic tree (Fig.1.6) is closest to AM fungi, but lacks the ability to form mycorrhizal symbiosis with higher plants (Gehrig et al. 1996). However, it is symbiotic with alga; for example, it is harbored intracellularly on the alga *Nostoc.* Spores produced by *Geosiphum* resemble that of *Glomus* even in terms of spore wall ultrastructure. Based on molecular analysis, Redecker (2002) opines that *Geosiphum* is perhaps an ancestor to mycorrhizal symbiosis that occurs on land plants.

3. C. Molecular Phylogeny of Ecotmycorrhizas

At present, both morphotyping and molecular techniques are in vogue to identify and classify ectomycorrhizas traced on root tips and soil. Sometimes, the two approaches may provide conflicting identifications of the symbiotic fungus. Several genotypes may be placed in a single morphotype. In some cases, morphotyping becomes inaccurate. Morphotyping was compared with PCR-RFLP analysis by Sakakibara et al. (2002). Specifically, mycorrhizas were classified using a detailed morphological approach and the same samples were subjected to PCR-RFLP analysis of the ITS region of rRNA gene repeat. Surprisingly, they noticed that in 93% of mycorrhizas analyzed, morphotypes would have been classified in the same way in either of the methods. Many of the morphotypes were positively identified using molecular methods. Major ECM fungi identified using a combination of both methods were *Rhizopogon, Lactarius, Russula, Cenococcum, Piloderma, Tuber*, etc. Sometimes, morphotyping did not identify ECM fungi such as *Piloderma* and *Rhizopogon* properly. Essentially, it means that morphotyping may be used for primary identification and followed by an accurate identification and classification using molecular approaches. The phylogenic aspects of some major ECM fungi have been dealt in detail in the following paragraphs.

Ectomycorrhiza-forming Mushrooms—Phylogeny of Agaricales

Several of the families belonging to order Agaricales are capable of ectomycorrhizal symbiosis. Perhaps, they are the most important ECM-forming mushroom fungi in most geographic locations. Moncalvo et al. (2000, 2003) have remarked that traditionally, taxonomy of mushrooms (Agarics) has been based on morphological characters subject to parallel evolution and phenetic plasticity. Emphasis on certain aspects, preoccupation on nomenclature and lack of a strong phylogenetic perspective have hampered their accurate classification. To date, the most comprehensive taxonomy of agarics is the one proposed by Singer (1986), which is also based on similarity and correlation of phenetic traits. Hence, Moncalvo et al. (2000, 2003) have attempted a classification of agarics, secotoids and puffballs by analyzing the large and small subunit ribosomal RNA, mitochondrial rDNA genes and mitochondrial protein-encoding genes. The molecular phylogeny of these mushroom-forming ECM fungi has also been studied by assessing the 5′ end sequences of nuclear large subunit RNA genes (nlsu-rDNA). Moncalvo et al. (2000) sequenced nearly 154 selected taxa that represented a wide range of families within Agaricales. To arrive at their phylogenetic positioning, weighted parsimony, maximum likelihood and distance methods were employed. Among them, weighted parsimony provided excellent estimates of phylogeny of Agaricales. Molecular phylogeny of the above agaricoid fungi corroborated either moderately or sometimes highly with previously known morphology-based classification schemes suggested by Singer (1986). It also supported the inclusion of two major grilled mushrooms *Boletus* and *Russula*. Amanitaceae, Coprinaceae, Agaricaceae, Strophariaceae, mycorrhizal species of *Tricholoma*, *Pleurotus*, and a few others comprise monophyletic groups under Agaricales. Whereas, Cortinariaceae, Hygrophoraceae, Tricholomataceae are examples of non-monophyletic groups. It is possible that in case of *Russula*, the two species *R. cremoricolor* and *R. silvicola* were identified as two different isolates based on red- and white-capped basidiospores. However, analyses of ITS sequences show that they are a single shape. Obviously, pilus color is variable in a population of *Russula*. This example highlights the pitfalls in relying on certain morphological traits for species identification. Microscopic traits may separate some species, but not all and that too accurately.

Amanita: A large set of database was developed for the ECM fungal genus *Amanita* by Drehmel et al. (1999). The molecular analysis involving nuclear-encoded large subunit rDNA sequences covered taxa from all sections of the current classifications of Singer et al. (1986). It included sections such as *Amanita, Amidella, Caesareae, Lepidellae, Mappae, Phalloideae, Ovigerae, Vaginatae, Vlaidae*. These studies confirmed that phylogenetic analysis based on the nuclear large subunit rDNA supports the previous classification based on morphology and biochemical tests that recognized nine distinct sections. Similarly, the taxonomic distinction of subgenera *Amanita* and *Lepidella* were also corroborated through rDNA analysis. In fact, Drehmel et al. (1999) have remarked that phylogenetic analysis among these nine terminal sections reveal several higher relationships, which again, are well supported by the morphological and biochemical traits. Figure 1.7 depicts a phylogenetic scheme that firstly recognizes *Amanita* and *Lepidella* as two subgenera; it then has four sections (*Amanita,*

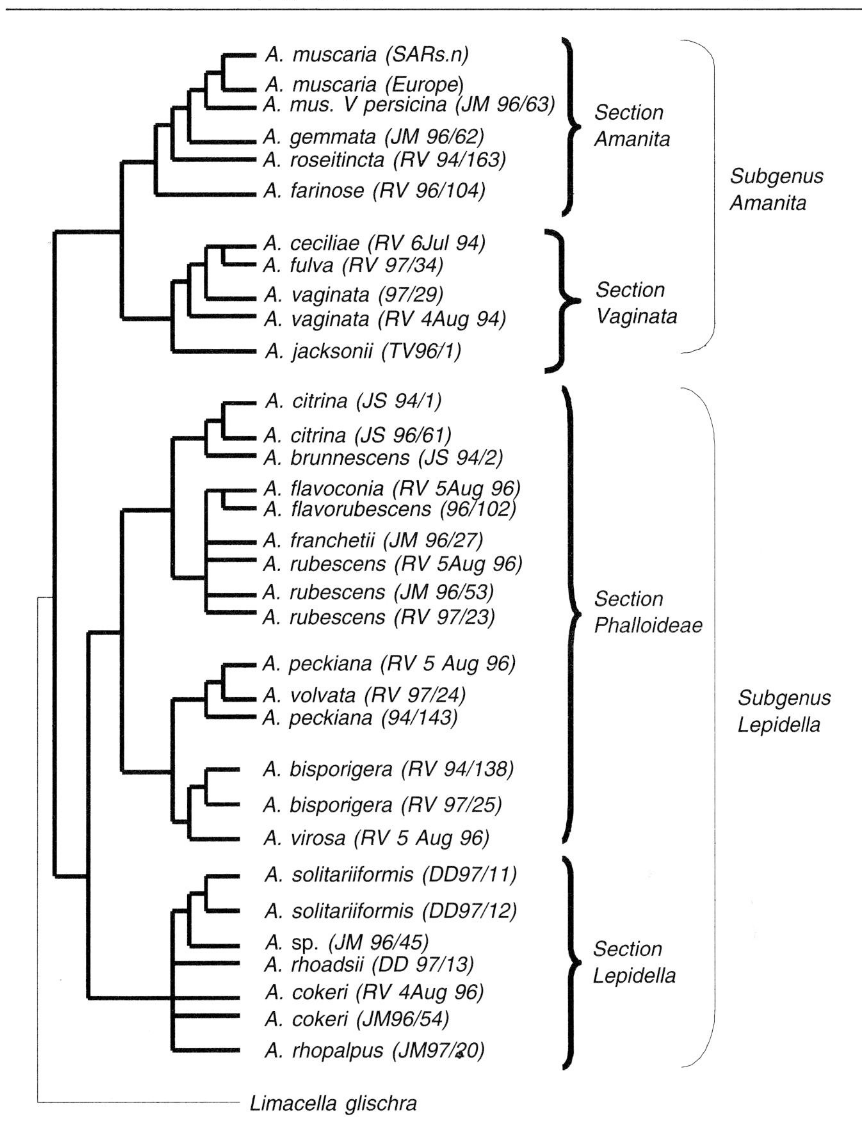

Fig. 1.7 A phylogram of genus *Amanita* developed based on the molecular analysis of large subunit ribosomal DNA sequences (*Source:* Drehmel et al. 1999).

Vaginata, Phalloideae, Lepidella), and seven subsections, namely *Amanita, Ovigerae, Amidellae, Caesareae, Phalloideae, Vaginata* and *Validae*. It also recognizes two series, namely, Mappae and Validae.

Hebeloma: The genus *Hebeloma* (Cortinariaceae, Agaricales) is a widespread ECM fungus in forests of Northern Hemisphere. It is mostly an early colonizer and forms symbiotic associations with members of Salicaceae, Pinaceae and a few other forest tree species. Aanen et al. (2000) made a phylogenetic analysis of this important ECM fungus based on nuclear ribosomal ITS sequences and cladistic methods. During phylogenetic analysis of sequences belonging to 51 collections and 39 taxa, *H. crustuliniforme* complex, a common ECM fungus was accorded special emphasis. Then, the relationships of other species/isolates were ascertained. According to them, the genus *Hebeloma* appears to be monophyletic. Although several thoroughly well-supported clades could be identified, certain basal relationships were unresolved or weakly authenticated. With regard to *H. crustuliniforme* complex, Aanen et al. (2000) inferred that it appears to be paraphyletic, comprising two clades and upto 17 intercompatibility groups. The second clade has uncertainties that are easily attributable to a high degree of recent speciation events, and many of them form ECM association with trees belonging to Salicaceae. In their later report, Aanen et al. (2001) have remarked that *Hebeloma* as a genus is taxonomically difficult to assess. Taxonomic position of many of its taxa is still unclear. Such taxonomic complexity is due to the occurrence of cryptic species. These cryptic species are populations isolated through reproduction and they behave as a biological species. They are not easy to differentiate. Aanen and Kuyper (1999) have reported over 20 biological species within the *H. crustuliniforme* complex occurring in northern Europe. Among them, at least two biological species belonged to *H. velutipes* (BSP 16 and 17). Within BSP 17, they found two divergent types of ribosomal ITS (ITS1 and 2). The two ITS types segregate in monokaryotic progenies of the dikaryon. These ITS types represent different alleles at homologous rDNA loci in the two nuclei. Analyses of polymorphism indicate potential problem for molecular identification of *Hebeloma* species based on ITS finger printing. Recently, Jany et al. (2003) have developed simple sequence repeat (SSR) markers from EST databases of *H. cylindrosporum*. These molecular markers allow unambiguous identification and scoring of this ECM fungus.

Pisolithus: *Pisolithus* is widespread in warm temperate regions. It forms ECM symbiosis with a wide range of woody plants. Regarding its phylogeography and phylogenetics, Martin et al. (2002) have stated that their rDNA ITS analysis revealed a total of 11 species in samples from across the world. Indeed, a strong phylogeographic pattern was noticeable. For example, *P. tinctorius* was frequent and dominated the trees in Holarctic belt, but were mainly associated with *Pinus* and *Quercus*. *Pisolithus arantioscarbrosus* occurred in restricted zones in eastern Africa on plants such as *Azfelia*. *Pisolithus albus*, *P. marmoratus* and *P. microcarpus* were localized in Australia on *Eucalyptus* and *Acacia*. In addition, two other species of *Pisolithus* were also restricted to Australian locations. Overall, there seems to occur a relationship between evolutionary/phylogenetic status of *Pisolithus* versus the biogeographical distribution of its common host. Hence, it was also inferred that an endemic host may dictate the phylogeography of *Pisolithus*. A different study by Arajo et al. (1999) also revealed considerable genetic variation among *Pisolithus* species. Isolates of *P. tinctorius* drawn from different locations on the globe were assessed using RAPD and RFLP methods. They were able to segregate them phylogenetically into two main groups. One group predominantly contained isolates of Brazilian origin and colonized *Eucalyptus* more frequently. The second cluster, containing *P. tinctorius* isolates from the Northern Hemisphere, was frequent on *Pinus* species, and they were clearly distinct in terms of the phylogenic positioning from those collected in locations of the Southern Hemisphere, particularly Brazil. A study by Gomes et al. (2000) utilized both ITS and mitochondrial DNA polymorphism to classify *Pisolithus* strains drawn from different continents. A consensus tree generated using sequence data of all *Pisolithus* isolates indicated the presence of a number of species among the isolates tested. Analysis of ITS sequence showed a 98.7% homology among isolates from the USA and France, 97% among isolates from Brazil and 76% among isolates from Australia, Brazil and Kenya. Cluster analysis based on MtDNA restriction fragments grouped the isolates into two distinct groups, which coincided with their host specificity and geographical origins.

About sixty-two *Pisolithus* isolates from New South Wales, Australia were assessed using random amplified polymorphism DNA (RAPD) to identify variation. Then, RFLP and ITS sequence analysis were conducted on a selected few isolates to generate a phylogenetic tree. Based on their analysis, Anderson et al. (1998) state that although groupings done on morphological considerations of carpophore and basidiophore of majority of isolates were consistent with bootstrap analysis based on molecular analysis, however, those for two isolates (cJO7 and WM01) differed. They believe that these two isolates may represent *Pisolithus* species distinct from the previously described Australian species of *Pisolithus*.

Rhizopogon: *Rhizopogon* (Basidiomycota, Bolatales) is a hypogaeus fungus that frequently forms ectomycorrhiza with Pinaceae. Over 100 different species of *Rhizopogon* have been identified. Grubisha et al. (2002) state that the greatest diversity of this genus is noticeable within the coniferous forests of North America. Despite being cosmopolitan in host range, it is predominant on Pine and Douglas fir (*Pseudotsuga menzesii*). A basic understanding about the phylogeny of *Rhizopogon* was developed around the reports by Lange et al. (1956); Smith (1971); as well as Smith and Zeller (1966). Smith and Zeller (1966), in fact, described several species of *Rhizopogon* from North America and Europe, and increased the number of species from 17 to 110. Grubisha et al. (2002) believe that because of the great diversity of its main host Pinaceae, the Pacific Northwest America is also, perhaps, the area with maximum divergence and evolutionary significance for the fungal symbiont *Rhizopogon*. They divided the genus into two subgenera, namely *Rhizopogonella* and *Rhizopogon*. The subgenus *Rhizopogon* was subdivided into four sections—Amylopogon, Rhizopogon, Fulviglebae and Villosuli. Such demarcations were based mainly on macroscopic and microscopic characters of sporocarps, color changes on the peridium and chemical reactions. Grubisha et al. (2002) have also pointed out that the placement of some species under Fulviglebae and Rhizopogon remained unresolved using the classification proposed by Smith and Zeller (1966). There is also a running hypothesis that genera *Suillus* and *Rhizopogon* are evolutionarily related (Theirs, 1984). Molecular evidences accrued via nuclear ribosomal large subunit (28s) DNA sequence also support the contention that *Suillus* and *Rhizopogon* are both monophyletic and taxonomically related.

Figure 1.8 depicts a phenogram of Rhizopogon developed on the basis of ITS1, ITS2 and 5.8S subunit nrDNA sequence (see Grubisha et al. 2002). Let us consider the section Rhizopogon in a slightly greater detail. The clade A comprises the genera *R. fuscorubens*, *R. luteus*, *R. occidentalis*, *R. ochraceorubens* and *R. succosus*. *R. succosus* and *R. luteus* share several morphological features. Molecular analysis has also shown that the two species share long insertions in the ITS1 sequences. Similarly, *R. ochraceorubens* and *R. fuscorubens* are closely related. The major difference is in rhizomorphs and peridium color. In the clade B, we encounter *R. subsalmonus* and *R. evadens*. *R. subsalmonus* does not stain red, but *R. evadens* turns red when cut, and the peridium stays white without yellow coloration. The clade C comprises

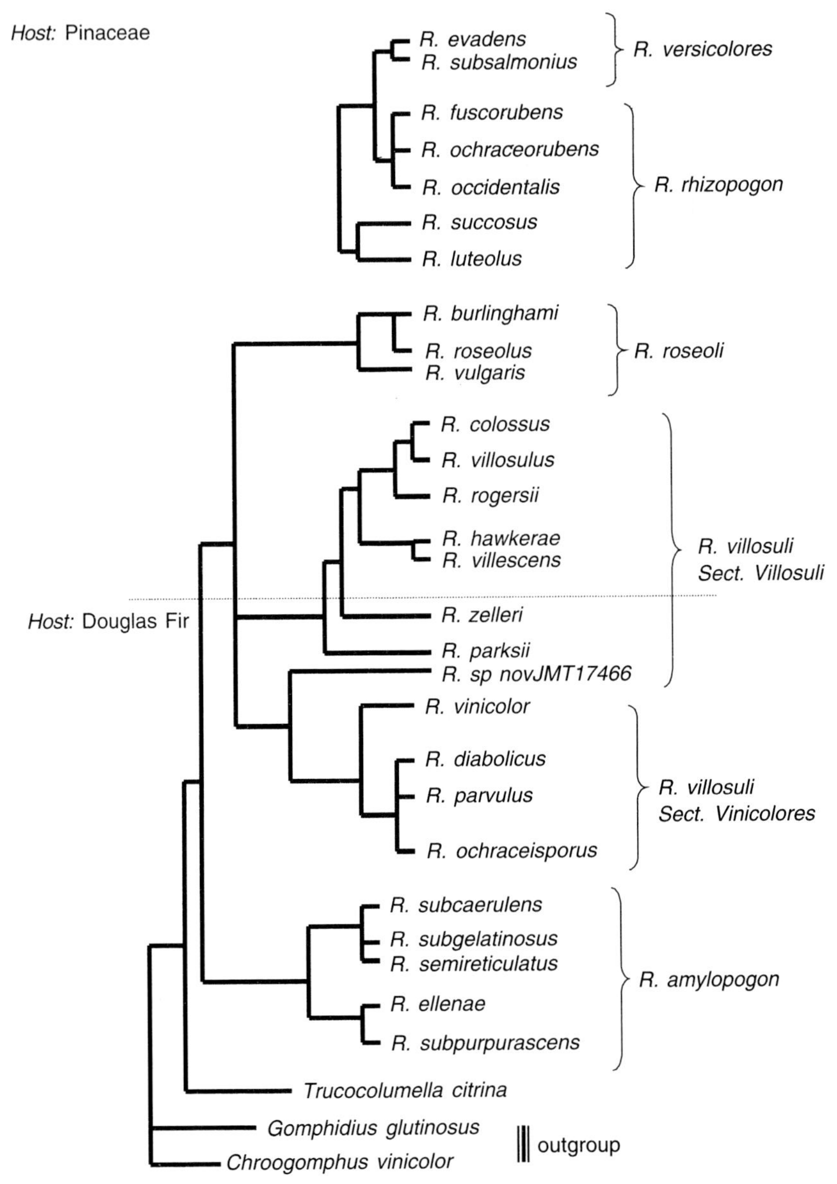

Fig. 1.8 A phylogram of the genus *Rhizopogon* based on ITS1, ITS2 and 5.8S Subunit nrDNA sequences (*Source*: Redrawn based on Grubisha et al. 2002).

R. burlinghensii, *R. roseoulus* and *R. vulgaris*. All these three species form mycorrhizal associations with *Pinus* species.

Next, the section Amylopogon is a monophyletic group and supports a clade exemplified by *R. ellenae* and *R. subpurpurascens*. Amyloid spores are an important phylogenetic trait. Reaction of peridium to KOH is a change in color from olive to green, blue, pink or red. These species possess a broad host range for mycorrhizal ability, but tend to confine to conifers. The section Fulviglebae consists of species such as *R. diabolicus*, *R. ochraceisporus*, *R. parvulus* and *R. vinicolor*. These species form a well-supported clade with bootstrap value 99. They share several morphological characters, including those of peridium. Section Villosuli comprises species such as *R. colossus*, *R. villosulus*, *R. regersii*, *R. hawkerae*, *R. villescens*, etc. A few species such as *R. colossus*, *R. hawkerae*, *R. reticulates* and *R. villosulus* were synonimized and clubbed as a single species *R. villosulus*. It was based on RFLP analysis of ITS rDNA (Martin et al. 1998). Incidentally, using similar molecular techniques, Grubisha et al. (2002) have prepared a revision of subgeneric level classification within subgenus *Rhizopogon*. It comprises subgenera *Rhizopogon, Amylopogon, Roseoli* (sections Roseoli and Fulviglebae), Versicolores, Villosuli and Vinicolores. Albeit, we may note that analysis of *R. roseolus* using ITS and 5.8S DNA sequences showed a high degree of variation among isolates drawn from the same collection. It resulted in classification patterns difficult to match with those proposed by previous authors (Martin et al. 2000). Recently, Kretzer et al. (2003) have reclassified the *R. vinicolor* complex into two distinct clades based on the ITS sequences and microsatellite loci. It is suggestive of two biological species, namely *R. vinicolor* and *R. vesiculosis*.

Ascomycetous Ectomycorrhizas

Cenococcum: *Cenococcum* is an ecologically important ECM fungus with worldwide distribution. It has a wide host range with over 200 gymnosperms and angiosperms as host. It survives and proliferates in hot, semi-arid and arctic environments. It adapts to pH range from 3.4 to 7.5. It grows slowly in culture medium, so a comparatively limited amount of taxonomic and genetic aspects are known. The difficulty in identification and phylogenic analysis of this ECM fungus is mainly because it produces only sterile mycelia and rarely sclerotial bodies, plus the morphological variations are not great among isolates (Sinohara et al. 1999; Trappe, 1977). It does not form asexual or sexual spores, which are the primary criteria for a classical taxonomic treatment of the fungus.

Based on morphological and ecological traits, the genus *Elaphomyces* (a hypogeous Ascomycete) is considered as the sexual state (teliomorph) of *Cenococcum geophilum*. In other words, *C. geophilum* is deemed an asexual state (anamorph) of *Elaphomyces*. No doubt, both *Elaphomyces* and *C. geophilum* are mycorrhizal. However, the molecular phylogenetic analysis of the 18S rRNA gene of *C. geophilum* conflicts with this hypothesis that *Elaphomyces* is the sexual state. Lubiglio et al. (1996) compared nucleotide sequence data for 5 isolates of *C. geophilum*, 3 of *Elaphomyces* sp. and 44 additional Ascomycetes, so that the phylogenetic status of *C. geophilum* could be understood better. Parsimony and distance analysis positioned *C. geophilum* as a basal and intermediate lineage between the two Laculoascomycete orders, Pleosporales and Dothidiales. *Cenococcum geophilum* had no sexual relatives among the Ascomycetes examined (Fig. 1.9; Lubiglio and Taylor, 2002)

In another study, Lobuglio et al. (1999) had earlier examined over 70 isolates of *C. geophilum* using RFLP data of entire repetitive units of rDNA and commented that it may represent more than one species. As we know, often rDNA ITS (ITS1, ITS2 and 5.8 S) genes analysis is done during molecular phylogeny. Shinohara et al. (1996) studied CgSSU (*C. geophilum* small subunit gene) intron sequences of rRNA and utilized it for taxonomic analysis at genus and species level. The aim was to understand the extent of variation in the secondary structures of ITS2 among *C. geophilum*

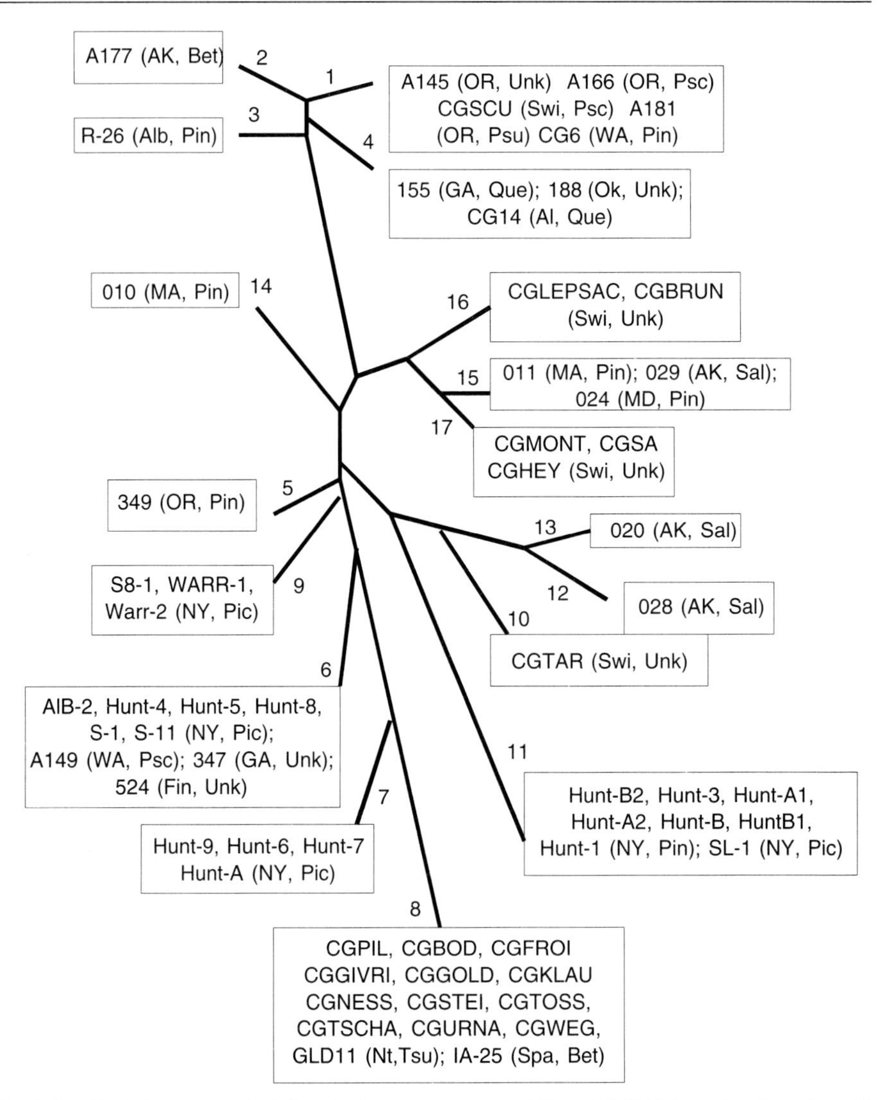

Fig. 1.9 A phylogram for isolates of *Cenococcum geophilum*. UPGMA-tree developed on 69 isolates of the ECM fungus based on restriction enzyme map of rDNA from ITS1, ITS2 and 5.8 S regions. (Source: Redrawn based on Shinohara et al., 1999). Information in the boxes relate to isolate, geographic origin and host. Note: Ak=Alaska; Al=Alabama; Alb=Alberta; Fin=Finland; GA=Georgia; MA=Massachussetts; NY=New York; OK=Oklahoma; OR=Oregon; Spa=Spain; Swe=Sweden; Swi=Switzerland; WA=Washington; Pin=*Pinus resinosa*; Pic=*Picea rubens*; Tsu=*Tsuga mertensiana*; Sal=*Salix rotundufolia*; Bet=*Betula nana*; Que=*Quercus alba*; Unk=Unknown.

isolates; evolutionary pressures on ITS sequences; and to examine whether broad geographic distribution and host range gets reflected in the genetic diversity assessed via ITS sequences and RNA secondary structure. Similarly, Portugal et al.

(1998) studied several isolates of *C. geophilum* using ITS1, ITS2 as well as amplified region of 18S gene and found polymorphism. Shinohara et al. (1999) have reported that ITS regions in *C. geophilum* are conserved and sequence similarity falls within the range of intraspecific variation found in other fungi. *C. geophilum* seems to possess one of the shortest ITS lengths, which are 90 and 140 bp for ITS1 and ITS2, respectively. In general, filamentous fungi have shorter ITS regions. With regard to secondary structure of ITS2 RNA, Shinohara et al. (1999) state that it is similar among all the isolates studied by them. They commented that high degree of ITS conservation is unexpected, perhaps not commensurate with their broad host range and worldwide distribution. They speculate that *C. geophilum* utilizes parasexuality in forming heterokaryons, instead of generating genetic variation via mating. *C. geophilum* hyphae may exchange nuclei, creating several versions of rDNA, but selective competition among nuclei/gene conversion may lead to homogenous rDNA repeats. Such a mechanism blends well for asexual fungi such as *C. geophilum*, because it allows genetic exchange in the absence of meiosis. Still, Shinohara et al. (1999) prefer to suggest that efficient means of dispersion and high adaptability are responsible for broad ecological, geographical and host range. Albeit, *C. geophilum* is a coherent taxonomic group and may represent only a single species.

Pezizales: Members of Pezizales are found in diverse habitats and plant hosts. A large number of species belonging to Peziza have been described including E-strain fungi. These E-strain fungi have been classified as *Wilcoxinia, Humaria, Sphaerosporella, Morchella*, *Geospora* and hypogeous fungi such as *Tuber*. Most of these fungal species are capable of ectomycorrhizal and/or ectendomycorrhizal association. They have a smooth, thin mantle that could be overlooked. Assessing the phylogenetic relationship among these hypogeous fungi is difficult because of extreme convergent morphology. Hence, molecular analysis of ITS region (Roux et al. 1998) and nuclear-encoded rRNA genes (Riccardo et al. 1998) were utilized to arrive at phylogenic positions of different families and species of Pezizales. A large portion of 18S rDNA (~1700 nt) was matched with known databases to infer phylogenies at the subgeneric level.

Truffles: Truffles are ectomycorrhizal with roots of trees belonging to oak, poplar, willow, hazel and several shrubs. Truffles are identifiable based on morphological traits of their fruit bodies. However, their identification may be a problem when found in a symbiotic phase, because several of their characteristic features may be lost. Mello et al. (2001) point out that with certain species such as *Tuber magnatum*, it is difficult to distinguish them morphologically from other related truffles. Therefore, they devised molecular methods based on ITS sequence to identify this truffle species when present along with several others. Later, Mello et al. (2002a) attempted to establish a rapid technique that types the truffles, especially *Tuber borchii* and other species, using PCR amplification and specific primers. The DNA isolated from fruit bodies, mycelia and mycorrhiza of *T. borchii* was amplified; later, the RFLP analysis of the amplified ITS region allowed discrimination of *T. borchii* from other truffle species. They also reported that primer pair TBA/TBB was specific for *T. borchii*. It helped in reliable and quick identification of the ECM fungus. Bertini et al. (1998) have analyzed white truffles (*Tuber borchii*) using PCR-RAPD techniques. They identified five different species. Later, two molecular markers specific for *T. borchii* were developed from the sequence of non-polymorphic RAPD fragment and from the regions flanking the 5′-3′ ends of a truffle gene. These markers helped in identification of *T. borchii* isolates using their fruit bodies, mycelia and mycorrhizas. Hence, in addition to phylogenetic placement, they could monitor and study life cycle of these fungi in the soil. Roux et al. (1998) studied the phylogenic position of a few European *Tuber* species such as *T. melanosporum, T. magnatum* and *T. uncinatum* that have commercial value. They used sequence variation in ITS region to classify these species. Rossi et al. (1999) selected four different mycelial strains of *Tuber brochii* and studied the DNA polymorphism. A very low genetic variability was noticed. The few polymorphisms found were sufficient to be used as strain-specific markers during phylogeny and other biotechnological applications. Iotti et al. (2001) have rightly pointed out that mycelia from many *Tuber* species grow slowly. Often, the mycelia are extracted from mycorrhiza/soil and tools to accurately identify their source were not available. Hence, they devized tests based on morphotyping and molecular analysis of mycelial samples from putative *Tuber* species. Morphologically, *Tuber* samples analyzed by them showed several similarities with regard to hyphal characters such as anastomoses, vesicle formation, hyphal granulation, etc. Using RFLP and sequence analysis of ITS region of mycelial samples, they could easily identify *T. maculatum, T. aestivum, T. macrosporum, T. rufum*, and *T. brumale.* Giomaro et al. (2002) used molecular approaches such as analysis of nucleotide sequence of ITS region and RFLP to *Tuber brumale* that was inoculated on to micropropagated *Quercus* and *Tilia* seedlings. They state that host plant had considerable influence on the morphological features on the truffle fungus. Considering the excessive variability noticed for morphological features on *Tuber brumale*, its identification based on molecular analysis seemed necessary. In order to ascertain the genetic variability and degree of relatedness of *Tuber uncinatum* with other *Tuber* species (e.g. *Tuber aestivum*), Mello et al. (2002b) utilized multi-locus approaches such as primed-PCR, and single locus markers such as nuclear and mitochondrial rDNA to compare the *Tuber* species. Taken altogether, *T. uncinatum* and *T. aestivum* clustered into two different groups. They noticed that the level of genetic polymorphism in *T. uncinatum* was higher than other *Tuber* species. The mitochondrial rDNA analysis was not able to differentiate between the morphologically related and unrelated truffles. They concluded that different molecular approaches plus morphological approaches could provide better judgment regarding phylogenetic relationships among *Tuber* species. *Tuber* species have also been characterized using morphological and molecular traits of their hyphae. Pure cultures of *T. maculatum, T. melanosporum, T. aestivum, T. macrosporum, T. rufum*, and *T. bromale* were grown to extract mycelia. Molecular identification tests such as RFLP and ITS region sequence were performed on mycelia of different isolates. *Tuber* species showed several common morphological traits such as hyphal anastomoses, vesicle formation and hyphal aggregation. Several other mycelial traits such as hyphal branch angle, septal distance, hyphal diameter and growth rate were also utilized to classify the *Tuber* species (Iotti et al. 2002). Amicucci et al. (2000) selected species-specific primers from the ITS region sequence and devized a were able to multiplex polymerize chain reaction (PCR) that aided identification of *Tuber* species. They were able to simultaneously identify or detect four different *Tuber* species namely, *T. magantum, T. borchii, T. maculatum* and *T. puberulum.* Two repeated DNA sequences specific to European strains of *T. melanosporum* were isolated by Paolocci et al. (2000). One of these denoted as SS14 is specific to

T. melanosporum. The second repeated sequence SS15 is specific to *T. melanosporum*. It is also useful to detect even Asian Black Truffle *T. indicum*.

3. D. Phylogeny of Ericaceous Mycorrhizas

The ectendomycorrhizas, commonly termed Ericaceous mycorrhizas, are worldwide in distribution. Ectendomycorrhizas are formed by plants belonging to families of the order Ericales, namely Ericaceae, Empetraceae and Epacridaceae (Watson et al. 1967; Cronquest, 1981; Thorne, 1992). These ectendomycorrhizas are frequent on heath and temperate forest species (Smith and Read, 1997; Xiao and Berch, 1996). Similarities between fungi that form ectendomycorrhiza, mainly among the ericaceous taxa, are suggestive of a monophylogenetic origin of the symbionts, especially host plant (Cairney, 2000). Details regarding the geological time frame for these ericoid mycorrhizas are poorly understood. Micro- and macro-fossils from the southern and northern America, Australia, Southern Gondwana and elsewhere in Pacific locations indicate that epacrids and other ericaceous mycorrhizas existed in Late Cretaceous period around 30 to 80 million years ago (see Cairney and Ashford, 2002). Analysis of fossil records of certain Ascomycetes that form ectendomycorrhizas suggest the occurrence of ectendomycorrhizas during the Early Cretaceous period, i.e. around 140 million years ago (Cullings, 1996). Cairney and Ashford (2002) have pointed that changes in biogeography, climatic patterns, and preferences exhibited by the symbionts could have caused the divergence among Ericaceous mycorrhizas.

Microscopic examination and molecular analysis of a large number of naturally occurring Ericoid associations have revealed that endophytes so far isolated are Ascomycetes or their anamorphs (Carney and Ashford, 2002). Many of these endophytes possess mycorrhiza-forming ability on Ericaceous plants. These fungi could be identified based on fungal coils inside the roots. However, it may not prove the mycorrhizal status and the functional benefits to plants (Read and Kearney, 1995). At times, basidiomycete mycelium too may be encountered in the root cells of epacrid hosts. Such basidiomycetes may even take part in reciprocal transfer of carbon and phosphorus, which needs authentication (Allen et al. 1989). Certain Ericaceous (epacrids) hosts may also show up arbuscular mycorrhiza like fungi, but these could be opportunistic, non-symbiotic infections (Reed, 1991, 1996; McLean and Lawrie, 1996).

Ascomycetes belonging to Helotiaceae, for example *Hymenoscyphus ericae* and anomorphic Myxotrichaceae, *Oideodendron* sp. are frequent on Ericaceous plants of Northern Hemisphere. In fact, the biology of this symbiotic fungus, *H. ericae*, has been studied in comparatively greater detail, and hence it is a model fungus to study ericoid symbiosis in laboratories. Zymograms of Ericaceous roots have generally revealed that Ericaceae and Epacridaceae roots naturally harbor a diverse array of ascomycetous endophytes. Molecular analysis has also confirmed that Ericaceous plants in Northern and Southern hemispheres typically house diverse endophyte taxa (Perrotto et al. 1996; Sharples et al. 2000). As in ectomycorrhizas (Dahlberg, 2001), only a small number of endophytic taxa are found to dominate the root space (Chambers et al. 2000; Midgley et al. 2001). Knowledge regarding the composition and intensity of different endophytes within an Ericoid root, no doubt, will be useful. On a broad scale, investigations on endophytes colonizing Ericaceae from Northern and Southern hemispheres have revealed the taxonomic similarity among them. The phylogenetic analysis of ITS region conducted by Sharples et al. (2000) has led to the identification of two strongly supported (100% boot strap support) clades. In their analyses, most endophytes from epacrids could be grouped to form one clad, along with a few endophytes that colonized Ericaceous hosts of Northern Hemisphere. The second clade comprised all *Hymenoscyphus* spp. and epacrid endophytes. From their study involving analysis of ITS2 nucleotide sequences of Heliotales clades, Monreal et al. (1999) reported that endophytes collected from different locations worldwide could be classified as part of a *Hymenoscyphus* like group. Further, the DNA sequence data suggests that *H. ericae* represents an aggregate of related isolates, and may not be a single well-defined taxa (Sharples et al. 2000; Reed, 2000; Vralstad, 2000). From a similar study utilizing 28s RNA gene sequences, Cullings (1996) has inferred that plant taxa with the ability to form ericoid symbionts could be grouped together. Gernand et al. (2001) made a phylogenetic analysis of Helotiales and Rhytismatales using small subunit ribosomal DNA sequence. They concluded that families such as Helotiacea were polyphyletic, but Rhytismataceae was paraphyletic. Phylogeographic studies conducted by Perrotto and Bonfonte (2002) were based on rDNA genes (ITS and IGS regions) of a large number of ericoid fungal isolates from locations worldwide. Their investigations were essentially aimed at elucidating the relationship, if any, between genetic/functional diversity versus the wide range of environments/locations that these ectendomycorrhizas inhabit. They suggested that the occurrence of some rare group-1 introns in several isolates may assume significance in phylogeographic studies. Berch et al. (2002) utilized traditional culturing and molecular characterization methods to assess ectomycorrhizas. They reported the occurrence of five putative and two polyphyletic assemblages of ericoid fungi in *Gaultheria* and other Ericaceous hosts. Using phylogenetic analysis of ITS2 sequences, they could confirm that these genotypes occurred in North America, Europe and Australia. The recovery rate and culturability of these endophytes were low, owing to contamination with a non-culturable basidiomycete resident on the root system.

Concluding Remarks

For a long time, the evolution of mycorrhizal phenomenon was studied predominantly utilizing fossil records and phylogenetics. During the recent past, application of molecular techniques has improved our ability to judge the evolution of mycorrhizal symbiosis. It is believed that AM endophytes occurred in the Devonian period. Evolutionarily, the order Glomales that colonized Triassic plants predates other types of mycorrhizas such as ECM. They seem to have descended from endophytes of algal precursors, whereas, ECM fungi belonging to Basidiomycetes and Ascomycetes are comparatively recent to evolve. Phylogenetic analysis using DNA sequence data suggests that wood-rotting agarics developed the ability to form ECM symbiosis during the Early Cretaceous period. Since we are dealing with symbiosis, evolutionary aspects of both fungus and plant need attention. With regard to plant, common bryophytes such as mosses are non-mycorrhizal. Fossil records and molecular analysis have proved that *Aglaophyton* are the earliest land plants that contained arbuscular mycorrhizas. Pteredophytes that dominated during the Silurian to Paleozoic era supported AM symbiosis. During Early Cretaceous period, both kinds of mycorrhizas, AM and ECM, became frequent on plants. Given this framework on the evolution of mycorrhizas, we may forecast that several more examples and details will be added in due course. There are few pertinent aspects that need attention. Firstly, what drives the evolution of two partners towards symbiosis? Host and fungal evolutionary genetics versus dynamics of environmental conditions during different periods need to be compared. Is it nutrients or other factors that dictated the evolution of mycorrhizal symbiosis? Evolutionary divergences of some plant orders and families towards being mycorrhizal or non-mycorrhizal need comparative analysis, especially with reference to molecular phylogeny. Do we find plant types whose ancestors might have alternated between being mycorrhizal for a certain length of history, to later became non-mycorrhizal and then switched again to symbiotic status? For example, are there Cruciferous species that were mycorrhizal at a certain point of time in their evolution? Analyses of such samples may lead us to interesting inferences on conditions that drive the partners towards symbiosis or otherwise. It is difficult to guess, but should we take to conjectures regarding the evolutionary pattern of mycorrhizas that may unfold in the future?

Regarding phylogeny of mycorrhizas, classical methods involving morphotyping have been very useful. Morphotyping has led to identification and classification of over 6000 ECM fungi and 160 AM fungi. Advent of molecular techniques, especially ITS-RFLP, has resolved many questions on phylogeny of mycorrhizas. Inferences from morphotyping and molecular analyses have generally coincided, but molecular techniques have proved to be more accurate. The current suggestion is to use both morphotyping and molecular analyses wherever feasible. There is a clear need to develop phenograms for all possible AM and ECM fungi and utilize them as a ready reckoner during identification and classification exercises. Phylogeography of mycorrhizal fungi is interesting and useful. It needs greater emphasis. In future, depending on the enthusiasm of mycorrhiza researchers and their trend to bifurcate or club fungal species, we may either encounter a higher or lower number of AM/ ECM species. Mycorrhizal phylogeny remains a dynamic aspect and promises to be so for a while.

References

Aanen, D.K. and Kuyper, T.W. 1999. Intercompatibility in the *Hebeloma crustuliniforme* complex in North Europe. Mycologia 91: 783-795.

Aanen, D.K., Kuyper, T.W., Boekhout, T. and Hoekstra, R.F. 2000. Phylogenetic relationships in the genus *Hebeloma* based on ITS1 and 2 sequences with special emphasis on the *Hebeloma crustuliniforme* complex. Mycologia 92: 269-281.

Aanen, D.K., Kuyper, T.W. and Hoekstra, R.F. 2001. A widely distributed ITS polymorphism within a biological species of the Ectomycorrhizal fungus *Hebeloma velutipes*. Mycological Research 105: 284-290.

Aldwell, F.E.B. and Hall, I.R. 1987. A Review of Serological Techniques for the identification of Mycorrhizal fungi. In: Mycorrhiza in the Next Decade. Sylvia, D.M, Hung, L.L. and Graham, J.H. (Eds). University of Florida, Gainesville, Florida, USA, pp. 305-307.

Aldwell, F.E.B., Hall, I.R. 1987 and Smith, J.M.B. 1983. Enzyme-linked immunosorbent (ELISA) to identify Endomycorrhizal fungi. Soil Biology and Biochemistry 15: 377-378.

Alia, R. and Dodd, J.C 1998. The screening of isolates of *Glomus etunicatum* for their effects on the growth of different plants and the presence of isozyme markers for detection *in planta*. Second International Conference on Mycorrhizas, Uppsala, Sweden, http://www.mycorrhizas ag.utk.edu

Allen, W. K., Allaway, W.G., Cox, G.C. and Valder, P.G. 1989. Ultra structure of mycorrhizas of *Dracophylum secundum* R.Br (Ericales). Australian Journal of Plant Physiology 16: 147-153.

Amiccuci, A., Guidi, C., Zambonelli, A., Potenza, L. and Stocchi, V. 2000. Multiplex PCR for the identification of white Tuber species. FEMS Microbiology Letters 189: 265-269.

Anderson, I.C., Chambers, S.M. and Cairney J.W.G. 1998. Molecular determination of genetic variation in *Pisolithus* isolates from a defined region in New South Wales, Australia. New Phytologist 138: 151-162.

Arajo, E., Elza, F., Eliana, A.G., Kasuya, M., Catarina, M. and Everaldo, G.B. 1998. Molecular characterization of *Pisolithus tinctorius* isolates by RAPD and rDNA PCR-RFLP. Second International Conference on Mycorrhizas, Uppsala, Sweden, http://www.mycorrhizas.ag.utk.edu

Avio, L. and Giovannetti, M. 1998. The protein pattern of spores of arbuscular mycorrhizal fungi: Comparison of species, isolates and physiological stages. Mycological Research 102: 985-990.

Bago, B., Bentivenga, S.P., Brenac, S.P., Dodd, J.C., Piche, Y. and Simon, L. 1998. Molecular analysis of *Gigaspora* (Gigasporaceae, Glomales). New Pytologist 139: 581-588.

Bentivenga, S.H. and Morton, J.B. 1994. Stability and heritability of fatty acids methyl ester profiles of glomalean endomycorrhizal fungi. Mycological Research 98: 1419-1426.

Bentivenga, S.P. and Morton, J.B. 1995. A monograph of genus *Gigaspora*, incorporating developmental patterns of morphological characters. Mycologia 87: 719-731.

Berch, S.M., Allen, T.R. and Berbee, M.L. 2002. Molecular detection, community structure and phylogeny of ericoid mycorrhizal fungi. Plant and Soil 244: 55-66.

Berch, S.M. and Trappe, J.M. 1985. A new species of Endogonecea, *Glomus hoi.* Mycologia 77: 654-657.

Berch, S.M. and Trappe, J.M. 1987. Revision of Trappe's 1982 Synoptic keys to genera and species of Endogone. USDA Forest Service, Northwest Forest and Range Experiment Station Report. Corvallis, Oregon, pp. 1-33.

Bertini, L., Agostini, D., Poteza, L., Rossi, I. Zeppa, S., Zambonelli, A., Stocchi, V. 1998. Molecular markers for the identification of the ectomycorrhizal fungus *Tuber borchii.* New Phytologist 139: 565-570.

Bougher, N.L. 1995. Diversity of Ectomycorrhizal fungus associated with *Eucalyptus* in Australia. In: Mycorrhizas for Plantation Forestry in Asia. Brudndrett, M., Dell, B., Malacjzuk, N. and Gog, M.D. (Eds). Australian Center for International Agricultural Research, Canberra, Australia, pp. 8-15.

Brundrett, M.C.2002. Coevolution of roots and mycorrhizas of land plants. New Phytologist 154: 275-304.

Brundrett, M., Bougher, N., Dell, B., Grove, T. and Malajczuk, N. 1996. Working with Mycorrhizas in Forestry and Agriculture. Australian Center for International Agricultural Research, Canberra, Australia, pp. 1-373.

Bruns, T.D. and Szaro, T.M. 1992. Rate and mode differences between nuclear and mitochondrial subunit rRNA genes in mushrooms. Molecular Biology and Evolution 9: 836-855.

Bruns, T.D., Vilgalys, R., Barns, S.M., Gonzalez, D., Hibbet, D.S., Lane, D.J., Simon, L., Stickel, S., Szaro, T.M., Weisberg, W.G. and Sogin, M.L. 1992. Evolutionary relationships within the fungi: Analysis of small subunit rRNA sequences. Molecular Phylogenetics and Evolution 1: 231-243.

Bruns, T.D., White, T.J. and Taylor, J.W. 1991. Fungal molecular systematics. Annual Review of Ecological Systematics 22: 525-564.

Buscot, F., Munch, J.C., Charcosset, J.Y., Gardes, M., Nehls, U. and Hampp, R. 2000. Recent advances in exploring Physiology and Biodiversity of Ectomycorrhizas highlight the functioning of these symbiosis in Ecosystems. FEMS Microbiology Reviews 24: 601-614.

Cairney, J.W.G. 2000. Evolution of mycorrhizal systems. Naturweissenschaften 87: 467-475.

Cairney, J.W.G. and Ashford, A.E. 2002. Biology of mycorrhizal associations of Epacrids (Ericaceae). New Phytologist 154: 305-326.

Chambers, S. M., Liu, G. and Cairney, J.W.G. 2000. ITS rDNA sequences comparison of Ericoid mycorrhizal endophytes from *Woollsia pungens.* Mycological Research 104: 168-174.

Clapp, J.P., Fitter, A. and Young, J.P.W. 1999a. Diversity of *Scutellospora* SSU rRNA sequences from woodland roots and spores. Second International Conference on Mycorrhizas, Uppsala, Sweden, http://www.mycorrhizas.ag.utk.edu

Clapp, J.P., Fitter, A. and Young, J.P.W. 1999b. Ribsomal subunit sequence variation with spores of an arbuscular mycorrhizal fungus, *Scutellospora* sp. Molecular Ecology 8: 915-921.

Clapp, J.P., Young, J.W.P., Merryweather, J.W. and Fitter, A.H. 1995. Diversity of fungal symbionts in arbuscular mycorrhizas from a natural community. New Phytologist 130: 259-265.

Cronquist, A. 1981. An integrated system of classification of flowering plants. Columbia University Press, New York, 328 pp.

Cullings, K.W. 1996. Single phylogenetic origin of Ericoid Mycorrhizae within the Ericaceae. Canadian Journal of Botany 74: 1896-1909.

Dahlberg, A. 2001. Community ecology of Ectomycorrhizal fungi. New Phytologist 150: 555-562.

Dalpe, Y., Plenchette, C., Frenot, Y., Gloaguen, J.C. and Strullu, D.G. 2002. *Glomus kerguelense* sp. nov a new Glomales species from sub-Antarctic. Mycotaxon 84: 51-60.

Dodd, J.C., Boddington, C.L., Rodriguez, A., Gonzalez-Chavez, C. and Mansur, I. 2000. Mycelium of Arbuscular Mycorrhizal fungi from different genera: Form, function and detection. Plant and Soil 226: 131-151.

Dodd, J.C., Rosendhal, J., Giovannetti, M., Broome, A. Lanfranco, L. and Walker. 1996. Inter- and intrageneric variation within the morphological similar arbuscular mycorrhizal fungi *Glomus mosseae* and *Glomus coronatum*. New Phytologist 133: 113-122.

Drehmel, D., Moncalvo, J. and Vilgalys, R. 1999. Molecular phylogeny of *Amanita* based on large subunit ribosomal DNA sequences: Implications for taxonomy and character evolution. Mycologia 91: 610-618.

Edwards, S.G., Fitter, A.H. and Young, J.P.W. 1997. Quantification of an arbuscular mycorrhizal fungus, *Glomus mosseae*, within plant roots by competitive polymerase chain reaction. Mycological Research 101: 1440-1444.

Franken, P. 1999. Trends in molecular studies of AM fungi In: Mycorrhiza. 2nd Edition. Verma, A. and Hock, B. (Eds) Springer Verlag, Berlin, Heidelberg, pp. 37-49.

Frostgaard, A., Tunlid, A. and Baath, E. 1996. Microbial biomass measured as Total Lipid Phosphate in soils of different organic matter content. Journal of Microbiological Methods 14: 156-163.

Gadkar, V., Adholeya, A and Satyanarayana, T. 1997. Randomly amplified polymorphic DNA using M13 core sequence of the vesicular arbuscular mycorrhizal fungi, *Gigaspora margarita* and *Gigaspora gigantea*. Canadian Journal of Microbiology 43: 795-798.

Gehrig, H., Schubler, A. and Kluge, M. 1996. *Geosiphum pyreforme*, a fungus forming endocytosis with *Nostoc* (Cyanobacteria) is an ancestral member of *Glomales*: Evidences by SSU rRNA analysis. Journal of Molecular Evolution 43: 71-81.

Gensel, P.G., Kotyk, K. and Basinger, J.F. 2001. Morphology of above-and below-ground structures in early Devonian (Pragian-Emsian). In: Plants Invade land: Evolutionary and Environmental Perspectives. Gensel, P.G. and Edwards, D. (Eds). Columbia University Press, New York, USA, pp. 83-102.

Gerdemann, J.W. and Nicolosn, T.H. 1963. Spores of mycorrhizal *Endogone* species extracted from soil by wet sieving and decantation. Transactions of British Mycological Society. 46: 235-244.

Gerdemann, J.W. and Trappe, J.M. 1974. The Endogonecea of the Pacific Northwest. Mycologia Memoir 5: 1-76.

Gernandt, D.S., Platt, J.L., Stone, J.K., Spatofora, J.W, Holst-Jensen, A., Hamelin, R.C. and Kuhn, L.M. 2001. Phylogenetics of Helotiales and Rhytismatales based on partial small subunit ribosomal DNA sequences. Mycologia 93: 915-933.

Giomaro, G., Sisti, D., Zambonelli, A. Amicucci, A., Cecchini, M., Comandini, O. and Stocchi, V. 2002. Comparative study and molecular characterization of ectomycorrhizas in *Tilia americana* and *Quercus pubescens* with *Tuber brumale*. FEMS Microbiology Letters 216: 9-14.

Giovannetti, M., Avio, L. and Salutini, L. 1991. Morphological, cytochemical and ontogenic characteristics of a new species of a vesicular-arbuscular mycorrhizal fungus. Canadian Journal of Botany 69: 161-167.

Giovannetti, M. and Gianinazzi-Pearson, V. 1994. Biodiversity in arbuscular mycorrhizal fungi. Mycological Research 98: 7015-715.

Giovannetti, M., Sbrana, C., Strani, P., Agnolucci, M., Rinauda, V. and Avio, L. 2003. Genetic diversity of isolates of *Glomus mosseae* from different geographic areas detected by vegetative compatibility testing and biochemical molecular analysis. Applied and Environmental Microbiology 69: 616-624.

Gobel, C., Hahn, A. and Hock, B. 1995. Production of Polyclonal and Monoclonal antibodies against hyphae from Arbuscular Mycorrhizal fungi. Critical Reviews of Biotechnology 15: 293-304.

Gomes, E.A., De Abreu, L.M., Burges, A.C. and De Araujo, E.F. 2000. ITS sequences and mitochondrial DNA polymorphism in *Pisolithus* isolates. Mycological Research 104: 911-918.

Graham, J.H., Hodge, N.C. and Morton, J.B. 1995. Fatty Acid Methyl Ester profiles for characterization of Glomelean Fungi and their Endomycorrhizas. Applied and Environmental Microbiology 61: 58-64.

Grubisha, L.C., Trappe, J.M., Molina, R. and Spatofora, J.W. 2002. Biology of the ectomycorrhizal genus*Rhizopogon*. VI. Re-examination of intrageneric relationships inferred from phylogenetic analysis of ITS sequences. Mycologia 94: 607-619.

Harrier, L.A., Wright, F. and Horker, J.E. 1998. Isolation of the 3-phosphoglycerate kinase gene of the arbuscular mycorrhizal fungus *Glomus mosseae*. Current Genetics 34: 386-392.

Hart, M.M. and Reader, R.J. 2002. Taxonomic basis for variation in the colonization strategy of arbuscular mycorrhizal fungi. New Phytologist 153: 335-344.

Hawksworth, D.L., Kirk, P.M., Sutton, B.C. and Pagler, D.N. 1995. Ainsworth and Bisby's Dictionary of the Fungi. 8th Edition. CAB-International, Wallingford, U.K. 448 pp.

Hijri, M., Hosny, M. vanTueinen, D. and Duelieu, H. 1999. Intra-specific ITS polymorphism in *Scutellospora castanae* (Glomales) is structured within the multinucleate spores. Fungal Genetics and Biology 26: 141-151.

Hosny, M., Hijri, M. and Duleiu, H. 1999a. Molecular genetics as a unique tool for identification of biodiversity and evaluating genetic exchange within endomycorrhizal species. Second International Conference on Mycorrhizas. Uppsala, Sweden, http://www.mycorrhizas.ag.utk.edu

Hosny, M., Hijri, M., Passieux, E. and Duleiu, H. 1999b. rDNA units are highly polymorphic in *Scutellospora castanae* (Glomales). Gene 226: 61-71.

Iotti, M. Amicucci, A., Stocchi, V. and Zambolini, A. 2001. Morphological and molecular characterization of the mycelium of different *Tuber* species in pure cultures. Third International Conference on Mycorrhizas, Adelaide, Australia, http:// www.mycorrhizas.ag.utk.edu.

Iotti, M., Amicucci, A., Stocchi, V. and Zambonelli, A. 2002. Morphological and molecular characterization of mycelia of some *Tuber* species in pure culture. New Phytologist 155: 499-505.

Jabaji-Hare, S. 1988. Lipids and fatty acid profiles of some vesicular arbuscular mycorrhizal fungi: Contribution to taxonomy. Mycologia 80: 622-629.

Jansa, J., Gryndler, M. and Matucha, M. 1999. Comparison of the Lipid profiles of Arbuscular Mycorrhizal (AM) fungi and soil saprophytic fungi. Symbiosis 26: 247-264.

Jany, J.L., Bousquet, J. and Khasa, D.P. 2003. Micro-satellite markers for *Hebeloma* species developed from expressed sequences tags in the ectomycorrhizal fungus *Hebeloma cylindrosporum*. Molecular Ecology Notes 3: 659-662.

Kendrick, B. 1992. The Fifth kingdom. Mycologue Publications limited, Waterloo, Canada, 85 pp.

Kjoller, R. and Rosendahl, S. 2000. Detection of Arbuscular Mycorrhizal fungi (Glomales) in roots by nested PCR and SSCP (Single-Stranded Confirmation Polymorphism). Plant and Soil 26: 189-196.

Kjoller, R. and Rosendahl, S. 2002. Arbuscular mycorrhizal fungi in roots from a Danish pea field determined by polyacrylamide gel electrophoresis of specific fungal enzymes.http://www mycorrhizas.ag.utk.edu

Kojima,T., Sawaki, H. and Saito, M. 2004. Detection of arbuscular mycorrhizal fungi, *Archaeospora leptoticha* and related species colonizing plant roots by specifc PCR primer. Soil Science and Plant Nutrition 50: 95-101.

Koske, R.E. 1985. *Glomus aggregatum* emended: A distinct taxon in the *Glomus fasciculatum*. Journal 77: 619-630.

Kretzer, A.M. Luoma, D.L., Molina, R. and Spatafora, J.W. 2003. Taxonomy of *Rhizopogon vinicolor* species complex based on analysis of ITS sequences and micro satellite loci. Mycologia 95: 480-487.

Lanfranco, L., Bianciotto, V., Lumini, E., Souza, M., Morton, J.B. and Bonfonte, P. 2001. A combined morphological and molecular approaches to characterize isolates of arbuscular mycorrhizal fungi in Gigasporaceae (Glomales). New Phytologist 152: 169-178.

Lanfranco, L., Perotto, S. and Bonfonte, P. 2002. ITS sequences from the arbuscular mycorrhizal fungus *Gigaspora margarita*. http://www.fgsc.net/asilomar/evopop.html. Abstract 111, p. 1.

Lanfranco, L., Dalero, M. and Bonfonte, P. 1999. Intrasporal variability of ribosomal sequences in the endomycorrhizal fungus *Gigaspora margarita*. Molecular Ecology 8:37-45.

Lanfranco, L., Wyes, P., Marzachi, C. and Bonfonte, P. 1995. Generation of RAPD-PCR primers for the identification of isolates of *Glomus mosseae*, arbuscular mycorrhizal fungus. Molecular Ecology 4: 61-68.

Largent, D., Johnson, D. and Watling, R. 1977. How to identify mushrooms to Genus III. Microscopic features. Madison River Pres inc. New York, USA, pp. 39-52.

Lange, M. 1956. Danish Hypogeous Macromycetes. Dansk Botanisk Arkiv 16: 5-84.

Larson, J., Olsson, P.A. and Jackobsen, I. 1998. The use of fatty acid signatures to study mycelial interactions between the arbuscular fungus *Glomus intraradices* and the saprophytic fungus *Fusarium culmorum* in root free soil. Mycological Research 102: 1491-1496.

Lobuglio, K.F., Berbee, M.L. and Taylor, J.W. 1996. Phylogenetic origins of the asexual mycorrhizal symbiont *Cenococcum geophilum* and other mycorrhizal fungi among Ascomycetes. Molecular Phylogenetic and Evolution 6: 287-294.

Lobuglio, K.F., Rogers, S.O. and Wang, C.J. K. 1999. Variation in ribosomal DNA among isolates of the mycorrhizal fungus *Cenococcum geophilum*. Canadian Journal Botany 69: 2331-2343.

Lobuglio, K.F. and Taylor, J.W. 2002. Phylogeny and population structure of asexual mycorrhizal fungus *Cenococcum geophilum*. http://www.fgsc.net.asilomar/mycorr.html.

Longato, S. and Bonfonte, P. 1997. Molecular identification of mycorrhizal fungi by direct amplification of micro-satellite regions. Mycological Research 101: 425-432.

Lloyd-McGilp, S.A., Chambers, S.M., Dodd, J.C., Fitter, A.H., Walker. and Young, J.P.W. 1996. Diversity of the ribosomal Internally Transcribed Spacers within and among isolates of *Glomus mosseae* and related mycorrhizal fungi. New Phytologist 133: 103-111.

Madan, R., Pankhurst, C., Hawke, B. and Smith, S. 2002. Use of fatty acids for identification of AM fungi and estimation of the biomass of AM spores in soil. Soil Biology and Biochemistry 34: 125-128.

Maia, L.C., Yano, A.M. and Kimbrough, J.W. 1996. Species of Ascomycetes forming ectomycorrhizas. Mycotaxon 57: 371-390.

Margulis, L. and Schwartz, K.V. 1988. Five kingdoms. An illustrated guide to the Phyla of life on earth. Freeman and Co., San Fransisco, USA, pp 238.

Martin, F., Diaz, J., Dell, B. and Delaruelle, C. 2002. Phylogeography of the ectomycorrhizal *Pisolithus* species as inferred from nuclear rDNA ITS sequences. New Phytologist 153: 345-357.

Martin, M.P., Hogberg, N. and Nylund, J. 1998. Molecular analysis confirms morphological re-classifications of *Rhizopogon*. Mycological Research 102: 855-858.

Martin, M.P., Karen, O., Nylund, J.E. 2000. Molecular ecology of Hypogeous Mycorrhizal fungi: *Rhizopogon roseolus* (Basidiomycetes). Phyton-Annales Rei Botanicae 40: 135-141.

McLean, C. and Lawrie, A.C. 1996. Patterns of root colonization in Epacridaceous plants collected from different sites. Annals of Botany 77: 405-411

Mello, A., Cantisani, A., Vizzini, A. and Bonfonte, P. 2002b. Genetic variability of *Tuber uncinatum* and its relatedness to other black truffles. Environmental Microbiology 4: 584-594.

Mello, A., Fontana, A., Meottto, F., Comandini, O. and Bonfonte, P. 2001. Molecular and morphological characterization of *Tuber magnatum* mycorrhiza in a long-term survey. Microbiological Research 155: 279-284.

Mello, A., Garnero, L., Lonato, S., Perroto, S. and Bonfonte, P. 2002a. Rapid typing of *Tuber brochii* mycorrhizae by PCR amplification with specific primers. http://www.fgsc.net /asilomar/ mycorr.html Abstract 52, p 1.

Midgely, D.J., Chambers, S.M. and Cairns, J.W.G. 2001. Diversity and distribution of fungal endophyte genotypes in the root system of *Woollsia pungens* (Ericaceae). Third International Conference on Mycorrhiza, Adelaide, Australia, http://www.mycorrhizas.ag.utk.edu/icoms/ icom3.html

Miller, O.K. 1982. Taxonomy of Ecto- and Ectoendomycorrhizal fungi. In: Methods and Principles of Mycorrhizal Research. American Phytopathological Society, St Paul, Mn, USA, pp. 91-101.

Molina, R., Massicote, H. and Trappe, J.M. 1992. Specificity phenomena in mycorrhizal symbiosis: community–ecological consequences and practical implications. In: Mycorrhizal Functioning: An interactive plant fungal processes. Allen, M.J. (Ed). Chapman and Hall, New York, pp. 357-423.

Moncalvo, J.M., Lutzoni, F.M., Rehner, S.A., Johnson, J.E. and Vilgalys, R. 2000. Phylogenetic relationships of agaric fungi based on nuclear large subunit ribosomal DNA sequences. Systematic Biology 49: 278-305.

Moncalvo, J.M., Vilgalys, R., Redhead, S.A., Johnson, J.E., James T.Y., Aime, M.C., Larson, E., Baroni, T.J., Thorn, R.J., Jacobsen, S., Clemenson, H. and Miller, O.K. 2003. One hundred and seventeen clades of Euagarics. Molecular Phylogeny and Evolution (in press).

Monreal, M., Berch, S.M. and Berbee, M. 1999. Molecular diversity of ericoid mycorrhizal fungi. Canadian Journal of Botany 77: 1580-1594.

Morton, J.P. 1988. Taxonomy of mycorrhizal fungi: Classification, nomenclature and identification. Mycotaxon 32: 267-324.

Morton, J.P. 1992. INVAM News Letters, 2:1-5, http://mycorrhizas.ag.utk.edu

Morton, J.P. 1993. Problems and solutions for integration of Glomelean taxonomy, systemic biology and the study of endomycorrhizal phenomenon. Mycorrhiza 2: 97-109.

Morton, J.P. 1995. Taxonomic and phylogenetic divergence among five *Scutellospora* species based on comparative development sequences. Mycologia 87: 127-137.

Morton, J.P. and Benny, G.L. 1990. Revised classifications of Arbuscular fungi (Zygomycetes): A new order, Glomales, two new suborders, Glominae and Gigasporinae, and two families, Acaulosporaceae and Gigasporaceae, with an emendation of Glomaceae. Mycotaxon 37: 471-491.

Morton, J.P. and Bentivenga, S.P. 1994. Levels of diversity in Endomycorrhizal fungi (Glomales, Zygomycetes) and their role in defining taxonomic and non-taxonomic groups. Plant and Soil 159: 47-59.

Morton, J.P., Bentivenga, S. and Bever, J. 1995. Discovery, measurement, and interpretation of diversity in arbuscular endomycorrhizal fungi (Glomales, Zygomycetes). Canadian Journal of Botany 73: s25-s32.

Morton, J.P., Bentivenga, S. and Wheeler, W. 1993. Germplasm in the International collection of arbuscular mycorrhizal fungi (INVAM) and procedures for culture development, documentation and storage. Mycotaxon 48: 491-528.

Morton, J.P., Bever, J.D. and Pfleger, F.L. 1997. Taxonomy of *Acaulospora gerdemanii* and *Glomus leptotichum*, synanomorphs of an arbuscular mycorrhizal fungus. Mycological Research 101: 625-631.

Morton, J.B. and Redecker, D. 2001. Two families of Glomales, Archeosporae and Paraglomaceae, with two new genera *Archeospora* and *Paraglomus* based on concordant molecular and morphological characters. Mycologia 93: 181-195.

Muller, M.M., Kantola, R. and Kitunen, V. 1994. Combining sterol and fatty acid profiles for characterizing fungi. Mycological Research 98: 593-603.

Olsen, G.J. and Woese, C.R. 1993. Ribosomal RNA—A key to phylogeny. FASEB Journal 7: 471-491.

Olsson, P.A., Baath, E. and Jackobsen, I. 1997. Distribution of an arbuscular mycorrhizal fungus between soil and roots, and between mycelial and storage structures as response to application studied by fatty acid signatures. Applied and Environmental Microbiology in 63: 3531-3538.

Olsson, P.A., Baath, E., Jackobsen, I. and Soderstorm, B. 1995. The use of phospholipid and neutral lipid fatty acids to estimate biomass of arbuscular mycorrhizal fungi in soil. Mycological Research 99: 623-629.

Paolocci, F., Rubini, A., Riccioni, C., Granetti, B. and Arcioni, S. 2000. Cloning and characterization of two repeated sequences in the symbiotic fungus *Tuber melanosporum*. FEMS Microbiology Ecology 34: 139-146.

Perrotto, R.S., Actis-Perino, E., Perugini, J. and Bonfonte, P. 1996. Molecular diversity of fungi from ericoid mycorrhizal roots. Molecular Ecology 5: 123-131.

Perrotto, S. and Bonfonte, P. 2002. Molecular and functional diversity of ericoid mycorrhizal fungi. http://www.fgsc.net/asilomar/mycorrh.html. Abstract 53, p.3.

Peyronel, B. 1923. Fructifications de l'endophyte a arbuscular et a vesicules des mycorrhizas endotrophes. Bull. Soc. Mycol. France 39: 1-8.

Peyronel, B. 1924. Specie de *Endogone* protuctrici di microrizi endotrofische. Bull. Mens. R. Staze. Pathol. Veg. Roma 5: 73-75.

Phipps C.J. and Taylor, J.M. 1996. Mixed arbuscular mycorrhizae from the Triassic of Antartica. Mycologia 88: 707-714.

Portugal, A. Martinho, P., Freitas, H. and Viera, R. 1998. Molecular characterization of some isolates of the ectomycorrhizal fungus *Cenococcum geophilum* by amplification nuclear ribosomal DNA. Second International Conference on Mycorrhiza, Uppsala, Sweden, http://www.mycorrhizas.ag.utk.edu.htm

Pringle, A., Moncalvo, J.M. and Vilgalys, R. 2000 High levels of variation in ribosomal DNA sequences within and among spores of a natural population of the arbuscular mycorrhizal *Acaulospora colossica*. Mycologia 92: 259-268.

Pryer, K.M., Schneider, H., Smith, A.R., Cranfill, R., Wolf, P.G., Hunt, J.S. and Sipes, S.D. 2001. Horsetails and ferns are monophyletic group and the closest living relatives to seed plants. Nature 409: 618-622.

Pyrozynski, K.A. and Dalpe, Y. 1989. Geological history of Glomaceae with particular reference to mycorrhizal symbiosis. Symbiosis 7: 1-36.

Redecker, D. 2002. Molecular identification and phylogeny of arbuscular mycorrhizal fungi. Plant and Soil 244: 67-73.

Redecker, D., Hijri, M., Duleiu.A. and Sanders, I.R. 1999. Phylogenetic analysis of a dataset of fungal 5.8S rDNA sequences shows that highly divergent copies of ITS spacers reported from *Scutellospora castanae* are of Ascomycete origin. Fungal Genetics and Biology 28: 238-244.

Redecker, D., Hijri, I. and Weimken, A. 2003. Molecular identification of arbuscular mycorrhizal fungi in roots: Perspectives and problems. Folia Geobotonica 38: 113-124.

Redecker, D., Morton, J.B. and Bruns, T.D. 2000. Ancestral lineages of arbuscular mycorrhizal fungi (Glomales). Molecular Phylogeny and Evolution 14: 276-284.

Redecker, D., Szaro, T.M., Bowan, R.J. and Bruns, T.D. 2001. Small genets of *Lactarius xanthogalactus*, *Russula cremoricolor* and *Amanita franchetsi*. Molecular Ecology 10: 1025-1034.

Redecker, D., Theirfelder, H., Walker. and Werner, D. 1999. Restriction analysis of PCR-amplified internal transcribed spacers of ribosomal DNA as a tool for species identification in different genera of the order Glomales. Applied and Environmental Microbiology 63: 1756-1761.

Reed, D.J.1991. Mycorrhiza in Ecosystems. Experientia 47: 376-391.

Reed, D.J. 1996. The structure and function of the Ericoid mycorrhizal root. Annals of Botany 77: 365-374.

Reed, D.J. 2000. Links between genetic and functional diversity—A bridge too far? New Phytologist 145: 363-365.

Reed, D.J. and Kearny, S. 1995. The status and function of ericoid mycorrhizal systems. In: Mycorrhiza: Structure, Function, Molecular Biology and Biotechnology. Varma, A. and Hock, B. (Eds) Springer Verlag, Berlin, Germany. pp. 499-520.

Remy, W., Taylor, T.W., Hass, H. and Kerp, H. 1994. Four hundred-million-year old vesicular arbuscular mycorrhiza. Proceedings of the National Academy of Science USA 91: 11841-11843.

Renzeglia, K.S., Duff, R.J. Nickrent, D.L., Garbary, D.J. 2000. Vegetative and reproductive innovations of early land plants: implications for a unified phylogeny. Philosophical transactions of the Royal Society of London., B355: 769-793.

Riccardo, P., Trevis, A., Zambonelli, A. and Ottonello, S. 1998. Molecular phylogeny of hypogeous Pezizales based on nuclear ribosomal DNA sequence analysis. Second International Conference on Mycorrhizas, Uppsala, Sweden, http://www.mycorrhizas.ag.utk.edu

Rodriguez, A., Dougall, T., Dodd, J.C. and Clapp, J.P. 2001. The large subunit ribosomal RNA genes of *Entrophospora infrequens* related to two different glomalean families. New Phytologist 152: 159-167.

Rosendahl, S. 1989. Comparisons of spore cluster forming *Glomus* species (Endogoneceae) based on morphological characterization and isozymes banding patterns. Opera Botanica 100: 215-223.

Rosendahl, S. and Taylor, J. 1997. Development of multiple genetic markers for studies of genetic variation in arbuscular mycorrhizal fungal communities. New Phytologist 130: 419-424.

Rossi, I., Zeppa, S., Potenza, L., Sisti, D., Zambonella, A., Stocchi, V. 1999. Intraspecific polymorphisms among *Tuber borchii* mycelial strains. Symbiosis 26: 313-325.

Roux, C., Sejelon-Delmas, N., Martins, M., Dargent, R. and Becard, G. 1998. Phylogenetic relationship between *Tuber melanosporum* and *T. indicum*, characterization of *T. melanosporum* specific genetic markers. Second International Conference on Mycorrhiza, Uppsala, Sweden, http://www.mycorrhizas.ag.utk.edu

Sakakibara, S.M., Jones, M.D., Gillespie, M., Hagerman, S.M., Forrest, M.F., Simard, S.W. and Durall, D.M. 2002. A comparison of ectomycorrhiza identification based on morphotyping and PCR-RFLP analysis. Mycological Research 106: 868-878.

Sanders, I.R., Alt, M., Groppe, K., Boller, T. and Wiemken, A. 1995. Identification of ribosomal DNA polymorphism among and within spores of the Glomales: Application to studies on the genetic diversity of arbuscular mycorrhizal fungal communities. New Phytologist 130: 419-427.

Sanders, I.R., Clapp, J.P. and Weimken, A. 1996. The genetic diversity of arbuscular mycorrhizal fungi in natural ecosystem—a key to understanding the ecology and functioning of the mycorrhizal symbiosis. New Phytologist 13: 95-101.

Sanders, I.R., Ravolanirina, F., Gianinazzi-Pearson, V. Gianinazzi, S., Lemone, M.C. 1992. Detection of specific antigens in the vesicular-arbuscular mycorrhizal fungi *Gigaspora margarita* and *Acaulospora laevis* using polyclonal antibodies to soluble spore fractions. Mycological Research 96: 477-480.

Schellenbaum, L., Gianinazzi, S. and Gianinazzi-Pearson, V. 1992. Comparison of acid soluble protein synthesis in roots of endomycorrhizal wild type *Pisum sativum* and corresponding isogenic mutants. Journal of Plant Physiology 141: 2-6.

Schenk, N.C. and Perez, Y. 1990. Manual for identification of mycorrhizal fungi. Synergistic Publications, Gainesville, Florida, USA, pp. 142-144.

Schenck, N.C. and Smith, G.S. 1982. Additional new and unreported species of mycorrhizal fungi (Endogoneceae) from Florida. Mycologia 74: 77-92.

Schmidt, E. and Oberwinkler, F. 1993. Mycorrhiza-like interactions between the achlorophyllus gametophyte of *Lycopodium clavatum* and its fungal endophyte studied by light and electron microscopy. New Phytologist 124: 69-81.

Schubler, A. 2000. *Glomus claroidium* forms and arbuscular mycorrhiza-like symbiosis with the hornwort *Anthoceros punctatus*. Mycorrhiza 10: 15-21.

Schwarzott, D., Walker, C. and Schubler, A. 2001. *Glomus*, the largest genus of the arbuscular mycorrhizal (Glomales) is non-monophyletic. Molecular Phylogeny and Evolution 21: 190-197.

Sharples, J.M., Chambers, S.M., Meharg, A.A. and Carney, J.W.G. 2000. Genetic diversity of root associated fungal endophytes from *Calluna vulgaris* at contrasting field sites. New Phytologist 148: 153-162.

Shinohara, M.L., Lobuglio, K.F. and Rogers, S.O. 1996. Group-1 intron family in the nuclear ribosomal RNA small subunit genes of *Cenococcum geophilum* isolates. Current Genetics 29: 377-387.

Shinohara, M.L., Lobuglio, K.F. and Rogers, S.O. 1999. Comparison of ribosomal DNA ITS regions among geographic isolates of *Cenococcum geophilum*. Current Genetics 35: 527-535.

Simon, L. 1996. Phylogeny of the Glomales: Deciphering the past to understand present. New Phytologist 133: 95-101.

Simon, L., Bousquet, J., Levesque, R.C. and Lalonde, M. 1993. Origin and diversification of endomycorrhizal fungi and coincidence with vascular land plants. Nature 363: 67-69.

Simon. L., Lalonde, M. and Bruns, T.D. 1992. Specific amplification of 18S fungal ribosomal genes from vesicular-arbuscular mycorrhizal fungi. New Phytologist 130: 419-424.

Singer, R.1986. The Agricales in Modern Taxonomy. 4th Edition, Koeltz Scientific Books, Koenigstein, Germany, pp. 185.

Smith, A.H. 1971. Taxonomy of Ectomycorrhiza forming fungi. In: Mycorrhizae. Hackskaylo, E. (Ed.). US Department of Agriculture Miscellaneous Publications, Washington D.C. pp. 1-8.

Smith, S.E. and Read, D.J. 1997. Mycorrhizal Symbiosis. Academic Press, London, pp. 486.

Smith, G.S. and Schenk, N.C. 1985. Two new dimorphic species in the Endogoneceae: *Glomus ambisporum* and *Glomus heterosporum*. Mycologia 77: 65-74.

Smith, A.H. and Zeller, S.M. 1966. A preliminary account of the American species of Rhizopogon. Memoirs of New York Botanical Garden 14: 1-178.

Soltis, D.E., Soltis, P.S., Chase, M.W., Mort, M.E., Albach, D.C., Zanis, M., Savolainen, V., Hahn, W.H., Hoot, S.B. Fay, M.F. Axtell, M., Swensen, S.M., Prince, L.M. Kreees, W.J., Nixon, K.C., and Farris, J.C. 2000. Angiosperm phylogeny inferred from 18s rDNA, rbcl, and atbp sequence. Botanical Journal of the Linnaean Society 133: 381-461.

Taylor, T.N., Remy, W., Hass, H. and Kerp, 1995. Fossil arbuscular mycorrhizae from the early Devonian. Mycologia 87: 56-573.

Taylor, T.N. and Taylor, E.L. 1993. The biology and evolution of fossil plants. Prentice Hall Inc., Engelwood Cliffs, New Jersey, USA, 356 pp.

Taylor, T.N. and Taylor, E.L. 1997. The distribution and interactions of some Paleozoic fungi. Review of Paleobotany and Palynology 95: 83-94.

Tehler, A., Farrris, J.S., Lipscomb, D.L. Kallejsro, M. 2000. Phylogenetic analysis of the fungi based on large rDNA data sets. Mycologia 92: 459-474.

Thaxter, R.1922. A revision of the Endogoneceae. Proceedings of American Academy of Arts and Sciences 57: 591-351.

Thiers, H.D. 1984.The sectoid syndrome. Mycologia 76: 1-8.

Thingstrup, I., Rozycka, M., Jaffrin, P., Rosendahl, S. and Dodd, J.C. 1995. Detection of the arbuscular mycorrhizal fungus, *Scutellospora heterogama* within roots using polyclonal antisera. Mycological Research 91: 1275-1232.

Thorne, R.T.1992. Classification and geography of flowering plants. Botanical Reviews 58: 225-348.

Trappe, J.M.1977. Selection of fungi for ectomycorrhizal inoculation in nursery. Annual Review of Phytopathology 15: 203-222.

Trappe, J.M. 1979. The orders, families and genera of hypogaeous ascomycetes (Truffles and their relatives). Mycotaxon 9: 297-340.

Trappe, J.M. 1982 Synoptic keys to the genera and species of Zygomycetous mycorrhizal fungi. Phytopathology 72: 1102-1108.

Trappe, J.M. 1987. Phylogenetic and ecological aspects of mycotrophy in the Angiosperms from an evolutionary standpoint. In: Ecophysiology of VA Mycorrhizal fungi. Safir, G.R. (Ed.). CRC Press, Boca Raton, Florida, USA, pp. 5-25.

Trappe, J.M. and Schenk, N.C. 1982. Taxonomy of fungi forming endomycorrhizae. In: Methods and Principles of Mycorrhizal Research. American Phytopathological Society, St Paul, Minnesota, USA, pp. 1-9.

Van Tuinen, D., Dulieu, H., Zeze, A. and Gianinazzi-Pearson, V. 1994. Biodiversity and characterization of arbuscular mycorrhizal fungi at the molecular level. In: Impact of Arbuscular Mycorrhizas on Sustainable Agriculture and Natural Ecosystems. Birkhauser, Basel, Switzerland, pp. 13-23.

Van Tuinen, D., Jacquot, E., Zhao, B., Gollotte, A. and Gianinazzi-Pearson, V. 1998. Characterization of root colonization profiles by a microcosm community of arbuscular mycorrhizal fungi using 25S rDNA-targeted nested PCR. Molecular Ecology 7: 879-887.

Vraltad, T., Forsheim, T. and Schumacher, T. 2000. *Picirhiza bicolorata* the ectomycorrhizal expression of Ericales. Proceedings of the Linnaean Society of London, London, UK. 178: 25-35.

Walker, C. 1983. Taxonomic concepts in the Endogonaceae: Spore wall characteristics in species descriptions. Mycotaxon 28: 443-443.

Walker, C. 1992. Systematics and Taxonomy of Arbuscular Mycorrhizal Fungi. Agronomie 12: 887-897.

Walker, C., Reed, L.E. and Sanders, F.E. 1984. *Acaulospora nicolsonii*: A new Endogonaceous species from Great Britain. Transactions of British Mycological Society 82: 360-364.

Walker, C. and Trappe, J.M. 1980. *Acaulospora spinuosa* sp. nov with a key to the species of *Acaulospora*. Mycotaxon 12: 515-521.

Walker, C. and Trappe, J.M. 1993. Names and epithets in the Glomales and Endogonales. Mycological Research 97: 339-344.

Warcup, J.H. 1985. Ectomycorrhiza formation by *Glomus tubiforme*. New Phytologist 99: 267-272.

Watson, L., Williams, W.T. and Lance, G.N. 1967. A mixed-data approach to Angiosperm taxonomy: the classification of Ericales. Proceedings of Linnaean Society of London 178: 25-35.

Wilmolte, A., Van der Peer, Y., Goris, A., Cheppele, S., DeBaere, R., Nelissen, B., Neefs, J.M., Hennerbert, G.L. and DeWatcher, R. 1993. Evolutionary relationships among higher fungi inferred from small sub-unit ribosomal RNA sequence analysis. Systematic and Applied Microbiology 16: 436-444.

Wipf, D., Bedelli, J., Munch, J.C., Botton, B and Buscot, F. 1996. Polymorphism in Morels: Isozyme electrophoretic analysis. Canadian Journal of Microbiology 42: 819-827.

Wright, S.F., Morton, J.D. and Sworobuck, J.E. 1987. Identification of vesicular arbuscular mycorrhizal fungus by using mono-clonal antibodies in an enzyme linked immunosorbent assay. Applied and Environmental Microbiology 53: 2222-2225.

Xavier, L.J.C., Xavier, I.J. and Germida, J.J. 2000. Potential of spore protein profiles as identification tools for arbuscular mycorrhizal fungi. Mycologia 92: 1210-1213.

Xiao, G. and Berch, S.M.1996. Diversity and abundance of ericoid mycorrhizal fungi from Northern Vancouver Island and impacts on growth *in vitro* of *Gaultharia shalon*. Mycorrhiza 9: 143-149.

Zambonelli, A., Giunchedi, L. and Pollini, C.P. 1993. An Enzyme-linked Immunosorbent Assay (ELISA) for the detection of *Tuber albidum* ectomycorrhiza. Symbiosis 15: 71-76.

Zeze, A., Duleiu, H. and Gianinazzi-Pearson, V. 1994. DNA cloning and screening of a partial genomic library from an arbuscular mycorrhizal fungus, *Scutellospora castanae*. Mycorrhiza 251-254.

Zeze, A., Sulistyowati, E., Ophel-Keller, K., Barker, S. and Smith S.E. 1997. Intersporal genetic variation of *Gigaspora margarita*, a vesicular arbuscular mycorrhizal fungus revealed by M13 minisatellite-primed PCR. Applied and Environmental Microbiology 63: 676-678.

2

PHYSIOLOGY AND CELL BIOLOGY OF MYCORRHIZAS

Physiology and cellular biology of symbionts are vital aspects of mycorrhizal symbiosis. They have a direct bearing on the sustenance and effectiveness of symbiosis. A large share of reports regarding AM physiology pertains to the life cycle, metabolic aspects and nutrient transfer at biotrophic interfaces. The bi-directional transport of nutrients such as C, N and P is a prominent issue, perhaps, the crux of symbiotic phenomenon. Obviously an adverse shift of equilibrium of nutrient exchange can turn symbiosis into pathogenesis, leading to a drain in nutrients from the host plant. Physiological analysis of plant growth improvement due to fungal partnership has been a priority, considering that it may have useful applications in practical agriculture and forestry.

Physiological interactions are unique to each type of mycorrhizal association, although some similarities occur with regard to nutrient translocation, exchange and metabolism. Accordingly, in this chapter, recent progress on cellular and molecular physiology has been delineated separately for each of the three types of mycorrhizas. The first section on physiology of AM symbiosis encompasses discussions on nuclear changes, plastid and microtubule rearrangements, spore dormancy, germination and germ tube formation, mycelial growth and sporulation. The phenomenon of microtubule reorganization in root cells and AM fungus aimed at accommodating the symbionts is deemed to be a crucial cellular event. The physiology of nutrient exchange between symbionts, especially carbohydrate and lipid storage and translocation and transfer at biotrophic interfaces has been emphasized in great detail. Lipid storage and movement between intra- and extraradical mycelium is considered to be a significant aspect of C metabolism in AM symbiosis. Nitrogen transport and metabolism in the AM fungal mycelium has also been included. Molecular biology of phosphorus transport and metabolism in both AM and ECM symbiosis has been discussed separately in Chapter 4. Knowledge on regulation of secondary metabolism in AM fungus is still rudimentary. Current status of progress on 'Yellow pigment' (mycorradicin) specific to AM fungi, isoprenoids, sterol and flavonoid biosynthesis has also been summarized.

The physiology of ECM fungi and their hosts is a complicated aspect. There are several aspects yet to be deciphered in detail, although innumerable reports are available on the cellular and molecular aspects of metabolism. A summarized view on the physiology of ECM establishment is followed by details on carbohydrate metabolism. Recent progress on storage of carbohydrates, sucrose transport, invertase activity and succrolysis, transfer of hexoses at symplastic interface has been highlighted. Nitrogen metabolism, especially mechanism of efflux and influx of N at biotrophic interfaces, kinetics of NH_4 and glutamate uptake are the other major topics discussed.

Compared with AM or ECM, research reports on physiology of ectendomycorrhiza are meager. Accordingly, a general idea regarding the structure, cytology and developmental physiology of ectendomycorrhiza has been presented.

1. Molecular Physiology of Arbuscular Mycorrhizal Fungi

1. A. Cellular Aspects of Arbuscular Mycorrhizal Symbiosis

Till date, a variety of techniques have been employed to study the anatomical features, developmental patterns and cytochemical changes during mycorrhizal symbiosis. Basic root clearing and staining methods, histological methods compatible with light microscopic analysis, immuno-fluorescence techniques and ultrastructural studies using transmission and scanning electron microscopy are some examples.

With regard to arbuscular mycorrhizas, morphological features external to plant roots include the sporocarps and/or ectocarpic chlamydospores in *Glomus, Sclerocystis* and *Acaulospora*. In addition to azygospore, soil-borne vesicles are also produced by the genera *Gigaspora* and *Scutellospora*. The external mycelium is dimorphic, with one portion of it consisting of 'permanent', coarse, thick-walled aseptate hyphae forming the major share. Rest of the mycelial phase comprises fine, thin-walled, highly branched, lateral hyphae.

Immediately after penetration of root epidermis (or root hairs), the fungal hyphae grow intra and intercellularly within the cortex. The proportion of intra-and intercellular hyphae partly depends on the host species (Brown and King, 1982). Intracellular hyphal system is generally composed of extensive loops and coils. Internally, AM infections are characterized by the production of thick-walled inter and intracellular vesicles which act as storage organs. Arbuscules are finely branched intracellular hyphae, with a primary function of bi-directional nutrient transfer between the fungal symbiont and the host plant. Arbuscules are produced progressively within the inner cortical cells including those close to steel, and a short distance behind the advancing hyphal tips. Actually, the penetrating lateral hyphae become the trunk of the arbuscule, which repeatedly branches dichotomously. The ultimate bifurcate terminal branches are usually less than 1.0 μm. More than one arbuscule may develop in a single cell of the host root. A mature arbuscule occupies a sizeable volume of the host cell. Arbuscules are dense with non-vacuolated cytoplasm during the early stages, but vacuolation increases with maturity. As arbuscules senesce, their branches collapse and form dense irregular masses. Senescence of arbuscules begins simultaneously at several tips of hyphal branches and proceeds towards the trunck. Ultimately, the entire arbuscule aggregates into a compact, lobed mass near the original point of penetration of the host cell. These dense aggregates comprise remnants of arbuscular walls, host cytoplasm and interfacial material. New arbuscules may replace the degraded previous ones. In all, the sequence of initiation, development and breakdown of arbuscules may require 4 to 15 days. Throughout this period, an intact host plasma lemma (peri-arbuscular membrane) surrounds it. Hence, the fungus is not in direct contact with host cell cytoplasm. An apoplastic interface occurs between the fungus and host cell. The host cell nucleus enlarges during the early stages of arbuscule development, but reverts to its normal size and position as the arbuscule deteriorates. Several unique cytoskeletal changes also occur with the formation of arbuscules. For example, changes in tubulin structure occur as a consequence of arbuscule initiation, development or senescence. In addition to tubulin genes, several other genes are induced or repressed preferentially in relation to arbuscule development.

The major function of arbuscule, especially the 'arum' type, is transfer of inorganic nutrients such as P, Zn, Cu and N from the fungus to plant root cells. In case of Pi, transfer mechanism involves a H^+ coupled symport that aids transport of P at apoplastic interface. The proton gradient developed due to acidification drives a H^+ ATPase located on peri-arbuscular membrane. Transport of Pi from the hyphae into the arbuscules may occur as polyphosphate. Basically, several plant/fungal genes may take part in nutrient transfer at the arbuscule. For example, Pi transporters such as *Lycopersicon esculentum* phosphate transporter (LePt), and *Medicago truncatula* phosphate transporter (MTPt) are vital in transport of Pi. These Pi transporters are generally upregulated in cells containing arbuscules. Details on several Pi transporter genes are available in Chapter 4. With regard to carbon transfer, we know that several sugar transporters are upregulated in arbuscular cells. For example, a sugar transporter MtSt1 from *Medicago truncatula* is upregulated in regions/cells with arbuscules and dense intercellular hyphae (Harrsion, 1996). Burleigh (2000) states that the fungal side of arbuscule function is still a bit obscure. There are several aspects yet to be understood. Some of the other functions attributed to arbuscules are nitrate assimilation. It is believed that nitrate assimilation may be mediated at arbuscules, mainly because nitrate reductase localized in them (Kaldorf et al. 1998). In addition, a large number of housekeeping genes are also expressed in arbuscules. A remarkable trait of arbuscules is their ability to degenerate after a certain lapse of time. In fact, a well-regulated turnover of arbuscules occur in host roots. In comparison, haustoria produced by pathogenic fungi (mildews, rusts) remain throughout, until the host root (or cell) dies. The turnover of arbuscules is indicative of the host plant's ability to regulate arbuscule formation. Within root cells containing arbuscules, plasma membranes of plant and fungus are separated by an interfacial apoplast. According to Guttenberger (2000), pH of this compartment between the symbionts is crucial for nutrient transfer to take place through arbuscules. Histochemical tests show that the interface is held to be acidic in nature.

The vesicles produced by AM fungi may be inter- or intracellular, terminal or intercalary, often thick walled. They are abundant in the outer cortical region. The cytoplasm within vesicles is moderately dense, multinucleate, and contains many small lipid droplets and glycogen particles. The cytoplasm becomes dense through condensation and volume of lipids increases. At maturity, lipid globules are conspicuous and fill the vesicles completely.

Nuclear changes in AM symbiosis: One of the earliest observations made about host nucleus relates to significant hypertrophy as a consequence of AM infection. Marked differences arise in the chromatin organization. It may either change to large strands of condensed chromatin (reticulate or chromocentric nuclei) or to decondensed chromatin (chromocentric or diffuse nuclei). The chromatin decondensation is characteristic of arbuscule containing cells, which indicates that fungal entry into host cell may induce transcription. Lingua (2001) reported that in three different plant species such as *Allium porrum*, *Pisum sativum* and *Lycopersicon esculentum,* modification of chromatin due to AM fungal invasion strongly increased metabolic activity of root cortical cells. In addition, fungal invasion of host cell may induce polyploidy. For example, in tomato, which is a plant with small genome but multiploid with 2n or 4n populations of nuclei, the presence of AM fungus induces polyploidy to 8n. Berta et al. (1998a) conducted flow cytometry and static cytofluorometry of cortical cells of tomato and confirmed that ploidy level in 90% of the colonized cortical cells was 8n and cells with 2n were almost absent. They suspected that AM fungal penetration induced polyploidy by acting on the passage between the post-synthetic phase (G2) and mitosis. In fact, enhanced transcription of genes relevant for transition to synthetic phase of cell cycle also supports this inference about induction of polyploidy. Further, Berta et al. (1998b) have stated that decondensation and ploidy induction may help in activating relevant genes and in copying the genome. A later report by Berta et al. (2000) indicates that nuclei of root cortical cells were larger with more condensed chromatin. A strong correlation between mycorrhizal colonization and polyploidization could be discerned. Concomitantly, cell respiratory activity determined using cytochemical assay of succinate dehydrogenase also increased. They believe that to a certain extent, mycorrhization of tomato roots may bring about changes in metabolic activity via polyploidization.

Under axenic conditions, the germ tubes emanating from the resting spores display rounded, migrating nuclei along the entire length. In hyphal specimens where apical septation had not been initiated, brighter fluorescent nuclei were noticed. Completely septate hyphae lack nuclei. Hyphae with incomplete septation may have both degenerating and healthy nuclei. These observations clearly indicate that random autolytic processes occur in the AM fungal germ tubes, in the absence of appropriate signals from host roots. In symbiotic state, thin runner hyphae usually show dimly stained nuclei along them, whereas, nuclei were clearly visible within thicker hyphae and their branches. When visible, nuclei appear anchored at regular intervals along the symbiotic extra radical hyphae. Nuclear migration occurs in pulses and through the hyphal central core. In symbiotic state, nuclei in the extra radical mycelium get distributed regularly and appear laterally, as if precise areas of coenocytic hyphae are under the control of an assigned nucleus (Berta et al. 1998b). The AM fungal hyphae also possess round-shaped nuclei. They are localized at the center and migrate along the hypha. The genome size, complexity and ploidy levels have been recently determined for AM fungi such as *G. intraradices*. Flow cytometric and DNA re-association tests indicate the genome to be haploid. The genome size was calculated to be around 16.54 Mb, comprising 88% single copy DNA, 1.6% repetitive sequences and 10.05% fold back DNA. The size of *G. intraradices* genome is said to be comparatively smaller than other AM fungi (Hijri and Sanders, 2004).

Plastids in Root Cells: Plastids are involved in a wide range of biochemical process in root cells. Some of them are fatty acid biosynthesis, nitrate reduction and its assimilation, starch and protein metabolism as well as secondary metabolism. Many of these primary metabolic functions are crucial to both host and AM fungus during establishment and maintenance of symbiosis. Secondary metabolites such as apocarotenoids are induced in the plastids by arbuscules. Fester et al. (2001a, b) studied these aspects about plastids and their functions in the arbuscular cells of tobacco roots. They utilized the epifluorescence and confocal laser scanning electron microscopy to analyze the plastids in mycorrhizal root cells. According to them, firstly, plastid surface and volume increases enormously in response to arbuscule formation. It is also accompanied by reorganization of plastids in host root cells. A few octopus-like or millipede-like plastids were also observed surrounding the nuclei within arbuscular cells whose function is unknown. These physical changes seem to be indicative of profound changes in metabolic activity. Interestingly, they observed that as arbuscules degenerated, plastids and their networks in host root cells also disintegrated. Clearly, at least some physiological interactions involving root plastids affect AM symbiosis.

Microtubules in Root Cells during Plant—AM Symbiosis: Developing a detailed understanding regarding the sequence of cellular events during the plant-AM symbiosis is crucial in many ways. At least a certain number of such cellular changes might be directly or indirectly related to regulation of nutrient exchange phenomenon between the two symbionts. No doubt, wide-ranging cellular changes occur as the AM fungus infects and arbuscules develop inside root cells. It includes fragmentation of vacuole, migration of nucleus, synthesis of membrane and cell wall components that surround the arbuscule, etc.

Let us now consider microtubules within plant root cells. Microtubules and actin filaments are important proteineceous cytoskeletal networks found in root cells of the host plant. They influence several cell functions such as cell elongation, shape, organelle transport, division, etc. (Gaiteman and Emons, 2000; Kost et al. 1999). We also know that both external and internal stimulus such as hormones, light, osmotic stress, mechanical force and others affect and reorganize microtubules. In turn, these changes in microtubules will influence cell functions, cell growth and developmental patterns of plant roots (Blancflor et al. 2001; Nick, 1998). In nature, both fungal root pathogens and symbionts (e.g. mycorrhiza) can influence these microtubules. In other words, the organization and function of cytoskeletal system within root cells is affected by AM/EM fungal colonization. Overall, there seems to be greater emphasis towards understanding the microtubules in plants that are in interaction with pathogens. However, there are only few reports that deal with microtubules in mycorrhizal symbiosis (Genre and Bonfonte, 1997; 1998; 1999). Recently, Douds (2002) has reviewed the progress in our understanding about cytoskeletal system in mycorrhizas and compared it with biotropic pathogens and symbionts such as *Rhizobium*. He opines that efforts to understand the cytoskeletal changes during establishment and functioning of AM symbiosis in greater detail have been hampered by technical difficulties of working with mycorrhizal material.

Recent studies with *Medicago-Glomus versiforme* association indicate that AM fungi colonize plant roots and induce changes in cytoplasmic organization, including reorganization of the microtubules. Extensive remodeling of microtubule cytoskeleton begins at an early stage of arbuscule development and continues till their collapse and senescence (Blancflor et al. 2001; Butehorn et al. 1999). Actually, in cortical cells containing arbuscules, microtubules reorganize to form complex arrays enveloping arbuscules, its branches as well as trunks to interlink them with nucleus. Development of such interconnections between host microtubules and fungus has also been reported in other symbiotic systems, such as Orchid-*Ceratobasidium* and *Pinus-Suillus bovinus* (Niini et al. 1996; Uetake and Peterson, 1997).

Microtubule reorganization in root cells of host is stage specific in AM symbiosis. Double staining root sections with anti-tubulin antibodies and WGA–Texas Red allows simultaneous evaluation of changes in both host cell microtubules and AM fungi hyphae. Such investigations have revealed that immediately after initial penetration/entry of the AM fungus into cortical cells, it undergoes terminal differentiation and profuse branching into an arbuscule. In due course, microtubules appear to form close link with hyphae. Again, in cells possessing only running hyphae but without arbuscule, the microtubules localize along the border of hyphae. In contrast, the microtubules are randomly distributed all across inside a non-invaded cell. The microtubule arrangement is also dependent on the stage of arbuscular development. With extensive branching and well-developed arbuscules, the microtubule aggregates become progressively less numerous, thinner and fragmented. However, as senescence sets in, arbuscule progressively depreciates in volume within cells, whereas cortical microtubules elongate. Blancflor et al. (2001) opine that such restructuring of microtubules in response to changes in arbuscules may also have implications on properties and composition of peri-arbuscular membrane. For example, the peri-arbuscular membrane is known to possess high ATPase activity, particularly surrounding fungal branches (Gianinazzi-Pearson et al. 1991).

Thus far, it is clear that microtubules are intricately linked to several functions in plant cells. Perhaps, signaling events leading to establishment of AM symbiosis, the changes in cell morphology and cytoplasmic architecture too might involve microtubules. Blancflor et al. (2001) have put forth certain evidences in support of this view. Firstly, microtubules of non-colonized root cells adjacent to arbuscule containing cells or intercellular hyphae were altered in organization. Such a rearrangement of microtubules in non-invaded cells could be evidence for signaling between the fungus that is in physical proximity of root cells, prior to actual penetration. A diffusible signaling molecule emanating from the neighboring cell (invaded) with arbuscule in it is a clear possibility. Such a signaling moiety could be of plant or fungal origin. Blancflor et al. (2001) propose that auxin and cytokinin are possible candidates as 'signal molecules'. These plant hormones are well known for inducing microtubule reorganization in plants. It is believed that such signaling, which leads to changes in microtubule patterns, may alter cell wall deposition, thus making cells more susceptible to future fungal penetration attempt.

1. B. Metabolic Aspects of Arbuscular Mycorrhizal: Pre-symbiotic Stages

Most AM fungal species form asexual chlamydospores in the soil, which are capable of the germinating in the absence of the host plant. These quiescent spores germinate in response to edaphic and environmental stimuli. However, they do not produce extensive mycelium and are physiologically incapable of completing their life cycle without functional symbiosis with a host plant. Giovannetti (2000) suggests that AM fungal development from spores to formation of a hyphal network involves a series of morphogenetic and biochemical events—spore germination, pre-symbiotic mycelial growth, hyphal branching, later formation of appresoria, infection and spread. Together, these represent early events in the life cycle of AM fungi (Fig. 2.1). However, at this point in discussion, only the early stages of AM fungal life cycle, in particular cellular and molecular events that trigger germination (release of dormancy), presymbiotic mycelial growth and expansion or senescence due to a lack host root are emphasized.

Spore Dormancy: Dormancy of AM fungal spores could be defined as the failure to germinate even when provided with congenial physical, chemical and other relevant

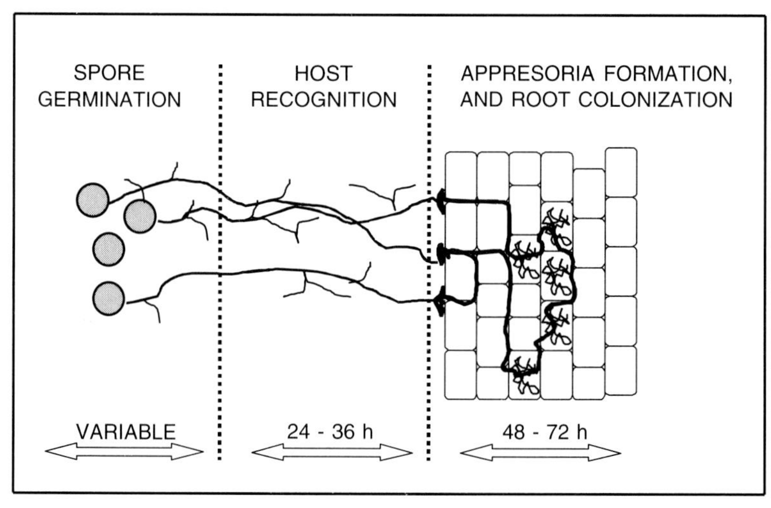

Fig. 2.1 A Time Scale depiction of Development of Arbuscular Mycorrhizal Symbiosis. *Note:* First step of spore germination may take a variable length of time, depending on spore dormancy. Host recognition is crucial. Earliest interaction may begin with hyphal differentiation and continue for 24 to 36 h. From physical contact to appresoria formation, hyphal spread and production of first few arbuscules may require between 48 and 72 h. (*Source:* Redrawn based on Giovannetti, 2000).

environmental factors. Given similar conditions, a non-dormant quiescent spore, having met the dormancy requirements in the general course, would germinate within stipulated period, say 3 to 4 days. Morphologically, it is difficult to distinguish between dormant spores and those ready to germinate. However, cytoplasmic differences between young and old resting spores can be identified. For example, dormant spores of *A. laevis* contain enlarged oil globules that restrict cytoplasm to small interstitial spaces (Maia and Kimbrough, 1998; Meier and Charvet, 1992; Mosse, 1970). Similarly, dormant spores of *Gi. margarita* contain cytoplasm interlaced with large lipid droplets (Sward, 1981a,b).

Spores may remain physiologically dormant for variable length of time (Fig. 2.1). Removal of such dormancy through storage has been confirmed with several AM fungal species (Giovanneti, 2000; Tommerup, 1983). For example, freshly harvested sporocarps germinated slowly, compared with those stored at 6°C for over 5 weeks. Similarly, storage at 10°C for 5 weeks improved the germinability of AM fungal spores. Spores may remain dormant, but be viable for one to several years. However, dormancy and viability of AM fungal spores also depend on the method and length of storage, temperature, moisture, etc. We may note that all species of AM fungi need not exhibit dormancy. For example, isolates of *Gi. gigantea* collected from sand dunes never displayed dormancy, but germinated immediately within 24 h. Again, *Gi. margarita* spores incubated on agar media germinated within 72 h. Essentially, spore dormancy within AM fungi constitutes a method to survive and overcome unfavorable conditions. Genetically, dormancy related traits will vary depending on the species and isolates of AM fungi, and we are yet to understand the biochemical and molecular aspects of dormancy in detail. Cold storage of AM fungal spores at 4°C for longer than 14 days increases germination. Longer cold storage generally preserved viability of spores better (Juge et al. 2002). In other words, coldness can affect physiological functions of AM fungal spores.

It is interesting to note that AM spore germination patterns vary; perhaps it is dependent on activation signals and the immediate environment encountered by germ tube/hyphae. For the first time, Juge et al. (2002) have clearly identified two types of germination. They believe that it may have ecological significance to the fungus. It may also have a direct bearing on the efficiency of a spore-based inoculum. "G'-type" germination is characterized by the main germ tube that arises from spores and radiate out in straight lines (Fig. 2.2). Apical dominance is noticed in spore branching. The other "g'-type" germination pattern that results after cold treatment at 4°C for <14 days produces curling type hyphae with numerous anastomoses points. Radiating hyphae are sparse. Juge et al. (2002) point out that "G-type" germination pattern may fetch successful root colonization more frequently, because of better reach to roots. It is a useful trait in terms of survival and proliferation in soil.

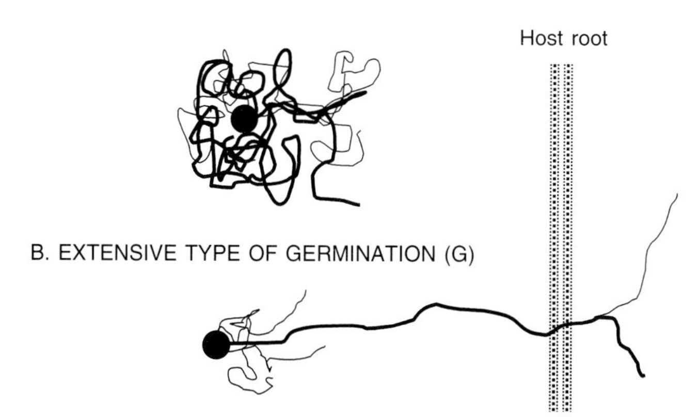

Fig. 2.2 Two types of spore germination patterns observed in *Glomus intraradices*. The 'g' type or restricted type allows comparatively less chances of encounter with host roots, whereas, 'G' type or extensive type of germination allows hyphae to cover comparatively longer distances rapidly; hence, enhances chances of encountering a congenial host root. (*Source:* Drawn based on depictions by Juge et al. (2002).

Stimuli that Trigger Spore Germination: The physical, chemical and microbiological conditions adequate to activate a quiescent spore and induce germination may vary. In nature, complex interactions among several factors such as pH, temperature, moisture, mineral and organic nutrients host roots, their exudates and microbial component triggers AM fungal germination. However, the effect of many of these individual factors on spore germination has been studied, mostly singly and in under controlled conditions.

The soil solution pH is an important factor that affects AM fungal spore germination. For example, a survey for AM fungal spores revealed that *A. leavis* is predominate in soils with low pH. Sometimes, it is the only AM species encountered when soil pH gets below 6.0. The congenial pH for germination of *A. leavis* is said to be 4 to 5. Any increase in congenial pH leads to a decrease in germinability, and it reaches less than 10% if soil pH is maintained between 6.5 to 8.0. Similarly, spores of

Gi. corolloidea and *Gi. heterogama* are known to germinate better in acidic conditions (pH 4 to 6). Generally, these fungi are less sensitive to acidic conditions in potting mix or soil. However, severe acidic pH (4 to 5) fails to trigger germination. In fact, in many AM fungal species, the pH requirement seems stringent. For example, *G. mosseae* spores were literally absent in soils of Columbia. On the other hand, spores of *G. mosseae* germinated better at pH 6 to 9, be it in soil or agar. Giovannetti (2000) opines that pH requirements for the germ tube production relates to the ability of different AM fungal isolates to survive and adapt to the environment.

Temperature is a crucial factor that influences proper germination of AM fungal spores. In general, spore germination may occur at temperatures ranging from 10°C to 34°C. If *A. leavis* spores germinated best at 15° C to 25° C, spores of *Gi. calospora* gave maximum germination at 10° to 30°C, and *G. caledonium* at 10 to 25°C. The optimum temperature requirements may depend on the soil environment to which a particular AM fungal isolate is acclimatized. For example, isolates of *Gi. corolloidea* and *Gi. heterogama* derived from tropical soils in Florida, germinated best at 34°C. Similarly, spores of a *G. mosseae* isolate from Northern latitudes germinated at 20°C, but failed to germinate at 34°C. It is obvious that an isolate derived from cold climate germinates better at lower temperatures. Incubation of AM fungal spores at temperature above 60 °C for 1 to 5 min is known to be lethal. The viability of *G. intraradices* and *G. mosseae* spores is nil when stored at 60°C. However, a certain isolate of *G. intraradices* is known to tolerate exposure to 45°C for 24 h (Daniels and Graham, 1976). In nature, sandy soils that reach temperatures above 45°C for longer durations of over 2 to 3 months, still harbor viable spores of *G. intraradices* (personal observations). The cellular and biochemical basis for tolerance AM fungi to extreme temperatures need to be understood in detail.

Soil moisture has variable effects on AM fungal spore germination. The soil matric potential between –0.5 and 2.20 MPa strongly inhibited germination of *G. intraradices*, *G. mosseae* and *A. longula* spores. Soil moisture, if held at field capacity, enhances germinability (Koske, 1981), whereas, certain isolates of *Glomus* tolerated drying and maximum germination could be observed even at –0.01 MPa. At the other extreme, spores from *G. epigaeum* germinated at moisture levels ranging from field capacity to soil saturation. It is not uncommon for AM fungi to experience alternate drying and wet conditions in the soil. AM fungal species encounter soils with low moisture as well as higher temperature within dry land ecosystems (Braunberger et al. 1996; ICRISAT, 1986).

In addition to soil moisture, nutrients may influence the spore germination and metabolic activity within the germ tubes and hyphae. Hepper (1983) observed that the germination of *G. mosseae* and *G. caledonium* spores was normal upto 30 mM PO_4 concentration in the medium, but at PO_4 concentrations above this limit, germination rate rapidly decreased. Similarly, spores of *Gi. margarita* germinated well upto 16 mM PO_4 in solution. However, in soil, for an increase in PO_4 concentration, germination of AM fungal spores progressively decreased. With reference to other nutrients, Daniels and Trappe (1980) reported that spore germination remained unaffected by increasing levels of NH_4NO_3 and K_2SO_4 concentration upto 200 ppm. On the other hand, presence of inorganic S compounds, such as thiosulphate in medium stimulated spore germination (Hepper, 1984). Giovannetti (2000) suggests that Mn, Cu, Zn and heavy metals (e.g. Cd) can inhibit spore germination. Sometimes, such responses to inorganic ions and heavy metals could be variable or erratic, depending on the AM fungal species, soil or incubating medium. Increasing salinity (NaCl conc.) inhibits both spore germination and hyphal growth. However, biochemical reasons for such metabolic suppression are yet to be deciphered.

Since AM fungal spores germinate in axenic culture systems devoid of roots, it is believed that roots may not be an essential trigger. The trigger for spore germination may not be entirely within roots. They may not possess factors that remove spore dormancy. However, we know that exudates from roots of different plant species induce germination and germ tube growth. Root exudates from non-host plants such as *Beta vulgaris* or *Raphanus sativus* did not influence AM fungal spore germination (Giovannetti and Sbrana, 1998).

In nature, a wide range of soil microbes, flora and fauna interact with AM fungal spores. Soil microbiota surrounding the AM fungal spores may exude substances that induce spore germination. Sterile soils are known to support much less germination percentage as compared to unsterilized ones. Giovannetti (2000) points out that although spores from many AM fungal species germinate in axenic systems, there is some positive effect from rhizosphere microflora. To quote examples, *Streptomyces orientalis* induced spore germination, *Klebsiella pneumonia* increased hyphal development, and *Trichoderma* spp. induced the growth of mycelia emanating from germinating spores of *G. mosseae* (Calvet et al. 1992). Contrary to the above reports, the presence of *Streptomyces cinnamomeous* in high numbers reduced germination of *G. etunicatum* spores. We are yet to know the biochemical basis for microbe mediate induction/suppression of AM fungal spores. There are suggestions that microbes residing on spore wall and those capable of chitin decomposition may aid better spore germination.

Molecular aspects of Spores and their Germination: The chlamydospores of AM fungi are large, ranging in size from 50 to 600 μm. In certain cases, such as *Gigaspora margarita,* the cytoplasm of large spore may contain two regions. The first region stores lipid, protein bodies and glycogen. The second region has a high number of nuclei, that are probably blocked at G0/G1 growth stage (Bionciotto and Bonfonte, 1993). Based on DNA content, the genome could be 10"8 to 10"9 nt, which has been effectively utilized in molecular phylogenic studies and in constructing genomic libraries (Table 2.1; Van Burren et al. 1999). Spore germination is initiated through increased cytoplasmic activity that brings about a transition from a quiescent state to a metabolically active condition. Some of the cytoplasmic events occur immediately as a consequence of imbibition. For example, in *A. leavis*, dense regions in cytoplasm harboring many dividing nuclei are seen in germinating spores. Similarly, highly condensed chromatin and prominent nucleoli are observable in *Gi. margarita* spores. These changes could be a prelude to rapid nuclear duplication as germination and hyphal extension begins. At this juncture, we may note that Burgraaf and Beringer (1987; 1989) reported lack of nuclear division during spore germination and immediately after. They based it on the fact that tritiated adenine did not incorporate into nuclear DNA, either during or immediately after germination. On the contrary, Bianciotto and Bonfonte (1993) and Becard and Pfeffer (1993) reported capability for DNA replication and nuclear division in germinating *Gi. margarita* spores. It was inferred using image analysis. During early stages, as germ tube and hyphae develop from spores, the nuclei are known to migrate from spore to the hyphae. In one instance, nuclei in spores dwindled from 2000 to 800, indicating streaming away of nuclei into hyphae. Giovannetti (2000) argues that streaming of nuclei is further authenticated

Table 2.1 Nucleic acid content in different tissues of *Glomus mosseae*

Tissue	*Amount (No)*	*RNA content (ng)*	*DNA content (ng)*
Ectocarpic spores	500	7.8	412
Chlamydospores	500	34.8	319
Sporocarps	100	37.2	323
Germinated spores	100	49.3	435
Extra-radical hyphae	1 mg	89.0	74
Mycorrhizal roots	1 mg	298	105
Uninfected control roots	1 mg	312	98

Source: Beilby and Kidby, 1980; Bionciotto and Bonfonte,1993; Van Burren et al. 1999.

by the recent finding regarding occurrence of cytoskeletal components, such as microtubule and microfilaments, in the hyphae emanating from germ tubes. These cellular components should be aiding the migration of nuclei and other organelles during active growth phase. Franken et al. (1997) identify yet another stage in spore germination/hyphal development, which is based on analysis of RNA accumulation patterns. For example, in *Gi. roseae*, the RNA content remains low/negligible immediately after isolation from soil, but one week after storage in water at 4°C, the RNA accumulates (2 μg/500 spores). In case of *G. mosseae*, the RNA content remains low/negligible for 3 months in storage but increases during germination. Franken et al. (2002) suggest that such RNA, stored during dormancy and spore activation, is effectively utilized for translation of different proteins, all through the hyphal development until reaching a host or beginning of hyphal retraction in the absence of a host.

Regarding biochemical processes, RNA and protein synthesis seem to begin within 35 min after imbibition (Beilby and Kidby, 1982). The tricarboxylic acid cycle, amino acid synthesis and gluconeogenesis were operative within 35 min after hydration. Increased ATP concentration by 45 min after germination is indicative of active respiratory system. It is obvious that protein synthesis should proceed as germ tubes are initiated and hyphae start appearing. Hepper and Smith (1979) confirmed this aspect by challenging spores with cycloheximide, an antibiotic that suppresses protein synthesis. Germ tube and hyphal growth ceases in the presence of cycloheximide. Further, use of radiolabeled amino acids, in particular leucine, added along with cycloheximide, confirmed the need for amino acid incorporation and protein synthesis during germination. At this point, we should also consider the presence of storage proteins inside AM fungal spores that may become available for germination and hyphal extension. Storage proteins do occur in AM fungal spores (Avio and Giovannetti, 1998; Samra et al. 1996). A decrease in native protein during spore germination is easily observable. Easily detectable amounts of ribosomal and mRNA are produced following imbibitions of *G. roseae* spores (Franken et al. 1997), which was required for hyphal growth. Fresh mitochondrial DNA was synthesized during germination (Hepper and Smith, 1979). Lipid synthesis and storage in AM fungal spores is common. Net degradation of storage lipids occurs during germination, with concomitant conversion and release of fatty acids and sterols (Beilby and Kidby, 1980). Ultrastructural studies indicate that lipids could be energy-rich compounds useful during germination and pre-symbiotic hyphal growth. Polyamine biosynthesis seems vital during spore germination and hyphal growth. Inhibitors of polyamine biosynthesis like cyclohexylamine decrease polyamine levels and interfere with germination of *Gi. rosea* spores and hyphal growth (Sannazzaro et al. 2004).

Spores of AM fungi germinate and produce hyphae seeking host roots to infect. The growing hyphae are incapable of extensive unlimited growth. In nature, it is common for germinated spores/germ tubes not to strike a host root. Actually, we know very little about metabolic changes that occur during spore adaptation to such unfavorable conditions. Firstly, there is arrested hyphal metabolism and growth if the germ tubes do not encounter host roots. Germlings are known to cease growth within 15 to 20 days. Details on vital metabolic pathways that function at slow rates, or those blocked, are not available as yet. One set of evidences indicates that exhaustion of spore reserves may automatically initiate reduction in the metabolic rate. However, Giovannetti (2000) states that not all the spore reserves are used up during germling growth. In most cases, the metabolic decline begins in 5 to 15 days after germination initiation. This is indicated by cessation of hyphal development, and retraction of cytoplasm, nuclei and other organelles from peripheral hyphae. A retraction septum results in the senescence of empty hyphal segments. Onset of senescence in peripheral hyphae is said to coincide with retardation of metabolic activity. Further, hyphae still not senesced, but situated in proximity to mother spore retained infectivity for upto 6 months. During this period, they possessed mild metabolic activity. This is said to be a biological mechanism through which AM fungal spores ensure better survival and infection of host roots.

At this juncture, we have to recognize the three different but prominent aspects of hyphal growth during AM symbiosis (Fig. 2.1). First, a asymbiotic or pre-symbiotic metabolic state, which is influenced by the mother spore and environment provided by soil/medium immediately after germ tube formation. Host roots may influence the pre-symbiotic hyphae only through diffusible chemical signals. Second, the intraradical hyphae and other structures that interact with the host, both morphologically and physiologically. Complex molecular regulations are to be expected between the intraradical hyphae and host root cell. Third, extraradical hyphae that are influenced both by the metabolic status of host root and external soil environment.

Metabolism in Pre-Symbiotic Mycelium: As stated above, pre-symbiotic mycelium of most species of AM fungi exhibit a certain degree of growth without physical contact with the host plant. Such an asymbiotic growth is easily observable in nutrient agar. Generally, addition of complex growth factors extends a limited hyphal growth, but it cannot augment perpetual independent extension of hyphae. The hyphal growths stops in 2 or 3 weeks, if hyphae do not reach a live root and successfully establish symbiosis. Video analysis has clearly shown that within the hyphal segments unable to reach the host root, the cytoplasmic contents retreat and the empty hyphae are abscessed by the formation of a septum. The inability to grow independently can be attributed to limitation in uptake and/or metabolism of carbon. Radio-tracer (^{14}C) studies indicate that hexose and acetate are absorbed and metabolized by pre-symbiotic fungi (Bago et al. 1999), and there is dark fixation of CO_2. Hence, CO_2-enriched conditions can enhance asymbiotic hyphal development (Becard et al. 1992). Pfeffer et al. (1999) suggest that lack of lipid biosynthesis is a major difference between asymbiotic and symbiotic mycelium, and it could be the limiting step in completing the life cycle. In this regard, Franken et al. (2002) have suggested that a detailed study of lipid metabolism by cloning relevant genes for fatty acid biosynthesis and analyzing their expression may provide us with some useful insight about their relevance to AM fungal growth and survival. With regard to nucleic acid metabolism, RNA display techniques have shown that RNA synthesized during early activation of spores suffices to maintain subsequent germination and asymbiotic development. Franken et al. (2002) argue that EST libraries constructed on activated spores should contain all genes necessary for pre-symbiotic hyphal growth. Studies by Requena et al. (2000), Lanfranco et al. (2000) and several other researchers indicate similarities in genes that code for proteins involved in basic metabolic processes such as translation, protein processing, transport processes, cell cycle, replication of chromatin, signaling, etc. Biologically, pre-symbiotic hyphae are programmed to grow linearly with apical dominance, but they change to strong branching patterns immediately after sensing a host root. Interestingly, this phenomenon of branching is not manifested when hyphae are in close proximity to roots of non-host species. We know that P-deficient plants tend to show quantitatively higher levels of AM colonization than those fertilized with P. Analysis of exudates from P-starved plants indicated the occurrence of a UV fluorescing compound that disappears after P fertilization or mycorrhization. This chemical principal, which is a salicylic acid derivative, effected such an elongation and branching of AM hyphae. Addition of salicylic acid to water agar is known to induce changes in RNA accumulation and pattern of transcripts in pre-symbiotic hyphae. Several other chemical compounds too have been studied for their effects on pre-symbiotic hyphal growth and physiology. Using nitrocellulose membranes, Giovannetti et al. (2000) separated the two symbionts and found that enhanced hyphal growth and branching were possible whenever fungus was in close proximity to roots. Using this method, root exudates and their fractions have been studied for their effects on morphology and growth pattern of presymbiotic hyphae placed in different media (Buee et al. 1998; Nagahashi et al. 1998). Certain fractions of root exudates were implicated to carry the diffusible fungal stimulants. Chemical compounds such as quercitin, naringenin or biochinin A also stimulated hyphal growth. The stimulatory effect of biochinin A, an estradiol could be suppressed by introducing an anti-estrogen compound (Poulin et al. 1997). Polyamines and/or salicylic acid derivatives from maize roots also exhibited positive effects on the development of presymbiotic hyphae. Vierheilig et al. (2001) studied the effect of chitinases on the growth and extension of hyphae derived from an in vitro culture of *Glomus mosseae*. Chitinase applied to the hyphal tip resulted in an inhibition of growth, lysis of apex and disarray in pattern of hyphal spread. However, chitinase applied to subapical parts of hyphae did not affect normal hyphal extension. Clearly, chitin deposition at the hyphal tip is crucial to extension of pre-symbiotic mycelium of AM fungi. During pre-symbiotic phase, host plant may release several critical metabolites that trigger fungal growth, including its branching and extension into root tissue. Buee et al. (2000) have identified one such fraction from exudates of carrot hairy roots that was highly effective on germinating spores of *Gi. rosae*, *Gi. margarita* and *Gi. gigantea*. According to them, such a chemical factor responsible for induction hyphal growth and branching is elaborated by all mycotrophic hosts, but not by non-host plant species. This chemical stimulant is yet to be characterized

Under natural conditions, soil microbes may influence the growth of pre-symbiotic mycelium. For example, the presence of plant growth-promoting rhizobacteria like *Bacillus subtilis* is known to down regulate the fatty acid oxidation gene GmFOX2 (Requena et al. 1999). Such down regulation of GmFOX2 is related to the induction of pre-symbiotic hyphal growth. Further, it is interesting to note that down regulation of corresponding proteins in *Neurospora crassa* also induce a similar mitogenic cycle cell/tissue growth. Yet another gene involved in *B. subtilis/G. mosseae* interaction encodes for GmTOR2, which is a protein involved in regulation of cell cycle and actin cytoskeleton patterns. The cell cycle controlling activity of the gene GmTOR2 is affected by rapamycin. However, it does not suppress spore germination, only germ tube extension and pre-symbiotic hyphal growth is suppressed. This observation supports the hypothesis that nuclear replication is not a prerequisite for germination, but necessary for continued pre-symbiotic hyphal growth (Bionciotto and Bonfonte 1993; Becard and Pfeffer, 1993).

Reaction to stimuli from soil/culture medium is an important physiological aspect of germ tubes or pre-symbiotic mycelium. The electrical properties of hyphal wall and cell membrane seem to influence this process. Assessment of transmembrane electrical potential of three different AM fungal species *Gigaspora margarita, Scutellospora calospora* and *Glomus coronatum* revealed that it ranges around –40 mV; it is low compared with other filamentous fungi. Addition of root extracts into the hyphal surroundings reduced the transmembrane electric potential further. Hence, it became more negative. Ayling et al. (2000) state that despite such low electric potential, high affinity P uptake is possible with germ tubes and external hyphae. Permeability of K^+ was also unaffected due to low electric potential at hyphal membrane. Further, they suspect that such interactions between plant and fungus at plasma membrane may play a more important role immediately during establishment of symbiosis.

1. C. Mycelial Growth of AM Fungi

Immediately after germination AM fungal hyphae grow forward, showing linear growth with a strong apical dominance. Cytoplasm and nuclei migrate along both directions inside the growing hyphae. These inferences on intrahyphal protoplasmic changes are based on photon-fluorescence microscopy and video-enhanced high-powered microscopy (Giovannetti 2000; Giovannetti et al. 2000). Hyphal branches are regular and right-angled to the main strand. Hyphae are generally thick-walled aseptate (coenocytic), 5 to 10 μm in thickness and multinucleate. The extension of mycelial network is dependent on individual spore and immediate environment. Hyphal growth is generally poor when compared with other fungi capable of rapid in vitro growth. Hyphal growth rate during early phase after germination was 2 μm day^{-1} for *G. caledonium*, reaching a length of 30 to 50 mm in 10 to 15 days (Logi et al. 1998). The presence of roots, root exudates or any other stimulant will obviously influence hyphal growth rate. In the absence of root factors, *G. mosseae* hyphae extended at a rate of 2.7 μm min^{-1}. In some fungi, the hyphal growth rate actually fluctuates around a mean and shows a pulsed growth. If it applies to AM fungi too, then factors such as host physiological/nutrient status, soil environment and fungal stage of growth may all influence such fluctuations. The total length of hypal network generated may also vary. For example, *G. clarum* hyphae extended to only 8 mm in 10 days, whereas *Gi. margarita* is known support rapid, extensive hyphal network reaching even 54 cm in 15 days when incubated under in vitro conditions (Douds, 2002). Hrselova and Gryndler (2000) have noticed that oligoamine such as spermine stimulated hyphal growth at concentrations about 1.5 μM L^{-1}. Addition of a metabolic inhibitor that suppresses polyamine synthesis caused significant decrease in the hyphal proliferation. This test confirms the stimulatory effect of spermines on hyphal growth. Growth of extraradical mycelium in soil is also dependent on the extent and kind of carbohydrate storage (Gavito and Olsson, 2003).

Giovannetti et al. (2000) have stated that ability of AM fungal spores to germinate in response to edaphic and environmental stimuli—especially hyphal growth that ensues immediately after—are independent of the presence or absence of a compatible host. In order to accrue evidence in support of the above hypothesis, they monitored mycelial elongation and protoplasmic flow-rates, including the bi-directional movement of cellular organelles using micro-chambers and video-enhanced light microscopy. The hyphal growth rate measured on cellophane sheets kept within petri dishes varied, depending on AM fungal species. It was 2.64 μm min^{-1} for *G. rosea*; for *G. caledonium* it was 1.97 μm min^{-1} and 2.7 μm min^{-1} for *G. mosseae*. In the absence of host-supported nutrient replenishments, the hyphal growth generally ceased within 5 to 15 days after germination. The cytoplasm, nuclei and cellular organelles firstly retract from peripheral hyphae. A retraction septum forms to separate viable portions from empty mycelium (Bago et al.1998; Cole et al. 1998; Logi et al. 1998). This leads to a progressive increase in proportion of empty mycelium, if AM fungal spores are incubated without hosts (Tommerup, 1984). In natural conditions the host plant, its physiological status and environment also influence the hyphal growth rate in vivo. Recently, Giovannetti et al. (2001) visualized monitored and quantified the hyphal growth in vivo, while still in symbiotic contact with host plant. Obviously, the hyphal growth rates were higher than in vitro. On average, the hyphae grew at 0.75 to 1,1 m day^{-1} (Table 2.2). The hyphal densities ranged from 10 to 40 mm mm^{-1} root length. Several other previous estimates have ranged from 1.6 to 142 mm mm^{-1} root length (Sanders and Tinker, 1973; Sylvia, 1988). Hyphal branching is again influenced by several factors related to the host and its environment. For example, studies by Nagahashi (2000) revealed that exposure to light induced primary germ tubes to branch into secondary hyphae. They reported that maximum branching was induced at low light intensity (10,800 μL s^{-1} m^{-2}), but multiple exposures alternating with dark and light periods did not influence the hyphal branching. Later, it was found that blue light and certain chemical signals could induce hyphal branching (Nagahashi and Douds, 2001). Buee et al. (1998) have shown that root exudates from compatible hosts can induce hyphal branching. The active molecule that induces branching is not a flavonoid, but is effective in very small quantities.

Table 2.2 Hyphal growth and physiological characters including anastomoses in *Glomus mosseae* found on three different host plant species

Mycorrhizal trait	*Allium porrum*	*Thymus vulgaris*	*Prunus cerasifera*
Root length (mm)	201.9	516.4	465.5
AM colonization (%)	54.4	46.7	64.4
Total hyphal length (mm/7d^{-1})	7471	5169	7096
Area covered by extra radical myceliium (mm^2)	2755	2525	2898
Hyphal length per total root length (mm mm^{-1})	40.2	10.1	15.9
Hyphal length per mycorrhizal root length (mm mm^{-1})	73.7	21.9	24.9
Number of hyphal branches mm^{-1} length of hypha	0.97	0.86	0.92
Hyphal density (mm mm^{-2})	27.2	2.07	2.61
Number of anastomoses mm^{-2}	1.31	1.09	1.34
Number of anastomoses mm^{-1} (length) of hyphae	0.46	0.51	0.51
Number of anastomoses per hyphal contact	0.75	0.78	0.64

Source: Giovannetti et al. (2001). *Note:* Values are means of 9 independent observations.

Purified root exudates from non-hosts did not induce hyphal branching. Branching and survival of branches noticed in this study were confirmed using natural fluorescence. Molecules that actually trigger hyphal branching and concomitant nuclear changes, if any, need to be ascertained.

Anastomoses, i.e. hyphal branches meeting to form a ring/loop or interconnection are common, both when grown in vitro and in natural soil. Literally, anastomoses means developing connective mechanisms. In the present context, such hyphal fusion as a phenomenon is supposed to aid the exchange of genetic material, both among genetically similar and dissimilar strains of AM fungi (Fig. 2.3). Giovannetti (2001) estimates that anastomoses occurs between 0.46 and 5.1 times mm^{-1}, depending on the host and soil type. Anastomoses are 4 to 5 times more frequent in natural conditions compared with hyphae radiating from spores incubated in medium. Analyses of anastomosed hyphae using epiflourescence microscopy revealed localized nuclei within the hyphal bridge. The viability and protoplasmic continuity in the anastomosed zone could be confirmed through histochemical tests for succinic dehydrogenase. Giovannetti et al. (1999) believe that occurrence of higher and larger number of nuclei in the extramatrical mycelium in the anastomosed zone is indicative of the flow of genetic information, in addition to nutrients. In AM fungi, anastomoses could be an important avenue through which genetic heterogeneity develops. Ecologically, hyphal fusions and anastomoses may represent a mechanism of adaptation. In nature, anastomoses may be a fundamental process, leading to the development of an intricate and non-finite fungal network connecting closely spaced plants (Molina et al. 1992; Read, 1997).

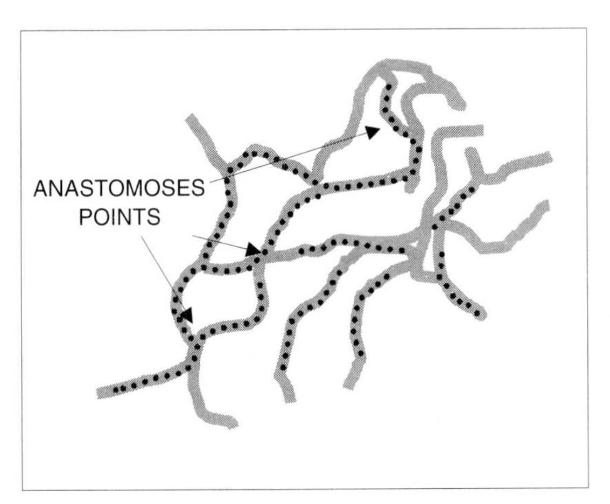

Fig. 2.3 A diagrammatic representation of extrametrical mycelium of an AM fungus showing anastomoses and nuclei. Note that nuclei flow across the mycelium at the points of anastomoses and fusion bridges. (*Source:* Based on a depiction by Giovannetti et al., 2001

Physico-Chemical aspects of AM Fungal Hyphae: We must realize that cell surface electrical properties of symbionts might be crucial during establishment of compatible associations and hyphal growth. Ayling et al. (1997, 1998) reported that as colonization progressed, AM fungal hyphae became more polarized, reaching values similar to that found on the walls of the host (leeke) root. Electrical potential differences and cytosolic pH were also affected. Obviously, such changes in electrical properties are attributable to biochemical changes in the fungal hyphae in response to invasion of host root. These changes in electrophysiology of hyphae could also influence nutrient uptake and transport at the root/fungal hyphae interface. Hence, appropriate advance knowledge about physico-chemical environment at the root cell-fungus interface may be crucial while inferring results from experiments on solute/nutrient transfer in either direction between symbionts. Briones et al. (2001) also studied the physico-chemical properties of AM fungal hyphae, especially their cell walls. They actually aimed at deciphering the cause for differences in metal-binding properties of AM fungal hyphae. High resolution acido-basic potentiometric titration and infrared spectroscopy revealed that wall surface of AM fungi is composed of four major functional groups, namely, carboxylic, amino acid, sulfonic, phosphate and phenolic. The pKa values ranged from 4 to 5.5 and from 8.5 to 9.6. The organic acid contents of AM fungal hyphae were two times higher than saprophytic fungus such as *Rhizopus arrhizus*. Metal sorption by AM fungal hyphae was generally by a release of Zn^{2+} and Ca^{2+}. Obviously, these surface mechanisms of AM fungal hyphae have an important place during nutrient absorption, especially mobility and availability in rhizosphere soil. Ayling et al. (2001) investigated the physico-chemical properties of growing germ tubes and extra radical hyphae of AM fungus *Gigaspora margarita* with a view to identify the cause for inability of germ tubes to absorb and metabolize nutrients. Transmembrane electric potential differences (p.d.) of ~40 mv were recorded for external hyphae. Thermodynamic calculations showed that despite such low values of p.d., efficient high-affinity P uptake is possible. Major differences in electrophysiological properties of germ tubes and extra radical mycelium were unlikely.

In temperate and boreal soils, the ability of extra radical hyphae to tolerate very cold and freezing conditions is crucial to AM fungal survival. Extraradical hyphae may lose viability and infectivity due to prolonged soil freezing (Addy et al. 1994, 1997). Acclimation to cold freezing via prior exposure to low temperatures is a possibility that has been examined. Addy et al. (1998) reported that following freezing at either –12°C or 5°C, more of AM fungal cultures preconditioned survived cold temperatures and remained in metabolically active state. Such acclimation is attributable to a range of physiological responses that include modification in the membrane structure and function. Specific mechanisms that impart freezing tolerance are yet to be known. However, in certain fungi, such freezing tolerance has been attributed to synthesis and accumulation of trehalose (Becard et al. 1991; Schubert et al. 1992). Trehalose is known to reduce the freezing point of cell contents and protect the membrane from damage.

1. D. Sporulation in AM Fungi

As stated above, aspects related to spore dormancy and factors triggering their germination have been investigated at least to a certain extent, whereas, our knowledge about the induction of sporulation and its biochemical basis in AM fungi is rudimentary. Sporulation is no doubt dependent on fungal species. We are yet to accumulate details on physiological and genetic basis for induction of sporulation. There are AM to fungal species that sporulate profusely but others may do so mildly.

Some AM fungi may not sporulate at all under certain conditions. Indeed, several factors/signals related to host, fungus and environment might actually influence the onset as well as the intensity of sporulation of an AM fungus. Host photosynthate and its flow may be a crucial factor determining the induction of higher levels of sporulation. Trials at ICRISAT, Hyderabad (India) showed that pruning the shoot system of host plant (grass) induces rapid and higher density of spores in the pot culture of *G. intraradices*. Pruning shoots suddenly reduces the supply of photosynthate to roots. Introduction of such unfavorable conditions in soil/medium may initiate sporulation. The molecular basis for such phenomenon needs to be understood. Addition of chitin may also induce sporulation. Gryndler et al. (2003) observed that chitin or even autoclaved *Fusarium oxysporum* used as chitin source induces sporulation in AM fungi. Jolicoeur et al. (2003) have tried to develop a model that describes the effects of inorganic P and sugar uptake by carrot hairy roots on spore production by *Glomus intraradices*. Caliberation of experimental data with expected values from the model suggested that intracellular Pi storage in host root controls 'decision switch' for initiation of sporulation. Sporulation is also dependent on the availability of soluble sugars in roots. In a study aimed at knowing the source of nitrogen and carbon to spores, Nakanol et al. (2001) observed that spore carbon in AM fungi was derived from plants, but spore N was mainly from the soil. In other words, host C status, or its supply in the medium in axenic cultures could be key to initiation of sporulation. Douds (2002) examined this aspect using monoxenic cultures of *G. intraradices* grown on transformed carrot roots in two compartment petri dishes. Glucose supply (200 mg L^{-1}) and replacement of gel was crucial to induction of sporulation. This treatment provided a three-fold increase in spore production per unit time and approximately 20,000 spores of *G. intraradices* could be harvested.

1. E. Carbohydrate Physiology of Arbuscular Mycorrhizas

In its asymbiotic phase, AM fungal spores and mycelia support substantial gluconeogenesis, mobilizing lipid stores to sustain the growth (Bago et al. 1999, 2000, 2003). Enzymatic analysis on asymbiotic spores and extraradical mycelia have also proved that tricarboxylic acid cycle, glycolysis and pentose phosphate pathway allow sufficient carbon fluxes within the fungus. As such, germinating spores and hyphae are known to imbibe hexoses from the medium. Fatty acid synthesis is not significant in asymbiotic phase. Therefore, Bago et al. (2000) argue that in such a situation, new spores may not be put forth by the asymbiotic AM fungus.

The carbon flow from plants to fungus was demonstrated early in 1970s. Sugars were suspected to be the main transport moiety for carbon (Bevege et al. 1975; Ho and Trappe, 1973). Woolhouse (1975) then proposed that host cortical cells release sugars at the plant-fungal interfaces. A passive efflux seemed to transfer sugars into the fungal symbiont. Such an efflux could be induced by the presence of AM fungus. Nuclear magnetic spectroscopy, isotope labeling and radio-respirometric analyses have all proved that intraradical fungal hyphae receive hexoses, later accumulate them. However, in the extraradical mycelium, major hexoses such as glucose, fructose, as well as mannitol and succinate are not traced in significant quantities (Pfeffer et al. 1999). Higher levels of invertase activity in different biotrophic associations, including AM mycorrhizas also proves that hexoses, such as glucose, are dominant carbon transport compounds. Bago et al. (2000) opine that in addition to passive flow of carbon, as an alternative mode, active H^+ mediated hexose transporter may also be operative. Such an energy-mediated transporter system does operate in ectomycorrhizas. As a consequence of C transport into fungal component and its metabolism, carbohydrate pools differ substantially between AM roots and non-mycorrhizal ones (Douds et al. 2000). Starch levels can be minimal or almost nil in the arbuscular cells. Invertase activity also differs between arbuscule containing cells and cortical tissue in non-mycorrhizal roots. We may also note that respiration rates are higher in mycorrhizal roots; hence photosynthates translocate towards the stronger sink developed in mycorrhizas. Invertase and sucrose synthase catalyse the first step in sucrose utilization by the symbionts. Time course studies have proved that these two enzymes play a role in transport of C (Ravnskov et al. (2001). Interestingly, quantitative expression of both invertase and sucrose synthase varied, depending on the AM fungal species. It has relevance to efficiency of C transfer from host to fungus. Recently, Schubert et al. (2004) have proved that activity of sucrose-cleaving enzyme is modified by AM fungus. Alkaline invertase activity was significantly higher in roots colonized by AM fungi. Actually, fungal colonization induces alkaline invertase in order to provide hexoses for the fungal symbionts and for development of colonized root cells.

Interestingly, tests with radiolabeled ^{14}C have proved that trehalose and glycogen pools increase in the intraradical mycelium of AM fungus, whereas, in ECM trehalose and mannitol accumulate. Trehalose and glycogen are easily detectable in both intra- and extraradical as well as germinating mycelium of AM fungus. Within AM fungus, cytoplasmic hexoses are also metabolized via pentose phosphate pathway. Occurrence of glucose phosphate dehydrogenase also supports this conclusion (Saito, 1995). Tricarboxylic acid cycle and glycoloytic pathway both operate in the intra and extra radically mycelium of the fungus. Substantial quantities of labeled hexose supplied by the host root are also utilized for lipid synthesis. Radiolabel ^{13}C patterns on glycerol and triacylglyceride molecules derived from hexose show that hexoses are converted to trioses and acetyl CoA. Later, an acetyl CoA carboxylase and fatty acid synthase complex supports synthesis of fatty acids and their elongation. Further, Bago et al. (2000) suspect that as in oleogenic fungi—where acetyl CoA used for tryglyceraldehyde is derived from citrate—a citrate lyase system could be functioning in the cytosol of AM fungus.

Rapid translocation of C from roots to external mycelium is equally crucial; because it directly influences growth of AM fungi in the soil. According to Bago et al. (2000) AM fungi receive hexoses and metabolize them to trehalose and glycogen in significant quantities. Therefore, carbohydrates are also suspected as a major avenue for C transport from intra to extraradical mycelium. However, occurrence of substantial flow of lipid bodies along the hyphae is to be noted. Evidences being accumulated prove that glycogen is translocated rapidly from intraradical to extraradical mycelium. It means that within AM fungi, C transport is mediated via two or three different modes. In their recent report, Bago et al. (2003) have restated that lipids synthesized from carbohydrates in the root could be yet another form of carbon that is exported to the fungus. Labeling patterns in glycogen as well as analyses of enzymes indicate that glycogen flux is also important in the symbiosis. Unlike ECM fungi, mannitol accumulation and cycling was absent within AM fungi existing in symbiotic state. Glycogen has four different roles within AM fungi; namely, sequestration of hexoses, long-term storage of carbohydrates in spores, translocation of carbohydrates from intraradical to extraradical mycelium and buffering of intracellular hexose levels throughout the life cycle.

The carbon compounds translocated from intraradical to extraradical hyphae are actively metabolized. Glycolytic enzymes assayed in extraradical mycelium were feeble in activity (Saito 1995). Instead, labeling patterns suggest a substantial gluconeogenesis via a glyoxolate pathway (Pfeffer et al. 1999). Occurrence of malate synthase and isocitrate lyase is indicative of glyoxolate cycle. As we know, lipids are also translocated into extraradical mycelium. Such lipid compounds are said to be utilized for anabolic processes in extraradical hyphae. Lipid flow into extraradical mycelium may also be utilized as respiratory substrate via tricarboxylic acid cycle. A pentose phosphate pathway seems to metabolize the carbohydrates received into the extraradical mycelium of the AM fungus. The carbon flux due to this pathway can be substantial, since enzyme measurements indicate higher activity. Ezawa et al. (2001a,b) assessed the glucose phosphorylation, the first step in glycolysis and pentose phosphate pathway (PPP). They found hexokinase that utilizes ATP, as a phosphagen was active in all components of AM fungi, namely spores, extra- and intraradical mycelium. However, polyphosphate glucakinase type activity that utilizes polyphosphate as phosphagen was negligible. Hence, it was concluded that glucose metabolism in AM fungi via glycolysis/PPP is mediated through hexokinase that depends on ATP as phosphagen.

In a further study, Johnson et al. (2002a) utilized in situ (CO_2)-C^{13} labeling to study the movement of carbon compounds in the plant-AM fungal continuum, right upto exudates from the extra radical mycelium in the soil. A technique involving mesh cores separated out the roots from the fungal hyphae effectively. The radiolabel movement into AM fungi peaked nearly 9 to 14 hours after ^{13}C was fed to shoots of the host plant, but it declined within 24 hours if roots and fungal mycelium were severed. Their calculations suggest that nearly 5 to 8% of carbon lost by plants was respired by the AM fungal mycelium in the first 21 h after labeling. Liming increased carbon fixation by plants; consequently greater quantity of C could be channeled to AM mycelium.

Yet another estimate by Johnson et al. (2002a, b) regarding short-term respiratory losses and accumulation of ^{14}C suggests that 70 h after labeling, mycorrhizal mycelium accounted for only 3.4% of ^{14}C initially fixed by the plant. Overall, they suggest that in field conditions, AM mycelia may provide rapid transit pathway for carbon flux from plants to the soil, then on to the atmosphere.

The reverse flow of carbon compounds, i.e. from fungus to a compatible symbiotic plant host has been a debatable issue for sometime. However, recent reports indicate that reverse flow of carbon compounds is possible in AM mycorrhizas. Some calculations indicate that nearly 0 to 10% of carbon in an AM root could actually be derived from the fungus through its previous connections to other host plants (Watkins et al. 1996; Graves et al. 1997). Such a situation is commonly encountered when physically closely situated plants are interconnected via AM fungal networks. Fitter et al. (1998) however, noticed that such carbon received from the AM fungus remained in the root of the recipient plant. Therefore, carbon received via hyphal interconnections could remain within the fungus, without being utilized by the plant host. It was also observed that vesicles and arbuscular cells in the recipient plant really accumulate larger amounts of carbon. This finding supports the view that carbon translocated in reverse into a second host is held within its roots. This aspect was further confirmed by the fact that radiolabeled carbohydrate fed to extra radical mycelium remained in the fungus, but was not detected in appreciable quantities in the recipient host's shoot system. Overall, the present view is that carbon-derived via reverse flow from the fungus into an interconnected second plant may not be of significant consequence to recipient host shoot growth. This aspect needs further study to understand the regulatory mechanisms involved in reverse flow of carbon within mycorrhizal systems. Such bi-directional flow of a nutrient is well reported in soil-plant relationships. Reverse flow of nitrogen in mycorrhizal symbiosis has also been reported (Johansen et al. 1993). Many of the recent reports from the University of California, Berkeley, regarding Monotropidae/ECM associations, also known as 'mycohetero-trophyic associations', clearly show that reverse flow of carbon and perhaps other nutrients, especially P, is a reality (see Lindahl et al. 2001). It is not a debatable issue any more. For such mycoheterotrophic plants, reverse flow of carbon via ECM fungus that is already connected to another carbon-rich host is ecologically a crucial component of survival and proliferation strategy. At this point, it should be made clear that bi-directional movement of C, P or N, as stated above, is totally different from *reciprocal transfer of C and P* discussed elsewhere. It refers to carbon flow from host to fungus versus P transport from soil/fungus to the host. It may help in calculating energy (carbon) costs incurred per unit P translocated.

1. F. Lipids in Arbuscular Mycorrhizas

First let us consider cytological aspects of lipid physiology in AM symbiosis. Carbon fluxes from plant roots to AM/ECM fungal mass existing below ground, and vice versa if any, can be vital to metabolic sustenance of symbiosis. Knowledge about lipid storage, movement and metabolic aspects that are intricately related to C dynamics within symbiotic systems seems to be equally important. Fungal lipid deposits (globules) are storage locations inside their thalli. Such lipid bodies are easily traceable in intra- and intercellular hyphae, extraradical spores and germ tubes. The AM fungus is known to acquire carbon from the plant host as hexoses, but at all stages of life cycle the carbon derived is stored mainly as triacylglycerol (TAG) (Jabaji-Hare 1988; Gasper et al. 1994, 2001). Generally, TAG is synthesized in the intraradical mycelium (IRM) and exported to be stored in extraradical mycelium, which is later utilized for anabolic processes and during germination. Lammers et al. (2001) and Bago et al. (2002) state that once TAG moves out of spore towards germ tube tips or from IRM to ERM, it may be stored, circulated around the fungal mycelium, catabolized via tricarboxylic acid or may serve as anabolic substrate entering the glyoxylate cycle.

Firstly, let us consider the distribution and movement of lipid in the germ tubes that emanate from AM fungal spores. Large lipid deposits are easily visible inside the germinating *G. intraradices* spores and germ tubes. In vivo microscopy reveals brightly stained lipids in the germ tubes of *G. intraradices* and *G. rosea*. Bago et al. (2002) have shown that such lipid bodies accumulate comparatively intensely near AM fungal spore, which progressively decrease farther away from the spore. Lipid globules are almost absent near the tip of the germ tube. They also performed a three-dimensional digital reconstruction of optical slices of fungal material to decipher the percentage of hyphal volume occupied by lipid globules. Within *G. intraradices* germ tubes close to spore, 15.6% hyphal volume was occupied by lipids. The volume of lipids decreased progressively, reaching only 0.3% at hyphal tips, whereas in case of

G. rosea, the gradient of lipid globules decreased from 4.6% nearby spore to nil at the hyphal tips. Bago et al. (2002) have also carried out excellent time-lapse studies, visualizing the movement of lipid globules along the germ tube. Their movies have shown that translocation of lipid droplets is rapid when situated closer to germinating spores. A few other interesting facts about lipid globules are that not all lipids in spore/germ tube are transported away—some remain static—but those in the middle region move rapidly. The movement of lipid is not smooth. It can be in pulses and is not unidirectional. Irregular movement of lipid globules is frequently observed in hyphal tips.

Inside the extraradical mycelium (ERM), fraction of volume occupied by lipid globules is greater than in germ tubes. The storage lipids may occupy upto 24% volume inside ERM closer to the root, but may be as low as 0.5% at growing hyphal tips, i.e. far away from roots. Investigation on ERM of *G. intraradices* and *G. margarita* using two-photon laser scanning microscopy revealed a similar pattern of lipid distribution. However, volume occupied by lipids was greater at 50% of *G. intraradices* hyphae. A three-dimensional reconstruction of gradient showed that lipid density is greatest at approximately 10 cm from apex in runner hyphae. Active cytoplasmic streaming and movement of lipids was clearly shown using movies based on Nile red stained AM fungal hyphae. Bago et al. (2002) reported that lipid bodies take irregular shapes as they move along the hyphae or reach hyphal membrane and cell wall edges. Movies prove that lipid bodies also moved against the cytoplasmic streaming. The fastest lipid movement occurred in runner hyphae. It is computed to be 4 $\mu m\ S^{-1}$ for *G. intraradices* and 8 to 11 $\mu m\ S^{-1}$ in *Gi. margarita.* Bago et al. (2000) remark that such lipid movement (cytoplasmic streaming) in the hyphae, in both symbiotic and asymbiotic portion, has greater relevance in C transport in AM symbiosis.

The physiological aspects of lipid transport in AM fungi are not yet clear. Understanding the cellular and biochemical routes of lipid transport in plant-AM fungus, and the rates of lipid movement within fungal thalli are vital. Cytochemical observations have shown that lipid bodies move with cytoplasmic stream, but some remain stationery and attached to cell membranes or hyphal wall. A few other lipid globules move in the direction opposite to cytoplasmic streaming. Bago et al. (2002) suspect that movement of lipid independent of cytoplasmic streaming could be associated with cytoskeleton, micro tubular or vacuolar arrays. These lipid transport 'tracks' occur all through the length of runner hyphae (Timonen et al. 2001; Timonen and Peterson, 2002). There are reports that alpha-tubulin and actin gene expression or protein formation matches acyl-COA dehydrogenase. If the above points are considered along with microscopic evidences on lipid movement in hyphae, AM fungi seem to possess a full array of cytoskeleton system to transport lipid along the runner hyphae. Carbon flux as lipids may depend on many factors, such as plant C status, lipid bodies in IRM, fungal demand for C in ERM, etc. No doubt, lipid translocation will depend on specific plant-fungus combinations. Bago et al. (2002) have reported that the quantity of TAGs translocated along hyphae may vary. For *G. intraradices*, it is approximately 0.26 μg TAG h^{-1} and that for *Gi. margarita* is 1.34 μg TAG h^{-1}. These observed rates of lipid bodies accounts for lipid accumulated in ERM, after due consideration to lipid recycled from ERM to IRM, and lipid consumed for fungal activity. An idea regarding such carbon fluxes as lipid bodies will be useful while judging the quantity of carbon moved from plant to fungus. Recently, Gavito and Olsson (2003) utilized signature fatty acid measurements in the extraradical mycelium and found that its exploratory/foraging activity in the soil, especially nutrient foraging, depended partly on plant carbon (lipid) and energy storage in AM fungal mycelium. The hyphal density measurements have shown that extraradical mycelium of AM fungi proliferated in response to amendments, lipid storage and energy reserves.

Monoxenic cultures of obligately biotrophic AM fungus have been utilized to study lipid metabolism. Recently, Fontaine et al. (2001a) have used bicompartmental culture plates, wherein sporulating extraradical hyphae can be obtained totally free of root, so that ^{14}C acetate can be provided as the precursor for lipid synthesis. These researchers recorded lipid synthesis in all three symbiotic stages studied by them, such as germinating spores, symbiotic fungus and fungus detached from host roots. In each case, the fungus was able to synthesize its own 1,3 diacylglycerols, triacylglycerols, phospholipids, sterols and free fatty acids de novo. Hence, Fontaine et al. (2001a) have clarified that 'obligate biotrophy' of AM fungi is not due to deficiency in the synthesis of any of the various major lipid classes. Almost similar inferences were reported by Gasper et al. (2001) regarding the ability of extraradical mycelium (ERM) to synthesize and hydrolyze its own acylglycerides. In fact, they observed a parallel relation between activity of acyl-CoA ligase (e.g. palmitoylCoA ligase) and hyphal development. They suggested that during hyphal growth, biosynthesis of triglycerides proceeds at a faster pace and prevails over degradation. In nature, the 'oleagenic' fungi accumulate high levels of lipids reaching upto 25% by dry weight (Jabaji-Hare, 1988). Technically, AM fungi can be called 'oleogenic'. Bago et al. (2002) point out that most other oleogenic fungi accumulated carbon as lipids, whenever nutrients such as N and P are limiting. But lipid accumulation in intraradical mycelium (IRM) of AM fungi does not follow this pattern of synthesis and utilization. The abundance of lipids in AM fungi is actually a compact form of C storage, which could yield calories throughout the length of hyphae as needed. In fact, Bago et al. (2002) believe that continuous streaming of lipid bodies may actually ensure continuous/uniform availability of energy to the fungus at all points in the hyphae. Storage lipids are synthesized and accumulated in IRM but not in ERM or germinating spores (Pfeffer et al. 1999). Hence, to attain uniform distribution, lipid movement occurs along with cytoplasmic steaming. Pfeffer et al. (1999) and several others have used radiolabeled ^{13}C glucose and $2H_2O_2$ in order to understand the aspects of lipid metabolism in the *G. intraradices*/carrot cell associations. These trials proved that triacyl glycerol (TAG) is synthesized in the IRM by the AM fungus, which then moves to ERM. There was no evidence for *de novo synthesis* of storage lipids in the ERM, but recycling of lipids from ERM back to IRM is a clear possibility.

^{13}C-NMR spectra obtained on fungal lipid extracts mostly indicate the predominance of TAGs. In the ERM of AM fungus, almost all fatty acid moieties are C16:1c11, but mycorrhizal roots possess a combination of fungal and plant TAGs (e.g. C18:2c9,11) (Pfeffer et al. 1999). In monoxenic cultures, P availability to extraradical mycelium affects metabolic activity and carbon distribution, especially lipid accumulation. In case of *G. intraradices*, nearly 18% of ^{13}C could be traced in the fatty acids (16:1wS) fractioned from hyphae. Such ^{13}C spectral differences allow estimation of lipids in each phase of the AM fungus, i.e. in IRM, ERM and mycorrhizal roots. The host and fungal fatty acids emit separate signals in the NMR spectrum. If ^{13}C-1,3 Glycerol was supplied to ERM, TAG in both mycorrhizal roots and ERM itself show radiolabel ^{13}C in glycerol but not in fatty acid moieties of TAG. If ^{13}C glycerol was given to mycorrhizal roots, TAGs in both ERM and mycorrhizal roots become highly labeled. Further, the ratio of labeling IRM/ERM was 1:1. When spores of *G. intraradices*, in other words, the asymbiotic phase of AM fungus was provided with ^{13}C glycerol, ^{13}C label could be traced in only glycerol moiety but not in fatty acids. If ^{13}C acetate was supplied to germinating spores, then neither glyceryl nor fatty acid moiety of TAG showed up in the radiolabel. This happens despite efficient utilization of acetate in anabolic processes of the fungus. Overall, radiotracer trials indicate a substantial movement of lipids from ERM and re-circulation of lipids from ERM back to IRM. It is questionable whether recycled lipid globules reach the same root via hyphae. The AM fungal hyphae are coenocytic and highly branched with interconnection to roots of other plants in vicinity. Hence, the lipid globule may translocate to root system of neighboring plants via a common mycorrhizal network (Bago et al. 2002).

1. G. Nitrogen Metabolism in AM Symbiosis

Nitrogen is absorbed by extraradical hyphae of AM fungi. This aspect was examined by Frey and Schuepp (1993) using ^{15}N tracer fed to *G. intraradices* existing in symbiotic state with maize seedlings. The experimental setup possessed 40 μM mesh that separated the hyphae from roots. From their observations on ^{15}N uptake, it is clear that much of the N traced in maize seedlings was actually garnered via extra radical hyphae. Depletion of ^{15}N noticed in the hyphal chamber further confirmed its transport via AM hyphae. Time course studies proved that higher levels of ^{15}N uptake occurred 10 to 15 days after application of ^{15}N $(NH_4)_2SO_4$. Mader et al. (2000) utilized a hydrophobic polytetrafluoroethylene (PTFE) membrane to inhibit mass flow and diffusion of mobile ions from a soil compartment loaded with 15N. This compartment was accessible to AM fungal hyphae but not to the roots. ^{15}N flux measurements proved that AM fungal hyphae translocated ^{15}N to the tune that mycorrhizal plants had 3 times higher amount of ^{15}N than non-mycorrhizal counterparts. They inferred that AM fungi explore and translocate substantial amounts of N from soil.

Interestingly, the extent of N transported by AM fungus depended on the species. They attributed it to genetic differences in patterns of hyphal spread and consequent volume of soil exploited. They suspected that differences in capacity for N uptake per unit length of AM hyphae also contributes to differences among species. Azcon et al. (2001) provided ^{15}N labeled nitrate to two different AM fungal species. Tracer analysis showed that nitrogen derived from ^{15}N radiolabeled fertilizer (NdfF) was 84 mg N kg^{-1} for *Glomus mosseae*, but 168 mg N kg^{-1} for *G. fasciculatum*. Clearly, AM fungal ability to absorb nitrate is dependent on substrate concentration. Timing of nitrate application also influences its uptake by the fungal hyphae.

Valentine et al. (2002) have reported some interesting facts regarding nitrogen forms utilized best by AM fungi and their effects on symbiosis. They fed cucumber roots with three different forms of N, namely NO_3, NH_4 and a mixture NH_4/NO_3, but each source provided the same N concentration (4 mM) in the medium. Plants provided with NO_3 supported the highest colonization, hyphal growth and possessed greater number of arbuscule per unit root length. However, NH_4^+ application supported comparatively lower levels of hyphal growth and vesicle formation. Hawkins and George (2001) too observed that NH_4-N did not support rapid hyphal growth. They utilized ^{15}N labeled NH_4^+ and fed it to wheat seedlings grown in hydroponics, wherein hyphal portions were kept compartmentalized. Data from their experimental system indicates that reduced hyphal length could be a direct effect of NH_4^+ on hyphal growth. In other words, nitrogen forms affected hyphal growth and metabolic activity. The nitrogen source supplied can influence root growth. As a result, it could affect nutrient exchange rate and growth of AM fungus, or it can have a direct influence on the growth of AM fungal hyphae. This aspect was examined using compartmentalized roots and hyphae of *Glomus mosseae*. Although NH_4^+ fed plants accumulated higher N than NO_3 fed counterparts, it was not favorably reflected in terms of hyphal densities or ^{15}N transport via hyphae. Hence. Hawkins and George (2001) suspected that the influence of NH_4 or NO_3 on hyphal growth could also be mediated via host root growth and its metabolic state.

Toussaint et al. (2001) analyzed N metabolism in AM fungus and its host using an in vitro culture system, which allows accurate control over experimental conditions. They found that among the key enzymes of N metabolism, glutamate dehydrogenase (GDH) activity increased in response to AM infection, but activity levels of glutamate synthase (GS) and nitrate reductase (NR) did not alter. Addition of NH_4NO_3 induced GDH, GS and NR activities to equal extent, in both mycorrhizal and non-mycorrhizal carrot root tissue. Accumulation of N was noticeable in extraradical mycelium and mycorrhizal roots, based on which they suggest that AM fungi may be partly mediating N transport and assimilation.

Mathur and Vyas (1996) examined several AM fungal species for their role in influencing the key enzymes of nitrogen metabolism. Among the fungi, *Scutellospora calospora* was more effective in inducing the key enzymes; namely GDH, GS and NR. Mycorrhizal infection leads to a more prominent induction of NADP-GDH. In fact, a three-fold increase in NADP-GDH was recorded due to colonization by *Scutellospora calospora*. Nitrate reductase activity may also be indicative of effectiveness of AM symbiosis. For example, Caravaca et al. (2003) found that root NR activity of host colonized with AM fungi such as *G. intraradices*, *G. mosseae* and *G. deserticola* were increased significantly compared to uninoculated control.

1. H. Secondary Metabolism Relevant to AM Fungal Symbiosis

Flavonoids are secondary metabolites with significant effect on spore germination, hyphal growth and root colonization by arbuscular mycorrhiza. Flavonoids are also considered as crucial compounds with role in signaling between the symbionts. A recent study by Larose et al. (2003) suggests that flavonoid synthesis in a host such as *Medicago* sp. is affected even before the AM fungus penetrates the root tissue, thus, emphasizing their role in signaling during pre-infection stages of development of symbiosis. Flavonoid accumulation in symbiotic roots is influenced by the stage and extent of colonization. AM fungal species (or isolate) can also influence the extent of flavonoid synthesized in the root tissue. Further, they suggest that understanding the effects of pattern of flavonoid synthesis and accumulation, on the symbiosis may be more important.

Regarding sterol synthesis, Fontain et al. (2001b) have shown that its synthesis does occur de novo in the mycelia of AM fungi, be it in germinating spore stage, detached mycelial stage or symbiotic state. They proved it using a combination of Ri TDNA transformed carrot cell culture held in symbiotic association with AM fungus *G. intraradices*. In all the three stages of symbiosis stated above, they traced at least 24 alkylated sterols, 24 methyl and 24-ethyl cholesterol. Ergosterol, a common sterol in most fungi was, however, not traced. It was concluded that AM fungus *G. intraradices* was capable of imbibing and utilizing exogenous acetate to synthesize sterols.

Hause (2003) suspects that secondary metabolism of the host and accumulation of specific metabolites may affect AM fungal establishment and functioning. For example, it is argued that jasmonites, which are known as regulators of specific plant responses to biotic and abiotic stresses, may also influence mycorrhizal fungal development. Accumulation of certain secondary metabolites seems to be due to mycorrhizal colonization. Examination of a number of tobacco and tomato species has proved that host plants accumulate several glycosylated C-13 cyclohexanone derivatives. In all, Maier et al. (2000) could identify eight derivatives by subjecting mycorrhizal root extracts to High-Performance Liquid Chromatography (HPLC) and spectroscopy. In contrast, accumulation of coumarins, scopoletin and its glucoside were suppressed as a consequence of AM fungal spread in roots. Obviously, certain rearrangements in secondary metabolic activity and accumulation of specific secondary metabolites occur due to AM fungal invasion. The biochemical causes and significance of such changes to both host plant and fungus, as well as to sustenance of symbiotic relationship, need to be understood. Spatial and temporal analyses of jasmonite accumulation in roots versus its relation with mycorrhizal colonization were studied. Jasmonite accumulation affects the mycorrhizal morphology. In a related study, Hause (2003) once again found that secondary metabolites such as 'mycorradicin', a typical yellow pigment encountered in mycorrhizal plants, actually localized itself around the arbuscules inside the host cells. Such a localization of mycorradicin did affect AM fungal development. In the later stages of symbiosis, mycorradicin was localized in vacuoles.

The yellow pigment frequently identified in intensely colonized roots is an acyclic C-14 polyene compound. This chromophore 10,10'-diapocarotene-10,10-dioic acid has been named 'Mycorradicin' because of its origin in mycorrhizal systems. A C-13 cyclohexanone glycoside, which is a carotenoid derivative, was also reported in mycorrhizal roots. Walter et al. (2003) state that precursor of both the above apocarotenoid compounds; a common carotenoid (xanthophylls) is affected by arbuscule development and degeneration. The biosynthesis of these apocarotenoids in the AM roots proceeds through a non-mevalonate pathway involving isopentenyl diphosphate biosynthesis. Transcripts for the key enzymes in this pathway are strongly elevated. Further, they believe that among the many systems specific to mycorrhizal symbiosis, it is possible to use 'yellow pigment' biogenesis and its regulation as a representative secondary metabolite. Molecular regulation of 'yellow pigment' may involve both fungal and plant genes, and its details need to be investigated. Fester et al. (2002) studied the stimulation of carotenoid metabolism in AM symbiosis. They suggested that the development of AM is correlated with accumulation of isoprenoids (acyclic ^{14}Cpolyene or 'mycorradicin') and cyclohexanone derivatives. Carotenoid profiling has revealed that zeta-carotene is accumulated in mycorrhizal *Zea mays* and *Medicago truncatula*. Interestingly, mycorradicin biosynthesis and mycorrhization both were impaired in maize mutants defective for carotenoid synthesis. It was concluded that mycorradicin biosynthesis is probably induced by mycorrhization, utilizing a C-40 precursor carotenoid. The induction occurs at transcriptional level.

Investigations by Walter (2003) have focused around isoprenoid biosynthesis in AM systems. Obviously, the purpose is to characterize such specific secondary metabolites, trace their biosynthetic pathways and understand the physiological significance to host-AM fungal partnership. Firstly, several apocarotenoids that accumulated in the mycorrhizal roots were characterized. Accumulation of apocarotenoids in AM systems seems ubiquitous. These apocarotenoids seem to integrate and complex strongly, since only alkali treatment can release them. Molecular analysis using EST probes for enzymes that operate within the carotenoid biosynthetic pathway revealed that enzymes in mevalonate-erythritol phosphate pathway are induced. The two key enzymes concerned here are 1-deoxy-D-xylose5-phosphate synthase and 1-deoxy-D-xyllulose 5-phophate reducto-isomerase. Fester et al. (1999) investigated the de novo synthesis of isoprenoids such as blumenin and consequent transient accumulation of hydroxy-cinnamate amides in mycorrhizal wheat and barley. Accumulation of these secondary metabolites began early in two-week-old seedlings. During the final stages of symbiosis, when massive amounts of spores were produced, these secondary metabolites occurred in trace amounts. They have also claimed that certain microbes termed mycorrhiza helper bacteria (MHB) were able to stimulate the synthesis of secondary metabolites such as blumenin in the mycorrhizal roots. Yet another inference derived from this study is that isoprenoid and phenylpropanoid biosynthesis is closely linked to the developmental stage and extent of mycorrhizal fungus in the host cereal roots.

Mycorrhizal Leek plants grown under hydroponic conditions accumulated higher quantities of phytohormones, namely auxin (Indole acetic acid) and zeatin riboside. To a certain extent, this effect could be mimicked by an addition of P into hydroponic system. Torelli et al. (2000) reported that such accumulation of phytohormones influences root architecture as well as mycorrhizal spread in root and soil. Similarly, inoculation with AM fungus induces indole 3-butyric acid (IBA) synthesis; as a consequence, free-IBA may increase in cortical cells of host root (Jutta and Kaldorf, 2000). Such an increase in free-IBA accumulation coincided with enhanced activity of auxin receptor, ABP1. It is also a well-known fact that IBA, at appropriate concentration, induces root growth and lateral root formation. This aspect was confirmed by the suppressive effect of IBA-inhibitor on root growth and lateral root formation. Shaul-Keinan et al. (2002) analyzed phytohormones in mycorrhizal tobacco seedlings using HPLC and radioimmunoassay. The AM fungal infection induced the accumulation of zeatin riboside-like and isopentanyl adenosine-like compounds in roots. It is clear that cytokinin status is influenced by AM colonization. Using gas chromatographic and mass spectrometric methods, they proved that during arbuscular stage of symbiosis, gibberlins, GA 1, GA 8, GA 19 and GA 20 were significantly abundant in mycorrhizal roots. Barker and Tagu (2000) have reviewed several metabolic aspects and role of phytohormones in mycorrhizal symbiosis in general. They have concluded that individual roles of various phytohormones, say auxins, gibberlins and cytokinins and their derivatives are specific and subtle biochemical controls exist. These aspects need to be understood and explained clearly.

2. Molecular Physiology of Ectomycorrhizas

Several ECM fungus-host combinations have been studied with regard to their development and physiological nature. To preserve brevity under this section, only a couple of host-fungus combinations have been discussed. Major emphasis is confined to carbohydrate and nitrogen metabolism. All the important aspects on physiology and molecular biology of phosphorus nutrition have been dealt with separately in Chapter 4.

2. A. Establishment of ECM Symbiosis: Some Physiological Aspects

Development of ECM involves a series of complex ontogenetic processes. The ECM roots are ensheathed by a mantle of dense fungal hyphae. Hartig net is characterized by profusely branching intercellular hyphae. A network of extraradical hyphae that develops in due course explores the nutrients and water in the soil. The development of the mantle itself involves a series of programed morphogenetic steps (Martin et al. 1997; 2001). Mantle formation begins in the apical region of root. Immediately after attachment to epidermal cells, hyphal layers aggregate to form a mat of several μm thickness. A proteinceous and polysaccharide matrix then enmeshes the mantle. A few other steps involved are cessation of fungal growth, initiation of lateral roots, aggregation of hyphae, radial elongation of epidermal cells, formation of hartig net and extension of hyphae into the soil. Martin et al. (2001) opine that subtle cellular and physiological changes that follow in ECM, actually extend the root functions. Mantle, along with intra and extraradical hyphae, aid nutrient uptake as well as carbohydrate flow from plant root to ECM fungus, thus making it a mutual association.

Herrmann et al. (2001) opine that although several different morphological and biochemical changes that occur during the early stages of ECM symbiosis has been studied, our efforts to understand the molecular changes at nucleic acid level has not been commensurate. In recent times, model in vitro micro-propagated systems of symbionts, such as *Quercus/Piloderma* or *Eucalyptus/Pisolithus* have been yielding useful data regarding molecular events during signaling and establishment stages of ECM symbiosis. Obviously, signaling and recognition in ECM are important aspects influenced by genetic constitution and physiological manifestations of both the symbionts. A stable symbiotic structure actually develops through a series of signaling events that trigger specific morphogenetic and physiological changes. Signal compounds play an important role during ECM formation. In case of *Pisolithus tinctorius*, hypaphorine is a major indole compound synthesized and excreted by the fungus. Hypaphorine overaccumulates as soon as the hyphae contact the host root surface. Hypaphorine concentrations within the ECM fungus may reach 1000 fold higher than indole acetic acid. Hypaphorine induces changes in host root hairs. Hypaphorine solutions, if applied to fungal colonies, induce net K^+ uptake and H^+ extrusion. Hypaphorine seems to regulate ion fluxes in the initial stages of ECM establishment, induce hyphal tip elongation and influence nutrient exchange (Thierry et al. 1998). This hypaphorine, which is not considered as an analogue of indole acetic acid, suppresses the root hair elongation, if present at 10 to 1000 μM concentration. It means that ECM fungi suppress host root hairs as the lateral roots and the fungus start developing. At this point, it is interesting to note that mycorrhizal helper bacterium (e.g. *Pseudomonas monteili*) stimulates hypaphorine production and increases the aggressiveness of the ECM fungal symbiont (Duponnois and Plenchette, 2003). A recent study by Ditengou et al. (2003) indicates that hypaphorine elaborated by ECM fungus *Pisolithus tinctorius* counteracts IAA activity and controls root hair elongation. Hypaphorine-actually induces changes in cytoskeletal changes. First signs of cytoskeletal alteration could be detected 15 min after hypaphorine treatment. Hypaphorine-treated root hairs do not display actin caps. Fine actin filaments are replaced by thick bundles, which extend from subapical region to the tip. Hypaphorine treatment reduced the number of actin filament bundles. The attachment of the fungus to host root and formation of symbiotic interface is said to be crucial to ECM symbiosis. We should note that immediately after recognition, cell surface changes that occur on the symbionts influence the interactions that proceed later. ECM fungi are known to produce a glycoprotein containing micro-fibril in response to its attachment to host roots. It helps the fungus in gaining a better anchorage. Next, the physical contact between symbionts is said to induce several biochemical changes, each serving as a key to unlock further differentiation in the mycorrhizal symbiosis (INRA, 2002).

Similar to other biotrophic fungal interactions, establishment of ECM too involves a range of cellular mechanisms. At this stage, fungal perception of the host cell surface environment is said to be important in developing a compatible association. Host-induced hyphal branching is an initial phenomenon, which is followed by the formation of adhesion pad on the root surface. The actual docking of the ECM fungus on to host root cells may involve several biochemical receptors at the mucilaginous surface, on cell walls and plasma membrane. A number of physiological changes such as alterations in protein composition of cell walls, extramatrical surface of hyphae, production of mannoproteins, alanine-rich acid polypeptides, induction of specific enzymes and hydrophobins occur during establishment stages of the ECM symbiosis. The development of Hartig net is an important aspect of ECM symbiosis. In some cases, a transient increase in phenylalanine ammonia lyase (PAL) activity and accumulation of phenolic compounds coincides with hyphal penetration between root cells during hartig net formation. It is believed that such cytochemical changes restrict and regulate hartig net formation and ECM fungal spread within the root cortex (Feugey et al. 1999).

Indole acetic acid (IAA) is key compound that regulates ectomycorrhizal development. Generally, vigorous mycorrhiza formers such as *Pisolithus* and *Laccaria* accumulate greater quantities of IAA. Production of IAA itself depends on tryptophan supply in the medium. Nylund et al. (1994) reported that even for IAA-over producing mutants of *Hebeloma*; tryptophan accumulation upto a certain level was a prerequisite, so that specific changes in root morphology could be effected during ECM formation. In the case, *Pinus pinaster/Hebeloma cylindrosporum* association auxin production firstly affects the growth polarity of cortical cells as the short roots are formed. Accordingly, Laurans et al. (2001) examined and compared the effects of 'auxin over-producing' and wild type *H. cylindrosporum* strains on root morphology during establishment of symbiosis. As expected, auxin over-producing ECM fungal isolates caused marked modifications inside roots by inducing 43% reduction in cortical cell elongation and a 35% increase in radial growth of short roots, whereas, wild type *H. cylindrosporum* could cause only 30% reduction in cortical cell elongation and 10% increase in radial growth of root cortex. Immunolocalization studies revealed no drastic changes in cytoskeletal aspects such as α-tubulin. Similarly, polysaccharide accumulation was not affected significantly by the use of auxin over producing ECM fungal strains. Hence, Laurans et al. (2001) inferred that root morphological changes that occur during development of ECM association are predominantly through changes in cortical cell/tissue growth pattern. Auxin transport inhibitors, such as 2,3,5 triiodobenzoic acid (TIBA) and 1-N-naphthylphthalamic acid (NPA) have been utilized to confirm the role of auxins in establishment of ECM symbiosis. Rincon et al. (2003) have reported that the presence of TIBA completely prevented the development of ECM structures such as mantle and Hartig net. It also suppressed the stimulatory effect of symbiotic fungus. In the presence of other inhibitor NPA, number of seedlings colonized by ECM fungus L. *bicolor* reduced drastically.

Both the auxin transport inhibitors counteracted the effect of exogenous auxin addition and/or ECM fungus on rhizogenesis. A more recent molecular analysis by Charvet-Candela (2002) also shows that auxin affects host plant gene expression during early mycorrhization steps. Products from an auxin upregulated gene Ppiaa88 isolated by them from a *Pinus pinaster* cDNA library, shares extensive homology with auxin inducible proteins. Cycloheximide did not suppress auxin induced Ppiaa88 upregulation. The Ppiaa88 transcript levels were enhanced if auxin-overproducing mutants of *Hebeloma cylindrosporum* were inoculated on to *Pinus pinaster.* They suggest that during auxin-mediated induction of changes in host root and early mycorrhizal development, a putative transcription factor such as Ppiaa88 could be important.

Generally, phenolic compounds present in Eucalypt root exudates are known to affect ECM fungal growth. Rutin, a specific phenyl glycoside (flavonol) component from root exudates stimulates hyphal growth in *Pisolithus.* Rutin is suspected to function as a flavonoid signal. A rutin gradient even at picomolar concentration can easily influence the orientation and elongation of *Pisolithus* hyphae towards the root tip. Perhaps it could be aiding infection and further physico-chemical interactions between symbionts (Lagrange et al. 2001).

In a study involving *Betula/Paxillus* association, it was found that glutamine, aspartate and asparagine pools were always lower in infected roots than in the non-infected ones. Citrate and malate were the two major organic acids detected. Aspartate aminotransferase, glutamine synthetase, NAD-dependent malate dehydrogenase and glucose 6-phosphate dehydrogenase activity were higher in infected roots. Quantitative analysis of these enzymes has clearly suggested that ECM formation leads to re-arrangement of metabolic pathways during early stages, especially after contact between symbionts. Such changes could also augment structural changes required during ECM formation.

2. B. Carbohydrate Physiology of Ectomycorrhizas

Most ectomycorrhizal fungi can be cultured comparatively easily on synthetic medium containing definite/single organic source. Consequently, we can assess the fungal ability to utilize different organic compounds as a source of energy. In natural conditions, organic compounds in root exudates are the most plausible sources of energy for microbes in rhizosphere and ECM fungi. Generally, root exudates contain low-molecular weight soluble sugars, carboxylic acids and amino acids. Previous tests indicate that ECM grew best on hexose (glucose, fructose, mannose), disaccharide (cellobiose, trehalose) or sugar alcohols (mannitol). ECM fungi also utilize galactose, maltose, sorbitol, dextrin, starch and glycogen. Fungal growth was not discernible if trioses, tetroses or pentoses or carboxylic acids were provided as the sole energy source.

The next best source of energy/nutrient that supports ECM fungal growth is the components derived from the host cell wall itself. Results from over 100 different ECM fungal species have supported the idea that pectins are not good carbon sources (Schaeffer et al.1997). Since the ability to degrade proteins and occurrence of proteases are ubiquitous in microbes, cell wall proteins could be a source of C/energy to ECM fungi. ECM fungi are capable of high protease activity (Bau and Barton, 1989; Guttenberger et al. 1994). Schaeffer et al. (1997) suspect that ECM fungi may even divert plant cell wall precursors away from the host by interfering with polycondensation (Fig. 2.4). Such diverted cell wall precursors could also be utilized to support ECM growth. Schaeffer et al. (1997) summarize that simple hexoses are easily the best C/energy sources for ECM fungi, but the significance of alternate C sources need to be understood in greater detail.

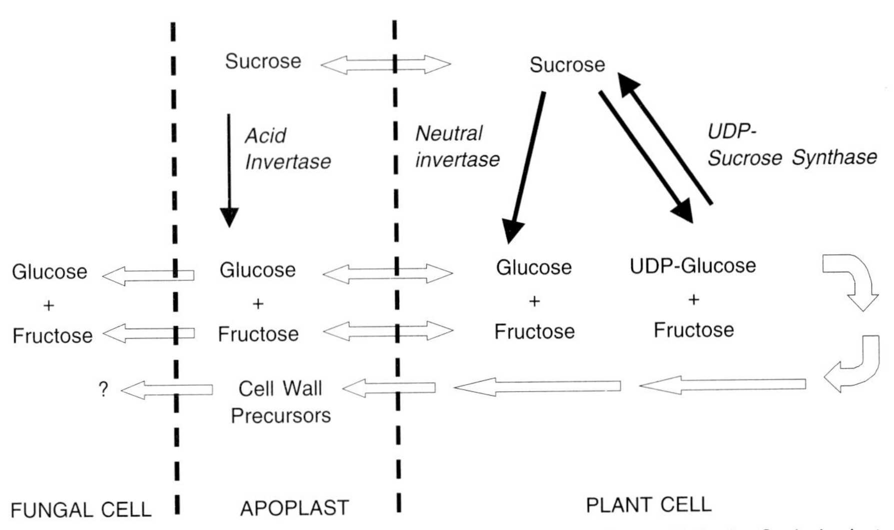

Fig. 2.4 Sucrolytic Enzymes in Ectomycorrhizal Associations. Role in Carbohydrate Translocation and Metabolism in Symbionts. (Schaeffer et al., 1997)

Sucrose is an important carbohydrate component in plants. It is the most common photo-assimilate transported across tissues, hydrolyzed and utilized by higher plants. In the present context, which is in conifers, transport of carbohydrates from leaves through phloem into root tips is primarily achieved as sucrose (Komor, 1983; Stanzel et al. 1988). Also, in conifers and other ECM hosts, sucrose is the photo-assimilate transported to roots, where phloem unloading occurs via passive leakage into apoplast, depending on the concentration gradient. Not just sucrose, but its monosaccharide components glucose and fructose are also traced in apoplast of the root. These observations suggest that sucrose may be the chief source of C/energy for ECM fungi. Naturally, the existence of an ECM fungal partner demands greater allocation of C to roots, so that it can be utilized to support mycorrhizal activity. Carbon distribution studies have proved that nearly 4 to 42 times more photo-assimilates are transferred to mycorrhizal roots compared with non-mycorrhizal roots (Cairney et al.1989). Similarly, 15 times more $^{14}CO_2$ was traced in the roots provided they were colonized with *Rhizopogon roseolus* compared with non-mycorrhizal seedlings.

Contrary to opinions, most, if not all ECM fungi lack the ability to utilize sucrose as the sole source of carbon (Salzer and Hager, 1991). However, as exceptions, there are reports that certain ECM fungi such as *Cenococcum geophilum*, *Tuber melanosporum*, *Rhizopogon* sp. and *Suillus* sp. could utilize sucrose. This sucrolytic activity attributed to these above situations could easily be due to pH of the medium, therefore, it is a non-enzymatic hydrolysis caused mostly by acidification. Hence, it is not directly attributable to ECM fungal enzymatic degradation of sucrose. Whereas, cell wall pH at hartig net ranges between 5 and 6; hence non-enzymatic sucrolytic activity may not occur in this zone. Most tests with ECM fungi have shown that ECM fungi utilize glucose and fructose, but lack a sucrose transport system. In other words, availability of sucrolytic enzymes/activity in the host is mandatory for ECM symbiosis to develop and the fungus to proliferate.

Invertases, enzymes responsible for sucrolysis occur in many fungi, including *Saccharomyces cerevisiae*, saprophytes such as *Phycomyces blakesleeanus*, *Botrytis cenerae* (see Salzer and Hager, 1991). Such invertases have not been detected in ECM fungi (Table 2.3). Sucrose incorporation into the growth media could not induce invertase activity in ECM fungi. Hence, if sucrose was the sole carbon, then ECM fungi tested ceased to put forth hyphal growth. However, such growth cessation was quickly overcome on supplying exogenous invertase. Further, liquid culture with sucrose as the sole source of C did not contain its hydrolytic products glucose and fructose, meaning that invertase was absent. Therefore, sucrose supplied into medium exogenously, or that available in the vicinity at hartig net under natural conditions can be utilized by ECM fungi, only if extraneous sucrolytic activity (or invertases) are present. It is believed that in nature, host cell root invertase plays the vital role in hydrolyzing sucrose and supplying glucose and fructose to ECM fungi (Table 2.3). It is also believed that apoplastic invertase plays a key role in sustaining the carbon/energy supply. Presence of an apoplastic invertase system in host root cells seems to be a prerequisite for ECM fungal proliferation.

Table 2.3 Invertase (sucrolytic) activity in the cell walls of Symbionts

	Invertase activity nKat g FW^{-1}	*SE*
Plant Host		
Picea abies	21.7	±3.3
Fungal Symbiont		
Amanita muscoria	0.041	±0.02
Hebeloma crustuliniforme	0.023	±0.05

Source: Salzer and Hager, 1991
Note: Hydrolysis of exogenously applied sucrose to suspension cultured cells of the spruce or fungal hyphae were measured at pH 4.6, 28°C. Values are averages from 3 to 8 experimental sets.

Knowledge about regulation of invertase activity in root cells and apoplastic space is crucial because it may determine the extent of carbon source available/utilized by ECM fungi. It is interesting to note that suspension cultured cells of *Picea* roots, when co-cultivated with ECM fungal hyphae, enhances the cell wall-bound invertase. Both the invertases tested by Salzer and Hager (1993a) were inhibited by fructose, but not by glucose. Optimum invertase activity occurred at pH 4.5, but decreased rapidly if pH was raised to 6.6. It is believed that as the ECM fungus draws away fructose for its growth, then invertase could be restored (re-induced), because it relieves the end product inhibition. Similar small changes in apoplastic pH caused by auxin produced by ECM fungus again, can induce cell wall-bound invertase (Salzer and Hager, 1993b; Gay et al.1989). In spruce roots, pH optimum for acid-invertase activity is 4.0, and at this pH neutral invertase activity is negligible. Schaefer et al. (1995) suggest that certain isozymes of invertase could be tissue specific. They remark that cortex specific distribution of acid invertase, neutral invertase and sucrose synthase is of interest because it may regulate the supply of sucrose, its hydrolysis and transfer in the hartig net (Fig. 2.4). In this particular example, invertase detected in mycorrhizal spruce roots ranged from 1 to 14 μmol sucrose hydrolyzed min^{-1} g^{-1} dw. This level of invertase is supposed to be reasonably higher when compared with other hosts. Understanding the influence of ectomycorrhiza formation on the quantity and quality of soluble sugars is important. There is no doubt that glucose and fructose are utilized as the sole C source by ECM fungi, but conversion of the monosaccharides into trehalose and mannitol is a clear possibility. Actually, trehalose is a major non-reducing disaccharide that serves as a carbon storage compound in many ECM fungi (Wisser et al. 2000). For example, in Spruce-*Amanita muscaria* association, trehalose was dominating (230 μmol g^{-1} dw), whereas *Cenococcum geophilum* accumulated mannitol (390 μmol g^{-1} dw) predominantly. Also, it is common to detect lower levels of soluble sugar in mycorrhizal roots compared to non-mycorrhizal roots. Main reasons suggested are 'dilution of sugars' in mycorrhizal roots, or ECM fungus acting as strong C sinks or due to the conversion of glucose and fructose to other sugars or sugar alcohols. Further, we should not discount the ability of ECM fungi to convert sugars into several other biochemicals. Some of these biochemicals may be important in setting up a gradient at apoplast so that carbohydrate transfer is effected from autotrophic partner to the heterotrophic fungus.

Now, let us consider some interesting facts about carbohydrate (sucrose/glucose) transfer in ECM fungi. They were derived from an experiment using protoplast cultures of ECM fungi, *Amanita muscaria* and *Cenococcum geophilum*. Hampp et al. (1995) observed an increased activity of sucrose synthase, which is a key enzyme involved in cytosolic formation of sugar in leaves. At the same time, formation and accumulation of fructose 2,6-bisphosphate decreased. These changes facilitate an increased formation of sucrose in mycorrhizal seedlings. Next, to understand the fate of sucrose that is transported into roots, they studied its metabolic aspects, utilizing the ECM fungal protoplasts. The protoplasts from *A. muscaria* and *C. geophilum* preferentially absorbed glucose (Km 1.25; Vmax 18 p mol 10^6 protoplasts min^{-1}), and no sucrose was transported. Further, histochemical analysis indicated that sucrose concentration became feeble, but trehalose and fructose bisphosphate was abundant in zones with intense fungal structures. These observations supported the view that invertase activity is confined to host tissue. In a related trial, addition of fructose 2,6 bisphosphate induced phosphofructokinase (PFK) in fungal tissue, but induced PPi dependent fructose-6 phosphate transferase (PFP). It means that the regulatory effects of fructose 2,6, bisphosphate may be pivotal in creating strong sink activity for carbohydrate transfer in mycorrhizal roots and the fungal symbiont (Hampp et al. 1995; Nehls et al. 2001). Hoffman et al. (1997) reported that amounts of cyclic AMP might also have regulator effects on the glycolytic pathway in the ECM fungus.

Buscot et al. (2000) argue that since carbon consumption rate of ECM fungi is around 30% of net photosynthesis by the autotrophic trees, smaller quantitites of C supplied via root exudates is unlikely to be significant. Earliest of the evidences showing that plant host supplies C to ECM fungi was provided through $^{14}CO_2$-based experiments (Melin and Nilson, 1957), wherein the radiolabel appeared in the hartig net within one day. Such carbon flow depended on sink strength and directed towards non-green areas such as roots and ECM fungi. ECM fungi are generally considered as strong sinks for C. Physiologically, key regulatory step seems to reside at sucrose synthesis in leaf cytosol, and the regulation is effected by fructose bisphosphate and sucrose phosphate synthase. Phosphorylation of this enzyme sucrose phosphate synthase results in its deactivation. In fact, many of the inducible/

inhibitory traits of sucrose phosphate synthase is known to be regulated by mycorrhization. For example, in spruce seedlings, phosphorylation of sucrose phosphate synthase was comparatively lower in the needles of mycorrhizal seedlings (Loewe et al. 1996, 2000). Certain other observations have shown that mycorrhization leads to higher amounts of fructose biophosphate, and increases sucrose phosphate synthase activity. It actually indicates better/higher ability to form sucrose consequent to mycorrhization. Clearly, mycorrhization induces increased formation of sucrose and net photosynthetic rates (Buscot et al. 2000). The allocation of sucrose, or other photo assimilates, and in general, C flow is regulated by the sink strength. In ectomycorrhizal plants, radiolabel tests have proved that younger the ECM fungus, greater the sink strength. A longitudinal analysis of spruce seedlings has indicated that sucrose concentration varies within the mycorrhizal plants, but fructose and glucose levels do not vary. Intense mycorrhizal activity reduces sucrose concentration; say in locations such as mid portion of young roots. Simultaneously, fungal-specific compounds such as trehalose and ergosterol accumulate in the fungal zone. We should note that very young or meristematic roots devoid of ECM fungus show no such gradient in sucrose concentration or the fungal components, trehalose and ergosterol.

After analyzing hundreds of ECM symbiotic partners, it is now clear that these fungi absorb glucose and fructose from the root system, but lack a transport system for sucrose at the apoplastic interface. Evidences indicate that ECM fungi can utilize sucrose, when it is hydrolyzed by cell wall-bound invertases. Roots contain invertases and one another sucrose splitting enzyme-sucrose synthase. Again, histochemical tests have shown that inveratase activity is predominant in root cortex, where mycorrhizal interactions are intense. Thus far, there is no evidence for the presence of sucrolytic enzymes in the ECM fungus (Buscot et al. 2000).

We should also know the fate of photosynthates transferred to ECM fungus. It is generally accepted that mycorrhizal roots synthesize trehalose and glycogen from glucose, and mannitol from fructose. In a specific case, Ineuchin and Wiemken (1992) traced higher levels of arbitol in vegetative mycelia of *Pisolithus tinctorius*. The accumulation of specific sugars and/or sugar alcohols seems to depend on the taxonomic class of ECM fungi. Wallenda et al. (1994) have shown that in basidiomycetous ECM such as *Hebeloma* or *Laccaria*, trehalose is the dominant soluble sugar. On the contrary, trehalose is only a minor sugar in ascomycetous ECM; instead, mannitol dominates. In situations comprising a tripartite combination of plant host, a basidiomycetous ECM and an ascomycetous ECM fungus simultaneously, the extent of trehalose and mannitol detected may be proportional to their activity in host roots.

So far, it has been accepted that sucrose is the major carbon-transport metabolite and its products, glucose and fructose, are most likely compounds that are transferred at the apoplastic interface. However, Guttenberger et al. (2003) point out that dicarboxylates that are common components in root exudates could also be the key carbon-transport metabolite. They have remarked that this possibility has been totally neglected for unknown reasons. Evidences are being accumulated to support a 'dicarboxylate hypothesis' for carbon transport in mycorrhizas. Genes encoding the respective transport proteins are being cloned and characterized by the group at Samuel Roberts Noble Foundation, Ardmore, in Oklahoma. Some of the major aspects such as molecular regulation of carbohydrate transfer, beginning with the carbohydrate supply to fungus; strategies that compensate carbohydrates with that drained to ECM fungus; mechanisms that generate strong monosaccharide sink in colonized plant roots; and the impact of apoplastic hexose concentration on fungal metabolism are also important (Nehls et al. 2001).

Lipids and sterols in ECM fungus *Pisolithus tinctorius* and its host *Pinus sylvestris* differ markedly when in an asymbiotic state. Immediately after the establishment of symbiotic association, saturated fatty acids in the neutral fatty acid fraction increased significantly and stayed in an elevated state. In due course, all of the saturated fatty acids appeared in the extraradical fungal mycelium and steadily increased in concentration. Clearly, lipids are transported from plant roots to fungus during symbiosis. In comparison, only a small fraction of plant-specific sterols were detected in extraradical mycelium. In view of these observations, Laczko et al. (2004) argue that lipid transfer from plant to ECM fungus might be an important mode of carbon transfer during ECM symbiosis.

2. C. Nitrogen Metabolism in Ectomycorrhizas

Both efflux and influx of NO_3 will influence net NO_3 uptake by ectomycorrhizas. The net NO_3 uptake rates are known to follow uniphasic Michalis-Menton kinetics in the concentration range between 10 μM to 1.0 mM external nitrate (Kreuzwieser et al. 2000). The NO_3 uptake proceeds linearly with increasing external NO_3 supply. Application of NO_3 along with regulators of NO_3 uptake such as NH_4 or glutamate decreases NO_3 uptake rates. Such an inhibitory effect seems to be due to the reduction of NO_3 influx and not due to the enhanced efflux. Sometimes, mycorrhization may lead to reduced influx, causing reduction in NO_3 uptake. Several extracellular hydrolytic enzymes produced by ECM and ericoid mycorrhizas have been characterized. The fate of soil proteins, peptides, and amino acids released has also been analyzed using appropriate radiolabels. We know that products of proteolytic enzymes are utilized by ECM fungi. However, the exact nature of nitrogenous compounds transferred at the interface between symbionts remains unclear (Chalot and Bruns, 1998).

The ECM fungi can translocate N from organic sources such as peptides, proteins and amino acids to their host plants (Finlay et al. 1992). In Alpine forests, soils are said to be rich in amino acids and ECM symbiosis is quite common on the native tree species. Hence, Lipson et al. (1999) argue that under natural conditions, N transport mediated by ECM fungi could be important. With a view to assess the extent of N transport achieved via ECM in the alpine meadows and forests, they devized experiments using nylon mesh screens that separate fungal hyphae from roots and fed them with ^{15}N-glycine. As expected, mycorrhizal plants absorbed greater quantities of ^{15}N compared to those with severed hyphae. Ectomycorrhizal tips on an average contained 4 to 5 fold more ^{15}N labels. In a different study conducted by Nasholm et al. (1998) ECM-infected pine and spruce seedlings absorbed at least 42% of N supplied as glycine. Tests with radiolabels indicated that ECM fungi such as *Cenococcum geophilum* utilize glycine as an N source, but not as C source. Generally, the N transfer from fungus to host seems to occur via glutamine, glutamate or alanine (Chalot et al. 1996); therefore, C drawn if any, from glycine, needs to be metabolized first and utilized indirectly. For example, glycine could be de-aminated to form an acetate and contribute C skeletons via glyoxglate cycle. If not, glycine could be converted via serine to pyruvate and utilized through tricarboxylic acid cycle. Using rhizoboxes that separate root and ECM fungi, Lipson et al. (1999) monitored labeled (^{15}N, ^{13}C) glycine uptake by mycorrhizal sedge plants. Mycorrhizas on an average transported 1.3% of the added ^{15}N label to plants. The ^{15}N enrichment was greater in mycorrhizal root tips compared to control roots. Blaudez et al. (1998, 2001) have shown that ECM fungi (*Paxillus involutus*) absorb exogenous carbon compounds such as [^{14}C] glutamate and [^{14}C malate). Mycorrhizal inoculation enhanced ^{14}C label uptake by 8 to 17 times over control roots. Sometimes it leads to profound alteration in the metabolic fate of these compounds. In the birch roots, these sources served as C skeletons for amino acid synthesis both in the symbiotic roots and in free-living ECM fungus.

The characteristics of glutamate uptake by ECM fungi were studied by Anderson et al. (2001) who assessed the kinetics of glutamate uptake in two Australian *Pisolithus* isolates. Total glutamate uptake did not follow Michaelis-Menton kinetics when the fungi were fed with glutamate at concentrations ranging from 0.5 m mol m^{-3} to 20 m mol m^{-3}. At glutamate concentrations above 5 mol m^{-3}, its uptake seemed to be through diffusion. However, at concentrations below 5 mol m^{-3}, an active component that followed Micehalis-Menton kinetics was observed. The Km and Vmax values for active glutamate uptake studied for several *Pisolithus* isolates ranged from 4.0 to 210 m mol m^{-3} and 80 to 637 n mol g^{-1} d wt min^{-1}, respectively. The pH optima for glutamate uptake ranged from 4.5 to 6.0, depending on the ECM fungal isolate.

Kinetics of ammonium ion uptake by ECM fungus *Paxillus involutus* was investigated using ^{14}C methylamine as an anologue of ammonium. One such trial by Javelle et al. (1999) has shown that apparent half-saturation constant (Km) and maximum uptake rate (Vmax) for the carrier-mediated transport were 180 μM and 380 nmol (mg dry wt)$^{-1}$ min^{-1}, respectively. The methylamine transport in *P. involutus* was dependent on electrochemical H^+ gradient. Generally, methylamine uptake was enhanced in the presence of fungal symbiont. Wallanda and Read (1999) have also investigated the uptake of amino acids various ectomycorrhizal systems. They utilized ^{14}C labeled amino acids and excized root tips of birch, spruce and pine to understand the kinetics of uptake. They state that all ectomycorrhizal systems absorbed N through a high affinity N transport system. Mostly, ECM roots possessed similar or higher affinity for N rather than control roots. Interestingly, the fungal isolate was a key factor determining the kinetic aspects of N uptake in ECM roots. Genetic variation for N acquisition is easily observable. Hence, they concluded that in nature, ECM fungal species composition influences net N uptake through differences in their inherent ability for N uptake.

Tree roots may respond differently to NH_4^+/NO_3 ratios existent in the soil. At high concentration of NH_4^+, certain species may accumulate excessive NH_4^+ in the root cytosol and exhibit high-velocity, low efficiency membrane fluxes of NH_4^+. In contrast, a species such as white spruce accumulates low cytosolic NH_4^+ but possesses much higher efficiency of NH_4^+ fluxes. It is inferred that the physiological response of ECM roots to NH_4^+/NO_3 ratios may even influence the pattern of tree distribution in a location. We know that ECM fungi are selective with regard to absorption of N forms. A step further, Emmerton et al. (2001a) have claimed that ECM can discriminate even N isotopes at the time of assimilation. Their tests involving labeled glutamate and glycine revealed that all N sources supplied were absorbed substantially, but isotopic preferences and patterns were different. ECM fungi showed net higher concentration of ^{15}N in their mycelium. With NH_4^+, ECM fungi showed preference for ^{14}N isotope. Clearly, in such situations, because of the isotope preferences, ^{15}N abundance measured on ECM fungi may not depict the exact extent of N assimilation. Emmerton et al. (2001b) confirmed such isotope preference while studying [^{14}N] and [^{15}N] discrimination by ECM and ericoid fungi found in subarctic zones. As we know, NO_3 fluxes into roots are dependent on its availability in the medium. At the same time, fluctuations of NO_3 availability are common in natural soils. Kohzu et al. (2000) reported that N isotope fractionation occurred during its transfer. In *Pinus* seedlings colonized with *Suillus granulatus*, mycelial mats possessed higher delta ^{15}N values and fractionation seems to occur during transport. With a view to understand the differential NO_3 uptake patterns, Gobert and Plassard (2002) measured its fluxes at the root surface using NO_3 selective microelectrodes. In N-starved conditions, maximum NO_3 uptake rates were reached in 3 days. Net NO_3 fluxes in non-mycorrhizal short roots increased two-fold due to a 3-day exposure to NO_3, but in mycorrhizal roots, NO_3 fluxes were uniform. According to them, mycorrhization may help plants with greater ability to use the fluctuating concentrations of NO_3 in the soil solution.

In a nitrogen fertilization experiment conducted in Germany, the nitrogen deposition in ectomycorrhizas was analyzed. Beckmann and Kottke (2001) quantified N accumulation in ECM fungus using electron energy loss spectroscopy. They observed that N content in cell granules was 53 times greater than that traced in vacuoles. All ECM isolates tested accumulated N in granules, but the extent of accumulation was dependent on ECM isolate. Generally, the number of granules per mm^2 of the hyphal mantle increased after N fertilization event in the plantation. Among ECM fungi examined, *Cenococcum geophilum* and *Xerocomus badius* showed high N-granule accumulation. Soil depth and horizon could also influence N-granule formation. Hence, in addition to mediating N-translocation, ECM fungi may play a crucial role in storing the N absorbed into tree roots. Thus, ECM too regulates N dynamics in the underground portions of temperate forest plantation.

An evaluation of ammonium metabolism in the ECM fungus *Tuber borchii* revealed that glutamine synthetase, glutamate synthase and glutamate dehydrogenase—all key enzymes of N metabolism are affected by symbiosis. In plant roots, glutamine synthase showed high specific activity and GDH activity was just detectable, whereas in free-living mycelia, glutamine synthase and NADH-GDH are easily traced (Pierleoni et al. 2001). Examination of affinity constants of substrates for NAD-dependent glutamate dehydrogenase from *Laccaria bicolor* has shown that it is mainly involved in the catabolism of glutamate, while NADP-GDH catabolizes the amino acid (Garnier et al. 1997). Grotjohann et al. (2000) experimented with axenic mycelia of ECM fungus *Suillus bovinus* provided with glucose as carbon source and four different N sources. They noticed that ammonia, nitrate and alanine produced similar weights of the fungus. Glutamine and glutamate as N source gave 60 to 80% more fungal mass compared with nitrate, but urea as external N source produced 35% less fungal material than nitrate. The NADH-dependent glutamate dehydrogenase exhibited high-aminating and low-de-aminating activities. Glutamate synthetase activity was also low. Aspartate amine showed similar affinities for glutamate and aspartate. Nitrate reductase was induced by the presence of substrate, but the induction of enzyme activity could be noticed only after mycelia had imbibed nitrate.

Kytoviita et al. (2001) have shown that elevated levels of CO_2 and ozone—two common atmospheric disturbances encountered in temperate forests—can actually affect nitrogen acquisition by *Pinus halepense* from its ectomycorrhizal symbiont. At 700 μmol mol^{-1} CO_2 and 200 nmol mol^{-1} ozone, the net N assimilation rate was reduced significantly. Nitrogen allocation to shoots increased in preference to roots due to ozone treatments. The higher shoot N concentration and altered carbon assimilation rates resulted in higher N use efficiency provided CO_2 levels were elevated. The ECM fungal growth in soil was also affected due to alteration in photosynthetic efficiency. Generally, the fungal activity seen in terms of N uptake was reduced due either of the treatments CO_2 and ozone. The carbon cost of symbiosis-mediated N uptake increased due to elevated CO_2 levels.

3. Developmental Physiology of Ectendomycorrhizas

At the outset, we may note that developmental physiology, cytology and molecular biology are least understood aspects within Ericoid mycorrhizas. Further, majority of inferences are based on field samples collected, wherein knowledge regarding fungal species, their physiological status and chronological stage of infection process are difficult to authenticate (McLean et al. 1998).

Generally, apical region of root is devoid of Ericoid fungus. In nature, extent of root infection encountered varies. Ericaceous hosts seem to support higher rates of colonization compared with Epacrids. The extent of extraradical mycelium produced too may vary (Cairney and Ashford, 2002). In nature, symbiosis can get initiated whenever a fungal hypha gains contact with a compatible region of hair root. In some Ericoid fungal species, an appresorium-like structure is formed immediately after root contact. We have accrued a certain degree of knowledge about biochemical controls for appresorium development in plant versus pathogen relationship, and even in plant-AM symbiosis, but molecular basis for regulation of appresorium formation in Ericoid mycorrhizas is least understood. Morphologically, immediately after appresorium formation, a narrow penetration of hyphae develops. It generally grows through the outer tangential wall and enters the periplasmic space of epidermal cells and widens into coils. In case of Epacrids (Ericaceae), it is interesting to note that invading fungi have simple septa and 'woronin bodies' indicating that these fungi are either Ascomycetes or their anomorphs (Steinke et al. 1996; Briggs and Ashford, 2001). A single penetration point with halo suggestive of cell digestion seems to be the rule with Ericaceous hosts. Under natural conditions, the ectendo-fungus encounters a mucilaginous layer that surrounds the root surface (Stienke et al. 1996; Briggs and Ashford, 2001). The amount of mucilage and its biochemical composition, especially the sugar moiety, may vary. The cell wall and mucilage above it may play crucial role in controlling the establishment of mycorrhizal association. The infective capacity of endophyte itself depends on the mucilage surrounding hyphae. The proportion of root system comprising ericoid mycorrhiza varies. Kemp et al. (2003) have reported that seasonal changes too affect extent on ericoid colonization in *Woollsia pungens*. It was lowest in April/May (50%) and highest in October (70%). They suggest that daily mean maximum and minimum temperature during a season could cause fluctuation in colonization.

Once inside root, the fungus develops a intracellular coil also called the "pelton". The pelton is itself intracellular but held outside the plasma membrane of the epidermal cell. Therefore, it is an apoplasmic structure within root cells. Such coils can enhance surface area of contact between the fungal hyphae ad cell protoplasm. After entering the root, the fungus may localize within the first colonized cell and spread very little into adjacent cells (Cairney and Ashford, 2002). Concomitantly, as the fungal hyphae enter root the production of mucilage ceases. However, fungal wall growth continues for a while and the 'interfacial matrix' that develops around hyphae/coils is said to serve in nutrient exchange. Unlike AM/ECM symbiosis, the root cell cytoplasm in ericaceous host degenerates before that of ectendo-fungus. In case of AM, the individual colonized root cells are known to out-line the fungal hyphae or arbuscule, and repeated colonizations are a clear possibility. In case of Ericoid mycorrhizas, both host root cell and coils may degenerate or be sloughed-off exposing the exodermis. Such exposed exodermis is not re-invaded by fungus. Such collapse of infected root cells have been reported in different ericaceous hosts such as *Lysinium ciliatum, Calluna vulgaris* etc. Ericoid mycorrhizas show up thickening of epidermal walls, but in certain Epacrids tangential and radial epidermal wall thicken excessively (Ashford et al. 1998). Sometimes, thickened walls are multilayered with spongy regions (Briggs and Ashford, 2001). In case of *W. pungens* the thickened wall is characterized by multi-lamellate secondary wall, with helicoidal cellulose and microfibrils. The epidermal cell wall regions possess net negative charge attributable to exposed carboxyl groups. Certain amounts of unesterified pectin also occur on walls. The *W. pungens* thick wall is known to be rich in galactose side chains, but not glucose and mannose. Histochemical analysis too revealed a high concentration of galactomannan (Briggs and Ashford, 2001).

Let us briefly consider the gross cytology of an ericoid mycorrhiza—*Pezizella ericae*, which is a good example of ectendomycorrhiza. Typically, distribution of *Pezizella ericae* is restricted to cells surrounding the central cylinder and the symbiotic relationship is highly compatible at the cellular level (Bonfonte-Fosolo et al. 1984). Light microscopic analyses reveal that *P. ericae* hyphae develop on the host (e.g. *Calluna vulgaris*) root surface and colonize the epidermal cells, without penetrating the stelar zone and apical meristematic tissue. Initially, in the presence of compatible host root, the outer wall of hyphae is typically a 'fibrillar sheath' that extends into mucigel, further leading to physical contact with outer wall layer of root cells. However, once the ericoid fungus penetrates into root cells, fibrillar sheaths disappear. In other ericoid mycorrhizal associations, the host root cell wall material closely surrounds the penetrating hyphae, but diminish as hyphal coils develop. Later, only scattered fibrils can be seen at the interfacial zones inside the host cell. Concurrently, the amount of cytoplasm and numbers of organelles such as mitochondria, plastids and dictyosomes increase. These organelles surround the developing fungal coils. At this stage, host plasma lemma surrounding the intracellular hyphae show up a three-layered structure. The host vacuoles contain electron-dense deposits.

We may realize that an ericoid fungus such as *Pezizella ericae* can also infect a few other hosts. For example, in *Trifolium repense*, the ericoid fungal infection did not result in an increase in cytoplasm nor the organelles, and hyphae were not surrounded by plasmalema. Instead, the cytoplasm of infected *Trifolium* roots cells degenerated, organelles broke down and disappeared in response to *P. ericae* infection. Disorganizaton of cell contents then leads to loss of tonoplast. In such non-ericaceous hosts, some time between 6 and 8 weeks, the fungus easily spreads into endodermis, central cylinder (stele), parenchyma and xylem vessels. Bonfonte-Fosolo et al. (1984) state that, in terms of cytological events, the constant presence of plant plasma lemma is a key factor that distinguishes specific interactions of *P. ericae* with a host or non-host plant species. However, they suggest that fibrillar sheath is produced by the fungus, both in response to host and non-host roots in the vicinity.

Nitrogen source: Cairney and Ashford (2002) opine that organic N available as soluble proteins and amino acids is a potential source of N to ericaceous mycorrhizal fungi. Endophytes on epacrids (Ericacea) utilized ^{15}N from labeled wheat straw incorporated into soil (Bell and Pate, 1994,1996). Similarly,^{15}N amino acid (e.g. glycine) supplied in solution was metabolized by mycorrhizal roots. Certain other studies have indicated that endophytes absorbed ^{15}N after it was mineralized by the regular soil micorflora, meaning that ericaceous fungi may not have a significant mineralizing effect on applied N. However, the ability to utilize BSA as the sole N source clearly demonstrates the ability to secrete extracellular proteases (Leake and Read, 1989). Several Helotiales mycorrhizal endophytes were shown to use NO_3 and NH_4^+ for growth, but were also capable of garnering N supplied as amino acids or simple proteins (Whittaker and Cairney, 2001). Among the 12 ericoid fungi tested, isolates of *W. pungens* showed significant preference to histidine and/or alanine, whereas isolates of *Hymenoscyphus* sp. preferred lysine over histidine as N source. Although preference to specific amino acid cannot be discerned, certain endophytes utilize basic amino acids, arginine and histidine poorly. We have yet to ascertain the form of N transferred into the host. Sometimes, amino acids are directly transferred to hosts by the ectomycorrhizal fungus. Certainly, ^{13}C-labeling experiments with ectomycorrhizas have proved that amino acid N is received by host plants through fungal endophytes (Smith and Read, 1997). Cairney et al. (2000) tested about 35 different isolates of ericoid fungi for their ability to utilize inorganic and organic N sources. Most isolates of *Hymenoscyphus ericae* preferred inorganic NH_4-N. They also found large intra specific variation for glutamine and BSA utilization. Soils from temperate forests and heath land are generally rich in phenolics that might impede easy utilization of organic N by plant roots and ECM fungi. To a certain extent, the wood-decomposing fungi as well as ECM fungi are known to degrade such phenolic substances. Tests with soluble phenolic compounds and insoluble phenolic lignin revealed that ECM fungi have only low abilities to degrade such phenolic substances, when compared with wood-digesting saprophytic fungi. Generally, ericoid fungi seem to possess better abilities to digest phenolic compounds encountered in forest soils (Bending and Read, 1997). Such phenol degradation ability in ericoid fungi was easily attributable to phenol-oxidizing enzymes. But the activity of phenol-oxidizing enzymes was not as intense as that traced in wood decomposing fungi. In a particular case, phenol oxidation by *Hymenoscyphus ericae* was associated with extra cellular O-polyphenol oxidase (tyrosinase) that possessed optimum activity at pH 5.5 and 30°C.

Concluding Remarks

Physiology of mycorrhizal systems is highly complicated, simply because it involves cellular and molecular manifestations of two different organisms and their interactions. With regard to fungal partner, dormancy is frequent in AM fungal spores. Together with formation of germ tubes and hyphal growth it forms a crucial face of fungal life cycle. AM fungal dormancy is variable. It is influenced by fungal species (isolate), several abiotic and biotic factors. Proper germination of spores is critical for successful contact with host roots. The type of germination, either 'g' or 'G' with different levels of apical dominance in hyphal growth, may dictate successful contact of fungus with congenial roots. We ought to investigate if this germination is genetic or it can be manipulated through external application of chemicals. This aspect may find application in practical inoculum preparation.

Progress on carbohydrate physiology of AM symbiosis has been significant during the past few years. It is due to adoption of molecular, cyto-immunological, microscopic and NMR techniques. Such a special attention has been attributed to the fact that 20 to 40% of photoassimilates are channeled to mycosymbiont. If extrapolated to global cropping and forestry zones, AM fungi may intercept flow of upto 5 billion tons of C per year. In plants, C transport is mediated mainly as sucrose. However, the mycosymbiont acquires carbon as hexoses and metabolizes it to trehalose and glycogen. In ECM fungi, trehalose and mannitol are C storage compounds. Both tricarboxylate and glycolytic pathways operate in AM fungi. Carbon metabolism via hexose monophosphate pathway is also a possibility. The extent and regulation of shifts between pathways needs investigation. Interestingly, C movement in AM symbiosis is bi-directional. Reverse flow of C compounds from compatible fungus to plant is a possibility. Sometimes, nearly 10% of root C could be derived via such reverse flow from an already mycorrhizal plant. Detailed studies on factors that regulate bi-directional flow C in mycorrhizal association, and extent of shift is equilibrium points should be understood accurately. It may actually suggest about alternations/shifts of a biotrophic association between symbiosis and pathogenesis.

Tryglycerides are observable throughout the life cycle of the fungus. Movement of lipids in germ tubes, intra and extraradical hyphae has been unequivocally proved using radiolabel and confocal microscopy. Carbon transport occurs in the form of glycogen. Sometimes, glycogen flux can contribute significantly to C transport. Molecular regulation of C transporters operating at symplastic interface is being investigated (see chapter 4). AM fungal preferences to C storage compounds, pathways of C metabolism and translocation need further study.

Interest in nitrogen metabolism in AM and ECM symbiosis is increasing and we may expect to learn more about it in the future. Extent of N translocation varies, depending on fungal species. The form of N utilized and transported by AM or ECM fungi could be important. Mycorrhizal fungi prefer NO_3-N. NH_4-N is suppressive on hyphal growth. Regulation of N metabolism in AM fungi is centered around key enzymes such as glutamate dehydrogenase, glutamate synthase and nitrate reductase.

Regulation of synthesis and degradation of secondary metabolism in symbiotic roots is a key topic being currently investigated. Mycorradicin, the yellow pigment is C-14 polyene compound regulated at transcriptional level. Investigations on mutants impaired in pigment synthesis may provide greater details on genetic controls. Detailed investigations are also required on phenolics, flavonoids and sterol biosynthesis specific to mycorrhizal fungi.

In case of ECM symbiosis, considerable details on carbohydrate physiology are available. Although translocation of photoassimilates in conifers and other hosts is achieved predominantly as sucrose, it is glucose and fructose concentration at the apoplastic space that is important. ECM fungi lack a sucrose transporter system at the apoplastic interface. Hence, invertases play a crucial role in C transfer utilization by ECM systems. ECM fungi may utilize a wide range of organic and inorganic sources of N, but the exact nature of N compounds translocated and at apoplastic interface and its regulation needs further investigation. Both efflux and influx of NO_3 influences nitrogen uptake. Considerable knowledge regarding the kinetics of uptake and utilization of major N compounds such as glutamate, glutamine, alanine are available. Glycine is also absorbed and metabolized via glyoxolate or TCA pathways.

ECM fungi vary with regard to inherent ability to absorb NH_4. Thus, physiological regulation of N uptake and assimilation to a certain extent is fungal dependent.

The role of IAA in regulating developmental physiology of ECM, short root formation and sustenance of symbiosis needs further investigation. Studies on molecular regulation of IAA formation and degradation rates in root and fungus, and its influence on ECM development requires greater emphasis. Auxin transport and its influence on cellular activity in ECM fungi is yet another aspect needing attention.

We may realize that establishment and functioning of AM/ECM fungus involves induction and repression of a large number of biochemical pathways and products. Meticulous biochemical and cellular interactions regulate the physiology of mycorrhizas. There has been a spurt in investigations on the physiology of mycorrhizas with the advent of easily applicable molecular techniques. Obviously, in due course, enormous amounts of details will be accrued that needs careful classification and utilization.

References

Addy, H.D., Boswell, E.P. and Koide, R.T. 1998. Low temperature acclimation and freezing resistance of extra radical VA fungal hyphae. Mycorrhizal Research 5: 582-586.

Addy, H.D., Miller, M.H and Peterson, R.L. 1997. Infectivity of the propagules associated with extra radical mycelium of two AM fungi following winter freezing. New Phytologist 135: 745-753.

Addy, H.D., Schefer, G.F., Miller, M.H and Peterson, R.L. 1994. Survival of the extraradical mycelium of a VAM fungus in frozen soil over winter. Mycorrhiza 5: 1-5.

Anderson, I.C., Chambers, S.M., Cairney, J.W. G. 2001. Characterization of glutamate uptake by two Australian *Pisoltihus* species. Mycological Research 105: 97-982.

Ashford, A.E., Allaway, W.G. and Reed, M.L. 1998. A possible role for the thick-walled epidermal cells in mycorrhizal roots of *Lysinema ciliatum* and other Epacridaceae. Annals of Botany 77: 375-351.

Avio, L. and Giovannetti, M. 1998. The protein pattern of spores of arbuscular mycorrhizal fungi: comparison of species, isolates and physiological stages. Mycological Research 102: 985-990.

Ayling, S.M., Edmonds, T., Smith, S.E. and Smith, F.A. 2001. Electrophysiology of vesicular arbuscular mycorrhizal fungi. Third International Conference on Mycorrhizas, Adelaide, Australia, http://mycorrhiza.ag.utk.edu.latest/icoms/icom3.htm

Ayling, S.M., Smith, S.E., Reid, R.J. and Smith, F.A. 1998. Changes in the wall potential of *Scutellospora calospora* associated with colonization *of Allium porum* roots are not accompanied by equivalent changes in the host. Canadian Journal of Botany 76: 153-156.

Ayling, S.M., Smith, S.E. and Smith, F.A. 2000. Transmembrane electric potential difference of germ tubes of arbuscular mycorrhizal fungi responds to external stimuli. New Phytologist 147:631-639.

Ayling, S.M., Smith, S.E., Smith, F.A. and Kobesil, P. 1997. Transport process at the plant-fungus interface in mycorrhizal associations: physiological studies. Plant and Soil 196: 305-310.

Azcon, R., Ruiz-Lozano, J. and Rodriguez, R. 2001. Differential contribution of arbuscular mycorrhizal fungi to plant nitrate uptake under increasing N supply to the soil. Canadian Journal of Botany 79: 1175-1180.

Bago, S. Azcon-Aguilar, C., Goubet, A. and Piche, Y. 1998. Branched absorbing structure (BAS): A feature of the extraradical mycelium of symbiotic arbuscular mycorrhizal fungi. New Phytologist 139: 375-388.

Bago, B., Pfeffer, P.E., Abubaker, J. Jun, J. Allen, J.W., Brouillette, J., Douds, D.D. Lammers, P.J. and Sachar-Hill, Y. 2003. Carbon export from arbuscular mycorrhizal roots involves the translocation of carbohydrates as well as lipid. Plant Physiology 131: 1496-1507.

Bago, B. Pfeffer, P.E., Douds, D.D., Brouillette, G., and Sachar-Hill, Y. 1999. Carbon metabolism in spores of the arbuscular mycorrhizal fungus *Glomus intraradices* as revealed by a nuclear magnetic resonance spectrometry. Plant Physiology 121: 263-271.

Bago, B., Pfeffer, P.E. and Sachar-Hill, Y. 2000. Carbon metabolism and transport in arbuscular mycorrhizas. Plant Physiology 124: 949-958.

Bago, B. Zipfel, W., Williams, R.M., Jun, J. Arreaola, R., Lammers, P.H., Pfeffer, P.E. and Sachar-Hill, Y. 2002. Translocation and utilization of fungal storage lipids in the arbuscular mycorrhizal symbiosis. Plant Physiology 128: 108-124.

Bago, B., Zipfel, W., Williams, R. M. and Piche, Y. 1999. Nuclei of symbiotic arbuscular mycorrhizal fungi as revealed by in vivo two photon microscopy. Protoplasma 209: 77-89.

Barker, S.J. and Tagu, D. 2000. The role of Auxins and Cytokinins in mycorrhizal symbiosis. Journal of Plant Growth Regulation 19:144-154.

Bau, K.W. and Barton, L.C.1989. Alkaline phosphatase and other hydrolyases produced by *Cenococcum gramiforme*, an ectomycorrhizal fungus. Applied and Environmental Microbiology 55: 2511-2516.

Becard, G., Doner, L.W., Rolin, D.B. Douds, D.D. and Pfeffer, P.E. 1991. Identification and quantification of trehalose in vesicular arbuscular mycorrhizal fungi in vivo ^{13}C NMR and HPLC analysis. New Phytologist 118: 547-552.

Becard, G., Doud, D.D. and Pfeffer, P.E. 1992. Extensive in vitro hyphal growth of vesicular arbuscular mycorrhizal fungi in the presence of CO_2 and Flavonols. Applied Environmental Microbiology 58: 821-825.

Becard, G. and Pfeffer, P.E. 1993. Status of nuclear division in arbuscular mycorrhizal fungi during in vitro development. Protoplasma 194: 62-68.

Beckmann, S. and Kotttke, I. 2001. Deposition of nitrogen in ectomycorrhizae of Norway Spruce (*Picea abies*) on NH_4^+ fertilized sites in the Black Forest, Germany. Second International Conference on Mycorrhiza, Uppsala, Sweden, http://www.mycorrhiza.ag.utk.edu

Beilby, J.P. and Kidby, D.K. 1980. Biochemistry of ungerminated and germinated spores of the vesicular arbuscular mycorrhizal fungus *Glomus caledonius*: Changes in neutral and polar lipids. Journal of Lipid Research 21: 739-750.

Beilby, J. P. and Kidby, D. K. 1982. The early synthesis of RNA, protein and some associated metabolic events in germinating vesicular-arbuscular mycorrhizal fungal spores of *Glomus caledonius*. Canadian Journal of Microbiology 28:623-628

Bell, T.L. and Pate, J.S. 1996. Nitrogen and phosphorus nutrition in mycorrhizal Epacridacea of South-West Australia. Annals of Botany 77: 389-397.

Bell, T.L., Pate, J.S. and Dixon, K.W. 1994. Response of mycorrhizal seedlings of SW Australian sand plain Epacridacea to added nitrogen and phosphorus. Journal of Experimental Botany 45: 779-790.

Bending, G.D. and Read, D.J. 1997. Lignin and soluble phenolic degradation by ectomycorrhizal and ericoid mycorrhizal fungi. Mycological Research 101:1348-1354.

Berta, B, Graziella, M. Fusconi, A.A., Anna, S.A., Sampo, S., Simonetta, T.A.. Perticoe, S, Sonia, S.E. Lingua, G., Guido, V. 1998a. Arbuscular mycorrhizal fungi induce polyploidization in tomato roots. ICOM II abstracts, Uppsala, Sweden, http://www.mycorrhiza.ag.utk.edu

Berta, G., Fusconi, A. M. Sampo, S., Lingua, G., Perticone, S. and Repetto, O. 2000. Polyploidy in tomato roots as affected by arbuscular mycorrhizal colonization. Plant and Soil 226: 37-44.

Berta, B, Zipfel, W., Williams, R. M., Webb, W. W. and Piche, Y. 1998b. In vivo observations on the behavior and fate of nuclei of two arbuscular mycorrhizal fungi grown under non-symbiotic and symbiotic conditions. Second International Conference on Mycorrhizas, Uppsala, Sweden, http://www.mycorrhiza.ag.utk.edu

Bevege, D.I., Bowen, G.D. and Skinner, M.E. 1975. Comparative carbohydrate physiology of ecto- and endomycorrhizas. In: Endomycorrhizas. Sanders, F.E., Mosse, B. and Tinker, P.B. (Eds) Academic Press, London, pp. 149-174.

Bianciotto, V. and Bonfonte, P. 1993. Evidence of DNA replication in an arbuscular mycorrhizal fungus in the absence of the host plant. Protoplasma 176: 100-105.

Blancflor, E.B., Zhao, L. and Harrison, M.J. 2001. Microtubule organization in root cells of *Medicago truncatula* during development of an arbusuclar mycorrhizal symbiosis with *Glomus versiforme*. Protoplasma 217: 154-165.

Blaudez, D., Botton, B., Dizengremel, P. and Chalot, M. 2001. The fate of [C^{14}]glutamate and [^{14}C] malate in birch roots is strongly modified under inoculation with *Paxillus involutus*. Plant Cell and Environment 24: 499-457.

Blaudez, D., Chalot, M., Dezengreimel, P. and Botton, B. 1998. Structure and function of the ectomycorrhizal association between *Paxillus involutus* and *Bentula pendula*: Metabolic changes during mycorrhiza formation. New Phytologist 138: 543-552.

Bonfonte-Fosola, P., Gianinazzi-Pearson, V. and Martinengo, L. 1984. Ultrastructural aspects of endomycorrhizae in the Ericaceae. 4. Comparison of infection by *Pezizella ericae* in host and non-host plants. New Phytologist 98: 329-333.

Braunberger, P.G., Abbott, L.K. and Robson, A.D. 1996. Infectivity of arbuscular mycorrhizal fungi after wetting and drying. New Phytologist 134: 673-684.

Briggs, C.C. and Ashford, A.E. 2001. Structure and composition of the wall in hair root epidermal cells of *Woollsia pungens*. New Phytologist 149: 219-232.

Briones, R., Mustin, C., Joner, G., Belgy, G. and Leyval, C. 2001. Chemical characterization of extra radical fungal walls of AMF and their metal binding capacity. Third International Conference on Mycorrhizas, Adelaide, Australia, http://mycorrhiza.ag.utk.edu/latest/icoms/icom3.html

Brown, M. F. and King, E. J. 1982. Morphology and histology of vesicular mycorrhizae. A. Anatomy and Cytology. In: Methods and Principles of Mycorrhizal Research. American Phytopathological Society, St Paul. Mn USA., pp. 15-21.

Buee, M, Nagahashi, G., Douds, D.D. and Becard, G. 1998. Branching signal or growth-promoting factor: In: Second International Conference on Mycorrhiza. Ahonen-Jonnarth, U., Danell, E., Franson, P., Karen, O., Lindahl, B., Rangel, I. and Finlay, S. (Eds). SLU services, Uppsala, p. 36, http://www.mycorrhiza.ag.utk.edu

Buee, M., Rossignol, M., Jauneau, A., Ranjeva, R. and Becard, G. 2000. The pre-symbiotic growth of arbuscular mycorrhizal fungi is induced by a branching factor partially purified from plant root exudates. Molecular Plant-Microbe Interactions 13:693-698.

Burgraaf. A.J.P. and Beringer, J.W. 1987. Nuclear division and VA-mycorrhizal in vitro growth. In: Mycorrhiza in the Next Decade. Sylvia, D.M., Hung, L.L. and Graham,J.H. (Eds) University of Florida, Gainesville, Florida, USA. p. 190.

Burgraaf. A.J.P. and Beringer, J.W. 1989. Absence of nuclear DNA synthesis in vesicular arbuscular mycorrhizal fungi during in vitro development. New Phytologist 111: 25-33.

Buscot, F. Munch, J.C., Charcoal, J.Y., Gardes, M. Nehls, U. and Hampp, R. 2000. Recent advances in exploring physiology and biodiversity of ectomycorrhizas highlight the functioning of the symbiosis in ecosystems. FEMS Microbiology Reviews 24: 601-614.

Burleigh, S. 2000. Cloning arbuscule-related genes from mycorrhizas. Plant and Soil 226: 287-292.

Butehorn, B., Gianinazzi-Pearson, V. and Franken, P. 1999. Quantification of β-tubulin RNA expression during asymbiotic and symbiotic development of the arbuscular mycorrhizal fungus *Glomus mosseae*. Mycological Research 103: 360-364.

Cairney, J.W.G. and Ashford, A.E. 2002. Biology of mycorrhizal associations of Epacrids (Ericaceae). New Phytologist 154: 305-326.

Cairney, J.W.G., Ashford, A.E. and Allaway, W.G. 1989. Distribution of photosynthetically fixed carbon within root systems of *Eucalyptus pilularis* plants ectomycorrhizal *Pisolithus tinctorius*. New Phytologist 112: 495-500.

Cairney, J.W.G., Sawyer, N.A., Sharples, J.M., Meharg, A.A. 2000. Intraspecific variation in nitrogen source utilization by isolates of the ericoid mycorrhizal fungus *Hymenoscyphus ericae*. Soil Biology and Biochemistry 32: 1319-1322.

Callow, J.A., Cappacio, L.M., Parish, G. and Tinker, P.G. 1978. Detection and estimation of polyphosphate in VA mycorrhiza. New Phytologist 80: 125-134.

Calvet, C., Barea, J.M. and Pera, J 1992. In vitro interactions between the vesicular-arbuscular mycorrhizal fungus *Glomus mosseae* and saprophytic fungi isolated from organic substrates. Soil Biology and Biochemistry 24: 775-780.

Caravaca, F., Algaucil, M. D., Diaz, G. and Roldan, A. 2003. Use of Nitrate Reductase activity for assessing effectiveness of mycorrhizal symbiosis in *Dorycnium pentaphyllum* under induced water deficit. Communications in Soil Science and Plant Analysis 34: 2291-2302.

Chalot, M., Bruns, A., Bolton, B. and Soderstrom, B. 1996. Kinetics, energetics and specificity of general amino acid transporter from the ectomycorrhizal fungus *Paxillus involutus*. Microbiology 142: 1749-1756.

Chalot, M. and Bruns, A. 1998. Physiology of organic nitrogen acquisition by ectomycorrhizal fungi and ectomycorrhizas. FEMA Microbiological Reviews 22: 21-44.

Charvet-Candela, V., Hitchin, S., Ernst, D. and Sandermann, H., Marmiesse, R., and Gay, G. 2002. Characterization of an Aux/IAA cDNA up regulation in *Pinus pinaster* roots in response to colonization by the ectomycorrhizal fungus *Hebeloma cylindrosporum*. New Phytologist 154: 769-777.

Cole, L., Orlovich, D.A. and Ashford, A.E. 1998. Structure, function and motility of vacuoles in filamentous fungi. Fungal Genetics and Biology. 24: 86-100.

Daniels, B.A and Graham, S.O. 1976. Effects of nutrition and soil extracts on germination of *Glomus mosseae* spores. Mycologia 68: 108-116.

Daniels, B.A. and Trappe, J.M. 1980. Factors affecting spore germination of the vesicular arbuscular mycorrhizal fungus, *Glomus epigaeus*. Mycologia 72: 457-471.

Ditengou, F.A., Raudoskoski, M. and Lapeyrie, F. 2003. Hypaphorine, an indole-3-acetic acid antagonist delivered by the ectomycorrhizal fungus *Pisolithus tinctorius* induces reorganization of actin and the microtubule cytoskeleton in *Eucalyptus globulus* ssp. *bicostata* root hairs. Planta 218: 217-225.

Dupponnois, R. and Plenchette, C. 2003. A mycorrhiza helper bacterium enhances ectomycorrhizal and endomycorrhizal symbiosis of Australian *Acacia* species. Mycorrhiza 13: 85-91.

Douds, D.D. 2002. Increased spore production by *Glomus intraradices* in the split-plate monoxenic culture system by repeated harvest, gel replacement and re-supply of glucose to the mycorrhiza. Mycorrhiza 12: 163-167.

Douds, D.D., Pfefer, P.E. and Sachar-hill, Y. 2000. Carbon partitioning, cost and metabolism of arbuscular mycorrhizas. In: Arbuscular Mycorrhizas: Physiology and Function. Kapulnik, Y. and Douds, D.D. (Eds) Kluwer Academic Publishers, Dordrecht, The Netherlands, pp. 140-151

Emmerton, K.S., Callaghan, T.V., Jones, H.E., Leake, J.R., Michelson, A. and Read, D.J. 2001a. Assimilation and isotopic fractionation of nitrogen by mycorrhizal fungi. New Phytologist 151: 503-510.

Emmerton, K. S., Callaghan, T. V., Jones, H. E., Leake, J. R., Michelson, A. and Read, D. J. 2001b. Assimilation and isotopic fractionation of nitrogen by mycorrhizal and nonmycorrhizal subarctic plants. New Phytologist 151: 513-524.

Ezawa, T., Smith, S. E. and Smith, F.A. 2001a. Differentiation of polyphosphate metabolism between the extra- and intraradical hyphae of arbuscular mycorrhizal fungi. New Phytologist 149: 555-563.

Ezawa, T., Smith, S.E. and Smith, F.A. 2001b. Enzyme activity involved in glucose phosphorylation in two arbuscular mycorrhizal fungi: Indication that poly-P is not the main phosphagen. Soil Biology and Biochemistry 33: 1279-1281.

Feng, G., Song, Y.C., Li, X.L., Christie, P. 2003. Contribution of arbuscular mycorrhizal fungi to utilization of organic sources of phosphorus by red clover in a calcareous soil. Applied Soil Ecology 22: 139-148.

Fester, T., Maier, W., and Strack, D. 1999. Accumulation of secondary compounds in barley and wheat roots in response to inoculation with an arbuscular mycorrhizal fungus and co-inoculation with rhizosphere bacteria. Mycorrhiza 8: 241-246.

Fester, T., Schmidt, D., Lohse, S., Walter, M. H., Giuliano, G., Bramley, P.M., Fraser, P. D. Hause, B., and Strack, D. 2002. Stimulation of carotenoid metabolism in arbuscular mycorrhizal roots. Planta 216: 148-154.

Fester, T., Strack, D. and Hause, B. 2001a. Reorganization of tobacco root plastids during establishment of the arbuscular mycorrhizal symbiosis. Third International Conference on Mycorrhiza, Adelaide, Australia, http://mycorrhiza.ag.utk.edu/latest/icoms/icom3.html

Fester, T., Strack, D. and Hause, B. 2001b. Reorganization of tobacco root plastids during arbuscule development. Planta 213: 864-868.

Feugey, L., Strulu, D.G., Poupard, P. and Simoneau, P. 1999. Induced defense responses limit hartig net formation in Ectomycorrhizal birch roots. New Phytol. 144: 541-547.

Finlay, R.D., Frostgard, A. and Sonnerfeld, A.M. 1992. Utilization of organic and inorganic nitrogen sources by ectomycorrhizal fungi in pure culture and in symbiosis with *Pinus contorta*. New Phytologist 120: 105-115.

Fitter, A.H., Graves, J.D., Watkins, N.K., Robinson, D. and Scrimgour, C.1998. Carbon transfer between plants and its control in networks of arbuscular mycorrhizas. Functional Biology 122: 447-454.

Fontain, J., Grandougin-Ferjani, A. Hartmann, M.A., Samholke, M. 2001a. Sterol biosynthesis by the arbuscular mycorrhizal fungus *Glomus intraradices*. Lipids 36: 1357-1363.

Fontain, J., Grandmougin-Ferani, A. and Sancholle, M. 2001b. Lipid metabolism of the endomycorrhizal fungus: *Glomus intraradices*. Competes Rendus De L Academie Des Sciences Serie 3 Life Sciences 324: 847-853.

Franken, P., Kuhn, G. and Gianinazzi-Pearson, V. 2002. Development and molecular biology of arbuscular mycorrhizal fungi. In: Molecular Biology of Fungal Development. Oseiwacz, H.D. (Ed.) Marcel Dekker, New York, pp. 325-348.

Franken, P., Lapopin, L., Meyer-Gauen, G. and Gianinazzi-Pearson, V. 1997. RNA accumulation and genes expresses in spores of the arbuscular fungus *Gigaspora roseae*. Mycologia 89: 295-299.

Franken, P. and Requena, N. 2001. Molecular approaches to arbuscular mycorrhizal functioning. In: Fungal Associations, Hock, R. (Ed.). Springer Verlag, Berlin, Heidelberg pp. 19-28.

Frey, B. and Schuepp, H. 1993. Acquisition of nitrogen by external hyphae of arbuscular mycorrhizal fungi associated with *Zea mays*. New Phytologist 124: 221-230.

Garnier, A., Berredem, A. and Botton, B. 1997. Purification and characterization of the NAD-dependent glutamate dehydrogenase in the ectomycorrhizal fungus *Laccaria bicolor*. Fungal Genetics and Biology 22(3): 168-176.

Gasper, M. L., Pollero, R. J., Cabello, M.N. 1994. Tryglycerol consumption during spore germination of vesicular-arbuscular mycorrhizal fungi. Journal of American Oil Chemical Society 71: 449-452.

Gasper, M.L., Pollero, R. and Cabello, M. 2001. Biosynthesis and degradation of glycerides in external mycelium of *Glomus mosseae*. Mycorrhiza 11: 257-261.

Gavito, M.E. and Olsson, P.A. 2003. Allocation of plant carbon to foraging and storage in arbuscular mycorrhizal fungi. FEMS Microbiology and Ecology 45: 181-187.

Gay, G., Roullon, Bernillon, J. and Favre-Bonvin, J. 1989. Indole acetic acid biosynthesis by the ectomycorrhizal fungus *Hebeloma heimale* as affected by different precursor. Canadian Journal of Botany 67: 2235-2239.

Geitman, A. and Emons, A.M.C. 2000. The cytoskeleton in plant and fungus cell tip growth. Journal of Microscopy 198: 218-245

Genre, A. and Bonfonte, P. 1997. A mycorrhizal fungus changes microtubule orientation in tobacco root cells. Protoplasma 199: 30-38.

Genre, A. and Bonfonte, P.1998. Actin versus tubulin configuration in arbuscule in arbuscule-containing cells from mycorrhizal tobacco roots. New Phtyologist 140: 745-752.

Genre, A. and Bonfonte, P. 1999. Cytoskeleton related proteins in tobacco mycorrhizal cells: r-tubulin and clathrin localization. European Journal of Histochemistry 43: 105-111.

Gianinazzi-Pearson, V., Smith, S.E., Gianinazzi, S. and Smith, E.A. 1991. Enzymatic studies on the metabolism of vesicular-arbuscular mycorrhizas. New Phytologist 117: 61-74.

Giovannetti, M. 2000. Spore germination and pre-symbiotic mycelial growth In: Arbuscular Mycorrhizas: Physiology and Function. Kluwer Academic Publishers, Dordrecht, Netherlands, pp. 47-68.

Giovannetti, M., Azzolini, D. and Citernesi, A.A. 1999. Anastomoses formation and nuclear and protoplasmic exchange in arbuscular mycorrhizal fungi. Applied and Environmental Microbiology 65: 5571-5575.

Giovanneti, M. and Citernesi, S. 1993. Time-course of appresorium formation on host plants by arbuscular mycorrhizal fungi. Mycological research 97: 1140-1142.

Giovannetti, M. Fortuna, P., Citrnesi, A.S., Morini, S. and Nuti, M. 2001. The occurrence of anastomeses formation and nuclear exchange on intact arbuscular mycorrhizal networks. New Phytologist 151: 717-724.

Giovannetti, M. and Sbarna, C. 1998. Meeting a non-host: The behavior of AM fungi. Mycorrhiza 8: 123-130.

Giovannetti, M. Sbrana, C. and Logi, C. 2000. Microchambers and video enhanced light microscopy for monitoring cellular events in living hypha of arbuscular mycorrhizal fungi. Plant and Soil 226: 153-159.

Gobert, A. and Plassard, C. 2002. Differential NO_3-dependent patterns of NO_3 uptake in *Pinus pinaster, Rhizopogon roseolus* and their ectomycorrhizal association. New Phytologist 154: 509-516.

Graves, J.D., Watkins, N.K., Fitter, A.H., Robinson, D. Scrimgeour, C. 1997. Intraspecific transfer of carbon between plants linked by a common mycorrhizal networks. Plant and Soil 192: 153-159.

Grotjoann, N., Kowallik, W., Huang, Y. and Den Baumen, A.S. 2000. Investigations into enzymes of nitrogen metabolism of the ectomycorrhizal basidiomycete, *Suillus bovinus.* Journal of Biosciences 55: 203-212.

Gryndler, M., Jansa. J., Hrselova, H., Chvatalova, I. and Vosatka, M. 2003. Chitin stimulates development and sporulation of arbuscular mycorrhizal fungi. Applied Soil Ecology 22: 283-287.

Guttenberger, M. 2000. Arbuscules of vesicular arbuscular mycorrhizal fungi inhabit an acidic compartment within plant roots. Planta 211: 299-304

Guttenberger, M., Compart, V., Spagale, S. and Hampp, R. 1994. Proteolytic activity in the ectomycorrhizal fungus *Amanita muscaria* and its mycorrhiza on *Betula pendula.* Plant Physiology supplement 105: 869.

Guttenberger, M., Martin, F. and Harrison, M. 2003. Carbon exchange between mycorrhizal symbionts: Are carbohydrates the whole story. Samuel Roberts Noble Foundation, Ardmore, Oklahoma, USA, Abstracts Plantbio.berkeley.edu/bruns/abstracts/guttenberg

Hampp, R., Scheuffer, C., Wallanda, T., Stulten, C., Johan, R. and Einig, W. 1995. Changes in carbon partitioning or allocation due to ectomycorrhiza formation in biochemical evidence. Canadian Journal of Botany 73: S 548-556.

Harrison, M.J. 1996. A sugar transporter from *Medicago truncatula*: Altered expression pattern in roots during vesicular arbuscular (VA) mycorrhizal associations. Plant and Soil 9: 492-503.

Hause, B. 2003. Cell biology of Mycorrhiza. Institute of Biochemistry, Halle, Germany. http://www.ipbhalle.de/english/institute/research/hause/introduction.html

Hawkins, H.J. and George 2001. N-15 nitrogen transport through arbuscular mycorrhizal hyphae to *Triticum aestivum* supplied with ammonium vs nitrate nutrition. Annals of Botany 87: 303-311.

Hepper, C.M. 1983. Effect of phosphate on germination and growth of vesicular-arbuscular mycorrhizal fungi. Transactions of British Mycological Society 80: 487-490.

Hepper, C.M. 1984. Inorganic sulphur nutrition of the vesicular arbuscular mycorrhizal fungus *Glomus caledonium.* Soil Biology and Biochemistry 16: 669-671.

Hepper, C.M. and Smith, G.A. 1979. Observations on the germination of Endogone spores. Transactions of British Mycological Society 66: 189-194.

Herrman, S., Kruger, A., Peskan, T., Oelmueller, R. and Buscot, F. 2001. An approach to elucidate events in the recognition of phase of ectomycorrhiza formation. Third International Conference on Mycorrhiza, Adelaide, Australia. http://mycorrhiza.ag.utk.edu/latest/icoms/icom3.html

Hijri, M. and Sanders, I. R. 2004. The arbuscular mycorrhizal fungus *Glomus intraradices* is haploid and as a small genome size in the lower limit of eukaryotes. Fungal Genetics and Biology 41: 253-261.

Ho, I. and Trappe, J.M. 1973. Translocation of ^{14}C from *Festuca* plants to their endomycorrhizal fungi. Nature 224: 30-31.

Hoffman, E., Wallenda, T., Schaefer, C. and Hampp, R. 1997. Cyclic AMP, a possible regulator of glycolysis in the ectomycorrhizal fungus *Amanita muscaria*. New Phytologist 137: 351-356.

Hrselova, H. and Gryndler, M. 2000. Effect of spermine on proliferation of hyphae of *Glomus fistulosum* an arbuscular mycorrhizal fungus in maize roots. Folia Microbiologica 45: 167-171.

ICRISAT 1986. Annual report 1985. International Crops Research Institute for the Semi-Arid Tropics. Patancheru, India, 53 pp.

Ineuchen, K ad Wiemken, V. 1992. Changes in fungus specific soluble carbohydrate pool during rapid and synchronous ectomycorrhiza formation of *Picea abies* with *Pisolithus tinctorius*. Mycorrhiza 2:1-7.

INRA 2002. INRA 2002. Molecular analysis of the Ectomycorrhizal differentiation—A research report. Institute National de Research Agronomique. Nancy, France, Myocrrhiza web page, pp. 1-3.

Jabaji-Hare, S. 1988. Lipid and fatty acid profiles of some vesicular arbuscular mycorrhizal fungi: Contribution to taxonomy. Mycologia 80: 622-629.

Javelle, A., Chalot, M., Soderstorm, B. and Botton, B 1999. Ammonium and methylamine transport by the ectomycorrhizal fungus *Paxillus involutus*. FEMS Microbiology Ecology 30: 355-366.

Johnson, D., Leake, J.R., Ostle, N., Ineson, P. and Read, D.J. 2002a. In situ (CO_2)-^{13}C pulse labeling of upland grassland demonstrates a rapid pathway of carbon flux from arbuscular mycorrhizal mycelia to the soil. New Phytologist 153: 327-334.

Johnson, D., Leake, J. R. and Read. D. J. 2002b. Transfer of recent photosynthate into mycorrhizal of an upland grassland: Short-term respiratory losses and accumulation of ^{14}C. Soil Biology and Biochemistry 34: 1521-1524.

Johansen, A., Jakobsen, I. And Jensen, E.S. 1993. Hyphal transport by a vesicular-arbuscular mycorrhizal fungi: Contribution to taxonomy. Mycologia 80: 622-629.

Jolicoeur, M. Bouchard-Marchand, E., Becard, G. and Perrier, M 2003. Regulation of mycorrhizal symbiosis: Development of a structured nutritional dual model. Ecological Modeling 163: 245-254.

Juge, C., Samson, J., Bastion, C., Vierhlig, H., Coughton, A. and Piche, Y. 2002. Breaking dormancy in spores of the arbuscular mycorrhizal fungus *Glomus intraradices*: A critical cold storage period. Mycorrhiza 12: 37-42.

Jutta, L.M. and Kaldorf, M. 2000. Arbuscular mycorrhizal fungi might affect host root morphology by increasing levels of indole-3-butyric acid. American Society of Plant Biologist Abs. No. 849, p. 43.

Kaldorf, M. Schmelzer, E. and Bothe, E. 1998. Expression of maize and fungal nitrate reducates genes in arbuscular mycorrhiza. Molecular Plant Microbe Interaction 11: 439-448.

Kemp, E., Adam, P. and Ashford, A.E. 2003. Seasonal changes in hair roots and mycorrhizal colonization in *Woollsia pungens*. Plant and Soil 250: 241-248.

Kohzu, A., Tateishi, T., Yamada, A., Koba, K. and Wada, K. 2000. Nitrogen isotope fractionation during transport from ectomycorrhizal fungi, *Suillus granulatus* to the host *Pinus densiflora*. Soil Science and Plant Nutrition 46: 733-739.

Komor, E. 1983. Phloem loading and unloading. In: Progress in Botany. Esser, K. Kubitzki, K., Ruge, M., Schuepf, E., and Ziegler, H. (Eds) Springer Verlag, Berlin, Heidleberg, pp. 68-75.

Koske, R.E. 1981. *Gigaspora:* Observations on spore germination of VA mycorrhizal fungus. Mycologia 73: 288-300.

Kost, B., Mather, J. and Chua, N.H. 1999. Cytoskeleton in plant development. Current Opinion in Plant Biology 2: 462-470.

Kreuzwieser, J., Stulen, I., Wiersema, P., Vaalburg, W., and Renennberg, H. 2000. Nitrate transport process in *Fagus-Laccaria* mycorrhiza. Plant and Soil 220: 107-117.

Kytoviita, M.M., Le Theic, D. and Dizengremel, P. 2001. Elevated CO_2 and ozone reduce nitrogen acquisition by *Pinus halepensis* from its mycorrhizal symbiont. Physiologia Plantarum 111: 305-312.

Laczko, E., Boller, T. and Wiemken, V. 2004. Lipids in roots of *Pinus sylvestris* seedlings and in mycelia of *Pisolithus tinctorius* during ectomycorrhiza formation: Changes in fatty acid and sterol composition. Plant Cell and Environment 27: 27-40.

Lagrange, H., Jay-Allguard, C. and Lapeyrie, F. 2001. Rutin, the phenylglycoside from eucalyptus root exudates, stimulates *Pisolithus* hyphal growth rate at Pico molar concentration. New Phytologist 149: 349-355.

Lammers, P., Jun, J., Abubaker, J., Arreola, R., Gopalan, A. Bago, B., Hernandez-Sebastain, C. Allen, J.W., Douds, D.D., Pfefer, P.E. and Sachar-Hill, Y. 2001. The glyoxylate cycle in an arbuscular mycorrhizal fungus: Carbon flux and gene expression. Plant Physiology 127: 1287-1298.

Lanfranco, L., Gabella, S. and Bonfonte, P. 2000. EST, a useful tool for studying gene expression in arbuscular mycorrhizal fungi. In: Papers of the Third International Congress on Symbiosis. Weber, H.C., Imhof, S. and Zeuske, D. (Eds). Philips University, Marburg, Germany, pp. 108-115.

Larose, G., Chenevert, R., Moutgalis, P. Gagne, S., Piche, Y. and Veirheilig, H. 2002. Flavonoid levels in roots of *Medicago sativa* are modulated by the developmental stage of the symbiosis and the root-colonizing arbuscular mycorrhizal fungus. Journal of Plant Physiology 159: 1329-1339.

Laurans, F., Pepin, R. and Gay, G. 2001. Fungal auxin overproducing affects the anatomy of *Hebeloma cylindrosporum-Pinus pinaster* ectomycorrhizas. Tree Physiology 21: 533-540.

Leake, T.R. and Read, D.J. 1989. The Biology of mycorrhizas in the Ericacea. 3. Some characteristics of the extracellular proteases activity of the Ericoid endophyte *Hymenoscyphus ericae*. New Phytologist 112: 69-76.

Lindahl, B., Finlay, R. and Olson, S. 2001. Simultaneous, bi-directional translocation of 32P and 33P between wood blocks connected by mycelial cords of*Hypholoma fasciculare*. New Phytologist 150: 189-194.

Lingua, G., Fusconi, A. and Berta, G. 2001. The nucleus of differentiated root plant cells: modifications induced by arbuscular mycorrhizal fungi. European Journal of Histochemistry 45: 9-20.

Lipson, D.A., Schadt, C.W., Schmidt, S.K. and Munson, R.K. 1999. Ectomycorrhizal transfer of amino acid nitrogen to the alpine sedge *Kobrusia myosuroides*. New Phytologist 142: 163-167.

Logi, C. Sbrana, C. and Giovannetti, M. 1998. Cellular events involved in survival of individual arbuscular mycorrhizal symbionts growing in the absence of the host. Applied and Environmental Microbiology 64: 3473-3478.

Loewe, A., Einig, W. and Hampp, R. 1996. Coarse and fine control and annual changes of sucrose-phosphate synthase in Norway spruce (*Picea abies*) needles. Plant Physiology 112; 641-649.

Loewe, A., Einig, W. and Hampp, R. 2000. Mycorrhization and elevated CO_2 both increase the capacity for sucrose synthesis in source leaves of spruce and aspen. New Phytologist 145: 565-574.

Mader, P., Veiheirlig, H., Treitwolf-Engel, R., Boller, T., Frey, B., Christie, P. and Wemken, A. 2000. Transport of ^{15}N from a soil compartment separated by a polytetraflouroethylene membrane to plant roots via the hyphae of arbuscular mycorrhizal fungi. New Phytologist 146: 155-161.

Maia, L.C. and Kimbrough, J.W. 1998. Ultrastructural studies of spores and hyphae of a *Glomus* species. International Journal of Plant Science 159: 581-589.

Maier, W., Schmidt, J., Nimitz, M., Wray, V. and Strack, D. 2000. Secondary products in mycorrhizal roots of tobacco and tomato. Phytochemistry 54: 473-479.

Martin, F., Duplesis,S., Ditengou, F., Lagrange, H., Voiblet, C. and Lapeyerie, F. 2001. Developmental cross talking in the Ectomycorrhizal symbiosis: Signals and communication genes. New Phytologist 151: 145-154.

Martin, F., Lapeyrie, F. and Tagu, D. 1997. Altered gene expression during ectomycorrhizal development. In: Mycota: Plant Relationships. Tudzenski,C. (Ed.) Springer Verlag, Berlin, pp. 223-242.

Mathur, N. and Vyas, A. 1996. Biochemical changes in *Zizphus xylopyrus* by VA mycorrhizae. Botanical Bulletin of Academia Sinica 37: 209-212.

McLean, C.B., Anthony, J., Sollis, R.A., Steinke, E. and Lawrie, A.C. 1998. First synthesis of Ericoid mycorrhizas in the Epacridaceae under axenic conditions. New Phytologist 139: 589-593.

Maier, R.I. and Charvet, I.1992. Germination of *Glomus mosseae* spores: Ultrastructural analysis. International Journal of Plant Science 153: 541-549.

Melin, E. and Nilsson, H. 1957. Transport of ^{14}C-labelled photo-assimilates to the fungal associate of pine mycorrhiza. Sven. Bot. Tidksr. 51, 166-186.

Molina, R., Massiote, H. and Trappe, J. 1992. Specificity phenomenon in mycorrhizal symbiosis: community-ecological consequences and practical implication. In: Mycorrhizal Functioning: An Integrative Plant-Fungal Progress. Allen, M.F. (Ed.) Chapman and Hall, New York, pp 357-432.

Mosse, B. 1970. Honey colored, sessile Endogone spores. 2. Changes in fine structure development. Archives. Mikrobiology 74: 129-145.

Nagahashi, G. and Douds, D. 2001. Blue light and chemical signals from host exudates synergistically stimulate hyphal branching of the AM fungus *Gigaspora gigantea*. Third International Conference on Mycorrhizas, Adelaide, Australia, http://mycorrhizas.ag utk.edu/ latest/ icoms/icom3.html

Nagahashi, G., Douds, D.P. and O'Connor, J. 1998. Fractioning of AM fungal branching signals from aqueous exudates of Ri T-DNA transformed carrot roots. In: Second International Conference on Mycorrhiza. Ahonen-Jonnarth, U., Donell, E., Fransson, P., Karen, O., Lindahl, O., Rangel, I. and Finlay, R. (Eds) SLU services, Uppsala, p. 125.

Nagahashi, G., Douds, D. and Buee, M. 2000. Light induced hyphal branching of germinated AM fungal spores. Plant and Soil 219: 71-79.

Nakanol, A., Takahashi, K., Koide, R.T. and Kimura, M. 2001. Determination of the carbon and nitrogen sources for arbuscular mycorrhizal fungi using stable isotopes. Third International Conference on Mycorrrhizas, Adelaide, Australia, http://mycorrhizas.ag.utk. edu/latest/ icoms /icom3.html

Napierala-Fillipiak, A., Warner, A., and Karolewski, P. 2002. Content of phenolics in mycorrhizal rots of *Pinus sylvestris* seedlings grown in vitro. Acta Physiologiae Plantarum 24: 243-247.

Nasohlm, T., Ekblad, A., Nordin, A., Geisler, R., Hogberg, M. and Hogberg, P. 1998. Boreal forest plants take up organic nitrogen. Nature 392: 914-916.

Nehls, U. Mikolajewski, S., Magel, E. and Hampp, R. 2001. Carbohydrate metabolism in ectomycorrhizas: Gene expression, monosaccharide transport and metabolic control. New Phytologist 150: 533-543.

Nick, P. 1998. Signalling to the micro tubular cytoskeleton in plants. International Review of Cytology 184: 33-80.

Niini, S.S., Tarkka, M.T. and Raudoskoski, M. 1996. Tubulin and actin protein patterns in *Scotspine (Pinus sylvestris)* roots and developing ectomycorrhiza with *Suillus bovinus*. Physiologia Plantarum 96: 186-192.

Nylund, J.E., Wallandr, H., Sundberg, B. and Gay, G. 1994. IAA-overproducer mutants of *Hebeloma cylindrosporum* mycorrhizal with *Pinus pinaster* and *Pinus sylvestris* in hydroponic cultures Mycoriza 4: 247–250.

Pfeffer, P.E., Douds, D.D., Becard, G. and Sachar-Hill, Y. 1999. Carbon uptake and the metabolism, and transport of lipids in an arbuscular mycorrhiza. Plant Physiology 120: 587-598.

Pierleoni, R., Vallorin, L., Sacconi, C., Sisiti, D., Giomaro, D. and Stocchi, V. 2001. Evaluation of the ezymes involved in primary nitrogen metabolism in *Tilia platyphyllos/Tuber borchi* ectomycorrhiza. Plant Physiology and Biochemistry 39: 1111-1114.

Poulin, M.J., Simard, J., Catford, J.G., Labrie, F., and Piche, Y. 1997. Responses of symbiotic endomycorrhizal fungi to estrogens and anti-estrogens. Molecular Plant-Microbe Interaction 10: 481-487.

Ravnskov, S., Wu, Y. and Graham, J.H. 2001. The influence of different arbuscular mycorrhizal fungi and phosphorus on gene expression of invertase and sucrose synthase in roots of maize. Third International Conference on Myocrrhizas, Adelaide, Australia, http:// www. mycorrhizas. ag. utk. edu/ icoms /icom3.html

Read, D.J. 1997. The ties that bind. Nature 388: 517-518.

Requena, N. Achatz, B., Mann, P. and Franken, P. 2000. Isolation of genes expressed during the early stages of arbuscular mycorrhizal fungi. In: Papers of the Third International Congress in Symbiosis. Weber, H.C., Imhof, S. and Zeuske, D. (Eds). Philips University, Marburg, Germany, p 176.

Requena, N., Fuller, P. and Franken, P. 1999. Molecular characterization of GmFOX2, an evolutionary highly conserved gene from the mycorrhizal fungus *Glomus mosseae*, down-regulated during interaction with *Rhizobium*. Mol. Plant-Microbe Interaction 12: 934-942.

Rincon, A., Priha, O., Sotta, B., Bonet, M., Le Tacon, F. 2003. Comparative effects of auxin transport inhibitors on rhizogenesis and mycorrhizal establishment of spruce seedlings inoculated with *Laccaria bicolor*. Tree Physiology 23: 785-791.

Saito, M. 1995. Enzyme activities of the internal hyphae and germinated spores of an arbuscular mycorrhizal fungus *Gigaspora margarita*. New Phytologist 129: 425-431.

Salzer, P. and Hager, A. 1991. Sucrose utilization of the ectomycorrhizal fungi *Amanita muscaria* and *Hebeloma crustuliniforme* depends on the cell wall-bound invertase activity of their host *Picea abies*. Botanica Acta 104: 439-449.

Salzer, P. and Hager, A. 1993a. Characterization of wall-bound invertase isoforms of *Picea abies* cells and regulation of ectomycorrhizal fungi. Physiological Plant Pathology 88: 52-59.

Salzer, P. and Hager, A. 1993b. Effect of auxins and ectomycorrhizal elicitors on wall-bound proteins and enzymes of Spruce (*Picea abies*) cells. Trees 8: 49-55.

Samra, A., Dumas-Gaudot, E., Gianinazzi-Pearson, V. and Gianinazzi, S. 1996. Soluble proteins and polypeptide profiles of spores of arbuscular mycorrhizal fungi. Interspecific variabilities and effects of host (myc^+) and non-host (myc^-) *Pisum sativum* root exudates. Agronomie 16: 709-719.

Sanders, F.E. and Tinker, P.B. 1973. Phosphate flow into mycorrhizal roots. Pesticide Science 4: 385-395.

Sannazzor, A.I., Alvarez, C.L., Menedez, A.B., Pieckenstain, F.L., Alberto, E.O. and Ruiz, O.A. 2004. Ornithine and arginine decarboxylase activities and effect of some polyamine biosynthesis inhibitors on *Gigaspora rosea* germinating spores. FEMS Microbiology Letters 230: 115-121.

Shaul-Keinan, O, Gadkar, V., Ginzberg, I., Grunzweig, J. M., Chet, I., Elad, Y., Wininger, S., Belausov, E., Eshed, Y., Arzmon, N., Ben-Tal, Y. and Kapulnik, Y. 2002. Hormone concentrations in tobacco roots change during arbuscular mycorrhizal colonization with *Glomus intraradices*. New Phytologist 154: 501-507.

Schubert, A., Allara, P. and Morte, A. 2004. Cleavage of sucrose in roots of soybean (*Glycine max*) colonized by an arbuscular mycorrhizal fungus. New Phytologist 161: 495-505.

Schubert, A., Wyss, P. and Weimken, A. 1992. Occurrence of trehalose in vesicular arbuscular mycorrhizal fungi and mycorrhizal roots. Journal of Plant Physiology 140: 41-45.

Schaeffer, C., Wallanda, T., Guttenberger, M. and Hampp, R. 1995. Acid invertase in mycorrhizal and non-mycorrhizal roots of Norway spruce (*Picea abies*) seedlings. New Phytologist 129: 417-421.

Schaeffer, C., Wallanda, T., Hampp, R., Salzer, P. and Hager, A. 1997. Carbon allocation in Mycorrhizae. In: Trees—Contributions to Modern Tree Physiology. Renenberg, H., Eschrich, W. and Ziegler, H. (Eds.). Backleys Publishers, Leiden, The Netherlands, pp. 397-407.

Smith, S.E. and Read, D.J. 1997. Mycorrhizal symbiosis. Academic Press, London, U.K. pp 605.

Stanzel, M., Sjohnel, R.D. and Komor, I. 1988. Transport of glucose, fructose and sucrose by *Streptanthus tortuoses* suspension cells. 1. Uptake at low sugar concentration. Planta 174: 201-209.

Steinke, E., Williams, P.G. and Ashford, A.E. 1996. The structure and fungal associate of mycorrhizas in *Leucopogan parviflorum*. Annals of Botany 57: 413-419.

Sward, R.J. 1981. The structure of spores of *Gigaspora margarita*. 1. The dormant spore. New Phytologist 87: 761-768.

Sward, R.J. 1981a. The structure of the spore of *Gigaspora margarita* 2. Changes accompanying germination. New Phytologist 88: 661-666.

Sward, R.J. 1981b. The structure of the spores of *Gigaspora margarita*. 3. Germ tube enlargement and growth. New Phytologist 88: 667-673.

Sylvia, D.M. 1988. Activity of external hyphae of vesicular arbuscular mycorrhizal fungi. Soil Biology and Biochemistry 20: 39-43.

Thiery, B., Huang, J. and Lapeyrie, F. 1998. Host plant stimulates the tryptophan betaine, hypaphorine, and accumulation in *Pisolithus tinctorius* hyphae during ectomycorrhizal infection. Fungal hypaphorine controls K^+ uptake, H^+ extrusion and root hair development. Second International Conference on Mycorrhiza, Uppsala, Sweden, http://www.mycorrhiza.ag.utk.edu.

Timonen, S. and Peterson, R. L. 2002. Cytoskeleton in mycorrhizal symbiosis. Plant and Soil 244: 199-210.

Timonen, S., Smith, F.A. and Smith, S.E. 2001. Microtubules of mycorrhizal fungus *Glomus intraradices* in symbiosis with tomato root. Canadian Journal of Botany 79: 307-317.

Tommerup, I. 1983. Spore dormancy in vesicular arbuscular mycorrhizal fungi. Transactions of British Mycological Society 81: 37-45.

Tommerup, I.C. 1984. Persistence of infectivity by germinated spores of vesicular arbuscular mycorrhizal fungi in soil. Transactions of British Mycological Society 82: 275-282.

Torelli, A., Trotta, A., Acerbi, L., Arcidiacono, G., Berta, G. and Branca, C. 2000. IAA and ZR content in leek (*Allium porrum*) as influenced by P nutrition and arbuscular mycorrhizae, in relation to plant development. Plant and Soil 225:29-35.

Toussaint, J.P, St-Arnaud, M. and Charest, C. 2001. Nitrogen assimilation in a mycorrhizal in vitro culture system. Third International Conference on Mycorrhiza. Adelaide, Australia, http://www mycorrhiza. ag.utk.edu/icoms/icom3.html

Uetake, Y. and Peterson, R.L. 1997. Changes in actin filament arrays in protocorm cells of the orchid species, *Spiranthes sinensis*, induced by the symbiotic fungus *Ceratobasidium corngerum*. Canadian Journal of Botany 75:1661-1669.

Van Buuren, M.L., Lanfranco, L., Longato, S., Minerdi, M.D., Harrison, M.J. and Bonfonte, P. 1999. Construction and characterization of genomic libraries of two endomycorrhizal fungi *Glomus versiforme* and *Gigaspora margarita*. Mycological Research 103: 955-960.

Valentine, A.J., Osborne, B.A. and Mitchell, D.T. 2002. Form of inorganic nitrogen influences mycorrhizal colonization and photosynthesis of cucumber. Scientia Horticulturae 92: 229-239.

Vierheilig, H., Alt-Hug, M., Weimken, A., Boller, T. 2001. Hyphal in vitro growth of the arbuscular mycorrhizal fungus *Glomus mosseae* is affected by chitinase but not by beta1,3-glucanase. Mycorrhiza 11:279-282.

Wallanda, T. and Read, D.J. 1999. Kinetics of amino acid uptake by ectomycorrhizal roots. Plant Cell Environment 22: 179-187.

Wallanda, T., Wingler, A., Schuefer, C. and Hampp, R. 1994. Fungus-specific soluble carbohydrates in ectomycorrhizas of Norway spruce with different fungal partners. Plant Physiology supplement. 105: 898A.

Walter, M.H. 2003. Molecular basis of Mycorrhizal symbiosis—A report. Institute of Plant Biochemistry, Halle, Germany, http://www.ipb-halle.de/english/institute/research/walter/introduction.htm

Walter, M.H., Fester, T., Hause, B., Hans, J. and Track, D. 2003. Molecular analysis of secondary metabolism in AM: The Yellow pigment story. COST Scientific meetings, Cologne, Germany, Abstract, p. 4.

Watkins, N.K., Fitter, A.H., Graves, J.D. and Robinson, D. 1996. Carbon transfer between C3 and C4 plants linked by a common mycorrhizal network, quantified using stable carbon isotopes. Soil Biology Biochemistry 28: 471-477.

Whittaker, S.P. and Cairney, J.W.G. 2001. Influence of amino acids on biomass production by Ericoid endophytes from *Woollsia pungens* (Epacridaceae). Mycological Research 105: 105-111.

Wisser, G., Guttenberger, M., Hampp, R. and Nehls, U. 2000. Identification and characterization of an extra cellular acid trehalose from the ectomycorrhizal fungus *Amanita muscaria.* Universitat Tubingen-Botanisches Institut reports (abstract) p. 23. <http:// www.unitubingen.de>/uni/bbp/lit/ Npip. htm

Woolhouse, H.W. 1975. Membrane structure and transport problems considered in relation to phosphorus and carbohydrate movement and the regulation of endotrophic mycorrhizal associations. In: Endomycorrhizas. Sanders, F.E., Mossea, B. and Tinker, P.B. (Eds). Academic Press London, United Kingdom, pp. 209-223.

3

GENETICS OF MYCORRHIZAL SYMBIOSIS: CLASSICAL AND MOLECULAR APPROACHES

Mycorrhizal symbiosis involves an intricate physiological and molecular interaction between two different organisms, plant host and fungus. Genetic constitutions of both symbionts are important, since they exert individual influence on establishment, sustenance and senescence of symbiotic relationship. In nature, we encounter host species, their genotypes and fungal isolates that are genetically predisposed to form symbiosis early. They may also possess the ability for higher levels of root colonization and facilitate better nutrient translocation and exchange, leading to greater growth benefits. At the other extreme, host species or mutants that simply suppress formation/functioning of symbiotic relationship at either early or later stages of symbiosis have also been identified. Initially, knowledge regarding genetic variation and mechanisms of inheritance of mycorrhizal traits was derived solely through classical approaches of selection and crossing. We know that genetic manipulation of host or fungus achieved using classical methods of crossing influences the mycorrhizal effect. However, current trend is to analyze the genetic aspects of symbionts using molecular techniques. In fact, as a result of this shift to molecular techniques, our knowledge about molecular regulation of symbiosis has improved enormously during the past 5 to 8 years.

With regard to AM symbiosis, the topics discussed in this chapter include:

(a) Availability of genetic variation for traits such as root colonization, P uptake and growth improvement, as well as hertiability of AM traits in host plants.

(b) Genetic loci in host roots that regulate early events of spore germination, root invasion and establishment in root cortex.

(c) Genetic variation and heritability of AM fungal characteristics such as spore morphology, colonization and P utilization.

(d) Knowledge about molecular regulation of AM symbiosis is comparatively recent. Molecular aspects including 'Symbiotic Regulatory (*SR*)' genes that are induced/repressed in response to AM symbiosis are discussed in detail. *SR* genes relevant to different stages of symbiosis such as initial signaling, presymbiotic, symbiotic and senescence are highlighted. *SR* genes that control nutrient transport and exchange have been discussed exclusively in chapter 4.

(e) Tripartite symbiosis within legumes involves comparatively more complex molecular interactions. Current knowledge on genetic loci, molecular lesions and other aspects relevant to establishment and functioning of tripartite symbiosis have been included.

With regard to classical genetics of ECM symbiosis, examples depicting genetic variation for relevant mycorrhizal traits, genetic crossing and progeny analysis have been discussed. Recent progress on molecular regulation of ECM symbiosis, especially *SR* genes and a few others affecting the cellular structure, primary and secondary metabolism has been emphasized.

1. Genetics and Molecular Regulation of Arbuscular Mycorrhizas

1. A. Host Genetic Variation for Mycorrhizal Symbiosis

In nature, plant genes play a vital role by modifying the effect of beneficial microorganisms, especially those involved directly in plant health, nutrient uptake, nitrogen fixation and plant growth promotion (Rengel, 2002; Smith and Goodman, 1999). In the present context, genetic constitution of plant-host could immensely influence the establishment, functioning and senescence of mycorrhizal symbiosis. One of the earliest evidences confirming the influence of host genetic component on AM colonization was forwarded by Krishna et al., (1984) based on a systematic analysis of a large number of genotypes of pearl millet from a germplasm bank at ICRISAT, Hyderabad, India. Genotype-dependent variation for AM colonization was shown across locations and seasons. The extent (percentage) of root colonized by natural AM flora in soil was a heritable trait in pearl millet. Additive effects were easily discernible for the extent of AM colonization among the F1 populations derived from selected crosses (Krishna and Lee, 1987). Such genotype-dependent variation for AM colonization has also been reported in other crops, for example in wheat (Hetrick et al., 1992, 1995, 1996), peas (Jensen, 1985), clover (Lackie, 1988), cowpea (Bertheu, 1980), trifolium (Eason et al., 2001), maize (Kaeppler et al., 2000), and prairie grasses (Wilson and Hartnett, 1998; see Tables 3.1, 3.2). In an effort to understand the genetics of AM colonization, root growth pattern and their relationship to P nutrition Nearly 400 pea genotypes were assessed in the field by Jensen (1985), who categorized these pea genotypes into groups with different levels of AM colonization. This study indicated the possibility of using AM colonization as a genetic trait during selection and breeding programs aimed at better nutrient uptake. White flowered pea varieties such as Lotta, Huka, Vretta or Lenka and varieties with colored flowers such as Marma, Minerva or Perdro were good sources of high root density and AM colonization, whereas, garden pea varieties tested were of less utility with reference to root growth or AM colonization improvement. Wilson and Hartnett (1998) evaluated nearly 95 different prairie grass species in Kansas, USA for interspecific variation in mycorrhizal symbiosis. They suggested that AM colonization and benefits in terms of biomass were generally higher in C4 grasses compared with cool season C3 species. Eason et al., (2001) recently tested 43 near isogenic lines (NILs) of *Trifolium repens* that were inoculated with *Glomus mosseae*. Individual lines exhibited a high degree of genetic variation for all the parameters recorded such as root length followed by mycorrhizal colonization and P uptake traits. According to them, genetically, most sensitive parameter was the root length. They suggested that some of these lines of *Trifolium* could help in deciphering aspects related to molecular and genetic control of AM symbiosis. Kaeppler et al., (2000) analyzed maize-AM fungal interaction with reference to genetics of P nutrition and growth responses. They reported that at least 10 regions in three chromosomes in maize possessed QTLs

Table 3.1 Genetic variation for AM fungal colonization in different crop species—examples

Crop species	*No. of genotypes examined*	*AM colonization range (%)*	*Reference*
Cowpea	2	21-78	Oliver et al. 1983
Medicago	10	4-13	Lakie et al. 1988
Peas	398	5-50	Jensen, 1985
Groundnut	30	17-39	Krishna and Williams, 1987
Pearl millet	30	25-56	Krishna et al. 1984a
Sorghum	24	23-67	ICRISAT, 1985
Wheat	14	0-13	Hetrick, 1992
Maize	6	6-19	Toth et al. 1984
	28	19-66	Kaeppler et al. 2000

Table 3.2 Heritability of mycorrhizal traits

AM fungal colonization (%)		*Phosphorus uptake (mg plant^{-1})*	
Genotype	**AM colonization**	**Genotype**	**Phosphorus uptake**
Female Parent			
5141A	51	81A	55
111A	37		
Crosses and Male Parent			
5141A × 631 P-3	51		
111A × 631 P-3	57	81A × ATP(T × P)554-5-(×)	162
631 P-3	37	ATP(T × P)554-5-(×)	59
5141A × 733 P-1	41	81A × SD2xE × B-1	95
111A × 733 P-1	54	SD2xE × B-1	92
733 P-1	38		
Controls			
WC C-75	38		145
MBH 110	40		66
CV (%)	15		18

Source: Krishna et al. 1984; Krishna and Lee, 1987

Note: Male parents 633 P-3 and 733 P-1 were utilized for studies on AM fungal colonization. Similarly, ATP(T × P)554-5-(×) and SD2 × E × B-1 are male parents used for investigations on heritability of P uptake. WC C-75 and MBH 110 are the popular pearl millet varieties used as controls for the experiment. Percentage colonization and P uptake data are averages from 3 trials each in 3 different locations.

that influence growth in low P situations, in the absence of AM fungi. A marker on chromosome 8 near the locus *phi 060* and a QTL each on chromosome 1 and 7 were directly responsible for P nutrition and shoot growth. The relevant QTL that determined the responsiveness to mycorrhiza also was located on chromosome 2 (see Table 3.3) wherein, a population denoted MO17 contributed the positive allele at this locus. Identification of these QTLs in maize further confirms the genetic basis for responsiveness of maize (plant host) to AM fungi. Genetic variation for growth in low P conditions ranged greater than six fold from the poorest genotype, and responsiveness ranged from 800% (cv B73) to 106% (cv MO17), averaging at 282% (Kaeppler et al. 2000). The genes relevant for enhanced P intake and growth responsiveness are known to express better in low soil-P conditions. In fact, there are apprehensions about using consistently high soil P levels during crop improvement procedures,

Table 3.3 Quantitative trait loci (QTL) controlling maize growth in the presence or absence of AM fungi

Trait	*Chromosome*	*Closest marker*	*Additive effect*
Shoot weight (M- P-)	1	umc 107a (200)	0.073
	7	umc 110 (68)	0.071
	8	phi 060 (98)	0.071
Shoot weight (M- P+)	10	npi 285 (28)	–0.46
Root volume (M+ P-)	10	phi 041 (0)	0.78
Root volume (M- P+	1	bnl 5.59 (114)	–3.40
Mycorrhizal colonization	4	php 20725 (16)	–1.4
Mycorrhizal responsiveness	2	umc 34 (64)	50.6

Source: Kaeppler et al. 2000
Note: M+ and M- denotes the presence or absence of AM fungi; similarly P+ and P- indicate P inputs. QTL analysis was conducted using data from 197B73 × MO17 recombinant inbred lines using composite interval mapping function of Plab QTL with LODof 2.0 The additive effect of each QTL is calculated as (mean of MO17 genotypic class minus mean of the B73 genotypic class divided by 2. Therefore, positive values indicate that B73 contribute the favorable allele for mycorrhizal responsiveness.

mainly because it may mask AM colonization under high soil-P conditions. However, it should be realized that during selection and breeding of most field crops, be it in high fertility conditions, there is no deliberate effort to reduce or delete AM colonization ability in roots. At high soil-P, there could be a mere physiological masking of AM in roots but without ill effects on its inheritance pattern as a genetic trait. At this point, it is worthwhile to note that suppressive effects of high P inputs on AM colonization can be overcome by applying certain chemicals such as formononetin (Fries et al. 1998).

In their recent review on applications and usefulness of classical genetic approaches to improve mycorrhizal performance, Barker et al., (2002) have befittingly remarked that during the past five years, classical genetic investigations are meager when compared with molecular analysis. A great deal of attention has been weaned away towards studying molecular aspects of AM symbiosis. Even among the reported studies involving classical breeding approaches, major focus has been on genetic analysis of plant hosts of AM symbiosis and the fungal partner in case of ECM symbiosis. Analyses of QTLs in AM/ECM that affect the capacity to colonize, and the quantitative genetic aspects of mycorrhizal responses has been attempted recently. In summary, there is a clear need to improve our understanding and ability to manipulate AM/ECM symbiosis for better results using classical genetic crossing methods, particularly where procedures are easier to carry out and are not time consuming. Rengel et al., (2002) have aptly stated that targeted efforts to select or breed genotypes for improved symbiosis, for example mycorrhiza, may bring yield improvement of crops cultivated in a wide range of environments, especially in regions with low soil fertility.

Phosphorus Efficiency of Asymbiotic and Symbiotic Plants: Genetic variation for P uptake and efficiency in plants has been studied and reported periodically for the past four decades. Several reviews and volumes are available on genetic variation and breeding for improvement in plant nutrition, that includes aspects on phosphorus (Clark, and Duncan, 1991; Duncan and Baligar, 1990; Gabelman and Gerloff, 1983;

Gourley et al., 1994; Krishna, 1997, Manske, 1990; Smith et al., 1992). The P efficiency reported in some cases involving flow culture/hydroponics or sterile soils that are devoid of micro flora including AM fungi, clearly refers to inherent genetic nature of plant species (e.g. Coltman, 1987). The P efficiency value denotes the rate of P uptake per time, or per unit P input or unit root length. Hence, in this case, the mycorrhizal component does not confound estimates of P efficiency of plant roots. Similarly, mycorrhizal P efficiency has been reported in some cases, wherein, extra radical hyphae of the mycorrhizal component have been exclusively studied using a separating mesh. Again, these values and inferences relate exclusively to mycorrhizal fungus. Experiments using non-fumigated fields, unsterile soil or with mycorrhiza inoculated sytems will definitely include P efficiencies of both AM fungus and plant root. For example, Rengel (2002) screened several wheat, triticale and rye genotypes for various components of P efficiency such as total P uptake, P uptake per unit root length, P uptake per unit P supply, shoot weight per unit P. Although not identified separately, P uptake efficiency values reported include mycorrhizal effects. Similarly, most studies reported include both plant and mycorrhizal P efficiency components (see Krishna, 1997). Obviously, genetic variation due to plant root and mycorrizal fungus need to be separated, wherever possible, using both experimental and/or statistical procedures. Consequently, complications in deciphering the genetic nature and inheritance patterns for P nutritional traits should also be removed and clearly depicted wherever possible. In a few cases, the inheritance of P uptake and utilization efficiency has been bifurcated into several components that contirubute to the total P efficiency of plants. For example, genetic variation and inheritance pattern of individual components such as root length, secondary roots, P translocation rates have been reported exclusively, without summing their effects. Similarly, we may have to separate the contribution of mycorrhizal component to P uptake efficiency. We should note that genetics of P nutrition with and without mycorrhiza could be mildly or drastically different. This needs to be ascertained.

The Spring-Wheat breeding program at CIMMYT (Mexico) is aimed at improving P uptake and utilization in order to make the crop sustainable even under low P fertility conditions. They screened wheat germplasm for a wide range of traits that contribute to P efficiency, including selection of component traits such as root proliferation rates, root hairs, acid phosphates activity, and P absorption related plant traits including AM fungal colonization. Phosphorus utilization traits such as plant growth rate, leaf surface, seed production and P translocation index were also assessed. According to Manske et al., (2000) under the prevailing soil fertility and environmental conditions, traits relevant to P uptake were a bit more important than those contributing to its utilization. Among the uptake traits, roots and their growth rate were more important and influenced the P uptake to a greater extent than AM fungal colonization and acid phosphatase activity. This example clearly enunciates the fact that in nature, genetic variation for P efficiency available both in plant and AM fungi are operable. We need to assess the germplasm and make a shrewd choice regarding the extent of importance to be provided to each component. Let us consider another example. It pertains to a tropical legume. Sanginga et al., (2000) evaluated a wide range of cowpea breeding lines cultivated in savanna zones of Nigeria. Depending on their ability to absorb and use P efficiently, cowpea germplasm from the International Institute of Tropical Agriculture, Nigeria, could be grouped into poor performers, P-responders, non-responders to P, P-efficient, or P-inefficient. As stated above, both plant and AM fungal traits might have contributed in different proportions to the net P-efficiency measured among the genotypes. At least in some cases such as IT81D-849, high correlation existed between P-efficiency and higher levels of mycorrhizal colonization (95%), although the extent to which AM fungus contributes to P-efficiency could not be known. At times, soil factors may markedly influence the extent of AM fungal colonization and P related benefits from symbiosis. Thingstrup et al., (2000) examined the effect of P fertilization on the contribution of AM fungi to P nutrition of the host crops. They noted that AM fungi transported less ^{32}P in high-P fertilized plots. The relative contribution of AM fungi to P nutrition was greater in low-P soils. High levels of P can be suppressive to AM fungi, and this effect may again depend on fungal species or its isolate. Phosphate-resistant AM fungal isolates are also known to occur in nature. In other words, we may have to assess the genetic variation of host/symbiont for extent of P contribution in different fertility regimes.

Genetic Control of Infection, Establishment and Senescence of Plant-AM Symbiosis

Mutants or genetic lesions have served useful purposes in understanding the genetic nature, deciphering the metabolic pathways and their regulation, cell/tissue functions within microbes and plants. However, there is a degree of difficulty, firstly in obtaining an appropriate host mutant, and secondly, in utilizing them to understand genetics and molecular regulation of plant-AM symbiosis. So far, mutants defective in plant-AM symbiosis (Amf-) have been utilized sparingly, although there are quite a few reports on this aspect (Table 3.4). In this regard, Marsh and Schultz (2001) have suggested that classical genetic approaches utilized in conjunction with genomic techniques can lead us towards understanding plant coded genes crucial to the establishment and regulation of Plant-AM symbiosis. Mycorrhiza specialists interested in the molecular aspects have identified and demarcated various stages of development of mycorrhizal symbiosis to suit their purpose of study. It generally begins with spore germination, followed by chemotaxis (pre-infection stage), appresoria formation on roots, penetration of epidermal cells, invasion of cortical tissue, formation of arbuscules, its development and senescence (Harrison, 1999; Marsh and Schultz, 2001; Parniske, 2000; Smith and Read, 1997). Similarly, Provorov et al., (2002) have identified five discrete developmental stages that consider both morphological and molecular genetic aspects of Plant-AM symbiosis. They are: *Pid*—preinfection development; *Apf*—appresoria formation; *Img*—intercellular mycelial growth; *Ard*—arbuscular development; *Myp*—mycorrhizal persistence. On occasions, *Pid* and *Apf* stages may not show up as discrete stages. Now, let us consider several of these mutants of plant host sequentially and in greater detail.

Stage 1: Germination of AM fungal spores, formation of germ tube, a certain degree of hyphal growth, and concomitant nuclear duplication but restricted to only a certain period of time—occur in soil/substrate independently of host roots (Harrison, 1999). The influence of host genetics on AM symbiosis begins earnestly right at this stage, when hyphal chemotaxis and recognition.of host/non-host are determined by biochemicals elicited by roots (Giovannetti et al. 1993, 1999). This stage is often termed the 'early signaling stage'. The biochemical regulation of quantitative/qualitative aspects of root exudations, consequent hyphal responses to differences in

Table 3.4 Plant-host mutants having genetic lesions that impair formation of arbuscular mycorrhizal symbiosis at different stages

Host (accession no)		*Allele*	*AM fungus*	*Reference*
STAGE 2				
Lack of Epidermal Penetration (*Pen-* mutants):				
Medicago truncatula	(C71)	dmi/dmil-1	NA	Catoria et al. 2000
	(B129)	dmi 1-2	NA	Penmetsa and Cook 1997
	(P1)	dmi2-1	NA	Catoria et al. 2000
	(TR 25)	dmi2-2	G. i., G.m.	Catoria et al. 2000
	(TR 26)	dmi2-3	G.I., G.m.	Sagan et al. 1995
	(TR25)	dmi3-1	G.i., G.m.	Catoria et al. 2000
Lycopersicon esculentum	(CV76R)	rmc	G.i, Gi.m, G.m	Barker et al. 1998
Pisum sativum	(P1)	a	G.i, G.m	Duc et al. 1989
	(P2)	a	G.i, G.m	
	(P3)	a	G.i., G.m	
	G(F4-1)(P53)	a/sym30	G.i, G.m	
	(F4-141)(P55)	c/sym19	G.i, G.m	Schneider et al. 1999
	(P4)	c	G.i, G.m	Duc et al. 1989
	P6	f	G.i, G.m	Duc et al. 1989
cv Rondo	K24	c/sym 19	NA	Weeden et al. 1990
cv Sparkle	NA	a/sym 30	NA	Gianinazzi-Pearson, 1996
	R 72	b/sym 9	G.i, G.m	Balaji et al. 1994
	NMU 1	c/sym 19	NA	Weeden et al. 1990
	NEU 5	c/sym19	NA	
	R 19	P/sym8	G.i, G.m	Albrecht et al. 1998
	E 140	P/sym 8	G.i, G.m	Gianinazzi-Pearson 1996
	R 25	P/sym 8	G.i, G.m	Balaji et al 1994
Vicia faba	Indiana 78	sym 1	G.i, G.m	Duc et al. 1989
STAGE 3				
Epidermal penetration only, with total lack of cortical invasion (*Coi* mutants)				
Lotus japonicus	282-287	Lj sym2-1	G.i	Wegel et al. 1998
	282-288	Lj sym 2-3	G.i	Schauser et al. 1998
	2557-1	Lj sym 3-2	G.i	Wegel et al. 1998
	282-227	Lj sym 4-2	G.i	Schauser et al. 1998
	EMS 1749	Lj sym 4-2	G.i, Gi.m	Bonfonte et al 2000
STAGE 4				
No invasion into inner cortical cells, only outer cortical cells are penetrated (*Ici* mutants)				
Lotus japonicus	mcbex	Lj sym 71-1	G.r-10	Senoo et al. 2000a
		Lj sym 71-1	G.r-10	
Colonization in inner cortical cell, without Arbuscule formation (*Arb* mutants)				
Medicago sativa		MN NN-1008	G.i., G.v	Bradbury et al. 1994
Phaseolus vulgaris	R 69	R 69	G.m, G.c	Shirtliffe and Vessey 1996
Pisum sativum	Ris Nod 24	S	G.m	Gianinazzi-Pearson 1996

Note: *Gi.m = Gigaspora margarita; G.i. = Glomus intraradices; G.c = Glomus clarum; G.m = Glomus mosseae; G.v = Glomus versiforme;* G.r-10 = Glomus R-10. Stage 1 mutants involve variation or genetic lacunae in chemotactic induction/suppression of AM fungal spore germination, germ table elongation and hyphae. Recent reports suggust such plant mutants occur. (see Gadkar et al. 2003) exudates have been understood feebly. Root exudates, volatile compounds and gaseous emissions are known to induce rapid and extensive hyphal branching in the root zone. On several occasions, specific phenolic and flavonoid bearing root exudates have been attributed to help in host recognition and induction of hyphal growth (Becard et al., 1992; Poulin et al., 1993; Sequiera et al., 1991; Tsai and Phillips, 1991). These phenolics/flavonoid compounds seem to be exclusively involved in attraction or induction/inhibition of hyphal growth. Their role in nutrition is minimal, and easily discounted because of the small quantities involved. The host/non-host recognition, binding to roots leading to initiation of appresoria may involve receptors. For example, Poulin et al., (1997) have reported the occurrence of AM fungal receptor capable of binding biochinin-A and estrogens. Overall, quantitative/qualitative regulation of host loci responsible for the root exudations and formation of receptors definitely influences the early stage of plant-AM fungal symbiosis. As yet, molecular regulation of root exudations/receptor functions and their direct influence on AM fungal infection has not been understood in detail.

Schwartz et al., (2002) have discovered a tomato mutant that resists infection via spores but not through an inoculum of *G. intraradices* containing extraradical mycelium. According to them, phenotype of this *myc*⁻ mutant (M161) was stable through nine generation and breeds true without being influenced by light intensity, temperature or other environmental parameters. The resistance to *G. intraradices* spores was specific because it formed a normal AM symbiosis with other AM fungi such as *G. mosseae* and *Gi. margarita*. During the trials, germination of *G. intraradices* was repressed to only 45 to 60% of the wild type rhizosphere. Schwartz et al., (2002) suspect that a single gene may control such repression of colonization by this specific AM fungus *G. intraradices* in tomato plants. The gene product responsible for specific repression of spore-mediated infection needs to be isolated and characterized because it may have useful applications. Recently, Shwartz et al., (2003) reported the isolation of yet another pre-mycorrhizal infection (*pme*) mutant named M20. Again, tomato mutants were resistant to infection by AM fungal spores. A significant reduction in AM fungal spores occurred whenever root exudates from mutant 20 were added. This tomato mutant M20 shared several of the physiological properties with mutant M161 discussed above; in addition, it has a more suppressive effect on spore germination. Characterization of root exudates indicated that inhibitory moieties are heat labile, bind to poly-venylpyrrolidone and not volatile. Gadkar et al., (2003) have ascribed suppression of AM fungal spore germination to inhibitory components of root exudates occurring *myc*⁻ mutants of tomato.

Stage 2: Chemotaxis of AM hyphae leading to encounter with plant roots is followed by the formation of approsoria and penetration of host roots. In addition to intricately close physical contact, this 'stage 2' of plant-AM symbiosis involves direct physiological/molecular interactions between AM fungus and host roots. Host mutations that hinder molecular interactions at this stage of the symbiosis are available. In fact, Marsh and Schultz (2001) have remarked that the vast majority of plant mutants isolated, so far, are actually defective at this stage of establishment of the symbiosis. Plants experience a genetic lesion that impairs or aborts appresoria formation. Also, the fungal ability to penetrate and colonize cortical tissue is curtailed. Such mutants are collectively denoted as *Pen*⁻ mutants (see Table 3.4). The *mcbip* mutant of *Lotus japonicus* is a good example of genetic locus that impairs/aborts appresoria. Similarly, the *MN-IN3811* mutant of alfalfa is known to suppress appresoria formation.

The details on molecular aspects of these lesions are not understood. Harrison (1999) suggests that *Pen*⁻ mutants could involve callose or phenolic deposition in cell wall that impedes appresoria formation, mainly because corresponding wild types do not show such depositions. She infers that a suppressor of defense response normally active in mycorrhizal wild type is perhaps mutated. Therefore, it induces a negative response on AM fungi, thus, suppressing the appresoria. As yet, the molecular nature of regulation of such putative fungal defense suppressor is unknown. Salicylic acid is known to accumulate at the beginning of penetration of root by AM fungus and appressoria formation. External addition of salicylic acid delays mycorrhization (Garcia-Garrido et al. 1998). Further, Medina et al., (2003) have reported that in transgenic tobacco with reduced ability to synthesize salicylic acid, mycorrhizal colonization was enhanced, whereas, transgenic CSA (constitutive salicylic acid) biosynthesis tobacco plants allowed only low levels of mycorrhization. In other words, breeding plant genotypes with varying levels of salicylic acid may help us manipulate mycorrhizal colonization levels attained by crop plants.

Stage 3: Penetration of epidermal cells followed by intraradical hyphal spread into the cortical tissue forms stage 3 of mycorrhizal symbiosis in the present context. A number of mutants with lesions on genetic loci that allow development of appresoria, but block hyphal growth prior to cortex invasion (*Coi*⁻) are available (Bonfonte et al., 2000; Parniske et al., 2000; Wegel et al., 1998). At least six different loci (*Ljsym*2, *Ljsym*3, *Ljsym*4, *Ljsym*5, *Ljsym*23 and *Ljsym*30) can result in *Coi*⁻ phenotype characterized by restricted hyphal development (Bonfonte et al. 2000). Sometimes, epidermal cell penetration leads to senescence/death of host cell and fungal hyphae. In general course, appresoria formation, followed by penetration of epidermal cells are sequentially followed by extensive spread of AM fungus inside the cortex. However, Marsh and Schultz (2001) suggest that we can encounter host mutants with varying levels of colonization of cortex. In certain mutants of *L. japonicus*, the hyphal invasion confines to outer layers of cortex, but it is blocked from invading inner cortical tissue (*Ici* mutants). These mutants differ from those with a block at the epidermis (*mcbee*) and exodermis (*mcbex*) (see Table 3.4; Senoo et al. 2000).

Stage 4: After hyphal spread, arbuscular development is a crucial stage that influences mineral nutrient and photosynthate exchange between plant host and AM fungi. Plant mutants that allow hyphal development/spread throughout the cortical tissue, but are defective for arbuscular formation (*Arb*⁻) are available. Such mutants may be helpful in providing better insights into molecular signals/control for arbuscular initiation. The *Ard*-mutants in pea allow arbuscular initiation. However, development of arbuscule is restricted, defective or at the least truncated. The molecular basis/ lesions that result in impaired arbuscular development are yet to be understood. Certain pea mutants such as *RisNod24*, again, do not allow complete formation and functioning of arbuscule, although initial stages of arbuscule development may proceed normally.

It is clear from the above discussions that genetic lesions/mutations at various stages during formation of AM symbiosis can result in *myc*⁻ phenotype in a plant host. If we may intend to reduce or eliminate mycorrhizal component in the agronomically elite varieties because of its carbon costs to host. In that case, transferring the *myc*⁻ trait can provide the required situation by enhancing host resistance to AM fungus. We lack knowledge about genes (loci), if any, that enhance susceptibility of host to AM and/or high root colonization. Mutants or polyploids with multiple copies of genes leadng to high AM colonization need to be traced, if they occur in nature.

1. B. Genetic Variation in AM Fungal Symbiont

Genetic variation within the fungal partner can be noticed for a wide range of traits. Firstly, AM fungal hyphae may genetically differ in terms of growth habit, extent of mycelial production, branching and frequency of anastomoses. Spore morphology also differs between species and their isolates. Genetic differences have also been recorded for traits such as AM fungal ability to colonize, extent of extramatrical mycelium, P nutritional benefits and growth promotion.

Variation in AM Fungal Spores: Genetic variation in AM fungal spores has been amply documented. Morphological traits of spores such as shape, size, volume, color, ornamentation, sporocarp formation may vary enormously even within a single species. Several of these spore traits may get transferred naturally to other coexisting isolates. As one of the examples, let us consider the variation available and mechanisms of inheritance of spore morphology in *Scutellospora pellucida*. We have to note that single spore cultures of *Scutellospora* may generate a large variation in spore morphology, which is heritable (Bever and Morton, 1999; Stahl et al., 1991; Sylvia et al., 1993). There is no doubt that the process of adaptation requires the existence of heritable phenotypic variation within a population. However, as such, we are yet to identify the sexual reproduction process, if any, within AM fungi that can generate genetic variation. In nature, AM fungi are multinucleate. Individual spores of *Scutellospora* may contain a large number of nuclei ranging from 2600 to 3850, which are genetically variable (Sanders et al., 1996; Pringle et al., 2000). Also, studies regarding the inheritance of morphological traits are limited. The developmental mechanism or even biochemical basis for the divergent spore morphology is yet unknown (Bever and Morton, 1999). Given the multinucleate nature of spores and hyphae, genetic drift may actually depend on effective generation time of nuclei within the coenocytic mycelium (Bentivenga et al. 1997). Bever and Morton (1999) suggest that regardless of developmental mechanism, spore morphology is a heritable and easily trackable trait, whose study may provide crucial insights into the genetic nature and inheritance mechanisms operative in AM fungi. With regard to spore shape of *S. pellucida* in particular, Bever and Morton (1999) suggest two possible mechanisms of inheritance. Assuming that genes encoding the spore shape are located on chromosomes within nuclei, one of the mechanisms may involve parental spores containing genetically identical nuclei (homokaryotic) or genetically different nuclei (heterokaryotic). Observed results seem to support the operation of a rare mechanism, wherein genetically diverse nuclei segregate through dividing hyphae, so that progeny spores also differ (see Fig. 3.1). Alternatively, hyphal fusion between diverse AM fungal isolates can allow reconstitution of genetic components and provide a basis for genetic variation in the spore shape. Segregation and subsequent random fusion may contribute a substantial variation noticed in spore morphology, or even other traits. Anastomoses are known to occur in AM fungi. However, hyphal fusion between AM fungi isolates as a means for gene transfer is yet to be documented, although it is known to occur with other groups of fungi. Variations in hyphal characteristics have also been noticed (Table 3.5; Doud et al., 2000), but information regarding its genetic components and heritability is yet to be accumulated. The extent of anastomoses also varies, depending on the AM fungal species and its isolates. As suggested above, it could be an important trait determining genetic transfer rates leading to heterokaryosis (see chapter 7). Obviously, we should study

GENETICS OF SPORE SHAPES

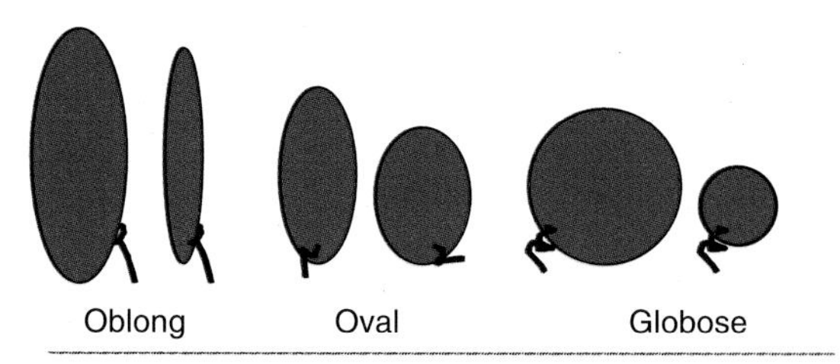

A. GENETIC SEGREGATION OF NUCLEI

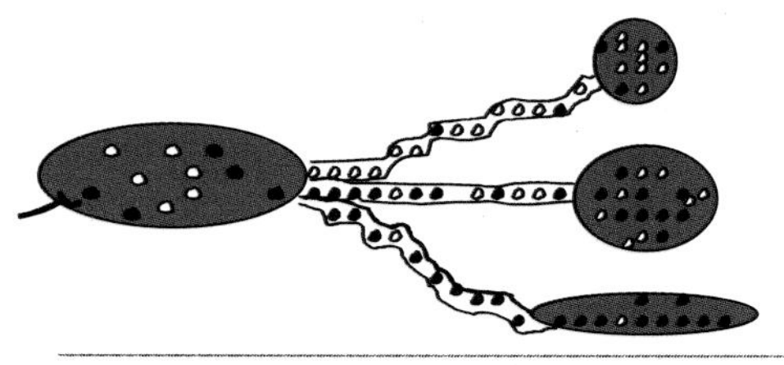

B. GENETIC MERGER OF NUCLEI

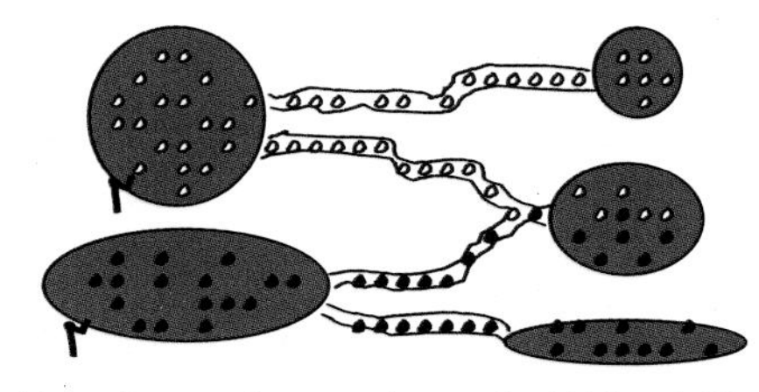

Fig. 3.1 Mechanisms of inheritance of spore shapes in AM fungi. Note: A. Depicts segregation of spore shapes. White nuclei code for round shape and dark ones code for an oblong spore shape. B. Depicts genetic merger and formation of hetrokaryotic spores. White nuclei code for rounded spores and dark ones for oblong shapes. (*Source:* Redrawn based on Bever and Morton, 1999.)

Table 3.5 Hyphal diameter of different AM fungi grown under similar conditions on *Peureria phaseoloide*s

AM fungal species	*Mean hyphal diameter (μm)*
Acaulospora morrowiae PHIL 11a	5.6 (3.7-9.1)
Acaulospora tuberculata BEG 4	3.7 (1.2-6.8)
Glomus manihotis BEG 112	7.9 (3.2-18.0)
Glomus manihotis CLB 1	4.9 (2.2-10.4)
Scutellospora heterogama BEG 35	
Pigmented hyphae	7.0 (4.2-9.9)
Hyaline hyphae	4.5 (1.6-7.2)
Scutellospora nigra	
Pigmented hyphae	5.4 (2.8-9.8)
Hyaline hyphae	3.2 (1.4-6.3)

Source: Dodd et al. 2000

Note: Values in parenthesis indicate range

the process of anastomoses in great detail and its relevance to gene transfer in AM fungi, especially considering that AM fungi are asexual in nature.

Variation for Colonization Ability: On a wider basis, variation in AM fungal ability to colonize a given host can be judged right at the taxonomic level. Hart and Reader (2002a) state that for a while, the colonization strategy of terrestrial AM fungal communities remained unclear. In their study with over 21 AM fungi belonging to different families, they found that isolates of *Glomus* were comparatively rapid colonizers and most extensive compared to *Acaulospora* or *Gigaspora*. Most *Glomus* isolates although extensive inside roots, they were slow to spread in the soil, if assessed based on the hyphal length. Taxonmic groupings were clearly correlated with variation in the ability/strategy to colonize different hosts. They concluded that genetic variation for colonization could be traced at the level of taxonomic groupings. This statement may hold good for at least few other traits of AM symbiosis. Interestingly, in their later study, Hart and Reader (2002b) found that variation in responsiveness among AM fungal families, to a certain extent, could be ascribed to size of the mycelium. For example, AM fungi with larger hyphae might transfer greater amounts of nutrients. We may, therefore, argue that a higher amount of colonization helps host plants to derive higher AM mediated benefits in terms of nutrient translocation from the soil. Graham and Abbott (2000) analyzed several AM fungal species for their ability to colonize wheat genotypes. Among the fungi, those consistently better in spreading into host roots were classified as 'aggressive AM fungi' and those with low percentage colonization ability were grouped as 'non-aggressive' fungi. From their trials, a few isolates from *Scutellospora calospora*, *Glomus invermaium*, *Acaulospora laevis* and *Gigaspora decipiens* were grouped as 'aggressive'. They averaged 50 to 89% root colonization on wheat, whereas, isolates of *Acaulospora* sp. and *Glomus* sp. that averaged only 10 to 19% were grouped as 'non-aggressive' on wheat. If we accept that higher levels of colonization are useful or a prerequisite for better plant response to AM fungus, then aggressive types could be preferred. This suggestion may not apply to all fertility conditions prevailing in the soil or growth medium.

Analysis of mycorrhiza-defective plants for any variation in their ability to form symbiosis may yield some useful results. Firstly, we may identify differences, if any, in the effect of host-genetic lesion on various AM fungal species. For example, the extent of suppression of colonization may vary, depending on the species or genotype of AM fungus. Gao et al., (2001) tested several AM fungi, such as *Glomus intraradices, G. mosseae, G. coronatum, G. versiforme, G.etunicatum, G. fasciculatum, Gigaspora margarita* and *Scutellospora calospora* by inoculating them onto mycorrhiza defective tomato seedlings. Interestingly, the development of *G. intraradices, G. etunicatum* and *G. fasciculatum* were arrested totally at the surface of the root. However, *G. mosseae, G. coronatum, G. margarita* and *S. calospora* frequently penetrated the root epidermis but colonization was rare. *G. versiforme* achieved relatively normal colonization. Obviously, AM fungal species have interacted differently to the defect in host roots. We should also understand the quantitative aspects of defect/expression of such genes. Conversely, will quantitative accentuation of certain genes induce higher levels of AM colonization? This aspect needs to be studied.

Fungal Variation for P Nutrition and Growth Response: There are several reports dealing with genetic variation among AM fungal species or their isolates for parameters such as ability to colonize, extent of colonization, P uptake and translocation and growth response. In fact, screening AM fungal cultures for their ability to colonize roots, improve P uptake and host growth is a very common exercise with mycorrhiza researchers. Most of the tests have shown the existence of genetic differences in AM fungi for the above traits. AM fungi may also vary with reference to their ability to utilize sparingly soluble P sources such as rock phosphates. The genetic components that determine P translocation and host growth are yet to be accumulated.

Let us consider a couple of recent examples on genetic variation in AM fungi. Rajan et al., (1999) screened several AM fungi on a forest species *Tectona grandis*. Considerable variation was noticed in symbiotic efficiency, but they found *G. leptotichum* to be the best among them in terms of colonization, plant growth and nutritional stimulation. In case of *Eleusine coracana*, among the six different AM fungal species tested, *Glomus caledonium* performed best, but a wide variation in terms of colonization, mineral uptake parameters and growth stimulation were clearly discernible (Tiwari et al. 1993). Recently, Clark (2002) assessed the variability of several AM fungi with regard to parameters such as colonization and mineral nutrition of a plant host (*Panicum virgatum*). Accumulation of mineral nutrients in the shoot system was generally higher in AM colonized plants, but varied depending on the fungal species. Mineral uptake per root length varied extensively among AM fungi. Plants colonized by *G. etunicatum* particularly had higher P, N, S, K, Mg, Zn and Cu. No doubt, the genetic and physiological reasons for such variation in induction of mineral uptake among AM fungal genotypes need to be understood in detail. It may be useful while devizing appropriate AM fungal inoculum.

With regard to host P nutrition and growth response, Graham and Abbott (2000) have made some interesting deductions based on their assessment of 'aggressive' and 'non-aggressive' isolates of AM fungi. It relates to variation in colonization versus P benefits derived by host depending on the P fertility of soils. They found that aggressive AM fungi spread all over into wheat roots and caused depletion of root sugar. It resulted in a growth depression. Concomitant P benefits were also not discernible if plant were cultivated under high P conditions. The extent of P removed from soil was not significantly higher in mycorrhizal plants having non-aggressive AM fungal isolates, whereas the use of aggressive AM fungal isolates resulted in higher colonization and induced better P uptake if cultivated in low P conditions. Nature has endowed us with AM fungi possessing wide genetic variation. Strains suited to different situations could be found and utilized.

Boddington and Dodd (1999) studied several species of AM fungi belonging to two genera: namely *Glomus* and *Gigaspora*, for genetic variation in host growth response and P nutrition. They aimed at understanding the underlying causes for variation in P nutrition. Some of their observations indicated that extraradical mycelium of *Gigaspora* accumulated polyphosphate, whereas it was not detected in the mycelium of *G. manihotis*. They suspected that mechanisms of phosphate metabolism and transport in the mycelium of AM fungi from different species or genera might have caused variations in uptake and transport. Life cycle strategies such as the formation of auxiliary cells in *Gigaspora*, excessive lipid storage in vesicles may also generate variation in P nutrition. There are possibilities of negative feedback mechanism between alkaline phophatase and polyphosphate in the extraradical mycelium of *Gi. rosea* being involved in variation in P uptake and translocation among species or its genotypes. Hart and Reader (2002b) found that to a certain extent, differences in mycelial size among AM fungi may contribute to a corresponding variation in responsiveness and net P translocation advantages derived by the host.

Recently, Burleigh et al., (2002) reported that genetic variation within AM fungi extends to many of the functional aspects of P transport and metabolism during symbiosis. They studied the variation in genes relevant to P transport in *Medicago* and Tomato. The seven AM fungal isolates tested differed with respect to magnitude of induction of expression of genes MtPT2 and LePT1 in the roots. Isolates of *Glomus mosseae* caused the greatest reduction of MtPT2 but the highest level of P uptake and growth stimulation, whereas, *Gigaspora rosea* induced MtPT2 to highest levels, but with low P and growth stimulatory effects. This study actually proves that genetic variation for certain traits such as P uptake and transport could be authenticated at the level of specific genes and their quantitative expression. This may have applied value while selecting AM fungal strains for inoculation on specific groups. Rapid screening tests based on mRNA levels of specific genes involved in P acquisition and translocation by AM fungi can really hasten the identification of useful AM fungal genotypes. Ravnskov et al., (2001) have identified variation in the activity of enzymes related to carbon metabolism and transport in AM fungi. They noticed differences in the expression of genes for invertase and sucrose synthase in maize plants colonized with isolates of different AM fungal species; namely, *G., intraradices, G. caledonium*. In other words, the extent of AM fungal influence on sucrose utilization in C importing tissues depends on fungal species. We should also be able to trace a genetic basis and variation for carbon cost of P translocation by AM fungi. Definitely, net P translocated per unit C consumed by fungus should be dependent on several genetic traits. Since, AM fungal activity involves expenditure of host carbon, it might be useful to assess variability appropriately by estimating net P translocation per unit carbon expended.

1. C. Molecular Regulation of Arbuscular Mycorrhizal Symbiosis

One of the main aims of studying molecular aspects of arbuscular mycorrhizal symbiosis is to understand the regulatory mechanisms involved in nutrient acquisition and its exchange between the symbionts. We hope that it may eventually help us in manipulating the symbiotic systems to our advantage (see chapter 4). In addition to genes that regulate nutrient acquisition and exchange, knowledge regarding the genes relevant to biochemical regulation of housekeeping functions and stimulation of plant growth are equally important. A generalized increase in metabolic rate through enhanced expression of genes related to growth, as well as hormone mediated stimulatory effects due to symbiosis (if any), should be understood in great detail.

Techniques: Firstly, let us consider the techniques that have been utilized to study the molecular regulation of AM (or even ECM) symbiosis. Franken (1999) suggests that to attain better insights into molecular phenomenon, both the symbiotic partners—plant and fungus, need to be biochemically assessed in time and space. At the minimum, a functional analysis of molecular regulation will require gene cloning and analysis of their expression, and a comparison of corresponding stages for physiological and morphological effects. In general, analysis of plant genes relevant to functioning of mycorrhizal symbiosis has been studied on more occasions, than AM fungal genes required to sustain the symbiosis. The prime reason is difficulty to culture AM fungus on synthetic growth medium. Since, ECM fungi are culturable molecular, analysis has been that much easier.

The different molecular genetic approaches utilized to study mycorrhizal symbiosis can be grouped into: (a) targeted approaches involving the screening of gene libraries, cloning, PCR with degenerated oligonucleotides; and (b) non-targeted approaches. Figure 3.2 depicts one of the flow charts utilized to assess the functioning of targeted protein-encoding genes of AM fungi. The non-targeted approaches involve isolation of up- and downregulated genes through differential screening of cDNA libraries. This method is based on the assumption that genes showing either induction or repression have bearing on a specific developmental stage. Firstly, RNA is extracted from two developmental stages, reverse transcribed and cloned as cDNA in two parallel experiments (Fig. 3.3). Both libraries are then screened for clones that hybridize to transcripts from one developmental stage but not the other. Obviously, if applied to early stages of mycorrhizal symbiosis, only plant genes will get scored, since fungal biomass is meager. However, if RNA from later stages is utilized, then fungal cDNA clones can be found (Murphy et al. 1995; Burleigh and Harrison, 1998).

Yet another technique is the differential RNA display (Fig. 3.4). In this case, RNA extracted from different tissues (mycorrhizal or non-mycorrhizal) is utilized as the template for PCR random primers. The amplification products are compared using gel electrophoresis. Fragments detected in one set but not the other are assumed to correspond to the differentially expressed genes, on which further analysis can be made. Using this technique, several plant and fungal genes have been identified

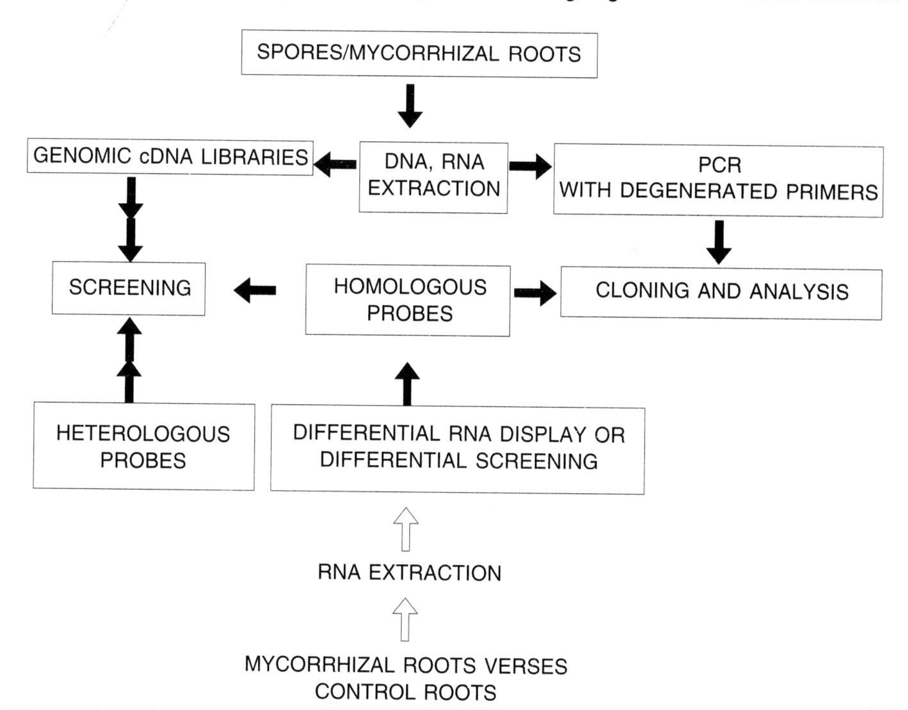

Fig. 3.2 Strategies to clone AM fungal genes. Mycorrhizal roots are used to extract nucleic acids and establish genomic or cDNA libraries. These libraries are then screened with heterologous probes. Alternatively, comparison of RNA populations of mycorrhizal and non-mycorrhizal roots by differential display or differential screening of cDNA libraries leading to homologous probes is possible. (*Source:* Franken,1999)

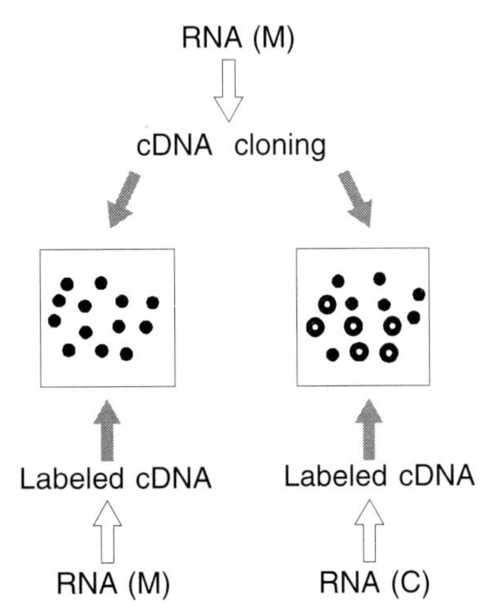

Fig. 3.3 Differential screening of a cDNA library. First, RNA is extracted from mycorrhizal roots (M) for cDNA cloning. Clones are then transferred on two membranes which are hybridized to labeled cDNA synthesized using RNA from mycorrhiza (M) or control roots (c). The presence of a signal after autoradiography is indicated by a *closed circle*, whereas its absence is indicated by *open circle*. The six clones shown that hybrize only to the mycorrhizal probe and probably correspond to genes that are expressed only in mycorrhiza and they could be of fungal origin. (*Source*: Redrawn from Franken,1999)

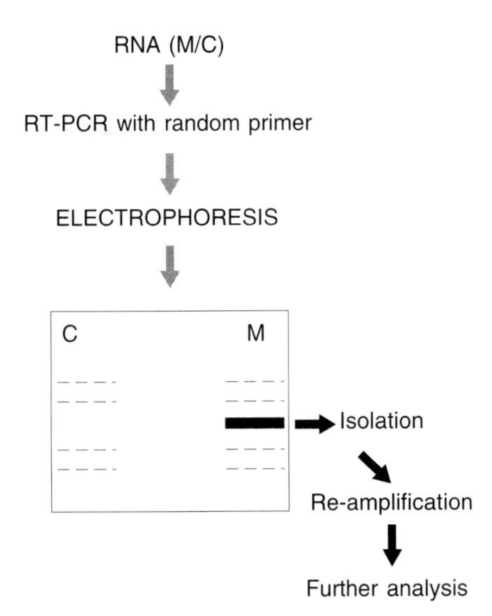

Fig. 3.4 Differential RNA display analysis Note: M = Mycorrhiza, C = Control roots. RNA extracted from mycorrhiza and control roots serve as template for RT-PCR with random primer. Genes that appear in the mycorrhizal sample correspond to genes expressed only during mycorrhization. These fragments are cut from the gel and re-amplified. (*Source*: Redrawn from Franken, 1999)

(Martin-Laurent et al., 1997; Voiblet et al., 2001). As a precaution, whenever mycorrhizal and non-mycorrhizal root fragments are assessed, fungal genes may not be the ones differently expressed during development. Proper analysis should, therefore, involve a comparison of samples from different stages of symbiosis or fungal life cycle (e.g. germinating spores, extra-radical mycelium, etc.). Franken (1999) further suggests that differential RNA display method is well suited for studying regulation of AM fungal genes that are induced or suppressed during symbiosis.

EST Libraries and Identification of Symbiosis Regulated (SR) Genes—An Example: Currently, EST library is available for several AM fungi. Regular biochemical kits are available to process the fungal tissue, uninfected plant roots or symbiotic portions (fungus plus roots). To quote an example, Stommel et al., (2001) constructed an EST library for *Gigaspora rosea*. The cDNA was synthesized, digested and cloned into vector lambda-ZAP express. About 1500 clones were obtained of which 1.5% carried inserts of the rRNA gene cluster. Overall, database searches for the genus *Gigaspora* revealed that 62% of the clones had similarities to already known sequences (Fig. 3.5). Further analysis of 100 randomly selected clones from the EST library indicated that 16 clones did not show homology with any known genes. Ten clones were probably involved in ribosomal RNA-transcribing polymerase-1, and those necessary for post-translational modification were identified. A second group of 9 clones had similarity with genes involved in DNA synthesis and cell cycle. Several of the remaining fragments represent genes that function during signal transduction, transport processes, respiration and fungal metabolic activity in general (see Table 3.6).

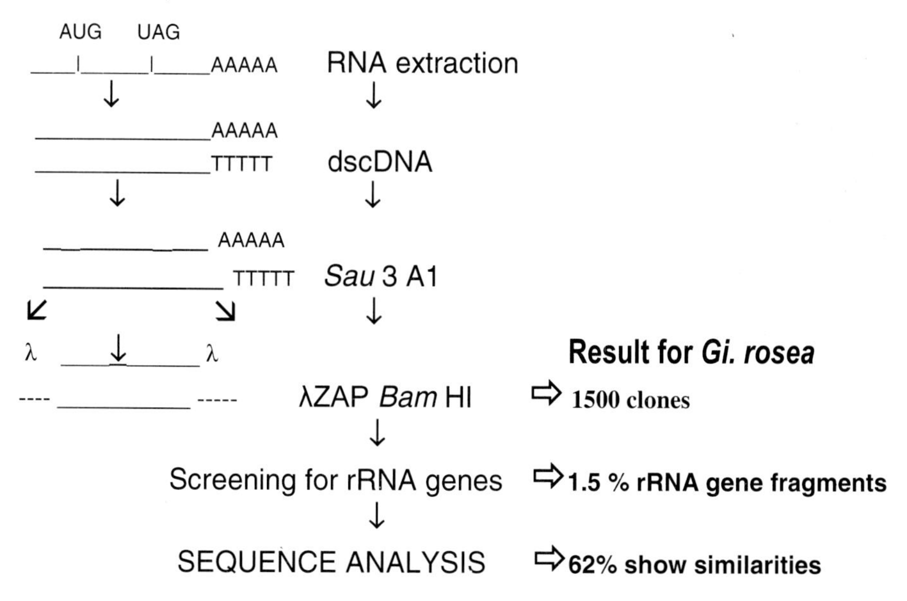

Fig. 3.5 Expressed Sequence tags (EST) library construction—an example. Double-stranded cDNA is synthesized from total RNA and digested with Sau3 A1. Clones are then screened for rRNA genes and inserts are sequenced. (*Source*: Redrawn from Stommel et al. 2001)

Table 3.6 Identification of cDNAs from an AM fungus *Gigaspora rosea* by comparing the homology to known proteins of the gene products deduced from sequenced ESTs—examples.

Protein class	*Size (bp)*	*Gene function*	*No clones*	*Identity/ length*
DNA synthesis and cell cycle	352	BRAHMA ortholog (DNA helicase superfamily 2)	3	33%/74
	517	ATP-dependent RNA helicase	1	74%/62
	359	Purine nucleoside phosphorylase	1	66%/42
	164	Ubiquitine precursor 1	1	100%/53
	208	Microtubule-associated protein	3	51%/39
Signaling	139	Nitrogen-activated protein (MAP) kinase	1	93%/46
	276	P13/14 kinase-like protein	1	43%/90
	115	Protein tyrosinase phosphatase	1	61%/34
	192	Histidine protein Kinase	1	52%/35
	182	Calcium P-type ATPase	1	64%/53
Transport	111	Vesicle transport v-snare protein	1	66%/36
	385	Efflux protein	1	28%/129
Respiration	107	NADH-oxidoreductase complex 1 subunit	1	45%/35
	117	Methyle transferase	1	45%/39
Others	343	Metalothionein	1	39%/48
	149	Homogentisate-1,2-dioxygenase	1	68%/29
	133	Malonyl CoA acyl carrier protein transcyclase	1	56%/23

Source: Stommel et al. 2001.

Tracing *SR* genes in ECM is an extended procedure involving accurate molecular analysis. Firstly, let us consider an example for construction of an EST library with an ECM (e.g., *Eucalyptus globulus/Pisolithus tinctorius*). In their study, Voiblet et al., (2001) utilized 4-day-old ECM samples that included fungus and root tissue. The RNA source material generally contained primary roots, root tips, meristematic tissue and root cap cells. The root samples drawn were all ensheathed by ECM fungal mantle, with differentiated inner and outer layers. Hartig net formation could be discerned by the hyphae growing between rhizodermal cells. Essentially, samples comprised all the early stages of ECM symbiosis. In these early stage samples, fungal transcripts accounted for approximately 65% of the total RNA extracted from ECM (Carnario-Diaz et al.1997). There are standardized procedures and few cDNA library construction kits. In the present case, a commercial kit 'Cap-Finder cDNA' (Clontech, Palo Alto, California, USA) was utilized by Voiblet et al. (2001). Through this procedure, high quality cDNA libraries containing a high proportion of full-length cDNA clones can be produced.

For a suppression subtractive hybridization (SSH), double-stranded cDNAs corresponding to mRNAs expressed in 4-day-old *E. globulus/P. tinctorius* ECM, free-living mycelium and uninfected roots were separately obtained. The cDNAs from free-living mycelium and roots were pooled, considering their probable concentrations in a 4-day-old mycorrhiza (Carnario-Diaz et al. 1997). The subtracted cDNAs were later primed, screened and designated appropriately (e.g. EgPtA, B, C, etc.). These SSH clones are purified, sequenced and utilized for homology comparisons and database construction (see Voiblet et al. 2001).

The key step during identification of *SR* genes is to estimate the expression profiles of differentially expressed genes in ECM (or AM). Later, we must pinpoint the mildly or highly induced/suppressed genes by comparing the mRNA levels in ECM tissue, free-living fungus and uninfected roots. In this study, Voiblet et al., (2001) utilized cDNA mini-arrays to investigate changes in gene expression profiles during the development of symbiosis. They studied around 480 selected ESTs from the ECM library. Obviously, among the array of cDNA tested, genes relevant for house-keeping functions, cell structure development, signaling and symbiosis regulation could be traced. A graphic analysis of gene (mRNA level) expression profiles can lead us to identify *SR* genes that are induced/suppressed, say, by 2.5 folds or 10 folds (see Fig. 3.6). Further, we can selectively analyze the *SR* genes induced in either plant root or fungus, also understand their relevance to specific functions such as signaling, cytoskeleton development, P transport, etc. (Fig. 3.6). By way of such techniques, several *SR* genes have been traced. Several more would be identified and studied in detail, in due course.

Reporter Genes: Genetic reporters are versatile molecular tools, through which several cellular functions such as gene transcription, protein and organelle localization can be studied. Such reporter genes have been traced and utilized for analysis of several fungi, including mycorrhizal symbionts. Bergero et al., (2003) have reviewed various aspects of AM/ECM fungal symbiosis that can be effectively analyzed using genetic transformation and introduction of reporter genes into AM fungi. They believe that the use of genetic reporters during molecular analysis of AM/ECM fungi may enhance our understanding about the biochemical functioning of the plant-fungal symbiosis.

Molecular Regulation of AM Symbiosis

Van Buuren et al., (1999) commented that for a while much of the investigation on molecular genetic aspects of AM symbiosis was comparable to plant versus pathogen interactions. Genes identified previously from interactions with pathogenic fungus became the major focus. They were traced in plant-AM symbiosis and then investigated. In contrast, non-defense genes that get accentuated at greater intensity were studied sparingly. This situation has no doubt changed, and several novel symbiotic genes are being thoroughly studied. Van Buuren (1999) reported three such genes induced during symbiosis between *Medicago*/*G. versiforme*. One of them is a xyloglucan endo-transglycosylation related protein; second, a putative arabinoglucan protein and the third, a homologue of initiator factor 3. These genes are not expressed in the plant host prior to AM fungal colonization, and are also not induced through phosphate application. Hence, it was concluded that induction of these genes was a direct consequence of fungal invasion.

While analyzing a tomato/*G. mosseae* symbiotic system, Karim et al., (1998) realized that for better understanding of the molecular regulatory aspects, firstly the relevant genes and their products should be known. It is also important to assess all three major types of changes that may occur as a consequence of symbiosis, namely the down and up regulation of some constitutive polypeptides, and thirdly, *de novo*

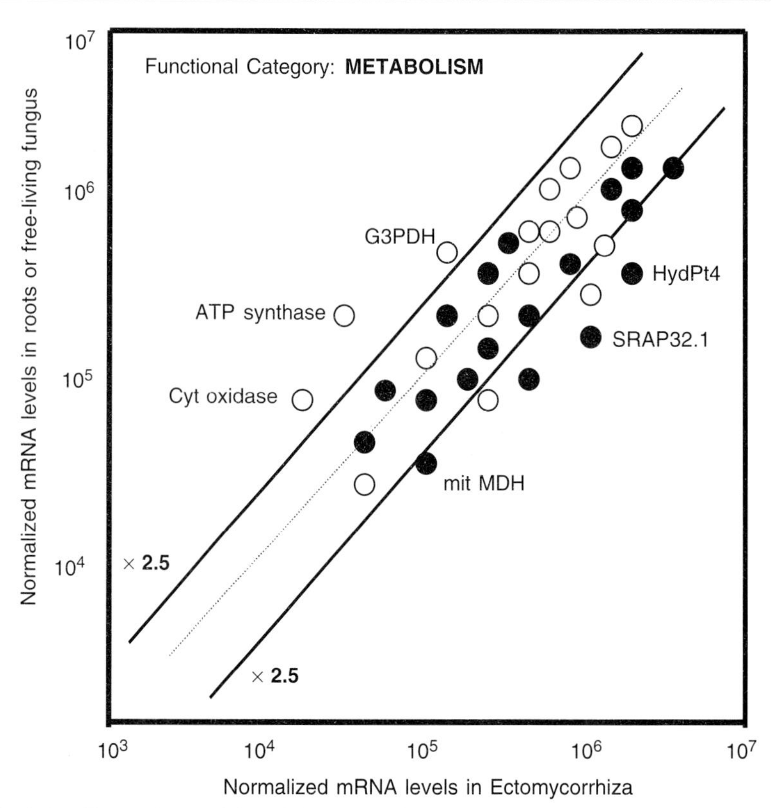

Fig. 3.6 Expression profiling of free-living *Pisolithus tinctorius*, non-mycorrhizal roots and 4-day-old mycorrhizal *Eucalyptus globulus* roots-genes involved in metabolism.
Note: The above example depicts induction of genes involved in metabolism. Similarly, induction/repression of genes relevant to cell signaling, host defense, housekeeping proteins and others have been profiled. A gene whose expression is unaffected will fall on the 45 degree line (gray line). Genes whose expressions are induced/repressed beyond 2.5 times the free-living host/fungus will fall outside the thick parallel lines on the graph—for example, G3PDH, SRAP32.1, ATP Synthase, Cytochrome oxidase and mitochondrial malate dehydrogenase have been shown above. So far, there are very few genes induced/repressed 10 folds beyond the free-living counterparts. They are G3PDH = Glucose phosphate dehydrogenase; HydPt4 = Hydrophobin gene; SRAP32.1 = Serine rich acid protein; MDH = Malate dehydrogenase. (*Source:* Modified and redrawn based on Voiblet et al. 2001)

induction of new polypeptides, if any. They prefer to denote such *de novo* polypeptides as 'mycorrhizenes', which seems synonymous with the protein products from 'symbiosis regulatory genes (*SR* genes)' discussed under ecotmycorrhizas within this chapter. The AM fungal origin of some of these polypeptides could be ascertained, but their significance to symbiotic relationship and plant growth stimulation is yet to be understood. The *SR* genes could also be of plant origin.

Symbiosis regulatory genes refer to those induced/suppressed so that symbiotic relation is established, maintained or terminated. They may be coded by host plant or fungus. In the normal course, several plant or fungal genes are expressed when these symbionts are in an independent state. Such genes are not termed symbiotic regulatory genes. The *SR* genes may actually relate to different stages of symbiosis. We can also classify them based on their functions (Fig. 3.7). In the present context, *SR* genes have been classified into those affecting pre-symbiotic stages, early symbiotic stage, cellular functions, primary and secondary metabolism, P nutrition, etc. During the past three years, several researchers have examined cDNA sequences and identified a wide range genes that are induced mildly or strongly, suppressed or those not affected due to AM symbiosis. These 'symbiosis regulated (*SR*)' genes code for enzymes, nutrient transporters, structural proteins, cell-cycle regulatory factors, etc. For example, Jun et al., (2002) aimed to discern the molecular strategies that AM fungi may adopt during symbiosis with host plants. They examined several ESTs that possessed homology with known enzymes such as trehalase, cystein synthase, transporters such as phosphorus and arsenate transporters, tubulins, actin, dynein, cell-cycle regulatory proteins and meiosis related proteins. These genes are known to significantly affect the establishment and functioning of plant-AM symbiosis. We may forecast that *SR* genes relevant to each and every aspect of

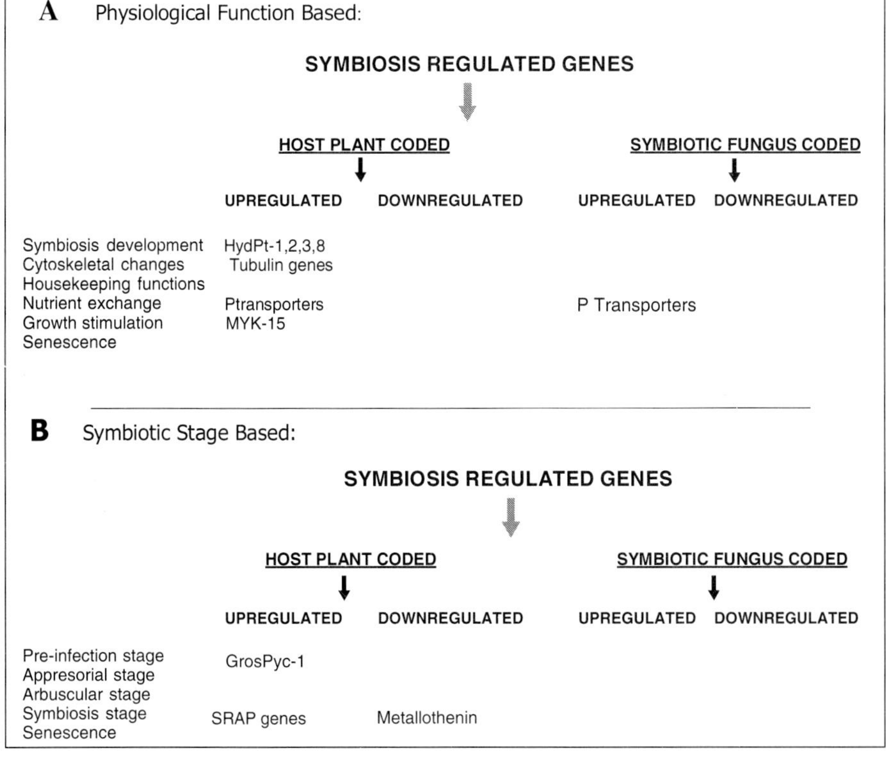

Fig. 3.7 Classification systems/charts for Symbiosis Regulated (SR) genes in AM/ECM association, with a few examples. Note: We may add any number of genes/gene families under each class, be it symbiotic stage related or function related, as and when identified/proved. Extent of induction of each gene or encoded protein can also be marked in parenthesis. It can be semi-quantitaively expressed as +, ++, +++ depending on extent of accentuation or as -, --, -- based on extent of depression.

symbiosis will be isolated and studied. Spatial and temporal changes in *SR* genes should also be given due importance (Delp et al. 2003). Hence, we may have to classify the *SR* genes appropriately, depending on the purpose of study and context. For example, some researchers have classified *SR* genes based on the extent of induction/suppression (Voiblet et al. 2001). There are many reports that categorize *SR* genes based on the stage of symbiosis that is affected most by their regulation. For example, *SR* genes induced in pre-symbiotic stages, early stages of infection, say appressoria formation, or those regulated at arbuscular stage or even at senescence of these structures (Tamsloukht et al. 2003 a,b). Yet another set of classification is based on the exact cellular and physiological functions affected by *SR* genes (Jun et al., 2002). For example, signaling aspects between plant root and fungus, cell-structural synthesis, enzymatic regulation of primary and secondary metabolism, nutrient transport functions, induction of host growth (Fester et al., 2002; Lammers et al., 2001). A very important aspect during analysis of *SR* genes is to classify them first as plant-coded or fungal-coded *SR* genes. Therefore, a first level classification should identify origin of *SR* gene (Fig. 3.7). In the following paragraphs, one such classification appropriate to the context and theme of this treatise has been narrated and discussed.

SR Genes of Pre-symbiotic Stages: We must realize that regulation of AM symbiosis begins much earlier than physical contact between the symbionts. It is already clear that fungal spores/hyphae receive and react to chemical signals emitted/ exuded by the compatible host roots. Tamasloukt et al., (2003a) point out that in nature, root exudates can change the growth pattern of the developmental stages of AM fungus while it is still in its asymbiotic state. It affects the apical dominance in asymbiotic hyphae and induces intensive branching in its pre-symbiotic state. In their study on pre-symbiotic hyphae of *Gigaspora rosea*, differential display analysis proved that specific changes in RNA accumulation pattern occur, in case the asymbiotic hyphae are provided with exudate preparation (from host plant) in the liquid medium. In order to identify genes relevant to pre-symbiotic stages of AM fungus, they adopted suppressive subtractive hybridization experiments. They reported that clustering of genes according to their sequences revealed that 'presymbiotic *SR* genes' were induced as early as 1 h after host root exudate treatment. These 'pre-symbiotic *SR* genes' were deemed to encode proteins that regulate signal transduction, mitochondrial development and DNA synthesis, as well as activity of other organelles in the AM fungus. Clearly, the presence of host root exudates induces genes that aid transition of AM fungus from asymbiosis to presymbiosis. Tamasloukt et al., (2003a) believe that such gene activation in asymbiotic hyphae may actually result in enhanced physiological activity and certain morphological changes that improve the chance for establishment of mycorrhizal symbiosis. To support this contention further, Tamasloukt et al., (2003b) have identified additional putative mitochondria related genes in the pre-symbiotic mycelium that were induced by the presence of root exudates. These genes were actually identified after subtractive hybridization, and they code for pyruvate carboxylase and mitochondrial ADP/ATP translocase. The gene *GrosPyc*1 that encodes pyruvate carobxylase was further cloned and quantified using its RNA accumulation. They also hypothesized that the presence of host root exudates in the vicinity of spores induces respiratory activity of AM fungal germ tubes/hyphae. Oxygen consumption and reducing activity of germ tubes/hyphae increased within 2 to 3 h, if root exudates were added to spore/hyphal incubation medium. Cytological observations have also indicated changes in mitochondrial morphology and its overall orientation in cytosol. At this stage, we may conclude that root exudates as a factor induces the expression of specific genes involved in respiratory activity and hyphal proliferation, thus augmenting the transition from asymbiotic to symbiotic phase, provided the stimulated hyphal growth leads it to a compatible host root. According to Tamasloukt et al., (2003b) this developmental switch via gene activation may need between 0.5 and 1 h, physiological and morphological effects could be discerned in about 5 to 8 h.

SR Genes of Early Stages of Symbiosis: Under this category, a wide range of genes relevant to the formation of appresoria, penetration and establishment of AM/ECM fungus can be considered. Several host mutants that repress early symbiotic stages are available. They should be assessed for subtle changes in gene expression.

SR Genes for Cell function and Structural Changes: Immediately after the establishment of a symbiotic relationship, cell structural and functional changes in both host and fungus may ensue. Several genes may get quantitatively induced or repressed. There are several recent reports on this aspect. Let us consider a few examples. Differential screening of mycorrhizal tomato roots yielded cDNA clone denoted as Le-MI-13. The le-MI-13 gene codes for a polypeptide that shows high amino acid sequence homology with a multigene family known as cullins. The Le-MI-13 accumulated whenever tomato roots were colonized with AM fungi (e.g. *G. mosseae*), but not when infected by a pathogen *Phytophthora nicotianae.* This suggested that upregulation of cullin-like gene Le-MI-13 is specific to AM fungal symbiosis. Induction of Le-MI-13 strongly correlated with nuclear polyploidization. Based on certain indirect evidences Tahir-Alaoui et al., (2002) speculated that Le-MI-13 might be involved in cell functions, especially cell-cycle regulation during symbiosis. The *SR* genes relevant to water transport in *Medicago truncatula* was studied by Krajinski et al., (2000). They screened for the presence of genes encoding tonoplast intrinsic proteins and identified a gene family that was induced during symbiosis. Injection of the cDNA into oocytes of *Xenopus* revealed that it encoded a protein called aquaporin (MtAQP1) that specifically facilitates water transport. The expression of MtAQP1 regulates cell functions, especially fluctuations in osmotic buffering in the highly compartmented vacuole of arbuscular cells.

A class-III chitinase gene, Mtchitinase III-3 was identified within the model system *Medicago truncatula-Glomus intraradicies* by Banonomi et al., (2001). Colonization by *G. mosseae* or *G. intraradicies* induced it specifically. Such an Mtchitinase III-3 gene was not induced in response to infection by different pathogenic fungi such as *Phytophthora megaspermae, Fusarium solani,* or *Ascochyta pisi.* MtchitinaseIII-3 was not induced even during nodule formation by *Rhizobium meliloti.* This situation contrasts with other chitinase isoforms related to plant-AM fungus association. Thus, Mtchitinase III-3 is expressed only in cortical root cells containing arbuscule, be it a developing or mature arbuscule. Obviously, cells without arbuscule were devoid of Mtchitinase III-3 activity. It is believed that mycorrhiza-specific chitinase in *M. truncatula* might exert its action in the space between fungal arbuscule and peri-arbuscular membrane. It may also be involved in modifying nascent chitin during the development of arbuscule. Perhaps, Mtchitinase III-3 is also involved in suppressing defense reactions within cells containing arbuscules (Salzer et al., 2000; see chapter 5). In summary, Banonomi et al., (2001) believe that Mtchitinase III-3 plays an essential role in suppressing a plant's defence responses towards AM fungus during the process of arbuscule formation. Several other *SR* genes relevant to cell structure and development have been reported, such as tubulin genes and cell wall related proteins (Voiblet et al., 2001; Stommel et al., 2001). Delp et al., (2003) recently reported spatial and temporal changes in certain *SR* genes relevant to metabolism in extraradical mycelium of *G. intraradices*. The fungal genes, Ginmyc1 and Ginhb1, were expressed only in extraradical mycelium and not in colonized roots of tomato. In contrast, a gene Ginmyc2 was induced both in extraradical mycelium and colonized roots. Their analysis showed that fungal-tubulin mRNA and protein both accumulated in greater amounts in the extraradical mycelium as the symbiosis proceeded. Its accumulation was relative to fungal rRNA. On the other hand, accumulation of tomato tubulin mRNA was more reduced in colonized roots when compared to non-mycorrhizal roots. Therefore, knowledge about differential spatial and temporal regulation of AM fungal genes seems crucial.

SR Genes Relevant to Primary Metabolism: Several *SR* genes may be involved in regulation of primary metabolic steps, especially carbohydrate conversion in mycorrhizal symbiosis. *SR* genes related to TCA cycle, pentose phosphate pathway, amino acid metabolism have not been investigated yet. Understanding gene expression and its regulation relevant for glyoxylate cycle is important. We know that fungal partner possesses sufficient quantities of carbon as lipid, and glyoxylate pathway is central to the flow of carbon in AM fungi. Radiolabel and NMR-based techniques revealed that carbon fluxes via glyoxylate cycle could be substantial (Lammers et al. 2001). Further, they reported that several cDNA clones with homology to gene sequences for isocitrate lyase and malate synthase were identified in *Glomus intraradices*. Quantitative real time PCR clearly proved that these genes are induced and their expression is significantly higher. Concomitantly, glyoxysome-like bodies observed under electron microscope were active. Several other genes that encode for relevant enzymes involved in primary metabolism and carbon fluxes via glyoxylate cycle were also isolated and identified by them.

SR Genes Relevant to P Nutrition: The *SR* genes relevant to P nutrition, especially transporters in roots cells and several others operative at the interface between symbionts have been discussed extensively in chapter 4.

SR Genes for Responsiveness: Classical approaches have already shown a genetic basis for enhancement of P nutrition and growth response. Induction or suppression of specific genes that regulate growth should be accurately identified. At times, classical approaches could be tedious and time consuming. Gaining detailed knowledge about genes relevant to responsiveness of a symbiotic system and their regulation are perhaps the most crucial aspects of genetic analysis. The ability to manipulate the expression of genes for mycorrhiza responsiveness is undoubtedly very useful. We have no idea regarding the number of *SR* genes involved in growth improvement. Most importantly, knowing their regulation in time and space could be a time consuming task. Recently, Fester et al., (2002) reported the identification and isolation of a small 'mycorrhiza responsive protein' (MYK15) in wheat roots that were colonized with mycorrhizal fungus *Glomus intraradices*. This protein occurred in abundant amounts in the root fractions supporting dense AM fungal colonization. The *MYK* gene codes for a protein with molecular mass 15kDA and an isoelectric point 4.5. The N-terminal sequence has high similarity to a peptide sequence deduced from EST clone derived from *Medicago truncatula* colonized by *G. intraradicies*. It contains a hydrophobic stretch of 24 amino acids predicted to form a transmembrane cc-helix and may correspond to cleavable domain. Involvement of MYK protein in growth response needs to be assessed. We may trace many more mycorrhiza responsive proteins. It is most important to decipher the regulation of genes responsible for these proteins and learn to manipulate them to advantage.

SR Genes Relevant to Secondary Metabolism: Regulation of secondary metabolism in arbuscular mycorrhizas is an important aspect, but it is yet to be understood in any great detail. Let us consider the few studies that have been reported. Isoprenoids are among the important secondary metabolites relevant to AM symbiosis. They are implicated in yellow pigments synthesized during AM symbiosis. Isopentenyl diphosphate, which is the precursor for isoprenoids, is synthesized via two separate pathways, one in cytosol and the other in plastids. The initial step in the plastid pathway is catalyzed by 1-deoxy-d-xylulose 5-phosphate synthase. Previously, only one gene was attributed to this step, but recently, Walter et al., (2002) identified two distinct genes: *MtDXS1* and *MtDXS2.* The DXS2 transcript levels are low in most tissues, but are strongly upregulated in root upon colonization by mycorrhizas. Its stimulation in mycorrhizal roots coincides with carotenoids and apocarotenoids. In a separate study on secondary metabolites, induction of genes relevant to jasmonite biosynthesis in a Barley/*G. intraradices* system was investigated by Hause (2003). They observed that an increase in jasmonates was accompanied with expression of jasmonate-induced protein (JIP23). Immuno-cytochemical tests and in situ hybridization indicates that expression of enzymes leading to jasmonate biosynthesis is cell specific and confines to arbuscular cells. Analysis of temporal expression patterns showed that transcripts accumulate 4 to 6 days after arbuscule formation. Hence, they conclude that jasmonate biosynthesis and endogenous increase transcripts of relevant genes may require a fully established symbiosis and interactive symbionts. It is possible that host carbohydrate transfer is a prerequisite for initiating a secondary metabolic cycle, leading to jasmonate biosynthesis.

Fungal Gene Products Overcome Host Genetic Defects: Mycorrhizal association may also be involved in correcting or regulating the effects of certain genetic defects in the host plants. AM fungal gene products may actually complement host functions and correct defects (genetic lesions) if any. For example, maize mutants defective in formation of lateral roots are also impaired with regard to sufficient levels of P absorption. It is interesting to note that Paszkowski and Boller (2002) have actually traced certain maize mutants defective in lateral root formation (Ltr1). These mutant plants exhibit decreased root hairs, a reduction in crown and brace roots, low P uptake ability and stunted growth. External P inputs overcome growth reduction and this effect can be mimicked by AM fungus *G. mosseae*. The AM symbiosis leads to induction of a new type of lateral roots that are bush-like in appearance. Clearly, AM fungi functionally complement Ltr1 mutation in maize. At this point, we must appreciate the fact that expression of several genes in host plant overcomes AM fungal defects, especially its inability for saprophytic cycle. Such a complementation actually ensures AM fungal growth and perpetuation. As a corollary, there may be several fungal genes whose products complement certain plant functions. We may even encounter situations where gene products from both the symbionts are obligately required for certain metabolic steps during symbiosis. These may be termed complementary genes obligate for symbiotic function. Complementation of gene products of symbionts may be operative in accomplishing quite a few symbiotic functions; they need to be traced and studied.

1. D. Plant Host, Rhizobium and AM Fungus: Molecular Regulation of Tripartite Symbiosis

Within the tripartite relationship, knowledge about molecular regulation of events that lead to the establishment of the two symbiotic systems on the same host, and molecular interactions among the three partners, i.e. Plant-Rhizobium-AM fungus is crucial. Essentially, molecular events that regulate nodule organogenesis and its function, as well as penetration and establishment of AM fungus in host roots and their interactions need to be thoroughly understood. Table 3.7 provides an overview of the comparable physiological events relevant to both microbial symbionts.

Table 3.7 Comparison between arbuscular mycorrhizas and legume-rhizobial symbiosis

Criteria	*Arbuscular mycorrhiza*	*Rhizobial symbiosis*
Analogous developmental stages		
Colonization of cortical tissues	Growth of infective hyphae from appresoria Formation of intercellular mycelium	Development of infection threads (inter- and intra-cellular)
Defense-like reactions in infected tissues	Synthesis of flavonoids, phenolics, peroxides, callose, modification of plant cell walls (regulated by symbionts)	
Development of subcellular symbiotic compartments	Arbuscules	Symbiosis
Differentiation of the infected plant cells	Abundant membrane structures, nucleus deformation, chromatin, decondensation, vacuole and placid reduction	
Differentiating properties		
Role of interaction for:		
Microbial Symbionts	Obligatory	Facultative
Hosts	Obligatory or facultative	Facultative
Specificity of interaction	Low	Pronounced
Mitotic reactivation of host cells	Absent	Involved
Histological differentiation of infected host tissues	Absent	pronounced
Systemic regulation of symbiosis	Not revealed	Evident
Endocytocis and formation of compartments	Not involved	Involved
Modified cell wall in subcellular compartments	Present	Absent
Connections between inter- and subcellular Compartments	Present	Absent
Stability of subcellular compartments	Low (days)	High (months)
Major plant-beneficial functions	Phosphate uptake from soil	N_2 fixation from air

Source: Provorov et al. 2002

While dealing with genetic systems controlling the development of both rhizobial and arbuscular symbiosis, it is important to recognize that a greater share of our current knowledge is based on analysis of pea mutants. Borisov et al., (2002) state that combined efforts in generating and isolating during the past two decades has yielded over 200 *Pisum* mutants, with lesions at a wide range of genetic points that are relevant to both rhizobial and arbuscular mycorrhizal partners. So far, at least 40 different genes relevant for tripartite symbioses have been identified. In case of nodulation, mutants with impairment in various developmental stages such as root hair curling (*hac*), infection thread initiation (*Iti*), infection thread growth inside the root hair (*Ith*), root tissue (*Itr*), nodule tissue (*Itn*), bacteriod differentiation (*BacD*) and nodule persistence (*Nop*) have been identified (Table 3.8). Simultaneously, many of these mutants were also blocked for arbuscular mycorrhizal symbiosis. Detailed investigations have led to identification of mutants with impairments predominantly at three different stages. They are: (a) formation of appresorium (*Myc*1); and (b) arbuscule formation (*Myc*2) and intensity of plant root colonization (*Myc*+/-) (Borisov et al., 2002; Marsh and Schultz, 2001: Parniske, 2000; Table 3.8).

Table 3.8 Pea (*Pisum sativum*) genes controlling developmental stages of root nodules and arbuscular mycorrhiza—some examples

Hair	*Infection thread Initiation*			*Bacteroid Initiation*		*Bacteroid differentiation*	
curling	*Itf*	*Ith*	*Itr*	*Itn*	*Idd*	*Bad*	*Nop*
sym8	sym7	sym2	sym5	sym33	sym40	sym31	sym13
sym9	sym14	sym36	sym34		RisFixA	sym32	sym 25
sym10	sym35						sym26
sym19	K24						sym27
sym30	KN1						FN1
	KN10						RisFixK
Myc1		Myc2			Myc+/-		

Source: Borisov et al. 2002.
Note: sym=symbiotic genes; Iti=infection thread initiation; Ith=infection thread growth in root hair; Itr=infection thread formation inside root tissue; Itn=infection thread growth inside nodule; Idd=infection droplet differentiation; Bad=bacteroid differentiation; Nop=nodule persistence; Myc-1=hyphal infection and appresorium formation; Myc-2 arbuscule formation; Myc +/-=Intensity of AM colonization.

The effect of some plant genes could be felt both on the rhizobial and fungal partner. We know that nodulation (*Nod*) factors such as lipo-chitoligosaccharides affect key developmental responses in leguminous plants. Using the nodulation defective mutants, Catoria et al., (2000) proved that *Medicago truncatula* defective for four different genes: namely *Dmi1, Dmi2, Dmi3* and *NSP* were pleiotrophic and firstly affected Nod factor responses. These genes were required for activated signal transduction, leading to root hair deformation, expression of nodulin genes and cortical proliferation. Interestingly, some of these Nod factors exhibited dual effects. Notably, at least mutations in any of the three genes; namely *Dmi1, Dmi2* and *Dmi3* also curtailed arbuscular mycorrhizal development. Hence, it was inferred that at least a few molecular steps are common to both rhizobial and mycorrhizal symbiosis. It is possible that signal pathways for the two symbioses overlap at different points in the molecular regulatory mechanisms. Parniske (2000) has suggested that some genes are common and exert similar effects on both AM fungus and nodule bacteria.

Evolution of these symbiotic regulatory genes for plant-AM symbiosis predates that for plant-nodule bacteria interaction. Therefore, at least part of the host genetic system controlling nodule development might have already originated during the co-evolution of ancient plants with AM fungi. For example, Kistner and Parniske (2002) reported an ancient receptor like-kinase involved in the transduction of signals between both the AM fungus and bacteria with host plant. They have suggested that it must be an ancient signal protein evolved originally with context to AM fungus, but one that became functional with rhizobia also during the later stages of evolution. Certainly, there are sets of ancestral genes that may have similar/overlapping effects on both the symbionts occurring on the plants (Provorov et al. 2002). However, they point out that there are also wide-ranging differences between AM fungal symbiosis and nodulation in plants (see Table 3.7). With regard to symbiotic regulatory genes, Provorov et al., (2002) state that mycorrhizal colonization induces *de novo* synthesis of proteins, which is equivalent 4 to 5% of the total proteins in the symbiotic system. Of these, some proteins are common to AM fungus and *Rhizobium*. Genes and their encoded proteins relevant to periarbuscular membranes, early nodulins (*enod2, enod11, enod12, enod40*) are common. There are *Nod* factors that can also stimulate AM development. It proves the presence of genes with effects on both AM fungal and rhizobial symbiosis. Jakobi et al., (2003) have recently stated that at least six genes control the development of both types of symbioses in pea. All of these genes, when mutated, determine a *Nod*-phenotype and two phenotypic traits in AM symbiosis. For example, a mutation in Sym8 or Sym9 that results in *Pen*- (lack of penetration) or *Myc* also leads to Nod-situation. Similarly, Lapopin et al., (1998) have reported that certain mutants defective in nodule development and functioning also show interesting variation in AM symbiosis. One of these nodule defective mutations *RisNod24* lacks the ability to form arbuscules, although other stages of AM formation proceed normally. Chaboud et al., (2002) found that early stages of infection such as appressoria formation is directly linked to expression of *MtENOD11* gene. Jackobi et al., (2003) have also pointed that certain glycoproteins are common to the functioning of the AM symbiosis and nodulation. Ruiz-lozano et al., (1999) found that at least seven defense genes were over expressed if pea roots were inoculated with *Glomus mosseae* or *Rhizobium leguminosarum* depending on the plant genotype. Importantly, early induction of similar plant defense genes in response to AM fungus and rhizobia reinforces the hypothesis that at least certain steps are common for both the types of root symbiosis. Both nodule bacteria and AM fungi are sinks to carbon sources in plant roots. Some genes related to carbohydrate transfer, for example one that encodes for sucrose synthase have been reported to aid both the symbionts in *Medicago truncatula* (Hohnjec et al. 2003). Analysis of expression pattern of the gene *MtSucS1* revealed that it is strongly induced in nitrogen fixing zones of the nodule. Equally so, its expression was traced intensely in cortical cells containing AM fungus. Its strongest expression was seen in cells possessing arbuscules. Unlike other members of sucrose synthase gene family, presence of nodule bacteria and AM fungi significantly altered its expression pattern. Hence, it was inferred that *MtSucS1* is specific to nodule and AM fungal symbiosis related generation of sink strength, and it may be aiding C transfer to both types of symbionts.

Let us consider nodulin genes associated with certain early stages of nodule organogenesis. The *enod40* gene in *Medicago truncatula* is known to induce accelerated nodulation if it over expresses. Increased primordial formation, proliferation response and enhanced root length are ascribed as effects of *enod40* over expression. In fact, root cortex of *enod40* transformed plants are generally more sensitive to nodulation signals (Charon et al. 1997; 1999). In a different study, Stachelin et al., (2001) observed that *enod40* also regulated both early nodulation events as well as mycorrhizal colonization. Plants overexpressing *enod40* exhibited stimulated AM fungal development. The overexpression of *enod40* clearly increased fungal development in the root cortex and induced higher proportion arbuscules. As a corollary, transgenic lines of *Medicago* with suppressed *enod40* transcripts did exhibit co-suppression of mycorrhizal colonization. Hence, Stachilin et al., (2001) have remarked that *enod40* might be a plant-coded regulatory gene that controls both nodulation and mycorrhizal symbiosis-related functions. Sinvany et al., (2002) reported a slightly more detailed analysis of *enod40* gene and its relevance to AM infection. They found that over expression of this gene could result in a three-fold increase in AM colonization. Application of cytokinin (benzyl purine) could induce overexpresson of *enod40* and enhance AM colonization. They found that over expression of *enod40* gene was actually controlled by a 35S constitutive promoter. The effect of *enod40* on AM colonization could also be seen in non-legumes cloned with this gene. Hence, they strongly suggested that *enod40* is relevant to both early nodulation and AM infection steps. Recently, Dey et al., (2004) have reported that *enod40*-mediated root growth response involves inhibition of ethylene biosynthesis. Ethylene supposedly inhibits *enod40* action both in legumes and non-legumes. On this same theme of overlapping effect, Starke et al., (2001) have made an interesting comment that while AM symbiosis is ancient evolutionarily and widespread across the plant kingdom, rhizobium is exclusive to leguminous species. Despite it, there is definite genetic convergence between the two symbionts whenever they coexist, particularly with reference to molecular regulation of development and functioning. They assessed the molecular basis for such a convergence by analyzing SYMRK (symbiotic receptor like kinase) gene in *Lotus* and *Pisum*. The SYMRK genes required for both rhizobial and mycorrhizal symbiosis is predicted to contain signal peptides, an extracellular domain corresponding to luecene-rich repeats, a transmembrane and intracellular protein kinase domain. At least one or more common signaling component seems to be involved in root nodule as well as AM fungal symbiosis.

Molecular interactions between the microbial symbionts may occur even without actual physical contact. In a trail using physical barrier that separates AM fungi and *Medicago truncatula*, Kasuta et al., (2003) demonstrated that a diffusible factor from germinating AM fungal spores is perceived without physical contact. The AM factor elicits expression of the nodulation gene MtNod11. Such a transgene induction occurs in the root cortex, with expression stretching from zone of root hair emergence to mature root hairs. It is interesting to note that diffusible factors from all three AM fungi used; namely, *Gi. rosea, Gi. gigantea* and *Gi. margarita* elicited a similar response by MtNod11 gene. However, exudates from pathogenic fungi such as *Phytophthora medicagenis* and *Fusarium solani* did not elicit a similar induction of *nod* gene. Obviously, molecular regulation of such a tripartite relation will be intricate and complicated. Presently, our knowledge about the finer aspects of molecular interactions of the three symbionts involved is still rudimentary.

The Ljsym Genes of Lotus: We know that plant mutants are excellent tools to understand the biochemical genetics. For much of the pioneering research on plant-

AM fungus-Rhizobium symbiosis, specific mutants from pea or medicago were utilized (Albrecht 1999; Duc et al., 1989; Marsh and Schultz, 2001; Borisov et al., 2002). However, Bonfonte et al., (2000) have remarked that within pea mutants used for AM/rhizobium studies, detailed investigations may become hampered because the genome is large sized. Instead, a model plant *Lotus japonicus* seems more amenable to genetic analysis of symbiotic genes, be it rhizobial or AM fungal association (Hagberg and Stougard, 1992). We should note that root epidermis of a legume species is an important physical checkpoint for symbiotic invasion in case of both AM fungus and rhizobium. In fact, root epidermal cells and their protruding root hairs form the first barrier. In case of lotus, *Mesorhizobium meliloti* reaches the root cortex via infection threads that invade through the root hairs. In a corresponding stage, the AM fungi gain access into roots via cortical cells, mostly by forming appresorium and separating two adjusting cells. AM fungi do enter the host directly, but follow a detour by first entering epidermal cells before piercing into the cortex. Molecular investigations too have revealed that in the phenotype of *Ljsym4* mutants, epidermal invasion is critical, and there are Ljsym4 alleles that affect epidermal cell reaction, leading to abortion of AM fungal infection. From a series of genetic analyses, Bonfonte et al., (2000) inferred that *Ljsym4* is a gene essential for both rhizobium and AM fungal symbiosis. Generally, in a symbiotic state, both fungal and rhizobial infections remain surrounded by an intracellular peri-microbial membrane. However, in an Ljsym4 mutant of Lotus, infection threads are not formed upon contact with *Mesorhizobium meliloti*, nor the peri-fungal membrane develops in response to *Gi. margarita* infection. Therefore, expression of *Ljsym4* is needed for accommodating both AM fungus and rhizobium in host cortical cells. It was argued that insufficient intracellular structural support might abort symbiosis. The root cells actually expand abnormally in response to invading hyphae. The epidermal cell death noticed was attributable to hypersensitive reaction. With regard to rhizobium in *Ljsym4*, the mutants nodulation trigger was suppressed well before infection and only root hair deformation and swelling were observed. As a consequence of arrested rhizobial growth and host cell death due to AM infection, both the symbioses are aborted. Bonfonte et al., (2000) have further reported that phenotypically also, the Ljsym4 mutants differ from those identified earlier. Firstly, the epidermal walls seem to provide signals and trigger for appresoria formation. It means that epidermal wall architecture in *Ljsym 4* mutants remain the same as in wild type. Appresoria were consistently formed because the necessary signals were available. In other words, Ljsym 4 mutants seem to carry a genetic lesion at a stage later to appresoria formation. Overall, although *M. meliloti* penetrate the tip of root hairs, and AM fungus gains access to epidermal cells, the expression of *Ljsym4* gene and its products are necessary to achieve the tripartite association. In particular, the host plant cells' accommodation program, allowing the establishment of both the symbionts needs Ljsym4 expression in lotus. The *Ljsym4* gene is a good example for genetic steps that overlap and are common to both symbionts. As already stated, there may be many genetic loci that are crucial to both the symbionts. With regard to genetic overlaps, Marsh and Schultz (2001) have constructed a hypothetical scheme, wherein there are several genes that overlap, meaning that a mutant/lesion at single locus affects the functioning of both AM fungal and rhizobial symbiont. These mutants would be phenotypically *Nod*$^-$ and *Myc*$^-$. A few examples of such genes with overlap effects are *Mtdm1,2,3*; *Ljsym 2,3,4*; *Ps sym 8,19*; *Ljmcbex*. Several Nod$^-$ *Myc*$^+$ mutants of legume have been identified, but *Nod*$^+$ *Myc*$^-$ have not been traced yet. Marsh and Schultz (2001) have also proposed that mycorrhiza-specific steps may be as abundant as nodule-specific steps. As a consequence, several parallel pathways may still operate separately, without overlapping during the establishment/functioning of each of the two symbioses.

Bastel-Corre et al., (2002) conducted a proteome analysis to study the molecular regulation of *SR* genes in the model plant *Medicago truncatula* that was challenged with both nitrogen fixer *Sinorhizobium meliloti* and AM fungus *Glomus mosseae*. Time course tests on root protein profiles indicated modifications, either up- or downregulation of several *SR* genes and a couple of newly induced polypeptides. They utilized matrix assisted laser desorption/ionization-time of flight mass spectrometry, as well as searches using previous databases to conclude that one of the polypeptides relates to *M. truncatula* hemoglobin. One other polypeptide could be easily attributed as elongation factor TU of *S. meliloti.* Among the rest, glutathione-s-transferase, a fucosidase, a myosin-like protein, a serine hydroxymethyl transferase and cytochrome-c-oxidase were also easily identifiable in mycorrhizal roots. In view of the above results, Bastel-Corre et al., (2002) have opined that proteome analysis could help in detecting molecular changes during the development and functioning of such tripartite symbiosis.

A recent review by Diouf et al., (2003) compared the steps in symbiotic establishment, structure and functioning and molecular aspects of different types of symbiosis. It concludes that there are appreciable similarities in molecular regulatory aspects of all three well-studied symbioses; namely, rhizobium, mycorrhizas and actinorhizas. Leghemglobin, for example, is expressed in all three types of symbiosis. Its regulation has similarities among the symbiosis types. It is easy to believe that both rhizobium and AM fungus do influence each other and interact biochemicaly within the plant host simultaneously while they exist in symbiosis. An interesting report by Requena et al., (1999) has suggested that even the free-living rhizobacteria may cause detectable changes in the gene expression of *G. mosseae* (Beg12). They concluded that AM fungus responds to the physiological status of host plant, but it equally receives stimuli from free-living rhizobacteria (e.g. *Bacillus subtilis* NR1) and other soil microbes and reacts with detectable changes in gene expression. More recently, Sanchez et al., (2004) found that profiles of plant gene induction by a fluorescent pseudomonad *Pseudomonas fluorescens* and a mycorrhizal fungus *Glomus mosseae* were similar. It supports the hypothesis that at least, some plant cell reactions are common to AM fungus and other beneficial organisms encountered in the rhizosphere. Obviously, there is much to learn about the specific changes in AMF gene expression, if any, in response to biotic alterations in the rhizosphere and soil environment.

2. Genetics and Molecular Regulation of Ectomycorrhiza

The existence of vast variability within and among ectomycorrhizal partners provides a good basis and potential to select and enhance performance of the symbiosis. Obviously, both classical and molecular approaches have been utilized to arrive at better symbionts, especially in terms of rapid establishment, nutrient uptake and growth stimulation (Martin et al. 1994). Regular selection procedures for plant host and ECM fungus are available. Gene transfer into fungal cells (protoplasts) has also been utilized to impart resistance against certain antibiotics and herbicides. In the present context, discussions on classical approaches involving genetic crossing, selection and breeding are confined to few examples. Genetic nature of some major

ECM fungi such as *Pisolithus*, *Laccaria*, *Cenococcum* and others has been dealt with.

2. A. Host Genetic Variation in Ectomycorrhizas

In a forest tree species, nutrient acquisition from the soil is partly attributable to the genetic variation in its overall growth rate, root proliferation rate and mycorrhizal component. Occurrence of intraspecific variations in roots and mycorrhizal component means that the benefits accrued in terms of nutrient absorption are host genotype dependent. Of course, environmental factors too affect nutrient acquisition by the symbionts. Genotypes of forest tree species vary widely with regard to mycorrhizal colonization in their roots (Dixon et al., 1987; Marx and Bryan, 1971; Rosado et al., 1994a; Tomkin et al., 1989; Tagu et al., 2001). Now, let us consider a few examples dealing with host genetic variation for ECM symbiotic ability. Tagu et al., (2001) inoculated parental clones and progeny of *Populus* species with ECM fungus *Laccaria bicolor* and evaluated mycorrhizal colonization. The data indicated clear variability in mycorrhizal colonization of F1 progeny, including many F1 progeny that possessed colonization levels greater than parental lines. This trend in mycorrhizal colonization represents a genetic basis and heterosis for mycorrhizal ability in the host species. It could be exploited to achieve a genetic gain in mycorrhizal colonization. Rosado et al., (1994b) analyzed the genetic variation in a *Pinus ellioti/Pisolithus tinctorius* association that is frequent in Canadian forests. The mycorrhizal traits that they studied, such as the number of ectomycorrhizas, percentage colonization of short roots and development of extraradical mycelium, varied significantly. All of these ECM traits were strongly inherited. Among them, percentage colonization possessed the highest heritability values (0.81), whereas, the other two traits evaluated, namely the number of ECMs and extramatrical mycelium showed inheritance values at 0.60 and 0.53, respectively. Rosado et al., (1994a) argue that genetic selection and improvement in mycorrhizal traits actually constituted a genetic gain for trees in terms of root quality and its performance (e.g. nutrient uptake) in nature. These inferences on host genetic influence on ECM colonization and net genetic gain in performance have been amply corroborated by similar studies on *Pinus/Pisolithus* association by Marx and Bryan (1971); *Eucalyptus/Pisolithus* (Tomkin et al.,1989); and in *Picea abies* (Tomkin et al., 1989). Based on their observations in the nursery with individual plant-host ECM fungus combinations and families, Rosado et al., (1994b) argue that since family heritability is higher than individual heritability, genetic gain in ECM traits of hosts could be improved better, if selections are performed on both the family and at individual level within selected families. Above all, a relatively tight host genetic control of ECM traits indicates a great potential for selecting individual genotypes of tree species with excellent ECM traits. This symbiotic trait can be used judiciously in tree breeding programs, if need be. Genetic variation for nitrogen uptake was estimated on six-week-old mycorrhizal *Picea abies* seedlings. Significant difference in nitrogen uptake and accumulation was discernable among the 30 families of *Picea abies* tested. Based on correlations between several physiological traits assessed, Mari et al., (2003) concluded that nitrogen content in the plant was a better indicator of growth response due to mycorrhization than the rate of nitrogen acquisition.

2. B. Genetic Variations in Ectomycorrhizal Fungi

Cairney (1999) points out that several thousand fungal species worldwide form ectomycorrhizal symbiosis with tree hosts. Hence, there is considerable interest in determining the genetic and physiological basis for their diversity with regard to nutrient mobilization and growth response advantages derived by the host tree species. However, only a few ECM fungal species have been examined for diversity with regard to nutrition and growth. Comparative studies of a single or few ECM fungal species may still be insufficient to judge the range of their functional capabilities. Clearly, extensive screening of well-defined fungal isolates is needed. On a single host species such as *Eucalyptus fastigata* found in Australian forests, over 100 different ECM fungal species were recognized by morphological tests. Over 109 ECM fungal species were differentiated from a single forest location using PCR-RFLP profiles of two genomic regions (nuclear rDNA ITS and mtLSU). According to Glen et al., (2001) intraspecific variation for one or both of the genomic regions was noticeable in one-third of the ECM fungal species identified.

Based on their data on the two genomic regions, Glen et al., (2001) surmized that ITS sequences are more conserved, but mtLSU seems to vary among the 23 ECM families tested. The above example relates more to phylogenetic variation. However, several other previous reports suggest that substantial inter- and intraspecific variation exist among ECM fungi. Such genetic variations pertain to a wide range of traits that determine the morphological, physiological and nutritional aspects of these fungi. Genetic variations for traits that contribute to host plant nutrition and growth stimulation have also been reported for ECM fungi such as *Amanita*, *Laccaria*, *Suillus*, *Hebeloma*, *Pisolithus, Tuber* species and several others (Bastide et al., 1995; Gay, 1993; Kendrick and Birch 1985; Marx, 1991; Sawyer et al., 2001, 2003; Sen, 1990). For example, recently, Sawyer et al., (2003) tested several genotypes of *Amanita muscaria* and found variation in their ability to utilize a range of inorganic and organic nitrogen sources. Genetic variation in ECM fungi has also been utilized to select and improve fungal strains superior in symbiotic performance (Kropp and Langlois, 1990; Marx, 1971; 1991). Details on genetic variation within several other ECM fungi, especially about their ability to colonize enhance P nutrition and growth is available in literature. Discussions are confined to only a couple of ECM fungi, considering the context and brevity of this volume.

Laccaria: Laccaria is a very common ECM symbiont found in the forests of Northern Hemisphere. It is easy to culture and study this fungus in the laboratory medium. Both monokaryotic and dikaryotic strains are capable of ECM symbiosis, but the extent of root colonization achieved varies. With both monokaryotes and dikaryotes the extent of colonization achieved, number of short roots and P translocation are genetically predetermined. Kropp et al., (1987) have identified aggressive monokaryons as well as those lacking the ability to form ECM symbiosis. In their study, ectomycorrhizal ability varied widely between no colonization to 63% of available short roots. Genetic analyses indicate that several component traits that influence the formation and functioning of ECM symbiosis are under polygenic control (Kropp, 1997). Production of indole acetic acid (auxin), nitrate reductase and glutamate dehydrogenase in monokaryotic *Laccaria, Hebeloma cylindrosporum* or *Pisolithus tinctorius* are under polygenic control, involving both additive and non-additive gene action (Debaud et al., 1995; Rosado et al., 1994b; Wagner et al., 1988). Now, let us examine the influence of genetic crossing in *Laccaria bicolor* on its ability to colonize and form symbiotic association of with *Pinus strobus*. In his study, Kropp (1997) first established the monokaryotic and dikaryotic strains of *Laccaria*. Dikaryotic lines were obtained by crossing two parental monokaryotic strains of known ECM ability, and then by isolating mycelium bearing clamp connections from the interface between the two colonies. All the monokaryons and dikaryons were tested and quantified for ECM ability on *Pinus strobus* seedlings contained in pouches. Figure 3.8 shows a simple protocol utilized by Kropp (1997) to generate mono- and dikaryotic F1 progeny. Relatively low ECM ability among F1 progeny of crosses between two monokaryons was attributed to cumulative allelic and non-allelic interactions, rather than to dominance of a single non-mycorrhizal gene. Quantitative analysis indicates additive and interactive effects (epistasis). It was concluded that mating between highly ectomycorrhizal *L. bicolor* need not generate F1 strains with high mycorrhizal ability. Instead, a sum of additive and epistatic gene actions may dictate its colonizing ability. In addition to colonization, the number of short root colonized, thickness of mantle, depth of Hartig net also varies, depending on ECM fungal genotype. Similar situation was noticed with monokaryons and dikaryons of *L. bicolor* tested on *Psuedotsuga menzesei* (Lumley et al. 1995). Dikaryotic strains colonized, formed multi-layered mantle, hartig net and established functional symbiosis with the host plant, whereas, only one monokaryotic strain formed functionally active ECM symbiosis, but most failed to infect. Instead, they showed up incompatible reactions and did not form ECM symbiosis. Yet another study by Kropp et al., (1987) suggests that loss in ECM ability of *Laccaria bicolor* is again variable and genetically determined. An isolate such as ss1 was non-mycorrhizal. Certain monokaryons even lost their ability to colonize rootlets. Dikaryotes reconstituted from compatible mating types were vigorous in ECM ability. Actually, dikaryons developed from vigorously growing competitive strains were highly mycorrhizal, but those those developed using weakly mycorrhizal monokaryons resulted in reduced ECM ability. At this point, we may note that ECM colonization is also a host genotype-dependent trait, which means that a certain amount of variation in colonization is also attributable to host plants' genetic constitution. Overall, the genetic mechanisms involved in causation of variation in ECM fungi, for example, in *Laccaria* is still not absolutely clear. In a study

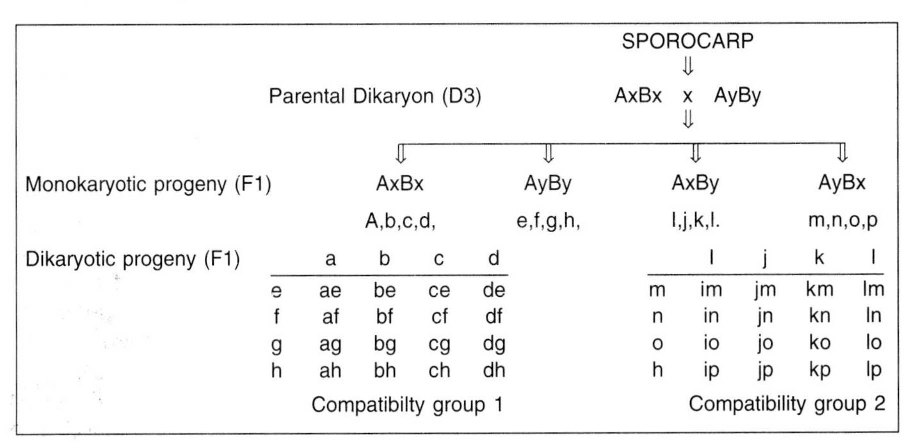

Fig. 3.8 Protocol for creating dikaryotic progeny from *Laccaria bicolor* strain D3. (*Source:* Redrawn from Kropp, 1997)

using *L. bicolor*, Bastide et al., (1995) found that in nature, more than one genotype of *Laccaria* colonizes a host to produce a genetically variable mycorrhizal mycelium. Such variation can permit competitive strains a better persistence, thus leading to changes in genetic composition in time and space. Evidences from RAPD analysis did not support the occurrence of dikaryon-monokaryon crosses; hence, the variation noted may have been due to presence of more than one dikaryon of *L. bicolor* on the root system of a single host plant. In *Laccaria bicolor*, multiallelic bifactorial mating systems are known to restrict the compatibility of sib mating. It is said that having multiple alleles may provide a high probability for crossing with unrelated mycelia. It may then promote out-crossing and genetic recombination, thus creating a genetic variation. The root colonization by monokaryons was variable, and ranged between 1.5 to 80%. Similarly, colonization by 32 dikaryotic (F1) progenies tested ranged from 30-84% of short roots. Variability was continuous, indicating that ECM colonization by *L. bicolor* is polygenically controlled. Kropp (1997) argues that the distribution of colonization rates indicated a complete dominance of genes for non-mycorrhizal character, and perhaps involves a single gene pair. Among the fungal crosses studied by Kropp and Fortin (1988), they could trace genomic interactions between mating types that affected (enhanced) colonization ability of *L. bicolor*. In certain crosses, a general reduction in percentage mycorrhizal colonization was easily discernible especially on mating poorly mycorrhizal monokaryons. Results from vigorous versus poor colonizers or vigorous × vigorous and other combinations showed that the inheritance of colonization ability is genetically complex involving major genes, polygenes and cytoplasmic factors. Hence, selection of appropriate fungal strains meant for genetic crossings/manipulation is crucial, in case higher levels of colonization and vigorous growth rates of symbionts are to be obtained.

Pisolithus: In a study aimed at identifying the genetic variability, Diez et al., (2001) analyzed polymorphism in ITS region of *Pisolithus* occurring on native and exotic Eucalyptus from Western Mediterranean region. Sequences for 17 *Pisolithus* isolates associated with native Mediterranean hosts were compared with databases in gene banks. They found that all *Pisolithus* strains associated with native Mediterranean host were clustering. The ITS analysis indicated the occurrence of several ecological species adopted to exploit different soil types, particular indigenous hosts, temperature regimes, etc. Interestingly, isolates derived from eucalypt plantation in Brazil, Kenya, grouped with Australian isolates. The wide host range of some of the exotic strains of *Pisolithus* seems to prevent inbreeding within the local species when they occur together. It also helps to overcome competition. A similar study by Gomes (2000) involved the assessment of genetic variability among the isolates of *Pisolithus* collected from different continents. Mitochondrial DNA analyses lead to identification of two clusters of isolates, which coincided with host specificity and geographic affiliation of samples.

The genetic nature and potential of particular ECM fungal isolate/strain utilized in the nursery for inoculation is important, because it can influence the symbiotic effects. There are indeed several genetically predetermined fungal traits that influence the efficiency of the establishment of the fungal symbiosis. In this regard, Rosado et al., (1994b) made an attempt to investigate the genetic variation available in *Pisolithus tinctorius* strains. They also aimed at assessing the heritability of different traits among the fungal crosses. They studied 16 full sib progenies of *P. tinctorius*. Both, inter- and intra-strain variations in the ability of the ECM fungus to colonize *Pinus ellioti* have been recorded (Lamahamadi and Fortin, 1991). The total genetic variation in colonization ability was attributable more to additive than to non-additive interactions. Rosado et al., (1994b) have also reported that that the general combining ability (GCA) for specific monokaryotic parent fungal lines which could be mated to yield strains able to colonize tree seedlings better than the average levels, whereas variation in traits such as mycelial growth rate or number of ectomycorrhizas was believed to be due non-additive genetic interaction. At this point, we may note that in basidiomycetes, superior performance of F1 offspring could be due to either non-allelic or allelic interactions such as dominance, epistasis, etc. Further, based on their analyses of dikaryons of *P. tinctorius*, Rosado et al., (1994b) concluded that fungal traits such as colonization, mycelial growth and ectomycorrhiza number could be improved by devizing appropriate fungal crossing/selection programs.

Cenococcum: Asexuality is a common feature among certain Ascomycetes. Morphologically, Ascomycetes lack the sexual reproductive ability and propagate clonally. Such clonally propagated organisms are sometimes thought to represent an evolutionary dead end, with least possibilities for genetic recombination and improvement. According to Lobuglio and Taylor (2002), their recent population genetic studies prove that recombination does occur within the life cycles of several putative asexual taxa of Ascomycetes, for example in *Cenococcum geophilum*. In nature, *C. geophilum* is a clonally propagated fungus. It lacks meispores as well as mitospores, and the propagules it produces are aggregations of mycelia called sclerotia. Lobuglio and Taylor (2002) made repeated observations on different populations of *C. geophilum* and concluded that clonal reproduction plays a very significant role in its perpetuation. However, genetic analysis of multilocus genotypes clearly indicates constant genetic recombination events within the populations. They suspect that a cryptic sexual state occurs during the life cycle of *C. geophilum*. Mitotic and para-sexual recombinations are other possibilities. Their phylogenetic tests did not yield a close sexual relative for *C. geophilum*. However, genetic differentiation and moderate gene flow were recorded on the two specific populations analyzed. Genetic variations available for traits such as colonization, P transport and growth responses in *Cenococcum* could also be identified and utilized appropriately in order to enhance the performance of inoculant strains.

2. C. Molecular Regulation of Ectomycorrhizal Symbiosis

A large number of ECM fungi have been subjected to molecular analysis, both in symbiotic and asymbiotic state. A vast body of useful information has been accrued and utilized by researchers and foresters. Let us consider some of the major advances with regard to molecular regulation of ECM symbiosis. A molecular analysis of *Eucalyptus/Pisolithus* model was conducted by Martin (2001) and Gibson (2002). They randomly selected and sequenced 650 cDNAs from the cDNA library of 4-day-old ECMs. Some 200 cDNAs were assessed by the suppression subtraction hybridization technique to pick cDNAs preferentially expressed in ECM symbiosis. Their analysis resulted in identification of 634 non-redundant expressed sequence tags (EST), which correspond to different genes. The gene expression profiles of these ESTs were investigated. Such profiling fetched surprisingly few genes with significant upregulation (induction) during symbiosis. Among the genes studied, only one of the cDNAs examined exhibited a 10-fold induction during early symbiosis. Twelve other genes were induced 5- to 10-fold over control. Among the genes induced, six were specific to the plant host, 6 others to ECM fungus *Pisolithus*, and 3 were fungal hydrophobin genes. There were several weakly up- or downregulated genes identified. There were also some strongly downregulated genes. According to Weimken and Boller (2002), despite such wide-ranging analysis of cDNAs, no gene qualified to be termed as *MYC* or *mycorrhizin* gene in the strict sense. In fact, these researchers remark that genes induced by 3 to 10 folds over individual controls have been traced, but none can be effectively utilized as positive or negative molecular markers for ECM symbiosis. For example even a 10-fold raise in hydrophobin gene expression in *P. tinctorius* is not specific to ECM symbiosis. There is no strict correlation, or preferential accumulation of hydrophobin in a tissue at any specific stage of symbiosis. Therefore, in order to obtain greater insights into regulation of gene expression in time and space, a time course molecular analysis of *Ecualyptus/Pisolithus* symbiosis was conducted. The sampling lasted from 0 to 28 days after inoculation with the fungus. Despite these analyses, genes exclusive to ECM symbiosis were not traced, even if they ever existed in *Eucalyptus-Pisolithus* model. They explained that firstly this model—although amenable to molecular analysis—might not be perfect enough to show up subtle changes in gene expression. Second, genes relevant to regulation of symbiosis and signal transduction could be expressed feebly at low levels. Hence, they were undetected or underrepresented in EST library selected. Thirdly, such molecular analysis includes a lot of host tissue that is non-mycorrhizal in content (e.g. central root cylinders), where the fungal tissue is not directly interactive with the host. These situations may conceal fine regulation of genes, which are specially induced/suppressed during symbiosis. Hence, such genes will be left undetected. To summarize, there seems to be a definite pay-off in discovering 'Myc or Mycorrhizin' genes, if any, and understanding its regulation. Factors that accentuate or suppress *MYC* or *Mycorrhizin* genes preferentially can then be studied in detail at the molecular level.

The Symbiosis Regulated (*SR)* genes of Ectomycorrhizas

As stated earlier for AM symbiosis, "Symbiotic Regulatory (*SR*)" genes relevant to EC symbiosis too may encompass aspects such as:

(a) Establishment of symbiosis (e.g. signal transduction, infection and hartig net formation),
(b) Cell structure and cytoskeletal network (e.g. Tubulin genes),
(c) Symbiotic functions (housekeeping proteins, nutrient transport genes such as sugar transporters, P transporters, etc.),
(d) Senescence (cellulotlytic factors), and
(e) Complementary genes (*SR* genes that complement host defects and vice versa).

Firstly, the development and functioning of ectomycorrhiza involves significant metabolic and morphological changes in roots and fungal hyphae. In nature, many of the molecular changes perhaps are common across wide taxa of ECM fungi and several tree species (Martin and Hilbert, 1991; Martin et al.,1997; Voiblet et al., 2001; Wiemken and Boller, 2002). We also know that ECM development alters gene expression in both the symbionts. Voiblet et al., (2001) have suggested that identification of a large number of such genes, expressed either exclusively or predominantly during symbiosis, will be useful in understanding the molecular regulation of ECM symbiosis better. Most importantly, we have to arrive at the basic/minimal set of genes whose expression is up- or downregulated in an ECM-plant symbiosis, if it happens to be a commonality across as many symbiotic pairs. They analyzed cDNA library from *Eucalyptus globulus–Pisolithus tinctorius* association and sequenced 850 cDNAs cloned randomly or those obtained through suppression subtractive hybridization (SSH). Their results indicated that nearly 43% of the ectomycorrhiza ESTs actually code for novel genes. Overall, the largest group of ESTs corresponded to genes encoding ribosomal proteins, translational regulatory proteins, elongation factor ubiquitine proteosome, etc. Cell wall proteins such as hydrophobins and symbiosis regulatory acidic polypeptide (*SR*AP 32) found on fungal cell wall were also abundantly detected (10%). In a symbiotic organ, say, hartig net or mantle where nutrient exchange occurs, significant proportion of genes (13%) coded for enzymes of primary and secondary metabolism (e.g. ATP synthase, glyderaldehyde-3-PO4 dehydrogenase, alanine aminotransferase). Transcripts involved in cell signaling and communication were also abundant (13%) in mycorrhizal root tissue. Next, Voiblet et al., (2001) also studied the profiles of highly expressed genes in the roots of Eucalyptus, free-living *Pisolithus* mycelium, and in symbiotic tissues. They utilized cDNA mini arrays assembled on selected 486 ESTs from ECM library. Most highly expressed genes belonged to housekeeping to structural genes, such as cell wall formation, carbon metabolism, ribosomal function, translation and some stress inducible protein genes. It is vital to identify the differentially expressed genes. The extent of up- or downregulation of these genes can be crucial to the functioning of symbiosis in many of its stages of development. Even before identification, we need to define mycorrhiza-induced or regulated (*SR*) genes, which are sometimes termed as 'differentially expressed' genes. In this regard, Voiblet et al., (2001) have regarded genes as mycorrhiza regulated (i.e. *SR* genes) only if the quantitative ratio of gene expression in mycorrhiza to that in free-living partner was >2.5. They have also stated that for *SR* genes such as hydrophobins, *SR*AP32 or certain others an estimated 2.5 times difference in signal intensity between mycorrhiza and free living depicts really a significant difference in transcript concentration. Using this criterion, they could classify nearly 45 genes as upregulated and 21 as downregulated of the 66 analyzed (Fig. 3.6; Table 3.9). In their opinion, the number of differentially expressed *SR* genes identified in their study is a conservative estimate, mainly because the morphogenesis and physiological activity of symbiotic partners involves intricate molecular regulation requiring activity of a very large number of genes, which are either up- or downregulated. Boffin et al., (1998) identified that NADP-dependent isocitrate dehydrogenase (NADP-ICDH) activity is induced during colonization of *Eucalyptus globulus* roots by *Pisolithus tinctorius*. The amino acid sequence of polypeptides encoded by the gene had similarity to plant NADP-ICDH (Eglcdh). Possibly, it is a plant cytosolic isoform induced due to ECM formation. RNA analysis showed that Eglcdh activity in roots was induced over 2 folds as a consequence of ECM fungal contact with roots. Induction and accumulation of Eglcdh was localized in the epidermis and ectomycorrhizal lateral roots. It is a good example for relatively mildly induced *SR* genes.

To study such symbiosis-regulated genes, Tagu et al., (1993) and Tagu and Martin (1995) devized differential screening techniques and an expressed sequence tags (EST) database using cDNA library. Using such methods, they estimated that nearly 50% of the fungal mRNA of ECM is regulated (up or down) by the formation of *Eucalyptus-Pisolithus* symbiosis. In yet another study, Kim et al., (1999b) reported an

Table 3.9 Differential expression (induction/suppression) of cDNAs in Ectomycorrhizas in comparison to free-living fungus and uninfected plant root. Examples of up- and downregulated genes in Ectomycorrhizas

EST clone ID	*Organism*	*Ratio Myc/Root*	*Database match (species)*
UPREGULATED GENES			
9C9	Fungus	12.3	Initiation factor eIF4A (*Schizosaccharomyces pombe*)
7C2	Fungus	7.8	Hydrophobin HYDPt-8 (*Pisolithus tinctorius*)
7E10	Fungus	7.5	No match
EgPtdB23	Fungus	2.6	No match
8D10	Fungus	2.5	Shp 1 protein phosphatase (*S. pombe*)
EgPtdB12	Fungus	2.5	*SR*AP-32-3 (*P. tinctorius*)
8E9	Plant	6.8	O-methyletransferase (*Prunus dulcis*)
6E3	Plant	6.0	Ubiquitine conjugating enzyme
3C3	Plant	2.7	Hypothetical protein T621.210 (*Arabidopsis thaliana*)
3C7	Plant	2.5	No match
DOWNREGULATED GENES			
EgPtd16	Fungus	–2.8	Hypothetical 30.9 kDA protein (*S. pombe*)
8A6	Fungus	–2.9	No match
7C5	Fungus	–3.6	mRNA maturase
7A4	Fungus	–12.1	Sm-like protein (*S. pombe*)
12D8	Fungus	–9.0	Ring box protein (*Homo sapiens)*
5C9	Fungus	–24.1	Rah1/Rad51 (*Coprinus cinereous*)
EgPtdB13	Plant	–2.7	Metalothionein (*Casuarina glauca*)
St144	Plant	–3.8	Absicic acid induced protein (*Ipomea batatis*)
12D4	Plant	–16.3	Zn-binding protein (*Arabidopsis thaliana*)

Source: Voiblet et al. 2001.
Note: Ratios are for normalized hybridization values of transcripts expressed in the symbiotic tissue and in the free-living partners. For greater details on procedures of identification and more examples on up- and downregulated genes, see above source.

in vitro system that allows the monitoring of changes in the expression of genes that code for proteins synthesized during the establishment of ECM symbiosis. Their system supports a conducive interaction between tree saplings and ECM fungus allows sampling at early stages during development of symbiosis. It is believed that sampling and examination at 0 to 24 h immediately after fungal infection is crucial to understand the subtle changes in gene expression. They identified genes that were expressed, not only during symbiotic initiation but even those throughout the different stages of symbiosis. It was inferred that cDNA clones represent genes induced and required both for initiation and sustenance of *Laccaria bicolor–Pinus rosea* symbiosis. Their investigation also revealed that certain genes were suppressed as the symbiosis progressed. In case of *L. bicolor*, Sundaram et al., (2001) reported isolation and characterization of a symbiosis-regulated gene *Lbras*, which belongs to *ras* family of genes known to be associated with signaling pathways controlling cell growth and proliferation. One such clone identified as RhythmA was homologus to Ap180-like vesicular protein. Immunolocalization revealed that it is present in the fungal cells around Hartig net (Sundaram et al., 2004). Similarly, Kim et al., (1998) have also reported the isolation and characterization of symbiosis-regulated gene from *Laccaria bicolor/Pinus resinosa* interaction. Homology comparison revealed that it might encode a protein involved in signal transduction, especially active during the formation and maintenance stages of ECM symbiosis.

Root colonization definitely leads to certain changes in gene expression. Beginning with fungal chemotaxis towards plant roots, penetration and establishment of symbiosis, it causes metabolic and structural alterations. Balasubramanian et al., (2002) argue that pre-infection stage of ECM symbiosis is crucial as molecular changes/ regulation occurring during this period sets in motion the establishment of later stages of symbiosis. They utilized an in vitro system to identify one of the symbiosis-regulated genes active during preinfection stage of ECM symbiosis between *Pinus resinosa* and *Laccaria bicolor*, namely malate synthase (*LbMS*). Malate synthase is involved in tricarboxylic acid pathway bypass, generating 4-C compounds for biosynthesis. They report that *LbMS* is regulated through interaction between the fungus and host, mainly through the presence of glucose and other carbon compounds. The upregulation of malate synthase is suspected to help the symbionts in utilizing 2C compounds derived from catabolic processes better, especially during preinfection and early stages of hyphal spread.

Now, let us consider a few specific *SR* genes relevant to the development of ECM symbiosis. Vioblet et al., (2001) report that symbiosis does not induce any mycorrhiza-specific gene. This is in consonance with the views expressed by Wiemken and Boller (2002), wherein they state that no gene identified by them is qualified to be denoted as *MYC* or *Mycorrhizin* gene. Instead, only a marked change occurs in the gene expression levels in symbiotic partners. A report by Duplessis and Martin (2002) suggests that for 11 to 23% of the genes analyzed at different stages of mycorrhiza formation between *Eucalyptus globulus* and *Pisolithus tinctorius*, the differential expression was greater than 2 folds. However, no mycorrhiza specific gene could be detected. Most of the upregulated fungal genes were related to cell wall, membrane proteins, and communication genes. In the host, a few downregulated genes were related to water transport. One of the highest increases in expression levels was traced for the gene elf4 from the mycosymbiont. The elf4 is a subunit of high molecular weight and is required as a single polypeptide chain for mRNA binding to ribosome. The upregulation of elf4 indicates that gene expression is regulated at translation initiation, ribosome and protein synthesis in *Eucalyptus-Pisolithus* symbiosis. Among other *SR* genes, several of the multigene *SR*AP32 and hydrophobins are strikingly upregulated during ECM symbiosis. The cell wall related clones of two different members of the *SR*AP family (type 1 and 2) were affected. Another set of fungal specific proteins whose genes are induced belong to hydrophobins, namely HydPt-1, HydPt-2, HydPt-3 and HydPt-8 (Voiblet et al. 2001). Induction of hydrophobin genes is known to help the development of hyphae in air pockets of the soil by repelling moisture and keeping them dry and non-wettable. There are also reports that *P. tinctorius* hyphae preferentially express a set of genes relevant to cell wall formation during symbiosis. This is known to aid fungal interaction with plant cell wall at the host-fungus interface ie apoplastic interface (Hardham and Mitchel, 1998; Martin et al.1999). The *SR* genes relevant to changes in cell structural aspects are mainly the proteins involved in membrane synthesis and function. Obviously, symbiosis induces certain changes in the membrane functions. In addition, there are several other *SR* genes involved in signal transduction and are upregulated. They form a sizeable fraction of ECM induced genes. They possess homology with GTPases, protein kinases, calmodulin, etc. Genes coding for elicitor-inducible isoflavine 7-O-methyletransferase and stress-inducible proteins (LT16B, SEND32) are confirmed *SR* genes with a transient increase in expression during plant stress/defense reactions in Eucalyptus roots (Wiemken and Boller, 2002). Kim et al., (1999 a, b) have identified LB-AUT7, a symbiosis-regulated gene expressed differentially 6 h after interaction between *Laccaria bicolor* and *Pinus resinosa*. The LB-Aut7p functionally complements its *Saccharomyces cerevisiae* homolog, which is involved in the attachment with autophagosomes to microtubules. Probably, this gene induces an autophagocy-tosis-like vesicular transport process during ectomycorrhizal interaction. Charvet-Candela et al., (2002) have reported that in *Pinus pinaster*, *Ppiaa88* is an auxin-induced upregulated gene involved in early stages of mycorrhization. If an auxin-overproducing strain of ECM fungus *Hebeloma cylindrosporum* is inoculated, it causes three-fold increase in the transcript levels of *Ppiaa88*. The kinetics of *Ppiaa88* transcript accumulation was closely linked to symbiosis establishment.

There are several *SR* genes that are downregulated as a consequence of ECM symbiosis. Some of the plant-coded genes such as metallothenin, ATP synthase, cytochrome oxidase, and zinc-binding protein are downregulated in the eucalyptus roots with ECM fungal component. A few downregulated genes were related to water transport. Vioblet et al., (2001) have opined that such large-scale cDNA profiling and identification of *SR* genes, which are either up- or downregulated, will provide a basis for precise molecular dissection of complex genetic regulation involved during plant ECM symbiosis. In future, we may, therefore, expect greater advance with regard to *SR* genes and their fine regulation at molecular level. In addition to the *SR* genes stated above, there are several other genes which are vital for maintenance of ECM symbiosis. Among them, genes involved in secondary metabolism, hydrophobin formation, tubulins, laccases, lignolytic enzymes have been investigated in greater detail. Let us consider a few examples.

Molecular Regulation of Short Roots: Short root development and hartig net formation is an important morphogenetic event during ECM symbiosis. Such short root morphogenesis, no doubt, involves regulation of several structural genes and cell function related genes. However, we are yet to decipher the details about the regulatory events crucial to short root formation. Investigations with *Pinus sylvestrus/ Suillus bovinus* associations has revealed that at least five proteins were upregulated and two were downregulated (Tarkka, 2001). The acidic short root-specific polypeptides traced were identified as class III secreatory peroxidases (PsyP1). The expression of PsyP1 was highest in short roots, which may have a role in reduction of short root elongation. With regard to structural and cytoskeletal proteins, Tarkka (2001) reported a change in alpha-tubulin proteins relevant to ECM formation. It is possible that novel alpha-tubulin (isotypes) guide the structural rearrangements of cytoskeleton required during hartig net formation. Tarkka (2001) also suggested that small GTPases, Cdc42 and Rac traced in *Suillus* probably regulate fungal morphogeneis by reorganizing actin cytoskeleton at the time of ECM formation.

Phenylalanine Ammonia Lyase (PAL) Genes: Phenylalanine ammonia lyase is a key enzyme of the secondary metabolism in most organisms, including ECM fungi. The PAL activity is required for both catabolic degradation and synthesis of phenyla lanine. The ECM fungus *Amanita muscaria* is not known to utilize phenylalanine as a nutrient source. Hence, in this case, understanding the regulation of *PAL* gene with reference to catabolic degradation of phenylalanine is of lesser significance. On the other hand, understanding the regulation of *PAL* gene has a direct bearing on synthesis of phenylalanine. Consequently, formation of several other secondary metabolites within the fungus could be influenced. Incidentally, *A. muscaria* fruiting bodies are known to accumulate high levels of glucose and phenolic compounds. As expected, these portions also exhibit high levels of PAL gene expression. In addition, hyphae of the hartig net and inner mantle that are exposed to high levels of monosaccharides may support enhanced levels of *AmPAL* expression. Nehls et al., (1999) studied this *PAL* gene of *A muscaria* in greater detail. Firstly, they identified the PAL gene using differential screening of cDNA library obtained from *Picea abies/ A. muscaria* association. A clone that codes for *PAL* was identified. It is 2311bp in length, codes for a protein that consists 740 amino acids, and its total molecular mass is 80167 Da. The PAL enzyme encoded by *AmPA*L gene is homologous with those derived from *Rhodotorula rubra* (Filpula et al., 1988) and *Rhodosporidium truncatula* (Anson et al. 1987). The *AmPAL* is said to be encoded by a single copy gene in the genome of *A. muscaria*.

We should note that sugar and nitrogen are two key nutrient components involved in ECM symbiosis that influence secondary metabolic activity. For example, expression of *PAL* gene in *Rhodotorula rubra* and *Rhodospora truncatula* is known to be repressed by sugar, but induced by the presence of phenylalanine. In the case of PAL gene from *Neurospora crassa*, phenylalanine induced its expression, but nitrogen input repressed it. A thorough examination of *AmPAL* gene expression patterns revealed that low monosaccharide concentration upto 2 mM strongly induces its expression, but high levels can repress its expression by 30 folds (Nehls et al. 1999). In their study, the *AmPAL* gene could have been regulated by glucose as such, or by compounds from its metabolism. Tests with glucose or its analogues (e.g. 3-O-methyle glucose) that are not processed further, induced *AmPAL* expression. However, exposure to 2-deoxyglucose—which is phosphorylated by hexokinase—resulted in 30-fold decrease in AmPAL expression. Overall, it is suspected that sugar regulates *AmPAL* expression in most ectomycorrhizal basidiomycetes, as in other saprophytes or ascomycetes may involve two independent mechanisms. Next, to assess the effect of nitrogen on *AmPAL* expression, Nehls et al., (1999) utilized nitrogen and glucose depleted fungal mycelia. In the absence of glucose, addition of aminoacids such as phenylalanine that is not utilized as nutrient source, enhanced *AmPAL* expression. Clearly, sugar and nitrogen can regulate genes relevant for secondary metabolism in an ECM association

A study slightly different from the one described above involved the differential screening of cDNA libraries of *A. muscaria*. It revealed a large number of upregulated and downregulated clones. Among them, Buscot et al., (2000) traced a clone that represents phenylalanine ammonia lyase (*AmPAL*). As such, abundant levels of *AmPAL* transcripts could be detected in hyphae grown at low external glucose in the ambient medium, but transcript levels were enhanced 30 folds in hyphal cultures provided with <2 mM glucose. They opine that regulation of *AmPAL* gene may be hexose kinase dependent or controlled via steps in glycolysis. Altered expression of *AmPAL* could also be effected by ammonium derived from amino acid catabolism. *AmPAL*, like PAL enzyme systems in plants and other ECM fungi, are related to defence function against pathogen. The expression of AmPAL is guessed to follow a gradient, depending on hexose availability in roots/hartig net. High levels of phenolics occur within ECM fungal cells exposed to soil (hyphal mantle), but comparatively low phenolics occur in the hartig net, where sugars are abundant and protective activity is required.

Hydrophobin Genes: The discussions above suggest that a wide array of differentially expressed genes occur in ECMs. Many of these genes are actually involved in initial signaling and maintenance of symbiosis. Therefore, accurate identification and characterization of these differentially expressed genes is required. We know that genes specific to mycorrhizal functioning may encode proteins which are involved in different aspects of the symbiosis such as signaling, nutrient transfer, or cell structure modifications. Hydrophobins are one such class of ECM fungal proteins, involved in cell surface interactions, cell adhesion and aggressive processes (Duplessis et al., 2001; Kershaw and Talbot, 1998; Tagu et al., 2001; Wessels, 1997). Hydrophobins are produced by Zygomycetes, Ascomyctes and Basidomycetes (Tagu et al., 2001; Wessels, 1997). Functionally, hydrophobins are vital for aerial hyphal development. Hydrophobins were first reported in *Schyzophylum commune* as protein excreted from hyphal tips. Aerial hyphae accumulate high levels of hydrophobins on their surface, and this protein layer is said to lower the surface tension, so that aerial hyphae can develop. Deletion of *sc3* gene that encodes for hydrophobin suppresses formation of aerial hyphae. Therefore, such fungal colonies lack aerial hyphae. Within the fungal life cycle, hydrophobins have been attributed a role in cell wall integrity, in covering spores and making them water repellant, and as elicitors of plant defense. Within ECMs in particular, hydrophobins have been detected in hyphal mantle. In the hartig net, every fungal cell was covered with hydrophobin (Wosten and Wessels, 1997). These hydrophobins, however, do not cover the entire hyphae; pores left in the layers allow oligosugars/oligopeptide diffusion (Mankel et al., 2002). Hydrophobin layer is known to help hyphal growth from a hydrophilic medium into air. Physically, a hydrophobin layer is a sheet of rod lets 10 nm in diameter and 100 nm in length that surrounds the hyphae. Mankel et al., (2002) studied hydrophobins in a very interesting pair of symbionts; namely *Pinus sylvestrus-Tricholoma terraeum*. In nature, trees may form symbiosis with a few different fungi. It is common to record 3 or 4 species on roots of single host. However, *Tricholoma terraeum* is known to be specific to *P. sylvestrus*. Most importantly, mycorrhization is required to produce mushrooms. Hydrophobins were suspected to have role in maintaining this specificity. In their study, Mankel et al., (2002) found that hydorphobins are produced in both aerial hyphae and hartig net of *Tricholoma terraeum* when in association with *Pinus sylvestrus*, but the hydrophobins were absent or feeble when this mycosymbiont was associated with incompatible plant host such as *Picea abies*. After analyzing the differential protein pattern and immuno-fluorascence in compatible and incompatible symbiont pairs, they concluded that hydrophobins on the ECM fungus might be involved in host recognition and specificity in ECM symbiosis. In a different study involving *Pisolithus-Eucalyptus* association, Duplessis et al., (2001) identified a new hydrophobin cDNA (*HYDPt-3*) in the symbiotic mycelium of the fungus. The sequence of this cDNA was highly divergent from the previously identified hydrophobin genes, such as the symbiosis regulated *HYDPt-1* or *HYDPt-2*. Expression analysis revealed that *HYDPt-3* is upregulated during symbiosis, and in contrast with pathogenic fungi changes in ammonium and glucose concentrations in the medium did not affect *HYDPT-3* expression. Hence, they believe that regulation of expression of *HYDPt-3* may be through other factors.

SRAP Genes: We know that establishment of ECM symbiosis, especially formation of mantle and hartig net requires aggregation of fungal hyphae. Obviously, cell surface interactions involving cell wall proteins are to be expected during these processes. In a trial aimed at understanding changes in cell wall proteins through different stages of *Eucalyptus-Pisolithus* symbiosis, Laurent et al., (1999) found enhanced synthesis (upregulation) of certain 31 and 32 kDa polypeptides. These were immunologically related to the fungus and were termed 'symbiosis-related acidic polypeptides' (*SR*AP). The transcripts for *SR*APs were upregulated, and the *PtSRAP-32* cDNAs actually encoded alanine rich proteins in the fungus. Further, these *SR*APs contained Arginine-Glycine-Aspartic acid, which are the portions involved in cell-adhesion. Immuno-electron microscopy of *SRAP,* have shown us that such polypeptides are frequent on the hyphal surface as well as on host root cell wall, provided the fungus *P. tinctorius* is already in a symbiotic state. The Pt*SR*AP gene transcripts increase during the formation of mantle. These observations seem to reinforce the idea that *SR*APs play a role in cell-cell adhesion and surface interactions during the establishment of ECM symbiosis. The gene transcripts for *SRAP32* possess sequences similar to those of integrins. Integrins are heterodimeric transmembrane receptor molecules.

Tubulin Genes: Morphological changes occur in the roots of host plant as a consequence of ECM symbiosis. This requires commensurate changes in the expression of genes encoding structural and cytoskeletal proteins. Diaz et al., (1996) tested this hypothesis by cloning and characterizing the alpha-tubulin cDNA (*EgTub A1*) from *Eucalyptus globulus.* They observed that induction of tubulin-gene was related to aggressiveness of fungal isolate used. For example, an aggressive isolate of *P. tinctorius 441* drastically upregulated the tubulin gene, but a less aggressive isolate *P. tinctorius 270* affected the tubulin gene expression less significantly. It was concluded that such changes in tubulin and cytoskeletal proteins was mostly related to cell structural modifications that should occur in host root cell in order to accommodate the fungus during symbiosis.

EgHyper genes: As stated earlier, identification and ascertaining the specific roles of *SR* genes is vital to unravel the molecular mechanisms involved in the maintenance of ECM symbiosis. These regulatory genes may affect either the primary or the secondary metabolic steps, or both. In this context, Nehls et al., (1998) have made an attempt to understand the relevance of auxin-specific secondary metabolites on *SR* genes. Firstly, they isolated *SR* gene 'EgHyper' from *Eucalyptus globulus-Pisolithus tinctorius* association. Its expression is largely confined to host root tissue. Its sequences coded for a protein (25.5 kDa) having homology with auxin-induced glutathione transferase. This EgHyper' was upregulated whenever *P. tinctorius* cell-free extracts were treated with indole3-acetic acid or 2,4-dichlorophenoxy acetic acid or tryptophan betain. Hence, they suspected that upregulation of EgHyper gene may be a reaction to morphological changes that occur in roots as the symbiosis proceeds.

Genes for Lignolysis: Genes for lignolytic activity are widespread among ECM fungi; perhaps they include all taxonomic classes of the fungal symbiont. Chen et al., (2001) screened a large number of ECM fungal species for lignin peroxidase (*LiP*) and manganese peroxidase (*MnP*) genes, using PCR and specific primers for known isozymes. Genes for *Lip* were detected in all the ECM fungi belonging to orders Agricales, Aphyllophorales, Boletales, Cantherallales, Hymenochaetales, Sclerodermatales, Stereales and Thelophorales. The *Mnp* was detected in *Cortinarius* sp. and three ECM Stereales taxa. The presence of lignolytic genes supports the saprophytic phase of the ECM fungi. It may also facilitate the access to nutrients associated with dead plant debris in the soil. For example, lignolysis can potentially increase C and N source to fungus and plant host. In other words, strict demarcation of a symbiotic ECM and its activity related to soil organic matter decomposition is difficult, even at the genetic level. Each type of activity of these fungi may be supplemental in nature, depending on the state of symbiosis and soil environment.

Mannoproteins: Analysis of protein extracts from *Pisolithus tinctorius*, using 2D polyacrylamide gel electrophoresis has shown the occurrence of several mannoproteins, which are accentuated in response to symbiosis. Tagu and Martin (1995) state that such mannoproteins are constituted by several isoforms and are covalently linked polymers located in the external layers of the cell wall. The synthesis of these isoforms is significantly downregulated during the early stages of ECM symbiosis. It is believed that the disappearance of protein moiety along with associated glycans is the first sign during recognition/interaction between the fungal symbiont and host. Its role is in avoidance of host defense, via suppression of surface glycans. A decrease in mannoproteins could also affect the nutrient transfer between symbionts.

Laccase Genes: Polyphenol oxidases are produced by a wide range of ECM and ericoid fungi. These enzymes include laccase, catechol oxidase and tyrosinase. Their substrate affinities may show considerable overlapping. Burke and Cairney (2002) have reviewed the aspect of stimulation of laccase and polyphenol oxidases by ecto- and ericoid mycorrhizas. Several of the ectomycorrhizal fungi belonging to Russulales screened by Chen et al., (2003) also showed laccase-like genes. Multiple genes for laccase were identified in *Piloderma* sp. The transcript levels of laccase-like genes were induced in the presence of high nitrogen in the medium, but were reduced if the nitrogen content was low. The induction of laccase-like genes noticed in ECM could have a potential role in nutrient mobilization and differentiation of hyphae.

Concluding Remarks

Thus far, search for genetic variation in mycorrhizal traits has provided significant results. Evidences gathered cover a large number of field and plantations crop species, in addition to several others. Similarly, genetic variation for symbiotic traits has been reported both in ECM and AM fungal species. Genetic analysis indicates that mycorrhizal traits are heritable. Additive gene action and heterosis seem to control the expression of these genes in the progeny. Therefore, wherever feasible, mycorrhizal traits could be improved using classical methods of selection and crossing, be it the fungus or the host.

Host mutants with lesions at various steps of infection, establishment and performance of symbiotic relations are available. Knowledge about genetic control of infection process in AM symbiosis is largely due to the availability of such *myc*$^-$ mutants. We may expect to discover several more *myc*$^-$ mutants with lesions at points that correspond to different stages of infection. It provides a better resolution about genetic control of infection process. Studies on biochemical and physiological consequences of these mutations to cellular functions need greater emphasis. We should also strive to find applications for these myc$^-$ mutants in practical agriculture and forestry. For example, if the focus is to improve the inherent ability of only plants towards better P uptake efficiency, then the influence of mycorrhiza on P nutrition and C drain can be effectively removed using myc$^-$ mutants. On the other hand, if correction of a particular mutation restores and improves P translocation drastically, then we should avoid occurrence of such mutants in the breeding stocks and elite varieties.

Recent progress in identifying, sequencing and characterizing genes relevant to AM and ECM symbiosis has been spectacular. The 'Symbiotic Regulatory gene' is a very useful concept. First, it helps us in focusing our investigation to only those genes whose expression is significantly altered due to symbiosis. Thus far, we have known that *SR* genes play a crucial role in establishment, maintenance and senescence of AM/ECM symbiosis. No doubt, understanding the molecular regulatory aspects of symbiosis in time and space may need more systematic and elaborate effort. Considering the large number of host-mycorrhizal fungal combinations that are being investigated and their inherent variation, it is time that we concentrated on a few selected specific symbiotic partners. Even then, by analyzing a large number of *SR* genes, we are actually attempting to decipher a complicated system at molecular level. The situation gets proportionately more complicated if a tripartite relationship of plant, mycorrhizal fungus and nodule bacteria or pathogen is involved. A single aspect such as colonization itself may involve a wide range of *SR* genes whose expression fluctuates due to a variety of factors related to host, fungus and environment. Such investigations invariably require a great deal of patience and perseverance. It might be worthwhile to systematically identify and classify the *SR* genes with regard to their origin (host or fungus), stage of symbiosis affected, cellular functions influenced. Figure 3.7 depicts a chart that helps in classifying *SR* genes suitably based on stages or functional aspects. Practically, in future, we will encounter a maze of gene induction/repression patterns and corresponding *SR* gene activity. Perhaps, we can focus further and confine analysis to a few important *SR* genes with the ability to influence symbiotic aspects drastically. Careful manipulation of *SR* genes may lead us to important applications in agriculture/forestry. They may hold the key for enhanced P uptake, translocation and growth stimulation of host species.

Since, commonly used ECM fungi are culturable, strain improvement and gene transfer through classical genetic approaches are routinely used in many laboratories. Where feasible, molecular gene transfer methods could help in accentuating useful traits in the ECM fungus. Some of these suggestions are not adoptable to AM fungi, since their sexual cycle is not known and they are not culturable in vitro without host support.

Overall, the prospect for research on genetic aspects of mycorrhizas is bright. Specifically, molecular approaches aimed at understanding the regulation of symbiotic effects may receive greater attention. More crucial is our ability to focus and find applications of genetic manipulation in practical situations.

References

Albrecht, C., Geurtz, R. and Bisseling, T. 1999. Legume nodulation and mycorrhizal formation: Two extremes in host specificity. EMBO Journal 18: 281-288.

Albrecht, C., Geurtz, G., Lapeyrie, F. and Bisseling, T. 1998. Endomycorrhizae and rhizobial *Nod* factors both require *Sym 8* to induce the expression of the early nodulin genes *PsNODS* and *PsENOD12 A*. The Plant Journal 15: 605-614.

Anson, J.G., Gilbert, H.J., Oran, J.D. and Minton, N.P. 1987. Complete nucleotide sequence of the *Rhodosporidium toruloides* gene coding for phenylalanine ammonia-lyase genes. Gene 58: 189-199.

Balaji, B., Ba, A.M., La Rue, T.A. Tepfer, D. and Piche, Y. 1994. *Pisum sativum* mutants insensitive to nodulation are also insensitive to invasion in vitro by the mycorrhizal fungus *Gigaspora margarita*. Plant Science 102: 195-203.

Balasubramanian, S., Kim, S. and Podila, G. K. 2002. Differential expression of a malate synthase gene during the preinfection stage of symbiosis in the ectomycorrhizal fungus *Laccaria bicolor.* New Phytologist 154: 517-527.

Bananomi, A., Wiemken, A., Boller, T. and Salzer,P. 2001. Local induction of a mycorrhiza specific class III chitinase gene in cortical root cells of *Medicago truncatula* containing developing or mature arbuscules. Plant Biology 3: 194-200.

Barker, S.J., Duplessis, S. and Tagu, D. 2002. The application of genetic approaches for investigations of mycorrhizal symbiosis. Plant and Soil 244: 85-95.

Barker, S.J., Stummer, B., Gao, L., Dispin, I., O'Connor, P.J., Smith, S.E. 1998. A mutant in *Lycopersicon esculentum* Mill. with highly reduced VA mycorrhizal colonization: Isolation and preliminary characterization. The Plant Journal 15: 701-797.

Bastel-Corre, G., Dumos-Gaudet, E., Poinsot, V., Dieu, M., Dierick, J.F., van Tueinen, D., Remacle, J., Gianinazzi-Pearson, V. and Gaininazzi, S. 2002. Proteome analysis and identification of symbiosis-related proteins from *Medicago truncatula* Geartn by two-dimensional electrophoresis and mass spectrometry. Electrophoresis 23: 122-137.

Bastide, P.Y., Kropp, B.R. and Piche, Y. 1995. Mechanisms for the development of genetically variable mycorrhizal mycelium in ectomycorrhizal fungus *Laccaria bicolor.* Applied and Environmental Microbiology 61: 3609-3616.

Becard, G., Douds, D.D., Pfefer, P.E. 1992. Extensive *in vitro* hyphal growth of vesicular-arbuscular mycorrhizal fungi in the presence of CO_2 and flavonols. Applied and Environmental Microbiology 58: 821-825.

Bentivenga, S.P., Bever, J.D. and Morton, J.B. 1997. Genetic variation of morphological characters within a single isolate of endo-mycorrhizal fungus, *Glomus clarum* (Glomacea). American Journal of Botany 84: 1211-1216.

Bergero, R., Harrier, L. A. Franken, P. 2003. Reporter genes: Applications to the study of arbuscular mycorrhizal (AM) fungi and their symbiotic interactions with plant. Plant and Soil 255: 143-155.

Bertheu, Y., Gianinazzi-Pearson, V. and Gianinazzi, S. 1980. Development and expression of endomycorrhizal associations in wheat: Evidence of varietal effect. Annals de Amelioration des Plantes 30: 67-75.

Bever, J.D. and Morton, J. 1999. Heritable variation and mechanisms of inheritance of spore shape within a population of *Scutellospora pellucida*, an arbuscular mycorrhizal fungus. American Journal of Botany 86: 1209-1216.

Boddington, C.L. and Dodd, J. C. 1999. Evidence that differences in phosphate metabolism in mycorrhizas formed by species of *Glomus* and *Gigaspora* might be related to their life cycle strategies. New Phytologist 142: 531-538.

Boffin, N.V., Hodges, V.M., Galvez, S., Balestrini, R., Bonfonte, P., Gadal, P. and Martin, F. 1998. Eucalypt NADP-dependent isocitrate dehydrogenase-cDNA cloning and expression in Ectomycorrhizae. Plant Physiology 117: 939-948.

Bonfonte, P., Genre, A., Facio, A., Martini, I., Schauser, L., Stougard, J., Webb, J. and Parniske, M. 2000. The *Lotus japonicus LjSym 4* gene is required for the successful infection of root epidermal cells. Molecular Plant-Microbe Interactions 13: 1109-1120.

Borisov, A.Y., Jacobi, L.M., Lebsky, V.K., Morzhina, E.V. Tsyganov, V.E., Voroshilova, V.A. and Tikhanovich, L.A. 2002. Genetic system controlling development of nitrogen-fixing nodules and arbuscular mycorrhiza. http://hermes.bionet.nsc.ru/pg/31/40.html

Bradbury, S.M., Peterson, R.L., and Bowley, S.R. 1994. Interactions between three alfalfa nodulation genotypes and two *Glomus* species. New Phytologist 119: 15-120.

Burleigh, S.H. and Harrison, M.J. 1998. A cDNA from the arbuscular mycorrhizal fungus *Glomus versiforme* with homology to a cruciform DNA binding protein from *Ustilago maydis.* Mycorrhiza 7: 301-306.

Burleigh, S. H. Cavagnaro, T. and Jackobsen, I. 2002. Functional diversity of arbuscular mycorrhizas extends to the expression of plant genes involved in P nutrition. Journal of Experimental Botany 53: 1593-1601.

Burke, R.M. and Cairney, J.W.G. 2002. Laccase and other polyphenol oxidases in Ecto and Ericoid mycorrhizal fungi. Mycorrhiza 12: 105-116.

Buscot, F., Munch, J.C., Charcoal, J.Y., Gardes, M., Nehls, U. and Hampp, R. 2000. Recent advances in exploring physiology and biodiversity of ectomycorrhiza highlight the functioning of the symbiosis in ecosystems. FEMS Microbiology Reviews 24: 601-614.

Cairney, J.W.G. 1999. Intraspecific physiological variation: Implications for understanding functional diversity in ectomycorrhizal fungi. Mycorrhiza 9: 125-135.

Carnero-Diaz, M.E., Tagu, D. and Martin, F. 1997. Ribosomal DNA internal transcribed spacers to estimate the proportion of *Pisolithus tinctorius* and *Eucalyptus globulus* RNAs in ectomycorrhiza. Applied Environmental and Microbiology 63: 840-843.

Catoria, R., Galera, C., DeBilly, F., Penmetsa, R.V., Journet, E.T., Maillet, F., Rosenberg, G., Cook, D. and Denaire, J. 2000. Four genes of *Medicago truncatula* controlling components of a *Nod* factor translocation pathway. Plant Cell 12: 1647-1665.

Chaboud, M., Venard, C., Defaux-Petras, G., Becard, G. and Barker, D.G. 2001. Targeted inoculation of *Medicago truncatula* in vitro root cultures reveals MtENOD11 expression during early stages of infection by arbuscular mycorrhizal fungi. New Phytologist 156: 265-270.

Charon, C., Johansson, D. Kondorosi, E., Kondorosi, A. and Crespi, M. 1997. *Enod* differentiation and division of cortical cells in legumes. Proceedings of National Academy of Sciences USA 89: 8906.

Charon, C., Sousa, C., Crespi, M. and Kondorosi, A. 1999. Alteration of enod40 expression modifies *Medicago truncatula* root nodule development induced by *Sinorhizobium meliloti*. Cell 11: 1953-1965.

Charvet-Candela, V., Hitchin, S., Ernst, D., Sandermann, H. Marmiesse, R. and Gay, G. 2002. Characterization of an Aux/IAA cDNA upregulated in *Pinus pinaster* roots in response to colonization by the ectomycorrhizal fungus *Hebeloma cylindrosporum*. New Phytologist 154: 769-777.

Chen, D.M., Taylor, A.F.S., Burke, R.M. and Cairney, J.W.G. 2001. Identification of genes for lignin peroxidases and manganese peroxidases in ectomycorrhizal fungi. New Phytologist 152: 151-158.

Chen, D.M. Bastias, B.A., Taylor, A.F.S. and Cairney, J.W.G. 2003. Identification of laccase-like genes in ectomycorrhizal basidiomycetes and transcriptional regulation by nitrogen in *Piloderma bysinum*. New Phytologist 157: 547-554.

Clark, R. B. 2002. Differences among mycorrhizal fungi for mineral uptake per root length of switch grass grown in acidic soil. Journal of Plant Nutrition 25: 1753-1772.

Clark, R.B. and Duncan, R.R. 1991. Improvement of Plant Nutrition through Plant Breeding. Field Crops Research 27: 219-240.

Coltman, R.R. 1987. Tolerance of tomato strains to phosphorus deficiency in root cultures. Horticultural Science 22: 1305-1307.

Debaud, J.C., Marmiesse, R. and Gay, G. 1995. Intraspecific genetic variation in ectomycorrhizal fungi. In: Mycorrhizal Structure, Function Molecular Biology and Biotechnology. Verma, A. and Hock, B. (Eds) Springer Verlag, Berlin, New York, pp. 79-113.

Dey, M., Complainville, A., Charon, C., Torrizo, L., Kondorossi, A., Crespi, M. and Data, S. 2004. Phytohormonal response in *enod40* over-expressing plants of *Medicago truncatula* and rice. Physiologia Plantarum 120: 132-139.

Delp, G., Timonen, S., Rosewarne, G.M., Barker, S.J. and Smith, S.E. 2003. Differential expression of *Glomus intraradices* genes in external mycelium and mycorrhizal roots of tomato and barley. Mycological Research 107: 1083-1093.

Diaz, E.C., Martin, F. and Tagu, D. 1996. Eucalyptus alpha-tubulin: cDNA cloning and increased level of transcripts in ectomycorrhizal root system. Plant Molecular Biology 31: 905-910.

Diez, J., Anta, B., Manjon, J.L. and Honrubia, M. 2001. Genetic variability of *Pisolithus* isolates associated with native hosts and exotic eucalyptus in the western Mediterranean region. New Phytologist 149: 577-587.

Diouf, D., Diop, T.A. and Ndoye, I. 2003. Actinorhizal, mycorrhizal and rhizobial symbiosis: How much do we know? African Journal of Biotechnology 2: 1-7.

Dixon, R.K., Garrett, H.E. and Stalzer, H.E. 1987. Growth and ectomycorrhizal development of loblobby pine progenies inoculated with tree isolates of *Pisolithus tinctorius*. Silvae Genetica 36: 240-245.

Dodd, J.C., Boddington,C.L., Rodrguez, A., Gonzalez-Chavez, C. and Mansur, I. 2000. Mycelium of Arbuscular mycorrhizal fungi from different genera: Form, function and detection. Plant and Soil 226: 131-151.

Douds, D.D., Pfeffer, P.E. and Sachar-Hill, Y. 2000. Carbon partitioning, cost and metabolism of arbuscular mycorrhizas. In: Arbuscular Mycorrhizas: Physiology and Function. Kapulnik, Y. and Douds, D .D. (Eds) Kluwer Academic Publishers, Dordrecht, Netherlands, pp. 325-347.

Duc, G., Trouvelot, A., Gianinazzi-Pearson, V. and Gianinazzi, S. 1989. First report of non-mycorrrhizal plant mutants (*myc-*) obtained in pea (*Pisum sativum* L.) and Faba bean (*Vicia faba* L.). Plant Science 60: 215-222.

Duncan, R.R. and Baligar,V.C. 1990. Genetics breeding and physiological mechanisms of nutrient uptake and use efficiency. In: Crops as Enhancers of Nutrient Use. Baligar, V.C. and Duncan, R. R. (Eds). Academic Press, London, pp. 1-29.

Duplessis, S. and Martin, F. 2002. Monitoring *Eucalyptus globulus* and *Pisolithus tinctorius* gene expression during the ectomycorrhizal symbiosis development using cDNA arrays and cluster analysis. Third International Conference on Mycorrhizas, Adelaide, Australia, http://www.mycorrhizas. ag.utk.edu/icoms/icom3.html

Duplessis, S., Sorin, C., Vioblet, C., Palin, B. Martin, F. and Tagu, D. 2001. Cloning and expression analysis of a new hydrophobin cDNA from the ectomycorrhizal basidiomycete *Pisolithus*. Current Genetics 39: 335-339.

Eason, W.R., Webb, K.J., Michealson-Yeates, T.P.T., Abberton, M.T. Griffith, G.W., Culshaw, C.M., Hooker, and J.E. Dhanoa, M.S. 2001. Effect of genotype of *Trifolium repens* on mycorrhizal symbiosis with *Glomus mosseae*. Journal of Agricultural Science 137: 27-36.

Fester, T., Kiess, M. and Strack, D. 2002. A mycorrhiza-responsive protein in wheat roots. Mycorrhiza 12: 219-222.

Filpula, D., Vaslet, C.A., Levy, A., Sukes, A. and Strausberg, R.C. 1988. Nucleotide sequence of gene for phenylalanine ammonia lyase from *Rhodotorula rubra*. Nucleic Acid Research 16: 11381-11385.

Franken, P. 1999. Trends in molecular studies of AM fungi. In: Mycorrhiza. 2nd edition. Verma, A. and Hock, B. (Eds) Springer Verlag, Berlin, Heidelberg, Germany, pp. 37-49.

Fries, L.L.M., Packovsky, R.S., Safir, G.R. and Kaminski, J. 1998. Phosphorus effect on phosphatase activity in endomycorrhizal maize. Physiologia Plantarum 103: 162-171.

Gabelmann, W.H. and Gerlof, G.C. 1983. The search for and interpretation of genetic control that enhance plant growth under deficiency levels of macronutrient. In: Genetical Aspects of Plant Nutrition. Saric, R. and Loughman, B.C. (Eds). Martinus Nijhoff Publishers, The Hague, pp. 380-394.

Gadkar, V., Shwartz, D.R., Nagahashi, G., Douds, D.D., Wininger, S., and Kapulnik, Y. 2003. Root exudates of *pmi* tomato mutant M161 reduces AM fungal proliferation *in vitro*. FEMS Microbiology Letters 223: 193-198.

Gao, L. L., Delp, G. and Smith, S. E. 2001. Colonization patterns in a mycorrhiza-defective mutant tomato vary with different arbuscular-mycorrhizal fungi. New Phytologist 151: 477-491.

Garcia-Garrido, J., Jose, M., Ikram, B. and Ocampo, J.A. 1998. Induction of salicylic acid in pea roots inoculated with the arbuscular mycorrhizal fungus *Glomus mosseae*. Second International Conference on Mycorrhiza, Uppsala, Sweden, http://www.mycorrhizas.ag.utk.edu

Gay, G., Marmeisse, R., Fouillet, P., Bouletreau, M., and Debaud, J.C. 1993. Genotype nutrition interactions in the ectomycorrhizal fungus *Hebeloma cylindrosporum*. New Phytologist 123: 335-343.

Gianinazzi-Pearson, V. 1996. Plant cell responses to arbuscular mycorrhizal fungi: Getting to the roots of symbiosis. Plant Cell 8: 1871-1883.

Gibson, G. 2002. Micro-arrays in Ecology and Evolution: A Preview. Molecular Ecology 11: 17-24.

Giovannetti, M., Sbarna, C., Avio, L., and Citernesi, A.S. 1993. Differential hyphal morphogenesis in arbusuclar mycorrhiza fungi during pre-infection stages. New Phytologist 125: 587-594.

Giovannetti, M., Sbarna, C., Citernesi, A.S. and Avio, L. 1999. Analysis of factors involved in fungal recognition responses to host-derived signals by arbuscular mycorrhizal fungi. New Phytologist 133: 65-71.

Glen, M., Tommerup, I. C., Bougher, N.L., and O'Brien, P.A. 2001. Interspecific and intraspecific variation of ectomycorrhizal fungi associated with Eucalyptus ecosystems as revealed by ribosomal DNA PCR-RLP. Mycological Research 105: 843-858.

Gomes, E. A., de Agreu, L. M. Borges, A.C. and de Araujo, E.F. 2000. ITS sequences and mitochondrial DNA polymorphism in Pisolithus isolates. Mycological Research 104: 911-918.

Gourley, C.J.P., Allan, D.L. and Russel, M.D. 1994. Plant nutrient efficiency: A comparison and suggested improvement. Plant and Soil 158: 29-38.

Graham, J.H. and Abbot, L.K. 2000. Wheat responses to aggressive and non-aggressive arbuscular mycorrhizal fungi. Plant and Soil 220: 207-218.

Hagberg, K. and Stougaard, J. 1992. *Lotus japonicus* an autogamous diploid legume species for classical and molecular genetics. Plant Journal 2: 487-496.

Hardham, A.R. and Mitchell, H.J. 1998. Use of molecular cytology to study the structure and biology of pathogenic and mycorrhizal fungi. Fungal Genetics and Biology 24: 252-283.

Harrison, M.J. 1999. Molecular and cellular aspects of the arbuscular mycorrhizal symbiosis. Annual Review Plant Physiology and Plant Molecular Biology 50: 361-389.

Hart, M.M. and Reader, R.J. 2002a. Taxonomic basis for variation in the colonization strategy of arbuscular mycorrhiza fungi. New Phytologist 153: 335-344.

Hart, M.M. and Reader, R.J. 2002b. Host plant benefits from association with arbuscular mycorrhizal fungi: Variation due to differences in size of mycelium. Biology and Fertility of Soils 36: 357-366.

Hause, B. 2003. Cell Biology of Mycorrhizas. Institute of Plant Biochemistry. http//:www.ipb-halle.de

Hetrick, B.A.D., Wilson, G.W.T. and Cox, T.S. 1992. Mycorrhizal dependence of modern wheat varieties, land races and ancestors. Canadian Journal of Botany 70: 2032 - 2040.

Hetrick, B.A.D., Wilson, G.W.T. and Cox, T.S. 1996. Mycorrhizal response in wheat cultivars: Relationship to phosphorus. Canadian Journal of Botany 74: 19-25.

Hetrick, B.A.D., Wilson, G.W.T., Gill, B.S. and Cox, T.S. 1995. Chromosome location of mycorrhizal responsiveness genes in wheat. Canadian Journal of Botany 73: 891-897.

Hohnjec, N., Perlick, A.M., Kuster, H. 2003. The *Medicago truncatula* sucrose synthase gene *MtSucS1* is activated both in the infected region of root nodules and in the cortex of roots colonized by arbuscular mycorrhizal fungi. Molecular Plant-Microbe Interactions 16: 903-915.

ICRISAT 1985. Sorghum Program—Annual Report, 1984. International Crops Research Institute for the Semi-arid Tropics. Patancheru, AP. India, pp. 51.

INRA 2002. Molecular analysis of the Ectomycorrhizal differentiation-a Research report. Institute national de Research Agronomique. Nancy, France, pp. 1-3.

Jackobi, L.M., Zubhkova, L.A., Barmicheva, L.M., Tsyganov, V.E., Borisov, A.V. and Thickanovich, I.A. 2003. Effect of mutations in the pea genes *Sym 33* and *Sym 40*. 2. Dynamics of arbuscular development and turnover. Mycorrhiza 13: 9-16.

Jensen, A. 1985. Evaluation of VA Mycorrhiza as a parameter in breeding field peas. In: Proceedings of the 6th North American Conference on Mycorrhiza (Ed.) Molina, R. Oregon State University, Corvallis, Oregon, USA, pp. 258.

Jun, J., Abubaker, J. Rehrer, C., Pfefer, P.E., Shachar-Hill, Y. and Lammers, P.J. 2002. Expression in an arbuscular mycorrhizal fungus of genes putatively involved in metabolism, transport, the cytoskeleton and cell cycle. Plant and Soil 244: 141-148.

Kaeppler, S.M., Parke, J.L., Mueller, S.M., Senior, L., Charles, S. and Tracey, W. F. 2000. Variation among maize inbred lines and detection of quantitative trait loci for growth at low phosphorus and responsiveness to arbuscular mycorrhizal fungi. Crop Science 40: 358-364.

Karim, B., Azcon-Aguilar, C. and Ferrol, N. 1998. Soluble and plasma membrane symbiosis related polypeptides associated with the development of arbuscular mycorrhizal in tomato roots. Second International Conference on Mycorrhizas, Uppsala, Sweden, http:// www.mycorrhizas.ag.utk.edu

Kasuta, S., Chabaud, M., Lougnon, G., Gough, C., Denarie, J., Barker, D.G. and Becard, G. 2003. A diffusible factor from arbuscular mycorrhizal fungi indices symbiosis specific MtENOD11 expression in the roots of *Medicago truncatula*. Plant Physiology 131: 952-962.

Kendrick, B. and Birch, S.M. 1985. Mycorrhizae: applications in agriculture and forestry. In: Comprehensive Biotechnology. Robinson, C.W. (ed.) Pergamon Press, Oxford, England, pp. 109-152.

Kershaw, M.J. and Talbot, N.J. 1998. Hydrophobins and repellants: proteins with fundamental roles in fungal morphogenesis. Fungal Genetics and Biology 23: 18-33.

Kim, S.J., Bernreuther, D., Thumm, M. and Podila, G.K. 1999a. LB-AUT7, a novel symbiosis-regulated gene from an ectomycorrhizal fungus, *Laccaria bicolor* is functionally related to vesicular transport and autophagocytosis. Journal Bacteriology 181: 1963-1967.

Kim, S., Hiremath, S.T. and Podila, G.K. 1999b. Cloning and identification of symbiosis regulated genes from the ectomycorrhizal *Laccaria bicolor*. Mycological Research 103: 168-172.

Kim, S.J., Zheng, J., Hiremath, S.T. and Podila, G.K. 1998. Cloning and characterization of a symbiosis-related gene from an ectomycorrhizal fungus *Laccaria bicolor*. Gene 222: 203-212.

Kistner, C and Parniske, M. 2002. Evolution of signal transduction in intracellular symbiosis. Trends in Plant Science 7: 511-518.

Krajinski, F., Biela, A., Schubert, D., Gianinazzi-Pearson, V., Kaldenhoff, R. and Franken, P. 2000. Arbuscular Mycorrhiza development regulates the mRNA abundance of MtAQP1 encoding a mercury insensitive aquaporin of *Medicago truncatula*. Planta 211: 85-90.

Krishna, K.R. 1997. Phosphorus efficiency of semi-arid dryland crops. In: Accomplishments and Future Challenges in Dry Land Soil Fertility Research in Mediterranean Area. Ryan, J. (Ed.). International Center for Agricultural Research in the Dry Areas. Aleppo, Syria, pp. 343-363.

Krishna, K.R. and Lee, K.K. 1987. Management of vesicular arbuscular mycorrhiza in tropical cereals. In: Mycorrhiza in the Next Decades. Sylvia, D.M., Hung, L.L. and Graham, J.H (Eds) 7th North American Conference on Mycorrhizas, Gainesville, Florida, USA, pp 43-45.

Krishna, K.R., Shetty, K.G., Dart, P.J. and Andrews, D.J. 1984. Genotype dependent variation in mycorrhizal colonization and response to inoculation of pearl millet. Plant and Soil 86: 113-125.

Krishna, K.R. and Williams, J.H. 1987. Vesicular arbuscular mycorrhiza and genotype effect on peanut symbiosis and growth. In: Mycorrhiza in the Next Decades. Sylvia, D.M., Hung, L.L. and Graham, J.H. (Eds). University of Florida, Gainesville, Florida, USA, p 48.

Kropp, B.R. 1997. Inheritance of the ability for ectomycorrhizal colonization of *Pinus strobus* by *Laccaria bicolor*. Mycologia 89: 578-585.

Kropp, B. and Fortin, J.A. 1988. The incompatibility system and relative ectomycorrhizal performance of monokaryons and reconstituted dikaryons of *Laccaria bicolor*. Canadian Journal of Botany 66: 289-294.

Kropp, B.R. and Langlois, C.G. 1990. Ectomycorrhizae in Reforestation. Canadian Journal of Forest Research 20: 438-451.

Kropp, B.R., McAfee, B.J. and Fortin, J.A. 1987. Variable loss of Ectomycorrhizal ability in monokaryotic and dikaryotic culture of *Laccaria bicolor*. Canadian Journal of Botany 65: 500-504.

Lackie, S.M., Pallardy, S.G. and Petterson, R.L. 1988. Comparison of colonization among half-sib families of *Medicago sativa* by *Glomus versiforme*. New Phytologist 108: 477-482.

Laurent, P., Vioblet, C., deCarvalho, D., Nehls, U., and Martin, F. 1999. A novel class of ectomycorrhiza-regulated cell wall polypeptide in *Pisolithus tinctorius*. Molecular Plant-Microbe Interactions. 12: 862-871.

Lamahamadi, M.S. and Fortin, J.A. 1991. Genetic variation of ectomycorrhizal fungi: Extra matrical phase of *Pisolithus* sp. Canadian Journal of Botany 69: 1927-1934.

Lammers, P.J., Jun, J., Abubekar, J., Arreola, R., Gopalan, A., Bago, B., Sebastian, C., Allen, J.W., Douds, D.D., Pfeffer, P.E., and Sachar-Hill, Y. 2001. Glyoxylate cycle in an arbuscular mycorrhizal fungus: Carbon flux and Gene expression. Plant Physiology 127: 1287-1298.

Lapopin, L., Gianinazzi-Pearson, V. and Franken, P. 1998. Derepression of genes involved in the late stages of arbuscular mycorrhiza development. Canadian Journal of Botany 73: s526-s532.

Lobuglio, K.F. and Taylor, J.W. 2002. Recombination and genetic differentiation in the mycorrhizal fungus *Cenococcum geophilum*. Mycologia 94: 777-780.

Lumley, T.C., Farquhar, M.L. and Peterson, R.L. 1995. Ectomycorrhiza formation between *Pseudotsuga menzeseii* seedling roots and monokaryotic and dikaryotic isolates of *Laccaria bicolor*. Mycorrhiza 51: 237-244.

Mankel, A., Krause, K. and Kothe, E. 2002. Identification of Hydrophobin genes that is developmentally regulated in the ectomycorrhizal fungus *Tricholoma terraeum*. Applied Environmental Microbiology 68: 1408-1413.

Manske, G.G.B. 1990. Genetical analysis of the efficiency of VA mycorrhiza with Spring Wheat 1. Genotype differences and a reciprocal cross between efficient and non-efficient variety. In: Genetic Aspects of Plant Mineral Nutrition. El-Bassam, N. (Ed.) Kluwer Academic Publishers, Dordrecht, Netherlands, pp 397-405.

Manske, G.G.B., Ortiz-Monasterio, J.I., Van Ginkel, M., Gonzalez, R.M., Rajaram, S., Molina, E. and Vlek, P.L.G. 2000. Traits associated with improved P-uptake efficiency in CIMMYT's Semi-Dwarf Spring Bread Wheat grown on acid Andisols in Mexico. Plant and Soil 221: 189-204.

Mari, S., Jonsson, A., Finlay, R., Ericson, T., Kahr, M. and Eriksson, G. 2003. Genetic variation in nitrogen uptake and growth in mycorrhizal and non-mycorrhizal *Picea abies* seedlings. Forest Science 49: 259-267.

Marsh, J.M. and Schultz, M. 2001. Analysis of arbuscular mycorrhiza using symbiosis-defective plant mutants. New Phytologist 150: 525-532.

Martin, F. 2001. Frontiers in molecular mycorrhizal research—Genes, loci, dots and spins. New Phytologist 150: 499-505.

Martin, F. and Hilbert, J.L. 1991. Morphological, biochemical and molecular changes during ectomycorrhiza development. Experientia 47: 321-331.

Martin, F., Lapayeri, F. and Tagu, D. 1997. Altered gene expression during ectomycorrhizal development in the Mycota. In: Plant Relationships. Lemke, P. and Caroll, G. (Eds) Springer Verlag, Berlin, pp 223-242.

Martin, F., Laurent, P., DeCarvalho, D., Voiblet, C., Ballestrini, R., Bonfonte, P. and Tagu, D. 1999. Cell wall proteins of the ectomycorrhizal basidomycetes *Pisolithus tinctorius*: Identification, function, and expression in symbiosis. Fungal Genetics and Biology 27: 161-174.

Martin, F., Tommerup, I.C and Tagu, D. 1994. Genetics of Ectomycorrhizal Fungi—Progress and Prospects. Plant and Soil 159: 179-181.

Martin-Laurent, F., van Tueinen, D., Dumas-Gaudet, E., Gianinazzi-Pearson, V., Gianinazzi, S. and Franken, P. 1997. Differential display analysis of RNA accumulation in arbuscular mycorrhiza of pea and isolation of novel symbiosis-regulated plant gene. Molecular and General Genetics 256: 37-44.

Marx, D.H. 1991. The practical significance of ectomycorrhizae in forest establishment. In: Ecophysiology of Ectomycorrhizae of Forest Trees. Marcus Wallenburg Foundation, Stockholm, Sweden, pp 75-79.

Marx, D.H. and Bryan, J.A. 1971. Formation of ectomycorrhiza on half-sib progenies of slash pine in asceptic culture. Forest Science 17: 488-492.

Medina, M. J. H., Gagnon, H., Piche, Y., Ocampo, J.A., Garrido, J. M. G., and Vierheilig, H. 2003. Root colonization by arbuscular mycorrhizal fungi is affected by the salicylic acid content of the plant. Plant Science 164: 993-998.

Murphy, P. J., Karakousis, A., Smith, S. E., Langridges, P. 1995. Cloning functional endomycorrhizal genes: Potential use for plant breeding. In: Biotechnology of Ectomycorrhizas. Stocchi, V. (Ed.) Plenum Press, New York, pp 77-83.

Nehls, U., Beguiristain, T., Ditengon, F., Lapayrie, F. and Martin, F. 1998. The expression of a symbiosis-regulated gene in Eucalypt roots is regulated by auxins and tryptophan betain of the ECM basidiomycete *Pisolithus tinctorius*. Planta 207: 296-302.

Nehls, U., Ecke, M. and Hampp, R. 1999. Sugar and nitrogen dependent regulation of an *Amanita muscaria* phenylalanine ammonia lyase gene. Journal of Bacteriology 181: 1931-1933.

Oliver, B., Berthaeu, Y., Deim, H.G. and Gianinazzi-Pearson, V. 1983. Influence de la variete de Vigna ungiculata dansl' Expression de tris associations endomycorrhiza varieties et arbuscular. Canadian Journal of Botany 61: 354-358.

Parniske, M 2000. Intracellular accommodation of microbes in plants: A common developmental program for symbiosis and disease. Current Opinions in Plant Biology 3: 320-328.

Paszkowski, U. and Boller, T. 2002. The growth defect of Lrt1, a maize mutant lacking lateral roots can be complemented by symbiotic fungi or high phosphate nutrition. Planta 21: 584-590.

Penmetsa, R.V. and Cook, D.R. 1997. A legume ethylene-insensitive mutant hyper infected by its rhizobial symbiont. Science 275: 527-530.

Poulin, M.J., Bel Rhild, R., Piche, Y., Chenevert, R. 1993. Flavonoids released by carrot (*Daucus carota*) seedlings stimulated hyphal development of vesicular-arbuscular mycorrhizal fungi in the presence of optimal CO_2 enrichment. Journal of Chemical Ecology 19: 2317-2327.

Poulin, M.J., Simard, J., Calford, J.G., Labrie, F. and Piche, Y. 1997. Response of symbiotic endomycorrhizal fungi to estrogens and anti-estrogens. Molecular Plant-Microbe Interactions 10: 481-487.

Pringle, A., Moncalvo, J.M. and Vilgalys, R. 2000. High levels of variation in ribosomal DNA sequences within and among spores of a natural populations of the arbuscular mycorrhizal *Acaulospora colossica*. Mycologia 92: 259-268.

Provorov, N.A., Borisov, A.Y. and Tikanovich, I.A. 2002. Developmental genetics and evolution of symbiotic structures in nitrogen-fixing nodules and arbuscular mycorrhiza. Journal of Theoretical Biology 214: 215-232.

Rajan, S.K., Reddy, B.J.D. and Bagyaraj, D.J. 1999. Screening of arbuscular mycorrrhizal fungi for their symbiotic efficiency with *Tectona grandis*. Forest Ecology and Management 126: 91-95.

Ravnskov, S., Wu, Y. and Graham, J.H. 2001. The influence of different arbuscular mycorrhizal fungi and phosphorus of gene expression of invertase and sucrose synthase in roots of maize. Third International Conference on Mycorrhizas, Adelaide, Australia, http://www.mycorrhizas. ag.utk.edu./icoms/icom3.html

Rengel, Z. 2002. Breeding for better symbiosis. Plant and Soil 245: 147-162.

Requena, N., Fuller, P. and Franken, P. 1999. Molecular characterization of GmFOX2, an evolutionarily highly conserved gene from the mycorrhizal fungus *Glomus mosseae*, down-regulated during interaction with rhizobacteria. Molecular Plant-Microbe Interactions 12: 934-942.

Rosado, S.C.S., Kropp, B.R. and Piche, Y. 1994a. Genetics of ectomycorrhizal symbiosis. 1. Host plant variability and heritability of ectomycorrhizal root traits. New Phytologist 126: 105-110.

Rosado, S.C.S., Kropp, B.R. and Piche, Y. 1994b. Genetics of Ectomycorrhizal Symbiosis. 2. Fungal variability and heritability of ectomycorrhizal traits. New Phytologist 126: 111-117.

Ruiz-lozano, J.M., Roussel, H., Gianinazzzi, S. and Gianinazzi-Pearson, V. 1999. Defense-related genes are differentially induced by a mycorrhizal fungus and *Rhizobium* sp. in wild type and symbiosis defective pea genotypes. Molecular Plant-Microbe Interaction 12: 976-984.

Sagan, M., Morandi, D., Tarenghi, E. and Duc, G. 1995. Selection of nodulation and mycorrhizal mutants in model plant *Medicago truncatula* Geartn after gamma-ray mutagenesis. Plant Science 111: 63 -71.

Salzer, P., Banonomi, A., Beyer, K., Vogali-langal, R., Aeschbacher, R., Lange, J., Wiemken, A., Kim, D., Cook, D.R. and Boller, T. 2000. Differential expression of eight chitinase genes in *Medicago truncatula* roots during mycorrhiza formation, nodulation and pathogen infection. Molecular Plant-Microbe Interactions 13: 763-777.

Sanchez, L., Weidmann, S., Brechenmacher, L., Batoux, M., van Tuinen, D.,Lemanceau, P., Gianinazzi, S. and Gianinazzi-Pearson, V. 2004. Common gene expression in *Medicago*

truncatula rots in response to *Pseudomonas fluorescens* colonization, mycorrhiza development and nodulation. New Phytologist 161: 855-863.
Sanders, I.R.M., Clapp, J.P. and Weimken, A. 1996. The genetic diversity of arbuscular mycorrhizal fungi in natural ecosystems—a key to understanding the ecology and functioning of the mycorrhizal symbiosis. New Phytologist 133: 123 -134.
Sanginga, N., Lyasse, O. and Singh, B.B. 2000. Phosphorus use efficiency and nitrogen balance of cowpea breeding lines in a low P soil of the derived savanna zone in West Africa. Plant and Soil 220: 119-128.
Sawyer, N.A., Chambers, S.M. and Cairney, J.W.G. 2001. Distribution and persistence of *Amanita muscaria* genotypes in Australian *Pinus radiata* plantations. Mycological Research 105: 966-970.
Sawyer, N.A., Chambers, S.M. and Cairney, J.W.G. 2003. Variation in nitrogen source utilization by nine *Amanita muscaria* genotypes from Australian *Pinus radiata* plantations. Mycorrhiza 13: 217-221.
Schauser, L., Handberg, K., Sandal, N., Stiller, J., Thykjaer, P.E., Nielsen, A. amd Stougaard, J. 1998. Symbiotic mutants deficient in nodule establishment identified after T-DNA transformation of *Lotus japonicus*. Molecular and General Genetics 259: 414-423.
Schneider, L., Handelberg, K., Sandal, N., Stiller, J., Thykjaer, P. E., Neilsen, A. and Stougaard, H. 1998. Symbiotic mutants deficient in nodule establishment identified after T-DNA transformation of *Lotus japonicus*. Molecular and General Genetics 259: 414-423.
Schwartz, D.R., Badani, H. Smadar. W., Levy, A. A. Galili, G., and Kapulnik, Y. 2002. Identification of a novel genetically controlled step in mycorrhizal colonization: Plant resistance to infection by fungal spores but not extraradical hyphae. Plant Journal 27: 561-569.
Schwartz, D.R., Gadkar, V., Wininger, S., Bendov, R., Galili, G., Levy, A.A. and Kapulnik, Y. 2003. Isolation of pre-mycorrhizal infection (pmi2) mutant of tomato, resistant to arbuscular mycorrhizal fungal colonization. Molecular Plant-Microbe Interactions 16: 382-388.
Sen, R. 1990. Intraspecific variation in two species of *Suillus* from Scots pine (*Pinus sylvestrus*) forests based on somatic incompatibility and isozyme analysis. New Phytologist 114: 607-616.
Senoo, K., Zakaria, S., Kwaguchi, M., Imizami-Anarku, H., Akao, S., Tanaka, A. and Obata, H. 2000. Isolation of two different phenotype of mycorrhizal mutants in the model legume plant *Lotus japonicus* after EMS-treatment. Plant Cell Physiology 41: 726-732.
Shirtliffe, S.J. and Versey, K. 1996. A nodulation (*Nod+/Fix-*) mutant of *Phaseolus vulgaris* L. has nodule-like structures lacking peripheral vascular bundles (*Pvb-*) and is resistant to mycorrhizal infection (*Myc-*). Plant Science 118: 209-220.
Sequiera, J.O., Safir, G.R. and Nair, M.G. 1991. Stimulation of vesicular-arbuscular mycorrhiza formation and growth of white clover by flavonoid compounds. New Phytologist 118: 87-93.
Sinvany, G., Kapulnik, Y., Winninger, S., Badani, H. and Jurkevitch, E. 2002. The Early nodulin enod40 is induced by and also promotes arbuscular mycorrhizal root colonization. Physiological and Molecular Plant Pathology 60: 103-109.
Smith, K.P. and Goodman, R.M. 1999. Host genetic variation for interactions with beneficial plant-associated microbes. Annual Review of Phytopathology 37: 473-491.
Smith, S.E. and Read, D.J. 1997. Mycorrhizal Symbiosis. Academic Press, San Diego, California, USA, pp. 453.
Smith, S.E., Robson, A.D. and Abbot, L.K. 1992. The involvement of mycorrhizas in assessment of genetically dependent efficiency of nutrient uptake and use. Plant and Soil 146: 169-179.
Stachilin, C., Charon, C., Boller, T., Crespi, M. and Kondorosi, A. 2001. *Medicago truncatula* plants overexpressing the early nodulation gene *enod40* exhibit accelerated mycorrhizal colonization and enhanced formation of arbuscules. Proceedings of National Academy of Sciences, USA, 98: 15366-15371.
Stahl, P.D. and Christensen, M. 1991. Population variation in the mycorrhizal fungus, *Glomus mosseae*: breadth of environmental tolerance. Mycological Research 94: 1070-1076.

Starke, S., Kistner, C., Yoshida, S., Mulder, L., Sato, S., Kanako, T., Tabata, S. Stougaard, J., Sczyglowski, K. and Parniske, M. 2001. A plant receptor-like kinase required by both bacterial and fungal symbiosis. Nature 417: 959-962.

Stommel, M., Mann, P. and Fraken, P. 2001. EST-library construction using spore RNA of the arbuscular mycorrhizal fungus *Gigaspora rosea*. Mycorrhiza 10: 281-285.

Sundaram, S., Kim, S.J., Suzuki, H., Mcquattie, C.J., Hiremath, S.T. and Podila, G.K. 2001. Isolation and characterization of symbiosis-regulated gene from the ectomycorrhizal fungus *Laccaria bicolor*. Molecular Plant-Microbe Interaction 14: 618-628.

Sundaram, S., Brand, J.H., Hymes, M.J., Hiremath, S. and Podila, G.K. 2004. Isolation and analysis of a symbiosis-regulated and RAS-interacting vesicular assembly protein gene from the ectomycorrrhizal fungus *Laccaria bicolor*. New Phytologist 161: 529-538.

Sylvia, D.M., Wilson, D.O., Graham, H.H., Maddox, J.J., Milner, P., Morton, J.B., Skipper, H.D., Wright, S.F. and Jarstfer, A. 1993. Evaluation of vesicular arbuscular mycorrhizal fungi in diverse plants and soils. Soil Biology and Biochemistry 25: 705-713.

Tagu, D., de Bellis, R., Belestrini, R., deVries, O.M.H., Picolli, G., Stocchi, V., Bonfonte, P. and Martin, F. 2001. Immunolocalization of hydrophobins HYDPT1 from the ectomycorrhizal basidomycete *Pisolithus tinctorius* during colonization of *Eucalyptus globulus* roots. New Phytologist 149: 127-135.

Tagu, D. and Martin, F. 1995. Expressed sequence tags of randomly selected cDNAs from *Eucalyptus* ECM by PCR assisted differential screening. Molecular Plant-Microbe Interaction 8: 781-783.

Tagu, D., Phthom, P., Cretin, C. and Martin, F. 1993. Cloning symbiosis-related cDNAs from *Eucalyptus* ECM by PCR assisted differential screening. New Phytologist 125: 339-343.

Tagu, D., Rampant, P. F., Lapeyrie, F., Frey-Klett, P., Vion, P. and Villar, M. 2001. Variation in the ability to form ectomycorrhiza in the F1 progeny of an interspecific poplar (*Populus* species) cross. Mycorrhizas 10: 237-240.

Tahir-Aloui, A., Lingua, G., Avrova, A., Sampo, S., Fusconi, A., Antoni, J., and Berta, G. 2002. A cullin gene is induced in tomato roots forming arbuscular mycorrhizae. Canadian Journal of Botany 80: 607-616.

Tamasloukt, M., Kluver, A., Becard, G. and Franken, P. 2003a. Plant-regulated gene expression of the arbuscular mycorrhizal fungus *Gigaspora rosea* during presymbiotic development. COST Scientific meetings, Cologne, Germany, Abstract, p 2.

Tamasloukt, M., Sejlon-Delmas, N. Kluver, A., Jauneau, A., Roux, C., Becard, G. and Franken, P. 2003b. Root factors induce mitochondrial related genes expression and fungal respiration during the developmental switch from symbiosis to presymbiosis in the arbuscular mycorrhizal fungus *Gigaspora rosea*. Plant Physiology 131: 1468-1478.

Tarkka, M.T. 2001. Developmentally regulated proteins in *Pinus sylvestrus* roots and ectomycorrhiza Ph.D. Thesis. University of Helsinki, Finland, pp 138.

Tiwari, L., Johri, B.N. and Tandon, S.M. 1993. Host genotype dependency and growth enhancing ability of VA mycorrhizal fungi for *Eleusine coracana*. World Journal of Microbiology and Biotechnology 9: 191-195.

Thingstrup, I., Kahiluoto, H. and Jakobsen, I. 2000. Phosphate transport by hyphae of field communities of arbuscular mycorrhizal fungi at two levels of P fertilization. Plant and Soil 221: 181-187.

Tomkin, C.M., Malajczuk, N. and McComb, J.A. 1989. Ectomycorrhizal formation by micropropagated clones of *Eucalyptus marginata* inoculated with isolates of *Pisolithus tinctorius*. New Phytologist 111: 209-214.

Toth, R., Page, T. and Castleberry, R. 1984. Differences in mycorrhizal colonization of maize selections for high and low ear leaf phosphorus. Crop Science 24: 994-996.

Tsai, S.M. and Phillips, D.A. 1991. Flavonoids released naturally from alfalfa promote development of symbiotic *Glomus* spores *in vitro*. Applied and Environmental Microbiology 57: 1485-1488.

Van Buuren, M.L., Maldonanado-Mendoza, E., Treiu, A.T., Blaylock, L.A. and Harrison, M.J. 1999. Novel genes induced during an arbuscular mycorrhizal (AM) symbiosis formed between *Medicago trucatula* and *Glomus versiforme*. Molecular Plant-Micorbe Interactions 12: 171-181.

Voiblet, C., Duplessies, S., Encelot, W. and Martin, F. 2001. Identification of symbiosis regulated genes in *Eucalyptus globulus–Pisolithus tinctorius* ectomycorrhiza by differential hybridization of arrayed cDNAs. The Plant Journal 25: 181-191.

Wagner, F., Gay, G. and Debaud, J.C. 1988. Genetic variability of glutamate dehydrogenase activity in monokaryotic and dikaryotic mycelium of ectomycorrhizal fungus *Hebeloma cylindrosporum*. Applied Microbiology and Biotechnology 66: 588-594.

Walter, M. H., Hans, J. and Strack, D. 2002. Two distantly related genes encoding 1-deoxy-D-xylulose 5-phosphate synthases: differential regulation in shoots and apocarotenoids accumulating mycorrhizal roots. Plant Journal 31: 243-254.

Weeden, N.F., Kneen, D.E. and LaRue, T.A. 1990. Genetic analysis of *sym* genes and other nodule-related genes in *Pisum sativum*. In: Nitrogen Fixation: Achievements and Objections (Eds) Gresshoff, P.M., Roth, L.E., Stacey, G. and Newton, W.E. Chapman and Hall, New York, USA, pp 323-330.

Wegel, E., Schauser, L., Sandal, N., Staougaard, J. and Parniske, M. 1998. Mycorrhiza mutants of *Lotus japonicus* define genetically independent steps during symbiotic infection. Molecular Plant -Microbe Interactions 11: 933-936.

Weimken, V. and Boller, T. 2002. Ectomycorrhiza: gene expression, metabolism and the wood-wide web. Current Opinions in Plant Biology 5: 355-361.

Wessels, J.G.H. 1997. Hydrophobins: Proteins change the nature of fungal surface. Advances in Microbial Physiology 38: 1-45.

Wilson, G.W.T., and Hartnett, D.C. 1998. Interspecific variation in plant responses to mycorrhizal colonization in tall grass prairie. American Journal of Botany 85: 1732-1738.

Wosten, H.A.B. and Wessels, J.G.H. 1997. Hydrophobins, from molecular structure to multiple functions in fungal development. Mycoscience 38: 363-374.

4

MOLECULAR BIOLOGY OF NUTRIENT EXCHANGE PHENOMENA IN MYCORRHIZAS

The nutrient exchange phenomenon between symbionts is a vital aspect of mycorrhizal partnerships. It has attracted the attention of several biologists. Clearer insights about nutrient transfer, particularly P, C, N, and a few other elements were possible mainly through the use of appropriate radiolabels and advanced physiological techniques. During the past five years, there has been a spurt in research activity on molecular aspects of nutrient exchange phenomenon in mycorrhizas. Several ingenious molecular and immuno-cytochemical techniques that allow us to isolate genes, clone them, and understand their localization patterns and regulatory aspects have been utilized. Knowledge about genes and proteins involved in the transport of major nutrients such as C, P and N has greatly improved during the recent past. Emphasis within this chapter is on the recent developments regarding cellular and molecular regulation of nutrient absorption, translocation and exchange within the mycorrhizal systems. Discussions have been confined to three major nutrients namely phosphorus, carbon and nitrogen.

1. Molecular Regulation of Phosphorus Nutrition in Mycorrhizas

1. A. Molecular Regulation of Phosphorus Nutrition in Arbuscular Mycorrhizas

We know that arbuscular mycorrhizal symbiosis is a mutualistic association, wherein P and C are reciprocally transferred between the symbionts. Phosphorus absorbed from the soil is immediately incorporated into cytosolic pool of the fungus. The excess P is transferred into the vacuolar system, where it is held mainly as ortho- and polyphosphate along with small amounts of organic phosphates. The vacuolar P is said to play a key role in the transfer of P into the plant. Long-distance transport of P from the site of its absorption in the soil to root cells via external mycelium is supposed to be accomplished by vacuolar components. Both protoplasmic streaming and tubular vacuoles are known to aid such P transport via the vacuolar system. Fungal arbuscules are deemed to be the site of transfer of P into apoplastic interface and then on to root cells. Arbuscules could also be the major site of exchange for C, N and other nutrients. There are innumerable reports dealing with the mechanisms of P acquisition and transport in AM. Despite this fact, there exist lacunae in our understandings which especially, regarding molecular and biophysical aspects of P transfer between the symbionts. These aspects are yet to to understood in greater detail (Ezawa et al. 2002). Within this chapter, a background about kinetics of P uptake by symbiotic plants has been provided, followed by a detailed discussion on the cytological aspects of P transport and accumulation and enzymes relevant to P transfer. Most importantly, recent knowledge on molecular regulation of P transport activity relevant to symbionts has been highlighted. Molecular aspects of carbon and nitrogen transfer have also been dealt with.

Let us first understand the available background knowledge about kinetics of phosphorus acquisition by plants. In comparison to other major nutrients such as N or K, P is an immobile element in natural soils. Hence, plants tend to root profusely, form extensive mycorrhizal hyphal network and reach out to absorb this element which is slow to diffuse towards them. Much of the P absorbed by plant roots is derived from the pool of 'available P' traced in soil solution. This inorganic P fraction (Pi) is actually absorbed rapidly by the roots. Physically, the P depletion zone in soil is confined to 1 to 2 mm surrounding the root cylinder (Fig. 4.1). In most natural conditions, kinetics of P removal by roots, and that replenished into the surroundings of root via diffusion do not tally. Depending on the physiological status of roots, the rate of P influx from soil could be 1000 folds greater than that replenished into the root zone. In addition, there are several factors such as soil texture, moisture, organic matter content that affect the P diffusion rates. In any case, inappropriately low rates of P replenishments result in dearth for P and a 'P zero zone' develops in the surroundings of the root. Obviously, concentration of P in soil solution within this 'P-zero zone' reaches well below the critical limits. Roots are then rendered

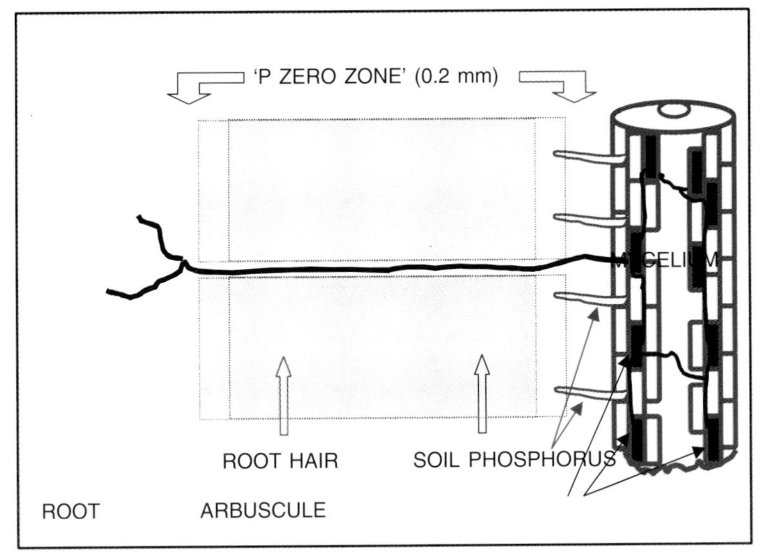

Fig. 4.1 Phosphorus uptake zones of roots, root hairs and mycorrhizal mycelium.
Note: 'Phosphorus depletion zone or phosphorus zero zone' develops as roots and root hairs absorb P at a faster rate than its replenishment through diffusion in the soil, whereas, AM fungal mycelium (blue) grow beyond the P depletion zone, garners extra P and translocate it to the roots. P unloading occurs predominantly in the arbuscular cells (blue). (*Source:* Modified based on Krishna, 2002)

physiologically incapable to acquire P that is available below the critical limit. Such critical limits vary, depending on the physiological status of roots, and their genetic constitution (see Harrison, 2003; Krishna, 1997, 2002a,b). Once a 'P-zero zone' develops, the rate of P acquisition by individual root segment is then *solely* determined by the P diffusion rate in the soil and the extent of impedence encountered, which thwarts P movement (Nye and Tinker, 1977). However, total P absorbed by plants is also influenced by several plant traits such as extent of rooting, its length, branching architecture, fine root density, hairs and mycorrhizal associations. These root adaptations or whole plant mechanisms overcome the restrictions placed due to physiological limits to P absorption by individual roots. The net advantage due to mycorrhizal symbiosis occurs mainly due to its ability to cross the 'P-zero zone' and explore P from a larger volume of the soil. The P influx rates suggested for mycorrhizal roots have varied. We are yet to decipher the causes. It could be related to differences in fungal species, its isolates, plant partner and experimental conditions. Tinker and Gildon (1983) calculated it to be closer to 10^{12} to 10^{15} mol sec^{-1} cm^{-1}. However, there are reports suggesting that P influx rates and kinetic parameters of mycorrhizal root segments differ from those known for non-mycorrhizal ones. The mycorrhizal roots possess two different absorption maxima for Pi. The P-binding sites, which relate to fungi, become active at very low Pi concentrations of 1 to 3 ppm (see Cress et al. 1979). Karunarathnam et al. (1986) have suggested that AM infection enhances the P-binding ability of soybean roots. In this case, AM enhanced the Vmax values of root segments without significantly affecting the affinity constant Km. Since then, there has been a constant effort to understand the molecular physiology of P nutrition in symbiotic roots, fungal hyphae and roots *sans* mycorrhizal hyphae.

Cytological Aspects of Phosphate Metabolism in Mycorrhizas: Translocation of phosphate from the soil to the host root through an efficient active mechanism involves the movement of polyphosphate granules containing vacuoles via cytoplasmic streaming (Gianinazzi et al. 1979, Marx et al. 1982). These vacuoles possess high alkaline phosphatase activity (Dexheimer et al. 1979; 1982), which perhaps plays a key role in PO_4 transport mechanism within AM fungal hyphae. Histochemical studies indicate that storage of phosphate as polyphosphate granules (metachromatic granules) is a common phenomenon across ecto-, ectendo and endomycorrhizas (Callow et al. 1978; Capaccio and Callow, 1982; Chilvers and Harley, 1980; Linglee et al. 1975; Straker and Mitchel, 1985). The molecular weights of such localized polyphosphates have varied, depending on the type of mycorrhiza. Ericoid mycorrhizas show up polyphosphates granules of molecular weight between 3000 to 7000, compared with AM fungal symbiosis that contain larger granules around 21,000 mol. wt (Callow et al. 1978).

Phosphate efflux from intraradical hyphae into cortical cells is a key step during P translocation via AM fungi. To study this aspect, Solaiman and Saito (2001) used enzymatically separated intraradical hyphae. They estimated the total P content, polyphosphates and efflux of P from such hyphae. The fungal P contents effluxed are substantial compared with the mere 2% biomass that it contributes to the root. Trials indicated that P efflux resulted in consequent reduction of polyphosphate in fungus. Therefore, P efflux away from intraradical hyphae could be coupled to polyphosphate hydrolysis. In fact, reduction of polyphosphate content in hyphae is indicative of P translocation to host root cells from the fungus. Perhaps, at this stage, we should also note the reverse flow of P from host cells to fungus; if any, along with its extent and regulation. Possibly, the P efflux reported so far, in many cases, is actually a difference between efflux and influx of P at root/hyphal interface. Methods that sharply distinguish and assess rates of efflux/ influx of P at the root cell-fungus interface are required. This crucial aspect needs proper biochemical assessment.

Cytochemical localization of phosphatase activity relevant to PO_4 transfer between fungal and root cells was studied by Marx et al. (1982), who state that ultrastructural organization of the arbuscule is well adapted to nutrient exchange and PO_4 transfer, in particular, is well achieved across membranes of living host/fungal cell. Their electron microscopic analyses indicated that localization of ATPase activities within cortical cells is identical in the three different AM-host combinations tested; namely, *Allium cepa-Glomus mosseae, A. cepa-G. fasciculatus and A. psuedoplatus-G. mosseae*. Further, the distribution of DES (Diethyl-Silbesterol)-sensitive plasma lemma-bound ATPase activity is modified, and becomes denser along the invaginations of the host membrane during the intracellular development of arbuscule. Compared with it, in the non-mycorrhizal cells, the activity of this enzyme gets localized along the cell periphery on plasma lemma. Within host cells possessing arbuscules/hyphae, the ATPase active is comparatively more intense. This is indicative of the involvement of ATPase in the active transfer of PO_4 between symbionts.

Gianinazzi et al. (1979) studied the cytochemical localization patterns of enzymes related to phosphate metabolism—namely, acid and alkaline phosphatase. No modifications were noticed with regard to acid phosphatase activity of host cells, either after AM fungal infection or within the period before senescence of arbuscules. However, within AM fungi, young and terminal parts of arbuscule branches possessed acid phosphatase activity. As the arbuscules branch profusely and mature, vacuolation increases but acid phosphatases activity progressively decreases.

In contrast, weak alkaline phosphatase activity confined to tonoplast is noticed at the early stages of arbuscules. Its activity increases and becomes intense in well-developed arbuscular and intercellular hyphae. However, alkaline phosphatase activity disappears as arbuscules collapse and senesce. Precise correlation between cytochemically traced alkaline phosphatase activity and mycorrhiza-specific alkaline phosphatase (MSAP) activity reported by Gianinazzi-Pearson and Gianinazzi (1978) may not exist. However, polyphosphate granules identified within vacuoles that play a role in phosphate transport mechanism within AM fungi, follow a cytochemical distribution similar to alkaline phosphatase activity. Hence, vacuolar phosphatase activity may be involved in the active mechanism of phosphate transport within AM fungi. Recently, Kojima et al. (1998) also studied histochemical aspects of alkaline phosphatase activity in the intraradical mycelium of an *Allium cepa/Gigaspora margarita* association. Hyphal phosphatase that appeared conspicuously under alkaline conditions was partially purified and characterized. It was also suspected that mycorrhiza-specific alkaline phosphatase plays a key role in phosphorus exchange in the mycorrhizal system.

Regulation of polyphosphate metabolism is an important aspect of AM symbiosis. Especially, the different enzymatic reactions, metabolic pathways and accumulation patterns of phosphates (polyphosphates) need to be understood. Ezawa et al. (2001) suggest that exopolyphosphatase activity of intra- and extramatrical hyphae may differ and this data should be ascertained. Their tests with marigold/*G. coronatum* proved that exopolyphosphatase activity in extra and intraradical hyphae differ, depending on the pH optima. Polyphosphate accumulation was pronounced in plants treated with high-P, but not in those that provided low P. The exopolyphosphatase activity in either intra- or extraradical hyphae did not alter, irrespective of polyphosphate status. They concluded that at least two different exopolyphosphatase type enzymes—which are expressed differently in intra- and extraradical hyphae—might function in AM associations. Their cytological assessments indicate that polyphosphate status of hyphae and/or root tissue could be indicative of a dynamic balance between its synthesis and hydrolysis.

In a study with monoxenic cultures of *G. intraradices,* Olsson et al. (2002) found that external phosphate concentration and its actual availability affected polyphosphate accumulation and its transport. Presence of motile tubular vacuoles in the hyphae suggested that, to a certain extent, P translocation might occur via this system. Tubular vacuole activity was significant when *G. intraradices* hyphae were in low P solution, but the vacuolar system did not function if high amounts of P were supplied to the medium. It is possible that tubular vacuole-mediated P transport is significant under low-P availability situations.

Phosphatases: Plant roots secrete extracellular phosphatases that can hydrolyze organic P. Most free-living soil microbes are also endowed with the ability to mineralize soil organic P, which occurs mostly as inositol phosphate or in phytates in the soil. Inorganic-P released is later utilized by the soil microbes for metabolism. Ectomycorrhizas are known to secrete extracellular phosphatase and utilize the organic P found in soil solution (Alexander and Hardy, 1981; Antibus et al. 1992). Clearly, such a physiological ability to mineralize soil organic P would augment P uptake by plant-AM fungus symbiosis. However, the unequivocal evidence of physiological ability of extraradical hyphae of AM fungi to secrete extracellular phosphatases has been lacking (Koide and Kabir, 2000). The mycorrhiza specific alkaline phosphatases (MSAP) were first reported long ago by Gianinazzi-Pearson et al. (1978, 1979). More recently, Negel et al. (2003) also have successfully identified mycorrhiza-specific alkaline phosphatases in *Glomus microaggregatum*/Leeke and *G. mosseae*/Leeke associations. Such alkaline phosphatases are active in the root. We have no idea regarding the extent of its excreation into soil, if any. The phosphatase activity estimated in many of the studies is attributable to both plant roots and AM fungus (Gianinazzi-Pearson and Gianinazzi, 1986) or to induction by the soil microbes (Krishna and Bagyaraj, 1985; Jaychandran et al. 1992; Joner and Jackobsen, 1995). The mycorrhiza-specific alkaline phosphateses could be of fungal origin, but no AM fungal alkaline phosphatase has been purified as yet (Negel et al. 2003).

In order to overcome such confounding arguments, Koide and Kabir (2000) utilized a sterile, in vitro plant-AM symbiosis system devized by Maldonado-Mendoza and Harrison (1999). Their study provides evidence for the ability of extraradical hyphae to secrete extracellular phosphatase that hydrolyzes organic P sources, namely 5-bromo-4-chloro-3 indolyl phosphate (BCIP) and phenolphthalic diphosphate (PDP). Such organic P hydrolyzing activity was surface bound or confined to patches on the root surface. The extraradical hyphae transferred sizeable quantities of P, which is not accountable by passive diffusion. Since the experimental system was sterile, involvement of soil microbes aided hydrolysis of organic phosphate could be ruled out. In this particular case, AM fungal ability to utilize organic P was lower by eight times, when compared with their preference to inorganic P (Pi). Still, considering that soil organic P is a sizeable fraction, this mechanism of organic P hydrolysis could be ecologically important (Koide and Kabir, 2000). On the other hand, a review by Joner et al. (2000b) suggests that quantitative importance of extracellular phosphatases released by extraradical mycelium of AM fungi is limited, considering the total phosphatase activity in soil. Joner and Johansen (1999) reported that in *G. intraradices*, 70% of the phosphatase activity measured on external hyphae was associated with hyphal wall and the rest to internal structures. Exuded phosphatases were not significant. They have argued that release of extracellular phosphatase activity is incidental in many soil organisms, including AM fungi. It has a limited beneficial effect to the host plant P nutrition.

At this juncture, our knowledge about regulation of extracellular phosphatases in AM fungi is still meager. Since, mycorrhization can enhance P concentration in root tissue; it may also reduce the phosphatase activity through feedback inhibition. Feng et al. (2002) conducted histochemical tests to visualize the release of extracellular phosphatase from extraradical mycelium. In their experimental setup, activities of acid and alkline phosphates and a corresponding depletion of organic P could be detected. Similar trends were observed for phosphatase and organic P when analyzed using transmission electron microscopy. Hence, it was concluded that AM fungi do release extracellular phophatase. Using monoxenic culture of carrot roots, Joner et al. (2000a) demonstrated that P supplied in organic form (^{32}P labeled AMP) was hydrolyzed by extraradical hyphae in the absence of other organisms and subsequently utilized as a nutrient source. Feng et al. (2003) provided external mycelia of *Glomus versiforme* with three different organic P sources (lecithin, RNA and sodium phytate) and an inorganic source (KH_2PO_4). At the initial stage of symbiosis, the three organic sources made a small contribution to plant P, meaning that only low amounts of P were hydrolyzed and mobilized by AM fungus. However, at later stages, P drawn from organic sources increased, reaching 71% of the total uptake. Therefore, in certain situations, secretion of extracellular phosphatases by AM fungi might mediate P hydrolysis and uptake appreciably. Arbuscular fungi are also known to mobilize sparingly soluble P sources. Release of organic acids, modifications in pH and solubilization of insoluble P sources could cause this situation. Evaluations by Yao et al. (2001) showed that external hyphae of *Glomus versiforme* mobilize Ca_2-P, Ca_8-P and Al-P better, but not Ca_{10}-P. Tracer ^{32}P analyses actually confirmed that plant roots and AM fungal hypha both garner the same fraction of P from the same type of P compounds. However, AM fungal hyphal network enhanced surface contact with P sources and explored a comparatively greater volume of soil.

Specific intracellular phosphatases identified among mycorrhizal plants are known to play a key role in the transformation of P compounds with in the roots, especially in the arbuscular cells. Biochemical aspects such as methods of extraction and purification, molecular weight, substrate specificity, factors influencing its activity and regulation have been studied (Ezawa et al. 1999; Gianinazzi-Pearson et al. 1978; Kojima et al. 1998; Negel et al. 2003). Ezawa et al. (2001) studied inhibitors and substrates for a mycorrhiza-specific alkaline phosphatase occurring in the arbuscules of *Glomus etunicatum* colonizing marigold (*Tagetes patula*). They extracted intraradical hyphae and arbuscules using cellulase and pectinase on the root tissue. Alkaline phosphatase was sensitive to beryllium, but acid phosphatase proved to be less sensitive. Inhibitory activity of beryllium could also be confirmed further, using particulate and soluble enzyme fractions. Both soluble and particulate enzyme hydrolysed phospho-monoester substrates such as glucose-6-phosphate, beta-glycerophosphate, trehalose-6-phosphate and glucose-1-phosphate. Pyrophosphates like ATP and polyphosphate were not hydrolyzed by these intracellular mycorrhiza-specific acid or alkaline phosphatases. Considering the kinetic properties of the particulate alkaline phosphatases, especially low Michalis-menton (Km) values for sugar phosphates such as glucose-6-phosphate and trehalose-6-phosphate, Ezawa et al. (1999) opined that it has a role to play in sugar and P metabolism in arbuscular cells.

Polyphosphate metabolism in mycorrhizal hyphae has direct relevance to P accumulation and transport within the mycorrhizal systems. More specifically, polyphosphatases in intra- and extracellular hyphae could be playing a crucial role in P translocation. A recent report by Ezawa et al. (2001) suggests that exopolyphosphatase activity may actually differ between intra- and extraradical mycelium. The pH optima for exopolyphosphatase activity may differ between extra- and intracellular hyphae. The substrate, polyphosphate occurred in hyphae given high P, but not those treated with low P. However, exopolyphosphatase activity and vacuolar acidity were relatively constant, irrespective polyphosphate status. More recent studies by Ezawa et al. (2004) have shown that rate of polyphosphate accumulation in extraradical mycelium is surprisingly rapid comparable to some of the polyphosphate-hyper accumulating microbes. They adopted an enzymatic system based on polyphosphate kinase/luciferase that is highly specific and sensitive to fluctuations in polyphosphate concentration in fungal mycelium.

1. B. Molecular Physiology of Phosphorus Acquisition and Transport in Plants *sans* Mycorrhizal Component

In most agricultural soils, P availability is either inherently low—if not, it is reduced due to chemical fixation into clay or organic fraction. As we already know, under such circumstances, plants manifest a range of physiological mechanisms such as secretion of organic acids, enzymes (phosphatases) or proteins to improve P acquisition. Phosphorus in soil is absorbed in ionic form by plant roots; later, it is loaded into xylem vessels and translocated into different tissues. During the process of acquisition, a gradient is encountered wherein P concentration in the plant tissue (roots) is higher than that in soil solution. In addition, most plant cells maintain a negative membrane potential. Therefore, an energy-aided active transport system becomes mandatory to drive P against the concentration gradient and into root cells (see Marschner, 1995). The regulation of P absorption and transport is still poorly understood (Delhaize and Randall, 1995). The kinetics of P acquisition is complex, depending on the variations in cell types and root tissues. However, kinetics of P uptake by plant root cells seems simpler, and easily comparable to single-celled protists such as yeast or *Bacillus*. Overall, there are two P uptake systems. One of them is a 'high affinity' inducible system that operates at lower levels of P (μM range) in the medium. The other, which is a low affinity constitutive system, functions at high levels of P (mM range) in the medium (Cress et al. 1978; Furihata et al. 1992; Schmidt et al. 1992).

***Pho* Mutants**: Mutations in P nutritional traits of plants can be useful in understanding the physiological and molecular aspects of P uptake and utilization. Mutations for P uptake (*Pho*) have been identified (Delhaize and Randall, 1995; Poirer et al. 1991). A P-deficient *Pho* mutant may have genetic lesion for absorption of P by roots, or at the site of P loading into xylem and translocation, or vein loading/ unloading in the leaves. Such *Pho* mutants may possess a lesion either at only one, two or all points of P transport. For example, *Pho1* described by Poirer et al. (1991) lacks a transporter protein responsible for P loading into xylem from roots. Whereas, Wu and Lefebvre (1990) reported a *Pho* mutant that lacks P transporter protein responsible for absorption of P from the soil solution. This mutation was traced to epidermal cells and a single recessive gene was attributed to this particular *Pho* locus. Several other examples of *Pho* mutants are available, which affect either the high affinity or low affinity P acquisition systems.

1. C. Phosphorus Transporters

Immediately after establishment of plant-AM fungal symbiosis, the extraradical hyphae absorb inorganic P (Pi) from soil. Phosphorus is then accumulated as polyphosphate (poly-P) in the vacuoles and transported along the hyphae. Hydrolysis of poly-P and release of Pi into symbiotic interface and apoplast are other crucial steps in P transport. In this regard, knowledge about P transporters is essential in order to gain a better uderstanding about the molecular basis of P transport and metabolism in AM symbiosis (Ezawa et al. 2001). Phosphorus transport throughout different tissues of the plant, especially translocation across membranes will need 'P transporters' (Chiou et al. 2001). For example, P concentration in xylem is high (~ 10 mM), and to load P into xylem in the roots, an active, energy-driven phosphate transport system is a requirement (Poirer et al. 1991). Regulation of P concentration within cells, i.e. influx/ efflux, involves P transport in and out of cells as well as vacuoles. Again, an active transport system mediates P uptake into cells and movement across tonoplast into vacuoles (Mimura, 1995). According to Chiou et al. (2001) our insights into molecular nature of the 'phosphate transporter genes' in plants began with cloning of these genes. The *PhT1* transporters are functional in the root tissue. For example, *LePT1* is a transporter from tomato roots active in P acquisition from soil solutes (Bucher et al. 2001). The *PhT1* family of transporter genes includes those relevant to the uptake of Pi from soil solution and redistribution in *Arabidopsis* tissue. These are H_2PO_4/H^+ symporters, upregulated in P-starved conditions. The efficient Pi uptake by cluster roots of *Arabidopsis* is attributed to such high affinity Pi transporters in the plasma lemma (Smith et al. 2003). The PhT2 represents a family of P transporters different from *PhT1* and exclusively active in the shoot system (Bucher et al. 2001). The *PhT2* gene in *Arabidopsis thaliana* reported by Daram et al. (1999) is a low-affinity transporter found exclusively in green tissue. It shares sequence similarity with sodium-coupled transporters from microbes *Neurospora crassa* and *Saccharomyces cerevisiae.* At present, several high-affinity transporters in plants which share sequence similarity with high-affinity, proton-coupled P transporter systems in *Saccharomyces cerevisiae* (Bun-ya et al.1991), *Neurospora crass* (Versau, 1995), and *G. versiforme* (Harrison and Van Burren, 1995) have been identified and cloned. For example, the high affinity P transporters from potato (*StPt1* and *StPT2*) (Leggewei et al. 1997); *Arabidopsis thaliana* (*AtPT1* and *AtPT2*) (Muchhal et al. 1996; Muchal and Raghothama, 1999; Smith et al. 1997); *Medicago truncatula* (*MtPT1* and *MtPT2*) (Liu et al. 1998b) and tomato (*LePT1* and *LePT2*) (Daram et al. 1998) share 75 to 80% similarity at the amino acid sequence level. Among the P transporters listed above, *MtPT1* and *2*, LePT1, StPT1, *2* and 3 (Rausch, 2001) complement with high-affinity transporter mutant of yeast, indicating clearly that these are functionally P transporters. Phosphate transporters such as *MtPT1, MtPT2, StPT1*or *StPT2* are downregulated as AM symbiosis develops; hence they may not be involved in P transport relevant to symbiosis (Harrison et al. 2002).

Rausch (2001) states that a P transporter such as *StPT3* that localizes itself predominantly in arbuscule-containing cells may help in transport of P derived from the fungus. In mycorrhizal soybeans, Bougoure and Dearnaley (2001) identified a set of P transporter genes that were upregulated and localized in arbuscular cells. All of these P transporters mentioned above are inducible through phosphate starvation and are expressed in root cells. So far, all of these high-affinity P transporters identified in plants are members of multi-gene families. It is interesting to note that among the rice phosphate transporter (*OsPT*) gene family analyzed, *OsPT11* was specifically induced in arbuscular cells (Paszkowski et al. 2002). Its activation was independent of nutritional status of rhizosphere. Infection by fungal pathogens such as *Rhizoctonia solani* and *Fusarium moniliforme* did not activate *OsPT11*, but colonization by AM fungus *G. intraradices* induced the gene. It complemented a defect in P uptake by yeast, and was phylogenetically related to a similar high affinity P transporter identified from *Solanum tuberosum*. Harrison et al. (2002) identified a phosphate transporter from *Medicago truncatula—MtPT4*, expressed exclusively in the arbuscular cells. Immunolocalization studies indicate that it is localized in periarbuscular fraction. The transport properties and spatial expression patterns of *MtPT4* indicate that it has a role in acquisition of phosphate released from the fungus in arbuscular cells. Its expression pattern is similar to the one described for *StPT3* by Rausch et al. (2001). Staining for *MtPT4* revealed that its activity was particularly intense on the membrane surrounding the fine branches of mature arbuscules, but absent surrounding the thicker hyphal trunk.

Location and Secretion of 'P Transporters': It is crucial to understand the tissue, cellular or subcellular location where high affinity P transporters are secreted and where they become active. It leads us towards a better understanding about spatial expression patterns and regulation of P transporter protein. In this regard, Marschner (1995) had predicted that such high-affinity mineral (e.g. P) transporters could be located in the plasma membrane of root epidermal cells or hairs, so that they are situated in proximity to soil solution P during the absorption process (Fig. 4.2). So far, phosphate transport proteins from several plant species have been isolated and studied (see Chiou et al. 2001; Daram et al.1998; Leggewei et al. 1997; Burleigh and Bechman, 2002). However, at present our knowledge about their secretion spots and spatial patterns are meager. Recently, Chiou et al. (2001) achieved a degree of success in deciphering the accumulation and spatial patterns of a P transporter protein (*MtPT1*) from *Medicago truncatula*. Carefully designed immuno-cytochemical tests and confocal microscopy revealed that *MtPT1* is predominantly distributed in the plasma membrane of root epidermal and root hair cells. Biochemical analysis of a comparable P transporter protein from tomato (*LePT1*) has revealed that its location/activity could be cell specific. In addition to epidermis and root hairs, LePT1 transcripts were also traced in root cap, stele and leaf parenchyma (Daram et al. 1999; Liu et al. 1998). Generally, *PhT2* family of P transporters are localized in green tissue and especially those involved in loading P into aboveground organs and leaves (Bucher et al. 2001).

While investigating the influence of tissue age on spatial patterns of MtPT1, Chiou et al. (2001) observed that its secretion/expression is detectable in both young and intermediate aged tissue, thus, indicating that P acquisition continues in both young and aged roots. Further, they suspect that although MtPT1 is traced in all epidermal and root hair cells, within each cell its distribution could be asymmetric. They believe

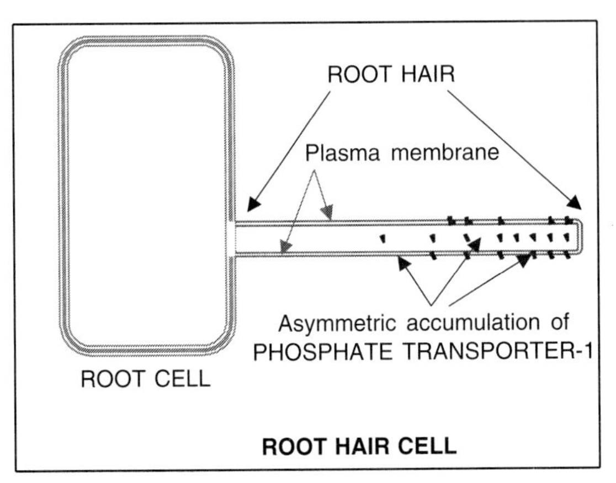

Fig. 4.2 A diagrammatic representation of asymmetric accumulation of Phosphate Transporter-1 of *Medicago truncatula* (MtPT-1) in the root hairs. Phosphate transporter accumulates at the tip of root hair, where P movement into root cell is predominant. *Note:* This example refers to the non-mycorrhizal plant. Therefore, root cells do not show mycorrhiza or mycorrrhiza coded transporters. (*Source:* Based on original) immunocytochemical analysis by Chiou et al. (2001)

that the tip of the root hair cells may be a preferred spot for MtPT1 to accumulate, because it may then attain close contact/proximity to P in the soil solution. Asymmetric distribution of mineral transporter proteins is not uncommon. The fact that root hairs contribute maximum towards P absorption (Ito and Barber, 1983) is an added evidence consistent with the current knowledge about location of P transporters in root hair cells (Fig. 4.2). Rausch (2001) examined the localization pattern of the P transporter *StPT3* and noted that *StPT3* mRNA was predominant in arbuscule-containing cells. Protein translated from it localized in the periarbuscular membrane, which is of plant origin and involved in the ultimate transport of P into plant. Benedetto et al. (2002) analyzed *Gigaspora margarita* EST with a view to identify the genes involved in P transport and metabolism. They traced Pi transporters with similarity to those from other AM fungi, especially *G. mosseae*. Considering AM fungi in general, they propose that several Pi transporters might be involved in P transfer and a few of them are surely expressed during the pre-symbiotic and/or extraradical fungal structures. Many other P transporter genes may be expressed during different stages of the symbiosis. Similar suggestions have been made by Requena et al. (2003).

Regulation of Phosphate Transporters in Plants (sans mycorrhiza): Knowledge of the genetic expression and physiological regulation of P transporters is important. The high affinity P transporter genes identified so far are regulated in response to P status of plants. Northern and Western blot analyses have indicated the correlation of both transcripts and protein expression patterns. The expression of P transporter is highest in plants starved of P, but negligible under high PO_4 conditions in solution. Such close correlation between transcript and protein expression patterns indicates that phosphate starvation induces rapid uptake of this element by the roots. It is normally accompanied by changes in Vmax but not Km. It is also interesting to note that P transporter levels, for example, MtPT1 decrease in response to colonization or roots by AM fungi. The MtPT1 gene, both its transcripts and protein expression are downregulated in the form of colonization levels of AM fungi intensifies. However, the underlying mechanism, as to why MtPT1 is not preferred in the mycorrhizal state is unknown. Perhaps it occurs as a routine 'take over' of phosphate-acquisition process, as the invading symbiotic AM fungi start to establish within plant roots. Overall, it is very clear that MtPT1 or similar high-affinity transporter genes in roots are not involved in P transport at plant cell-AM fungal interfaces or in arbuscular cells. They are functional only in non-mycorrhizal roots, and aid the uptake of P from soil solution. Burleigh and Bechman (2002) point out that not all P transporters are regulated upwards. For example, expression of MtPT2 within mycorrhizal roots of *Medicago* is inversely related to P status within the shoots. It means that P supply in plants affects gene expression in the AM fungus. The downregulation of such P transporters could also be influenced by carbon demand of the symbionts.

1. D. Molecular Aspects of Phosphorus Acquisition by Plant-AM Fungal Symbiosis

Molecular regulation of P transport from soil-fungus interface through the hyphae and its delivery into root cells is one of the most significant aspects of mycorrhizal symbiosis. However, our knowledge about molecular aspects of P transfer in mycorrhizal systems is meager, and much remains to be elucidated (Gianinazzi-Pearson, 1996; Harrison, 1997; Maldonado-Mendoza et al. 2001; Harrison et al. 2002). There is no doubt that phosphorus acquisition and transport occurs both directly, via roots and through plant-AM fungal symbiosis. Gianinazzi-Pearson (1996) believes that AM fungal-mediated P transport is a preferred P acquisition mechanism in the natural environment. The contributions of AM fungus-mediated P acquisition and delivery system within the plants could be significantly higher than root alone (Pearson and Jackobsen, 1993a). Reports by Chiou et al. (2001) and Liu et al. (1998) indicate that high affinity root phosphate transporters are downregulated in the root, and that plants prefer to depend on AM fungal P transporters when in a symbiotic state. At this point, there is need to identify the timing of shift to AM fungal P transport system, the extent of P transported by AM fungal versus 'root only' system, and the factors that regulate this shift between the two systems of P acquisition and transport. For example, in a fungicide-treated plant, how quickly will the 'root only' system of P transport system take over from plant—AM fungal P transport system? In nature, perhaps both fungal and plant-encoded P transporters genes may be expressed simultaneously, but at different proportions, depending on the P status of individual root segments and surrounding zones in the medium. It will be worthwhile to investigate the fluctuations in quantitative expression of host and AM fungal-coded P transporter genes, but first, accurate procedures will be needed (Fig. 4.3).

Using an in vitro arbuscular system for *G. versiforme* (St Arnaud et al. 1996), a phosphate-transporter gene (*GvPT*) was identified and cloned by Harrison and Van Buuren (1995). Later, using a slightly modified AM fungal in vitro culture developed on *M. truncatula* roots with *G. intraradices*, Maldonado-Mendoza et al. (2001) analyzed in greater detail the gene expression and regulation of *GiPT*. This *GiPT* from *G. intraradices* is obviously a homolog of *GvPT* (*G. versiforme*) derived from a related AM fungus. Within such AM fungal systems, P transporter (*GiPT*) transcript levels increase in the extraradical hyphae in response to the supply of micromolar levels of phosphate in the surrounding environment. Induction of *GiPT* is rapid and its increases

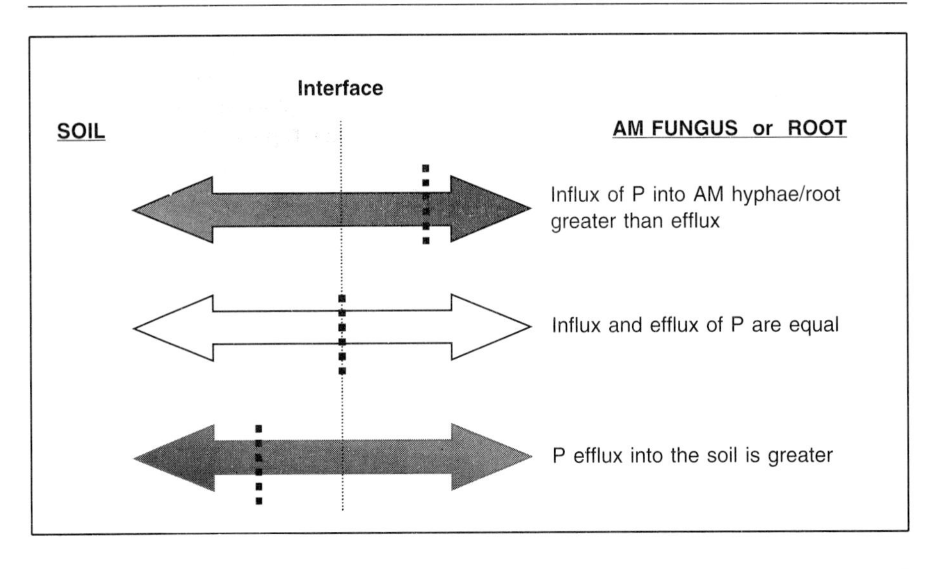

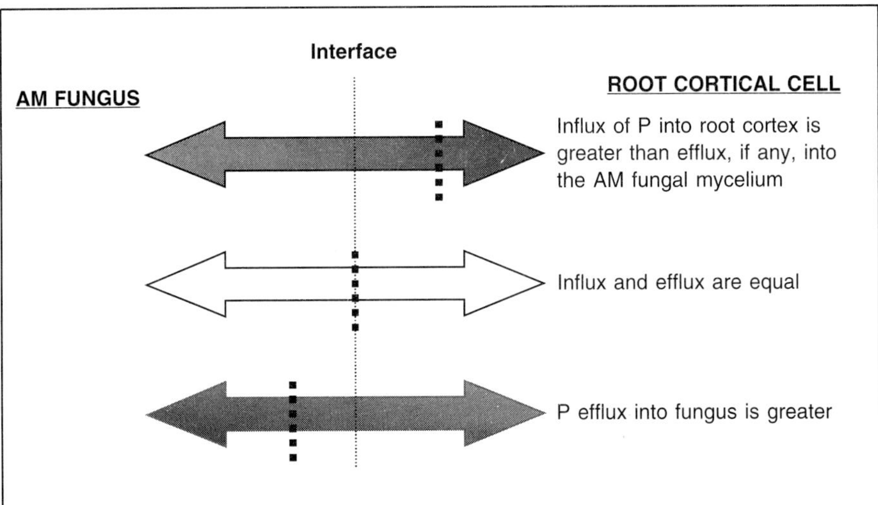

Fig. 4.3 The Bi-Directional of Movement of Phosphorus and Equilibrium Points. Top: P translocation at the interface between the soil and AM/ECM fungal mycelium. Bottom: P translocation at AM/ECM fungal hypha and root cortical cell interface. *Note:* Hairline is indicative of mid-point of the bi-directional translocation phenomenon; ▪▪▪▪▪ Bold dotted lines indicate possible locations of the point of equilibrium in each case.

can be discerned within 3 h. The magnitude of increase in the expression of P transporter gene is reported to be consistent. Maldonando-Mendoza et al. (2001) found that for a treatment with 35 μm phosphate, *GiPT* transcripts increased by 10 to 30 folds within 24 h. Further, they have confirmed that expression of *GiPT* subsides to nil, if P is absent or added at a very high concentration, say at mM levels. Clearly, genetic elaboration and regulation of P transporters are directly linked to the level of phosphate in the medium (soil/agar) surrounding the extraradical hyphae. Phosphate concentrations as low as 1 to 5 μM are sufficient to induce expression of *GiPT*. A depletion in phosphate that occurs due to its utilization by roots can shut down *GiPT* expression, but its continued expression is dependent on consistent supply of phosphate. Further, a rapid decline in P transporter transcripts in extraradical hyphae in response to depletion in phosphate in the medium implies that *GiPT* transcripts turnover quickly. Maldonando-Mendoza et al. (2001) point out that such P transporter (e.g. *GiPT*) is to be considered as high-affinity P transporter. It is consistent with the fact that *GiPT* is induced when phosphate concentration is low, a situation quite frequently encountered in the soil solution. Knowledge about the chemical nature of an inducer is crucial. Vanadate, a transition-state anologue of phosphate, also induces *GiPT* expression to the same level. Hence, phosphate itself could be the inducer of *GiPT* in the extraradical mycelium. At present, we lack information on the manner in which *G. intraradices* senses phosphate, whether it senses external or internal P level, or a P-bearing metabolite such as ATP, phospholipid, sugar-PO_4, etc.

Increase in *GiPT* transcripts is generally accompanied by a decrease in phosphate concentration in the medium. Conversely, high P in the medium shuts off *GiPT* expression. However, P uptake and transport by extraradical mycelium continues to occur even at high phosphate concentrations in the medium, despite an absence of *GiPT* transcripts. This clearly indicates the presence of another P transporter system, perhaps via a low affinity P transporter that operates under such conditions. For example, Thomson et al. (1990) predicted the occurrence of a low affinity P transporter system that shows Km in the order of 10 mM. Such low-affinity P transporters are known to occur in *S. cerevisiae* and *N. crassa*.

After analyzing and comparing the information on GiPT with P transporters from other microbes, Maldonando-Mendoza et al. (2001) suggest that regulation of GiPT is similar to high-affinity P transporter from *Saccharomyces cerevisiae* and *Neurospora crassa*. The genes *PHO* 84 in yeast and PHO 5 in *N crassa* were derepressed under P-limiting conditions, but repressed if high phosphate levels were encountered in the medium. Such high-affinity P transporters also occur in roots, which are not mycorrhizal (i.e. plant origin) (Furihata, 1992), and their regulation is similar to high-affinity P transporters from AM mycelium. Overall, the expression of high-affinity P transporter genes, their activation, derepression or repression seems to be controlled by a '*PHO* regulon', be it in plants, microbes or mycorrhizal fungus in the present context.

In a symbiotic state, P transport activity of AM fungi could be influenced by phosphate in the soil solution as well as P status of the host plant. The discussion above, so far, relates only to regulation of P transporter expression in response to phosphate concentration in the medium soil solution. However, experimentation by Maldonando-Mendoza et al. (2001) has revealed that if the P status of the host plant is held high through a pre-treatment with 35 μM P solution, then *GiPT* is induced but very feebly. The *GiPT* transcripts are either undetected or negligible in roots pretreated with P, compared with those not supplied with P previously. Definitely, the P status of root and/or intraradical fungal structures affects *GiPT* gene expression. Maldonando-Mendoza et al. (2001) suggest that depending on source-sink effect, P efflux into the root cell at arbuscule might be regulated by P acquisition rate by the extraradical hyphae (see Fig. 4.3). Alternately, a P status related/induced signal from plant might be involved in regulating GiPT. It is not surprising to note that extraradical hyphae from 'phosphate pretreated plants' cease to absorb P. Clearly, such a suppression of P acquisition by AM hyphae is dependent on the host P status. However, the magnitude/extent of regulatory influence on P transporters by the plant P status, and that by P in soil/medium is not easy to quantify. We lack information on mechanisms, especially feedback effects. Perhaps a gradient of P across root tissue/fungal hyphae regulates the induction/suppression of P transporters. At this juncture, we also lack information on reverse flow of P, if any, from 'P-pretreated' plants (roots) to AM/EM fungi.

With regard to nutrient acquisition from the soil, the net nutrient uptake by plant roots is generally discerned as the difference between the total influx and efflux due to passive and active mechanisms. We can think of a similar situation at host cell-AM hyphae or arbuscular interface, wherein the net P transported by AM fungi will actually be the difference between the total P transported by AM hyphae (i.e. P efflux from AM fungus to host cell) minus that drained from the plant to hyphae. The regulation of P drain from plant root to AM fungus, if any, could be mediated either passively or actively. Observation by Lindahl et al. (2001) using fungal mycelia of *Hypholoma fasciculare* supports the views on the bi-directional movement of P. They utilized radiolabeled ^{32}P and ^{33}P in combination with electronic autoradiography to monitor the bi-directional movement of P. Actually, significant amounts of P were transported in both the directions simultaneously between two root blocks connected by mycelial cords. Therefore, several reports about the extent of P uptake into roots via AM/ECM fungal mycelia may actually refer to the difference between the bi-directional movements of P in those conditions. Movement of P into roots from fungus, therefore, means that P flow into roots is greater than efflux (reverse flow) from roots into fungal mycelium. The equilibrium point for P flow in most of these cases definitely favors the transport of higher quantities of P into roots from fungal mycelium (Fig. 4.4).

1. E. Vacuolar Transport of Phosphorus

Phosphorus storage and transport within the AM fungal hyphae are crucial aspects of symbiosis, which need to be deciphered in detail. Accurate knowledge regarding transport of P along the AM hyphae holds the key to our ability to understand several other related aspects of P exchange in mycorrhizal systems. In this regard, tubular vacuoles may have a role in P transport, particularly within cells, and to longer distances along the fungal hyphae (Ashford and Allaway, 2002; Uetake, 2001). Reports by Hyde and Ashford (1997) and Cole et al. (1998) indicate that vacuole systems of AM fungi form extensible tubules, but they may or may not interconnect larger, less motile, more spherical components depending on the location in mycelium and environmental conditions. Such motile tubular vacuoles exist in all three types of mycorrhizas, i.e. ecto-, ectendo and arbuscular mycorrhizas (Ashford, 2002). Basically, fungal vacuoles are dynamic multifunctional organelles. Ashford (2002) opines that such a vacuole system in AM fungi practically serves as an elongate cellular subcompartment that forms a continuum for transport of nutrients, especially P over longer distances within hyphae (Fig. 4.5). For example, we encounter extensive tubular vacuole network in *Gi. margarita*, which are present in the germ tubes, intra- and extraradical hyphae. There are many reports suggesting that, generally, nutrients (including P) are held in a relatively higher concentration in vacuoles. Phosphorus accumulated as polyphosphate in fungal vacuoles is transported and released to plant root cells at high rates. This may require energy-mediated cytoplasmic transfer

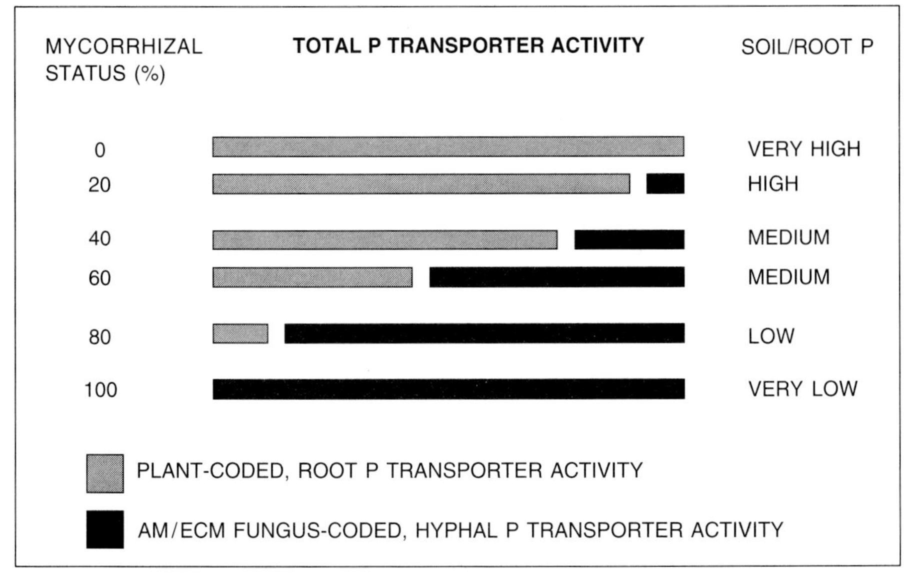

Fig. 4.4 Regulation of expression of plant-coded and Mycorrhizal fungal-coded phosphate transporter in symbiotic roots—A Model. Extent of fungal colonization and soil/ root P status regulates P Transporter Activity in plant root and fungal mycelium. Very high soil or P status is known to reduce mycorrihzal fungus ard represses its P transport activity. Under low soil P or P deficiency in roots, mycorrhizal roots depend entirely on AM/ECM fungal P absorption activity.

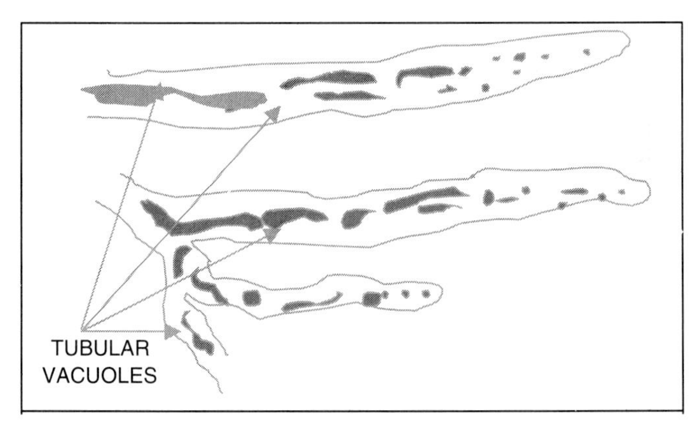

Fig. 4.5 A diagrammatic representation of Tubular Vacuoles in the fungal hyphae. They are suspected to help in long distance P transport within fungal mycelium.

system (Cox et al. 1980; Cox and Tinker, 1976). However, according to Ashford (2002), presence of a motile tubular vacuolar system throughout the mycelium of *Gi. margarita* offers an alternative P transport mechanism. It is suspected that P-bearing compounds (e.g. polyphosphate) that are common in AM fungal vacuoles may get transported via bulk flow along the lumen of vacuolar tubules. Further, such a transport mechanism avoids membrane activity and recycling at the point of delivery. Vacuolar movement, which is autonomous, may either retard or hasten P transport rates along the entire length (pathway) or a portion of hyphae. At this juncture, it is useful to note that a microtubule inhibitor such as benomyl inhibits P transport in *G. caledonium* (Larson et al. 1996), meaning that tubular vacuoles are involved in P transport. According to Ashford (2002), there are a few aspects that need clarification. For example, what form does the polyphosphate take in the vacuoles? Is it precipitated as granules? A few other points to ponder are: (a) Do we encounter a gradation/range in the extent of tubular vacuoles in different species or isolates of ECM/AM fungi? (b) Are there EM/AM fungal mutants devoid of tubular vacuole mediated P transport system? (c) Is vacuole-mediated transport directly related to symbiotic efficiency (P benefits)? What do we know about vacuolar P transport in plant-fungal pathogen or in plant–AM fungus systems with nil P benefits? (d) Do we encounter a reverse flow (or bi-directional) of P in tubular vacuolar? In that event, P measured at the host root cell is the difference of bi-directional flow of P? (e) To what extent does the tubular vacuole mediated transport mediate transfer of other mineral nutrients such as S, Ca, Zn, etc.? Recently, Declerck et al. (2003) have shown that elements such as caesium could be absorbed by extraradical mycelium and translocated either via tubular vacuolar system or by cytoplasmic streaming. Using root organ cultures, they demonstrated the uptake and accumulation of radiocaesium (^{137}Cs) in hyphae of *G. lamellosum*.

1. F. ATPases in AM symbiosis

Phosphorus uptake from the soil by plant roots is an energy-driven process facilitated by H^+ATPases. The ATPases generate a proton motive force across plant membranes, enabling the co-transport of phosphate and protons into the root cell. They play a key role in establishing the electrochemical gradient required for transfer of nutrients across the plasma membrane. A similar process occurs in mycorrhizal roots. The mediation of bi-directional nutrient transfer at the symbiotic interfaces between fungal and plant partner in mycorrhizal symbiosis is crucial to the sustenance of symbionts. Membrane proteins regulate such nutrient movement between symbiotic partners. Ferrol et al. (2001) have stated that among nutrients, it is most important to understand the regulation of membrane transport proteins responsible for movement of carbon, nitrogen and phosphorus across the membranes. Requena et al. (2003) have aptly suggested that modification of gene expression patterns is necessary to bring about certain physiological changes, in order to maintain bi-directional movement of nutrients in plant-AM fungus. In this regard, H^+ATPases are known to play a key role in establishing electrochemical gradient needed to transport nutrients across the plasma membrane. Application of molecular techniques has enhanced our understanding about plasma membrane H^+ATPases, a family of enzymes that drive active secondary transport systems. The genes-encoding H^+ATPases have been isolated, sequenced and their regulation have been deciphered in several mycorrhizal fungal/plant combinations. However, molecular basis for bi-directional transfer of nutrients during AM symbiosis seems as yet unclear.

Bago et al. (2002) isolated a fungal cDNA clone that codes for H^+ATPase-*GmPMA1* from *G mosseae*. Detailed analysis proved that this H^+ATPase possessed homology with a previously identified ATPase gene *GmHA5*. The *GmPMA1* and *GmHA5*, both were developmentally regulated genes. Expression levels of both *GmPMA1* and

GmHA5 genes were maintained throughout the intraradical development of the fungus. It was upregulated during asymbiotic development of the fungus. Its expression level did not change during the establishment of symbiosis, unlike *GMHA5,* that was upregulated during recognition stages and as the fungus entered root tissue through appresoria. Phosphate inputs into medium induced *GmHA5*, but sucrose addition had a negative effect. Further, Requena et al. (2003) propose that different isoforms of H^+ATPases and their molecular regulation might actually be more important during establishment and functioning of plant-AM fungal symbiosis. In yet another study, Nuria et al. (1998) cloned P-type H^+ATPase from *G. mosseae*. Amplified PCR products were sequenced and matched with plasma membrane ATPases from different sources. Some of them, especially the ones from other fungi shared 85% homology in nucleotide sequences, while certain others shared 40 to 50%. H^+ATPases were identifiable in all five partial genes coding for putative plasma membrane. They suggested that in mycorrhizas, there are several genes coding for membrane H^+ATPases, just like in other organisms. Two subfamilies of ATPase genes could be discerned in *G. mosseae*. Gene-specific primers for two sets of genes are available, so that these genes could be cloned and an expression pattern understood using RT-PCR. The plasma membrane H^+ATPase from *Medicago truncatula* (*Mtha*) is yet another example of enzyme involved in nutrient translocation at the symbiotic interface (Krajinski et al. 2002). A family of three genes (*Mtha1-3*) coding for H^+ATPases were identified. ATPase genes *Mtha1* and *Mtha2* were constitutively expressed in roots and remained unaffected by AM, whereas, transcripts of *Mtha1* could be detected only in AM tissues. In situ hybridization techniques revealed that *Mtha1* RNA accumulates only in cells containing AM fungal branches or arbuscules. Gianinazzi-Pearson et al. (2000) observed that in transgenic tobacco plants colonized with AM fungi, H^+ATPase is active in the plasma membrane around arbuscules but absent in plant mutants defective in arbuscule development. A family of seven genes coded such H^+ATPase activity. Immunocytochemical tests revealed that H^+ATPase protein was localized around peri-arbuscular membrane in the cortical cells. It was absent in non-colonized cortical cells. Studies on regulation of the seven H^+ATPase genes showed that two genes, *pma2* and pma4 were induced in arbuscules containing cells but not in non-mycorrhizal cortical tissue. According to them, these two H^+ATPases are synthesized *de novo* in the peri-arbuscular zone through selective induction corresponding (*pma*2 and *pma*4) genes.

Rosewarne et al. (2001) aimed at understanding the influence of developmental stage of mycorrrhizas on expression of H^+ATPases. Hence, they examined the expression genes for seven different *Lycopersicon esculentum* H^+ATPases (*LHA1-7*) in mycorrhizas occurring at different stages of development. They reported that only three genes—LHA1, 2 and 4 were significantly induced in roots. Phosphorus nutrition had no influence on the expression of these genes. Two H^+ATPases genes *LHA1* and *LHA2* were least affected by mycorrhizal colonization. The *LHA4* activity was conspicuous in arbuscule containing cortical cells. It is indicative of the involvement of *LHA4* gene in P acquisition from fungus. Ferrol (2002) too found that expression of H^+ATPases (*LHA*) genes in tomato plants colonized with AM fungus *G. mosseae* were differentially regulated, depending on tissue and stage of symbiosis. In summary, several ATPase isozymes are involved in nutrient acquisition and transfer between the symbionts. Their roles are specific, depending on the developmental stage of mycorrhiza and root tissue. Phosphorus transfer in mycorrhizal fungal portions is an important aspect of its cell physiology, because it directly relates to the generation of biochemical energy required for several vital functions.

1. G. Phosphate Transport in Ectomycorrhizas

In addition to cytochemical microtechniques (Ling-lee et al. 1975), energy-dispersive X-ray microanalysis can also improve our understanding about localization and distribution of phosphorus within roots colonized by ectomycorrhizal fungi. Bucking and Heyser (1999) analyzed both mycorrhizal and nonmycorrhizal roots of *Pinus sylvestris* using energy-dispersive X-ray spectroscopy. They report that in non-mycorrhizal Pinus roots, phosphate distribution and its accumulation is regulated by external P supply. Phosphorus absorption and accumulation in root cells is enhanced by a supply of NH_4^+. A phosphate starvation restricts P localization within cytoplasm. If P supply is sumptuous, then vacuoles in cortical cells accumulate P appreciably. It is argued that localization of P in the cytoplasm during a low-P stress situation is aimed at protecting the metabolic activity with P supply despite an existing deficiency in the environment. Regarding mycorrhizal plants, Bucking and Heyser (1999) noted that infection of pine seedlings with *Suillus bovinus* improved the P nutrition and removed the effects of P starvation. Phosphorus localization became more uniform. Whenever external P supply was high, mycorrhizal effect on localization of P was masked or not discernable. On the other hand, intracellular localization and accumulation of P in fungal portion was unaffected by external P supply, but dependent on regulation of polyphosphate formation. Application of ammonium phosphate caused an increase in P accumulation in the cytoplasm and in the hyphae of hartig net. They concluded that P accumulation in the cytoplasm (at hartig net) regulates the translocation of P from fungus to root cells at the interfacial apoplast.

Micro-autoradiography has also provided some useful insights about phosphate localization and movement at the hartig net and apoplastic interfaces of ECM. In their trials with *Populus alba*, Bucking and Heyser (2001) found that Pi distribution was not homogeneous in the root tissue. Instead, a sizeable fraction of labeled ^{33}P gets localized along the longitudinal axis. The fungal sheath and hartig net possess greater amounts of ^{33}P label in the median parts of the roots rather than in the apical or basal zones. Hence, it is imperative that phosphate uptake and transfer occurs in this zone of the root. The Pi is mainly translocated at the hartig net and interfacial apoplast. The Pi is distributed through stellar tissue into the meristematic zones of shoot system, such as young leaves that act as the major sinks. Such Pi translocation is somehow related to carbon supply. If photosynthesis is obstructed or reduced, then Pi translocation rates decrease proportionately. Micro-autoradiographic analysis has also showed that fungus possesses a good ability to attract labeled ^{14}C. The labeled carbon too localizes at median zone in the root, not near apex or base. In this nutrient zone, cortical cell nuclei show an increased sink strength for ^{33}P, indicating that median zone of roots support brisk metabolic activity during bi-directional P transfer. Bucking et al. (2002) have postulated that phosphate demand by the host regulates Pi translocation. The bi-directional movement of Pi and other nutrients, say, carbon does occur through the same interface between the ECM fungus and host tree root.

Ashford et al. (1999) have confirmed that high levels of P occurred in the vacuoles of sheath and Hartig net of *Pisolithus tinctorium* in ectomycorrhizal association with

Eucalyptus pilularis. These vacuoles also possess high levels of potassium. Sulphur too occurred in higher levels in P-rich vacuoles than in the cytoplasm. Vacuoles are also rich in S containing amino acids/peptides. Hyphal tips may contain P-rich vacuoles, but are not rich in S content (Cole, 1998). Vacuoles of hartig net and sheath generally possess greater quantities of P than in root cells. We may realize that root cell vacuoles are larger but variable; hence, the total P held per vacuole or cell varies (Ashford et al. 1999).

In order to assess the factors that influence phosphate accumulation, Gerlitz and Gerlitz (2002) conducted comparative in vivo ^{31}P NMR tests on the ECM fungus *Suillus bovinus* grown in pure culture. Firstly, they assessed the effect of monovalent cations such as Li^+, Na^+, K^+, or Rb^+ provided at 10 mM. The experimental set up allowed constant estimation of polyphosphate or phosphate accumulation and its kinetics. The effects of monovalent cations on polyphosphate accumulation varied, depending on the individual monovalent cation applied to the pure culture. Compared with a control equated to 100%, intracellular polyphosphate accumulation ranged from 94 to 119, again, depending on the monovalent cation. They opine that the effect of individual monovalent cation was dependent on their position in the periodic chart. Whenever K^+ was exchanged for one of the treated monovalent cations, phosphate metabolism returned to normal. The increasing content of mobile polyphosphate was attributable to higher energy demand. In terms of mechanism of action, Gerlitz and Gerlitz (2002) state that increasing depolarization of cell membrane through application of different ions inhibited Pi accumulation in the following order: Rb^+, $<K^+$, $<Na^+$, and $<Li^+$. The intracellular poly P content also controlled the Pi accumulation. Nuclear magnetic resonance (NMR) methods were utilized to study the accumulation and trafficking of phosphate in ectomycorrhizas. For example, Martins et al. (1999) studied the symbiosis established between *Castanea sativa* and *Pisolithus tinctorius*. The ECM fungus accumulated large quantities of polyphosphates during the rapid growth phase. While both mycorrhizal and nonmycorrhizal roots possessed orthophosphates, only mycorrhizal roots accumulated polyphosphates. The polyphosphate accumulation proceeded briskly for three months, but the levels of orthophosphates decrease in mycorrhizal roots. Hence, they concluded that such ECM systems might be relying more on polyphosphate accumulated by the ECM fungus.

Phosphorus translocation in short ranges, say from substrate hyphae to the root tissue is accomplished as polyphosphate. A set of high-affinity phosphate transporters mediates P absorption from soil solute into ECM fungal hyphae. In *Tricholoma vaccinium,* two symporters—one active under acidic and the other in neutral pH was reported by Kothe et al. (2002). Phosphorus starvation induced expression of P transporters. Franco-Zorilla et al. (2004) opine that induction and regulation of phosphate-starvation-responsive genes requires a sophisticated system that senses fluctuation in internal and external P*i* concentration. They believe that a transcription factor (PHR1) plays a central role in the regulation of P transporter genes in response to internal and external P status. Such P-starvation induced gene responses seem to be modified by hormonal signaling, especially cytokinins.

2. A. Molecular Regulation of Monosaccharide Transfer in Ectomycorrhizas

A carbon (glucose) sensing system is a necessity, so that key functions of transfer and utilization of carbon by mycorrhizal partners can be effectively regulated.

Mechanisms that sense carbon levels and generate appropriate signals should be operative within mycorrhizas. Discussions under this section are confined to monosaccheride transport in ECM and are highlighted by considering a couple of examples.

We already know that a disaccharide such as sucrose—that might be easy to reach in vascular bundles—is not utilized as such by the ECM fungus. Instead, monosaccharides like glucose and fructose are the preferred C sources that get transported across from host plant to fungus at the apoplastic interface (at hartig net). The regulation of monosaccharide transport at the apoplastic interface is mediated partly by a 'monosaccharide transport system' of the fungal partner. In *A. muscaria, AmMST1* protein is one such active monosaccharide transport system (Nehls et al. 1998; Weise et al. 2000). The protoplasts *of A. muscaria* too possess a monosaccharide transporter activity. With *AmMST1,* the Km value for glucose was determined at 0.4 mM and 4mM for fructose. A strong inhibition of fructose uptake occurred in the presence of glucose. In *A. muscaria*, basal activity/expression of *AmST1* gene is maintained when glucose concentration is still below 5 mM, but above this threshold a four-fold increase in *AmMST1* gene transcripts were easily discernable. A similar enhancement in AmMST1 activity occurs even with *A. muscaria* existing in symbiotic state with *Picea abies* or *Populus truncatula*. Clearly, *AmMST1* in mycorrhizas (*A. muscaria* protoplasts) is sugar regulated and depends on monosaccharide concentration at plant/fungus interface. According to Buscot et al. (2000) the AmMST1 uptake ability (transport activity) gets saturated at 5mM glucose concentration. The fungal monosaccharide transporter system seems to sense the saturation levels of monosaccharide. In symbiotic associations, the presence of additional photo-assimilates may trigger additional transporter proteins. The *AmMST1* activity enhances gradually, be it in symbiotic fungus or protoplasts. In fungal cultures, it requires 18 to 24 h to discern a significant change in AmMST1 gene expression from constitutive levels to maximum inducible activity. The explanation offered is that in soil, the growing mycelia are exposed generally to low concentration of carbohydrate, because higher C concentrations are sporadic and limited in time and space. In symbiotic state, fungal mycelia are exposed to a continuous supply of plant-derived carbohydrates at the apoplastic interface, but mycelia residing in the soil do not experience the same levels of monosaccharides. Buscot et al. (2000) believe that ECM fungus adapts to rapid fluxes of monosaccharides at the other apoplastic interface, by converting them immediately into storage sugars such as trehalose, mannitol or glycogen. Overall, the extended lag phase before induction of *AmMST1* gene expression in response to higher levels of monosaccharide seem to be a natural adaptation of the ECM fungus.

According to Buscot et al. (2000) fungal gene expression is regulated in response to the supply of glucose by the host. Their argument is based on the fact that analogs such as 3-O methyl glucose, which is not metabolized, or 2-desoxyglucose which is phosphorylated could mimic the glucose effect in vivo. At this juncture, we should note that hexose transporter gene is not the only gene responding to glucose availability or supply by ECM fungus. Differential screening of cDNA libraries of *A. muscaria* and mycorrhiza formed between *A. muscaria* and *Picea abies* has shown that a range of clones are up-regulated if hexose concentration is > 5 mM threshold. Weise et al. (1999, 2000) state that at the apoplastic interface, both symbionts, i.e. fungal and root cells compete for available monosaccharides. The kinetic property of monosaccharide transporter is also distinguishable based on the utilization rates of hexose substrates. For example, the monosaccharide transporter from *A. muscaria*

(*AmMST1*) distinguishes the pyronose in galactose by stearic hindrance at C-4; hence it transports comparatively lower levels of galactose.

We may realize that hexoses formed at the root/fungus interface are equally accessible to both fungal and plant monosaccharide transporters. As stated above, monosaccharide transporter from the ectomycorrhizal fungus *Amanita muscaria* has been characterized. It gets upregulated, leading to increase in hexose uptake capacity of the fungal partner. To characterize host (*Picea abies*) monosaccharide transporter, nucleotide sequences from known plant hexose transporters were matched (Nehls et al. 2000). A full-length clone from plant host denoted as *PaMST1* was identified. The entire cDNA codes for a protein with 513 amino acids. It revealed the best homology with H^+/monosaccharide transporters from *Arabidopsis*, *Saccharum* and *Ricinus*. The PaMST1 was expressed highly in the hypocotyls. However, mycorrhiza formation leads to a decrease in the *PaMST1* expression in spruce seedlings. The *PaMST1* gene, along with fungal monosaccharide transporters, seems to play crucial role in carbon allocation between symbionts.

In a *Populus-Amanita* system, ECM fungal sheaths separated from the root but still containing hartig net was analyzed for expression of sugar transporter genes in response to hexose supply. Expression of both the monosaccharide transporter gene (*AmmST1*) and phenylalanine ammonia lyase (*AmPAL*) followed patterns typical of low sugar availability in fungal sheath. It was clear that fungal sheath possessed typically low hexose, but hartig net was richer in hexose concentration (Nehls et al. 2001). Bucking and Heyser (2001) argue that a stringent barter occurs with regard to carbohydrate versus phosphate transport. Phosphate acquisition and transfer by ECM fungus is dependent on carbohydrate supply. Shading, which reduces $^{14}CO_2$ supply to ECM, resulted in reduction in PO_4 absorption. Radiolabeled studies with *Populus tremmela-Laccaria laccata* model using $^{14}CO_2$ and ^{32}P too indicate a direct relationship between C supply and P absorbed.

2.B. Molecular Regulation of Lipid Metabolism in AM Fungus

The physiological and cellular aspects of carbon transport and its regulation have already been dealt with in Chapter 2. Hence, under this section, facts regarding the molecular aspects of lipid translocation are discussed.

Lammers et al. (2001) searched for sequence homology among 420 randomly selected clones of cDNAs from *G. intraradices*, and they identified partial sequence homology to known acyl-COA dehydrogenase sequence. The deduced amino acid sequence was aligned with those from known acyl-COA from *Neurospora*, *Pseudomonas* and *Homo sapiens*. The N-terminal amino acid of the fungal acyl-COA dehydrogenase matched with that of *Neurospora* but not with human or *Pseudomonas* acyl-COA dehydrogenase sequence. A further examination of 291 ESTs from germinating spores of *G. intraradices* has revealed a sequence with strong homology to fatty acyl-COA ligase (acyl-COA synthase) and several others with polyketide synthases. The sequence obtained for acyl-COA dehydrogenase from *G. intraradices* matched with those from acyl-COA dehydrogenase of other fungi by over 37% in a stretch of 329 amino acids. Hence, Bago et al. (2002) inferred that this sequence might be responsible for coupling coenzyme A to fatty acids from TAGs in *G. intraradices*. Using real time reverse transcription (RT)-PCR techniques, Bago et al. (2002) demonstrated the expression of this putative acyl COA dehydrogenase in the

ERM and germinating spores. It compared with those of β-tubulin and rRNA, in almost equal quantities (see Table 4.1).

Table 4.1 Expression of genes relevant to metabolism and movement of lipids in the extraradical mycelium and germinating spores of *Glomus intraradices*

Gene	*Extraradical mycelium*		*Germinating spore tissue*	
	–RT	*+RT*	*–RT*	*+RT*
Acyl-COA Dehydrogenase	<50	8300 (870)	<50	2700 (310)
B-Tubulin	<50	6700 (2600)	<50	2800 (710)
rRNA	55 (3)	$6.6 (2.0) \times 10^8$	102 (6)	$5.6 (0.74) \times 10^8$

Source: Bago et al. 2002.
Note: Triplicate real time RT-PCR was performed on 6 ng of DNAse 1 treated total RNA isolated from the two tissues. Table values are copy numbers with standard deviation shown in parentheses.

The putative acyl-COA dehydrogenase catalyzes the first step in β-oxidase of fatty acid to acetyl-COA, and it occurs in peroxisomes/glyoxisomes. In yeasts and higher plants, B-oxidation is mediated by acyl-COA oxidase, but in filamentous fungi such as *Neurospora, Aspergillus* or the AM fungus *G. intraradices,* acyl-COA oxidation is entirely catalyzed mainly by dehydrogenase rather than oxidases. The glyoxylate cycle is very active in both ERM and germinating spores of *G. intraradices* (Lammers et al. 2001). Further, Pfeffer et al. (1999) suggest that lipid exported to RM or that stored in spores is the carbon sourced for anabolic processes. Therefore, acyl-COA dehydrogenase could be playing a role in carbon flux into anabolic processes, and maybe in the supply of acetyl-COA destined for oxidation in the TCA cycle.

The regulation of lipid transfer protein (LTP) in rice seedlings is affected by AM fungal colonization. Blilou et al. (2000) report that transcript levels of LTP increased when *Glomus mosseae* formed appressoria and penetrated the root epidermis, but decreased at the onset of extensive intercellular colonization of root cortex. Histochemical test indicated that LTP was associated with appressoria formation and penetration area. The induction of LTP coincided with a transient increase in the expression of a phenylalanine ammonia lyase gene. It also coincides with a transient increase of salicylic acid in the mycorrhizal roots. Both *LTP* and *PAL* genes were also implicated with plants' defense response. In plants, LTPs are known to participate in membrane biogenesis and regulation of intracellular fatty acid pools. They are also implicated in adaptation to various environmental conditions, cutin formation and defense responses.

3. Molecular Regulation of Nitrogen Exchange in Mycorrhizas

3. A. Molecular Regulation of Nitrogen Exchange in Ectomycorrhizas

In nature, host plants can be strong sinks for amino acids available in mycorrhizal fungi, be it AM or ECM. Ectomycorrhizal fungi do assimilate organic nitrogen from soil, which can then be transported to the plant host. Recently, Frommer et al. (2003) selected an ECM combination of *Hebeloma cylindrosporum/Pinus pinaster* to investigate organic nitrogen transporters. These organic-N transporters were cloned by complementation with yeast mutants deficient in amino acid and peptide uptake systems. Putative transporters were isolated by selection of transformed yeast on amino acids and peptides as sole nitrogen source. Frommer et al. (2003) also identified *Hebeloma* cDNA (*HcGAP1*) with strong homology to amino acid permease of yeast and filamentous fungi, and two other cDNAs (*HcPTR2A* and *B*) with strong homology to peptide transporters of yeast and filamentous fungi. These organic transporters were characterized using ^{13}C-Asp and ^{3}H-peptide. Overall, it was suggested that organic-N transporters such as *HcGAP*, *HcPTR2A* and *HcPTR2B1* could be involved in the transfer of amino acids and peptides from the soil and translocation to the host plant.

Chalot et al. (2002a, b) investigated the kinetics, energetics and specificity of amino acid transporter from the ECM fungus *Paxillus involutus*. They reported that uptake of glutamate, aspartate, and alanine by the ECM fungus followed simple Michaelis-Menton saturation curves, and the Km ranged from 7 μM for alanine to 27 μM for glutamate. Further, pH dependence and inhibition by protonophores indicated that a proton symport mechanism was operative during amino acid uptake by *Paxillus*. This particular amino acid transporter recognized a wide range of substrates, resembling the general amino acid permease from yeast. They suspect that this amino acid transporter from *Paxillus* may be essential for nitrogen transport at the interface between plant host and fungus, particularly during the early stages of ECM symbiosis.

Kleber and Nehls (2003) aimed at understanding the influence of N supply on expression of an amino acid transporter in *Amanita muscaria* (AmAAP1). Their results suggest that nitrogen status of the fungal cell may regulate *AmAAP1* gene expression, which is a similar situation to the one observed in Ascomycetes. The N status of cells is presumably sensed via glutamine synthetase/glutamine concentration. To confirm it further, they isolated the promoter region of *AmAAP1* (1 kb) using electrophoretic mobility shift assay. The nitrogen-dependent protein binding could be observed in a promoter region between 250 and 540 bp upstream of the translation start. Next, detailed analysis of this region led to identification of two binding sites, 20 and 40 bp in length, thus confirming the nitrogen-dependent expression of the *A. muscaria AmAAP1* gene.

Ammonium transporters: Ammonium or ammonia transporters from a wide range of microbes have been isolated, characterized and their regulatory aspects have been deciphered. Generally, the encoded proteins of ammonium transporter (*AMT*) gene family vary in size from 391 to 622 aminoacyl residues (see UCSD, Protein database). In ectomycorrhizas, appropriate expression of ammonium transporter genes is crucial for nitrogen uptake activity by their extraradical hyphae. It definitely influences nitrogen nutrition of the ECM symbiosis. The mechanism of energy coupling, if any, to methyl-NH_2 or NH_3 uptake by *AMT* protein is as yet unclear. A proton motive force (pmf), cytoplasmic K^+ and ATP have all been ascribed. The ATP dependence involves glutamine synthase reaction. NH_4^+ uniport driven by pmf, energy-independent NH_3 facilitation, and NH_4^+/K pump are proposed as possible mechanisms. Many organisms are said to contain multiple homologues of the *AMT* gene. The generalized transport reaction catalyzed by members of *AMT* gene family is as follows:

$$NH_3 \text{ (out) or } NH_4 \text{ (out) } \{+ K^+ \text{ (in)}\} \underset{\text{(ATP)}}{\overset{\text{pmf}}{\rightleftharpoons}} NH_3 \text{ (in) or } NH_4 \text{ (in) } \{+ K^+ \text{ (out)}\} \quad (1)$$

$$CO_2 \text{ (in)} \rightleftharpoons CO_2 \text{ (out)} \quad (2)$$

Mantanini et al. (2002) have reported the isolation of a high-affinity ammonium transporter from the ECM fungus *Tuber borchii*, an ascomycetous truffle. The polypeptide encoded by this gene named TbAMT1 functionally complements ammonium uptake-defective yeast mutants. It bears sequence similarity with ammonium transporters of *Saccharomyces* (*MeP*) and *Arabidopsis* (*AtAMT1*). The *TbAMT1* is a high affinity NH_4^+ transporter with Km for ammonium at 2 ìM. It is specifically up regulated in response to nitrogen deprivation. Both NH_4^+ and NO_3 seem to be effectively utilized by this transporter found in *T. borchii* mycelia. Hence, it should be playing a key role in acquiring N from soil in to hyphae.

Ammonium is the most abundantly available inorganic N source in soils. It is absorbed by ECM fungi. Its transport across the plasma membrane is mediated by transporters. Chalot et al. (1998) characterized a specific ammonium transporter in ECM fungus *Paxillus involutus*. They cloned and sequenced it using yeast primers. Sequence homology tallied with yeast NH_4^+ transporter genes *mep 1, 2* and *3*. Several other transporters involved in absorption and utilization of organic N sources by ECM fungi were also investigated and reported. Recently, Javelle et al. (2003a) have cloned and characterized genes for an ammonium transporter (*AMT1*) and ammonium metabolizing enzymes glutamine synthase (*GLNA*) and glutamate dehydrogenase (NADP-*GDH*) from *Hebeloma cylindrosporum*. The gene products of AMT1 from *H. cylindrosporum* effectively overcomes hyphal growth defect of *Mep* mutant of *Saccharomyces cerevisiae*. The *AMT1* reported here has much higher affinity than the previously known *AMT2* and *AMT3* genes of the *AMT/Mep* family in *Hebeloma*. The *AMT1, AMT2* and *GDHA* are all downregulated in response to exogenously supplied ammonium or glutamate. These genes are also subjected to nitrogen repression in *Hebeloma* sp. Exogenous NO_3 may not downregulate *AMT1* in N-starved conditions. It is suggested that glutamine is the main effector of *AMT1* and *AMT2* repression, whereas intracellular NH_4^+ controls GDHA repression and its effect is independent of glutamine or glutamate concentration. The ammonium transport activity, to a good extent, is controlled by intracellular NH_4^+. Javelle et al. (2003b) have proposed that such high-affinity ammonium transporters from ectomycorrhizas also act as ammonium sensors, which is essential for hyphal differentiation and growth.

Anapleurosis and Amino Acid Transfer: The ability of ECM fungi to draw inorganic-N in the soil and transport N-bearing solutes is well known (Chalot and Bruns, 1998; France and Reid,1984). Buscot et al. (2000) state that in natural soil, when nitrogen mineralization rates are low, nitrification is too slow; therefore, the inorganic N supply could be limiting and ammonium-N may dominate. They believe that in such conditions, ECM fungus can access organic nitrogen available in soil as well as increase NH_4^+-N uptake via mycelial network in soils. The ECM fungi grow better utilizing NH_4^+N rather than NO_3-N. Further, regulation of amino acid formation and transport across apoplastic interface seems crucial. In this context, dark fixation of CO_2, which induces amino acid formation via anapleurosis, seems to shift between both symbionts. In plants, phosphoenol pyruvate carboxylase (PEPC), which is a key anapleurotic enzyme, is downregulated by mycorrhiza under optimum NH_4^+ ion availability. Micro-histochemical tests indicate that such downregulation of PEPC is greater in areas with intense fungal activity and symbiotic interaction. It implies that in symbiotic association, ammonium assimilation and related anapleurotic carboxylation is shifted from plants to the fungal partner. This mechanism may augment the transport of amino acids such as glutamine, aspergine and others from fungus to host (Chalot and Bruns, 1998). The cDNA encoding the fungal amino acid transporter has been identified in the ECM fungus *A. muscaria* (AmAAP1). We should note that in the absence of a nitrogen source, expression of AmAAP1 increases ten-fold. Studies on the expression of AmAAP1 gene in yeast mutants revealed it to be a high-affinity transporter with a broad substrate range. It can transport wide range of amino acids with Km values between 22 μM for histidine and upto 100μM for proline (Buscot et al. 2000).

In plants, excess supply of nitrogen diverts photo-assimilates away from formation of storage sugars or transport carbohydrates such as starch or sucrose, respectively, into producing amino acids and proteins. The main regulatory steps involved occur at fructose 1,6 bisphosphatase (FBPase) and phosphoenol pyruvate carboxylase (PEPC). The FBPase is inhibited by an accumulation of fructose 1,6 biphosphate. Buscot et al. (2000) found that needles of spruce provided with higher levels of inorganic N, shift towards a 6-fold increase in protein synthesis related to PEPC activity, while accumulation of fructose bisphosphate was three-fold more than control plants. Therefore, starch contents become depleted. Hence, depleted levels of photosynthetic carbon fixation and an increased level of nitrogen supply cause a shift from gluconeogenesis; then the starch or sucrose formation will decrease. A decrease in sucrose formation will reduce C delivery to sinks such as mycorrhizal roots and fungus. In other words, it affects the plant-ECM fungus negatively. In general, it is clear that gene expression related to hexose metabolism, supply of CO_2 and photo-assimilates could lead to better fungal growth in the soil. On the other hand, excessive N supply under C limitation will divert photo-assimilates towards amino acid synthesis rather than sucrose formation (Tarnau et al. 2000). Hence, hexose concentration at hartig net depletes, again leading to a suppression of hexose related fungal genes. Under such conditions, fungal spread in the root may halt, which is indicated by a decrease in fungal metabolites such as mannitol, ergosterol and trehalose. Therefore, a surplus N supply will negatively impact on fungal gene expression and the net benefit derived by the plant will decrease in the long run (Buscot et al. 2000).

Nitrate Assimilation and Regulation: In natural conditions, ECM fungi aid nitrate assimilation by seedlings. The extent of nitrate assimilated may be influenced by several factors. There are many reports indicating that assimilation of nitrate is regulated on the basis of substrate. In view of such variation, Plassard and Gobert (2001) studied firstly, the NO_3 transport operating at low substrate concentrations (< 100 μM) in the *Pinus pinaster* roots and *Rhizopogon roseolus* thalli. Regulation of nitrate assimilation was studied in the presence of NO_3, NH_4 and glutamate in the solution. Their observations indicate that in non-mycorrhizal plants NO_3 uptake measured at 50 μM KNO_3 was slowly inducible and it reached a maximum rate in 3 days. In contrast, NO_3 uptake was not inducible with exposure to NO_3 (1 mM) in *R. roseolus* thalli. Similar results were observed when NO_3 uptake was measured in mycorrhizal and non-mycorrhizal short roots. When NH_4 or glutamate were added, NO_3 uptake was strongly inhibited in host roots and fungal hyphae. They summarized that in symbiotic roots of *Pinus pinaster*, a constitutive high-affinity transport system (HATS) and NO_3 inducible HATS has inhibited by NH_4 and glutamate.

Jargeat et al. (2002) have reported that transcription of a nitrate reductase gene isolated from a symbiotic basidiomycete fungus *Hebeloma cylindrosporum* does not need induction by nitrate. These researchers cloned and studied the transcriptional regulation of two nitrate reductase genes, namely *NAR1* and *NAR2*. The *NAR1* gene codes a polypeptide (908 amino acids) that complements nitrate-deficient mutants of *H. cylindrosporum* upon transformation. It demonstrates that protein encoded by the gene *NAR1* functions as nitrate reductase. The *NAR2* codes for protein which is non-functional. It is deemed ancestral and duplication of *NAR1*. In wild type strains of *H. cylindrosporum*, *NAR1* is repressed by high NH_4^+ concentration in the medium. Its transcription is induced in the presence of NO_3 or organic N sources such as urea, glycine or serine. Recently, Jargeat et al. (2003) have cloned and characterized genes that code for nitrate reductase and a nitrate transporter in the ectomycorrhizal basidiomycete *Hebeloma cylindrosporum*. Occurrence of nitrate assimilation gene clusters was actually demonstrated by the divergent transcription and their link to previously cloned nitrate reductase. The fungal nitrate transporter polypeptide (NRT2) is characterized by 12 transmembrane domains. It has both a long putative intracellular loop and a short C-terminal tail. These structures distinguish the fungal high affinity nitrate transporters from those of plant origin. Presence of nitrate and availability of organic-N also induced nitrate transport gene cluster, whereas, ammonium repressed the nitrate transporter gene.

Guescini et al. (2001) have attempted to understand the nitrogen regulation in symbiotic tissue of plants colonized with *Tuber borchii* by cloning and examining the transcriptional regulation of nitrate reductase gene *(NR)*. This *NR* gene was actually isolated and identified using a lambda EMBL4 genomic library. The predicted protein pattern had a high sequence similarity with nitrate reductase from several other organisms. The coding region of *TbNR1* is 2787 nucleotides in length and codes for a protein containing 929 amino acids (Guescini et al. 2003). According to them, nitrate reductase that catalyzes conversion NADPH-nitrate to nitrite plays a pivotal role in regulating nitrogen flux. The *NR* gene is substrate inducible. The *TnBr1* is expressed to be eight folds greater in mycorrhizal roots compared to free-living mycelium. Nitrate inputs resulted in higher levels of enzyme activity, but if ammonium was supplied as nitrogen source, then NR activity ceased. Similar observations were made on nitrate reductase from *Pisolithus tinctorius*. It was induced in the presence of nitrate but repressed by ammonium (Aouadj, 2000). As a consequence, *Pisolithus* grew better on nitrate medium than on a medium supplemented with ammonium. In addition, a shift of ECM fungus from a medium containing nitrate to ammonium and vice versa, resulted in the appearance of NR activity quickly in a few minutes, but the disappearance of activity took several hours. Enzymatic analysis has revealed that NADPH-dependent nitrate formation increase in ECM roots, indicating that it is involved in nitrate assimilation. Further, higher levels of amino acids such as glutamate, glutamine and asparagine in symbiotic tissues compared free mycelial controls suggested that these amino acids may serve to transfer nitrogen to the host plant.

Montanini et al. (2003) point out that nitrogen retrieval and assimilation is a crucial aspect of ectomycorrhizal symbiosis. In this regard, glutamate synthetase is thought to be a key regulatory enzyme. Hence, these workers studied the regulation of gene cluster responsible for glutamate synthetase (*TbGS*) in mycorrhizal ascomycete *Tuber borchii*. The *TbGS* mRNA is encoded by a single-copy of the gene in *Tuber* genome. It is upregulated in N-starved mycelia and returns to basal levels of expression, whenever different forms are supplemented. The most effective substrate was nitrate. The TbGS is supposed to be controlled at the pre-translational level. Comparative analysis indicated that among the related enzymes such as glutamate dehydrogenase and glutamate synthase, *TbGS* mRNA was the key responder to N-starvation, and was most abundant under N-limiting conditions. Regulatory effects of N-starvation were also observed in the symbiotic hyphae and fruit bodies of *Tuber borchii*, in addition to symbiotic roots.

In comparison to ECM, reports on the molecular aspects of N transport in AM symbiosis are meager. We are yet to understand numerous details. With regard to AM symbiosis too, it is important to know the molecular regulatory aspects of nitrogen metabolism in both the host and AM fungus. It is possible that AM fungus aids the host plants in nitrate assimilation in different ways. Kaldorf et al. (1998) investigated this aspect using PCR techniques. The gene coding for nitrate reductase (NR) apoprotein from the AM fungus *Glomus intraradices* and maize were amplified, cloned and sequenced. Northern Blot analysis showed that mRNA levels for maize nitrate reductase gene was lower in roots of mycorrhizal plants when compared with non-mycorrhizal ones, whereas, the fungal gene was transcribed in mycorrhizal roots. Nitrate formation catalyzed by NR was mainly NADPH-dependent in roots of AM colonized plants. It is consistent with the fact that NRs of fungi prefer NADPH as reductant. Further, in situ hybridization exhibited that fungal mRNA was conspicuous in arbuscules, but was not detected in vesicles. Clearly, transcripts of NR gene from host and AM fungus are expressed differentially.

Concluding Remarks

We may agree to view 'root biology' as 'mycorrhizal biology', since mycorrhizas are manifested at almost all times on 90% of plants on the earth. However, accurate estimates of nutrient transport attributable to host and fungal component *per se*, at a point of time, in an individual mycorrhizal root is not yet possible. Numerous reports on the uptake of P and other nutrients are approximations derived via computation. We lack a rapid technology to exactly measure P or other nutrients intercepted and translocated by fungal thalli and that by 'root *sans* mycorrhizal component' in a naturally growing plant. Assessment of extraradical mycelium provides a vague idea about the relevance of fungal mediation of nutrient transport. In addition to physical spread of fungal hyphae, knowledge about the physiological regulation of nutrient transport proteins is crucial. Effectiveness of symbiosis, surely, depends on regulation of nutrient transporters in fungus and 'root *sans* mycorrhizal fungus' and their individual contributions. Hence, clear distinctions regarding genes coded by plant and/or fungus are required while discussing the molecular biology of nutrient transport and exchange in mycorrrhizal plants. So far, nutrient transporter genes from quite a few plant-mycorrhizal pairs have been characterized. Chalot et al. (2002b) have aptly remarked that in due course, as we analyze nutrient transporters from genomes of different mycorrhizal fungal and plant combinations, clearer knowledge regarding their molecular functions should become available. Mainly, we ought to understand the mechanisms by which fungal and plant cells obtain, process and integrate information regarding nutrient levels in the external environment and plants' demand. A hypothetical model depicting the regulation of plant and/or fungal coded P transporter activity in relation to P in the ambient medium is shown in Figure 4.3. Some time in the future, a rapid molecular technique that estimates nutrient transporter activity and pinpoints it as due to plant or fungal coded genes is a necessity. This feat is easier said than done. Such a technique can help us to pick fungal isolates and plant genotypes that are physiologically efficient in nutrient transfer and exchange.

Vacuolar transport plays a key role in P movement for longer distances along the hyphae of mycorrhizal fungi. Further details on factors that influence vacuole formation, P accumulation and transport are required. Genetic variations, if any, for vacuolar transport of P should be identified. It may have relevance to P transport efficiency of AM/ECM fungal isolates. Are there any fungal isolates that lack ability to form vacuoles and transport P through them? Studies on genetic and molecular control of vacuole formation and vacuolar transport of P need a great deal of attention.

Regulation of gene expression related to extracellular phosphatases, poylphosphate accumulation and movement, production of mycorrhiza specific alkaline phosphatases, accumulation of P*i* in biotrophic interfaces within hartig net or arbuscular cells, role of ATPase isozymes are a few items that should be investigated in greater detail.

Molecular biology of nitrogen nutrition, especially transport and exchange in mycorrhizas has gained in importance during the past several years. Genes for ammonium transporters in few plant-fungal combinations have been sequenced, cloned, and their regulation understood, using yeast mutants lacking amino acid transporters. Obviously, they play a crucial role in sensing nitrogen source and concentration. However, details on energy coupling, ATP source and NH_4/K pump are as yet unclear. With regard to N assimilation, nitrate reductase (NR) gene clusters are known to operate in mycorrhizal fungi. Presence of nitrate induces NR genes, but NH_4^+ suppresses nitrate reductase genes in AM/ECM fungi. Information on the regulatory aspects of N uptake and assimilation is comparatively meager for AM mycorrhizas. Recently, genes coding nitrate reductase in *G. intraradices* have been sequenced. There is need to characterize NR genes from a few other plant-AM fungal combinations in order to arrive at generalizations. In situ hybridization and immuno-cytochemical tests have shown that mRNA levels for nitrate reductase are high in arbuscules, but not detectible in vesicles. Clearly, there are several aspects related to N exchange at biotrophic interfaces in AM /ECM that needs investigation.

Regulation of genes relevant to monsaccharide transport system is crucial in any mycorrhizal combination. A strict barter between C and P exchange seems to be regulated at the interface. Therefore, it decides the net physiological efficiency of the bi-directional movement of nutrients. In case of ECM, a monosaccharide transport system operative at apoplastic interface is regulated by glucose availability in the medium. In nature, mycorrhizal fungi sense fluctuations in C flux from host. Excess is stored as trehalose, mannitol or glycogen. Genes for monosaccharide transporters from different ECM/AM fungi have been sequenced, characterized and cloned. Currently, the major preoccupation of mycorrhiza researchers seems to be sequencing, characterization, cloning and understanding the regulatory aspects of nutrient transporters. We may expect to learn more about these aspects in the near future.

References

Alexander, I.J. and Hardy, K. 1981. Surface phosphatase activity of Sitka spruce mycorrhizas from serpentine site. Soil Biology and Biochemistry 13: 301-305.

Aouadj, R., Es-Sgaouri, A. and Botton, B. 2000. A study of the stability and properties of nitrate reductase from the ectomycorrhizal fungus *Pisolithus tinctorius*. Cryptogamie Mycologie 21: 187-202.

Antibus, R.K., Sinsabaugh, J. and Linkins, A.E. 1992. Phosphate activities and phosphorus uptake from inositol phosphate by ectomycorrhizal fungi. Canadian Journal of Botany 70: 794-801.

Ashford, A.E. 2002. Tubular vacuoles in arbuscular mycorrhizas. New Phytologist 154: 545-547.

Ashford, A.E. and Allaway, W.G. 2002. The role of motile tubular vacuoles in mycorrhizal system. Plant and Soil 244: 177-187.

Ashford, A.E., Vesk, P.A., Orlavich, D.A., Markovina, A.L. and Allaway, W.G. 1999. Dispersed polyphosphate in fungal vacuolesin *Eucalyptus pilularis/Pisolithus tinctorius* ectomycorrhizas. Fungal Genetics and Biology 28: 21-33.

Bago, B., Zipfel, W., William, R.M. Jun, J., Lammer, P.J., Pfeffer, P.E. and Sachar-Hill, Y. 2002. Translocation and utilization of fungal storage lipid in the arbuscular mycorrhizal symbiosis. Plant Physiology 128: 108-124.

Benedetto, A., Lanfranco, L. and Bonfonte, P. 2002. Phosphate metabolism in arbuscular mycorrhizal fungi: A molecular approach. Proceedings of Phytopathology Conference, CIRAD, Aussols, Savole, France, p 34.

Blilou, I., Ocampo, J. A. and Garcia-Garrido, J. M. 2000. Induction of LTP (lipid transfer protein) and PAL (Phenylalanine ammonia lyase) gene expression in rice roots colonized by the arbuscular mycorrhizal fungus *Glomus mosseae*. Journal of Experimental Botany 51: 1969-1977.

Bougoure, D.S. and Dearnaley, J.D.W. 2001. Investigations of phosphate transporter expression in the soybean AM symbiosis. Third International Conference on Mycorrhiza, Adelaide, Australia, http://www.mycorrhiza.ag.utk.edu.icoms/icom3.html

Bucher, M., Rausch, C. and Daram, P. 2001. Molecular and Biochemical mechanisms of phosphorus uptake into plants. Journal of Plant Nutrition and Soil Science 2: 209-217.

Bucking, H. and Heyser, W. 1999. Subcellular compartmentation of elements in non-mycorrhizal and mycorrhizal roots of *Pinus sylvestris*: An X-ray microanalytical study. 1. The distribution of phosphate. New Phytologist 145: 311-320.

Bucking H. and Heyser, W. 2001. Micro-auto-radiographic localization of polyphosphate and carbohydrates in mycorrhizal roots of *Populus tremmela* × *Populus alba* cross and the implications for transfer process in ectomycorrhizal associations. Tree Physiology 21: 101-107.

Bucking, H., Kuhn, A.J., Schroeder, W.H. and Heyser, W. 2002. The fungal sheath of ectomycorrhizal pine roots: an apoplastic barrier for the entry of calcium, magnesium and potassium into the root cortex. Journal of Experimental Botany 53: 1659-1669.

Bun-Ya, M., Nishimura, M., Haroshima, S. and Oshima, Y. 1991. The *PHO* 84 gene of *Saccharomyces cervesiae* encodes an inorganic phosphate transporter. Molecular and Cell Biology 1: 3229-3238.

Burleigh, S.H. and Bechmann, I.E. 2002. Plant nutrient transporter regulation in arbuscular mycorrhizas. Plant and Soil 244: 247-251.

Buscot, F., Munch, J.C., Charcoal, J.C., Gardes, M., Nehls, U. and Hampp, R. 2000. Recent advances in exploring physiology and biodiversity of ectomycorrhizas highlight the functioning of the symbiosis in ecosystems. FEMS Microbiology Review 24: 601-614.

Callow, T.A., Cappacio, L.C.M., Parisch, G. and Tinker, P.B. 1978. Deletion and estimation of polyphosphate in vesicular arbuscular mycorrhizas. New Phytologist 80: 125-134.

Cappacio, L.C.M. and Callow, J.A. 1982. The enzyme of polyphosphate metabolism in VA mycorrhizas. New Phytologist 91: 81-91.

Chalot, M., Lerat, S., Javelle, A., Wipf, D., Jacquot, J. and Botton, B. 1998. Molecular and biochemical characterization of transporter for nitrogenous compounds in the ectomycorrhizal fungus *Paxillus involutus*. Second International Conference on Mycorrhiza, Uppsala, Sweden, http:/www.mycorrhiza.ag.utk.edu.html

Chalot, M. and Brun, A. 1998. Physiology of organic nitrogen acquisition by ectomycorrhizal fungi and ectomycorrhizas. FEMS Microbiology Review 22:21-44.

Chalot, M., Brun, A., Botton, B. and Soderstorm, B. 2002a. Kinetics, Energetics, and Specificity of a general Amino acid Transporter from the Ectomycorrhizal fungus *Paxillus involutus*. http://www.mycorrhiza.ag.utk.edu

Chalot, M., Javelle, A., Blaudez, D., Lambillote, R., Cooke, R., Sentenac, H., Wipf, D., Botton, B. 2002b. An update on nutrient transport processes in Ectomycorrhizas. Plant and Soil 244: 165-175.

Chilvers, G.A. and Harley, J.L. 1980. Visualization of phosphate accumulation in beech mycorrhizas. New Phytologist 84: 319-326.
Chiou, T.J., Liu, H. and Harrison, M.J. 2001. The spatial expression patterns of a phosphate transporter (MtPT1) from *Medicago truncatula* indicates a role in phosphate transport at the root/soil interface. Plant Journal 25: 281-293.
Cole, L., Orlavich, D.A. and Ashford, A.E. 1998. Structure, function and motility of vacuole in filamentous fungi. Fungal Genetics and Biology 24: 86-100.
Cox, G., Moran, K.J., Sanders, F., Nokolds, L. and Tinker, P.B. 1980. Translocation and transfer of nutrients in vesicular arbuscular mycorrhizas. 3. Polyphosphate granules and phosphorus translocation. New Phytologist 84: 649-659.
Cox.G. and Tinker, P.B. 1976. Translocation and transfer of nutrients in vesicular arbuscular mycorrhizas 1. The Arbuscular Mycorrhizas and phosphorus transfer: a quantitative ultra-structural study. New Phytologist 77: 371-378.
Cress, W.A., Thronbiery, G.D. and Lindsay, B. 1979. Kinetics of phosphorus absorption by mycorrhizal and non-mycorrhizal tomato roots. Plant Physiology 64: 484-487.
Daram, P., Brunner, S., Person, B.L., Amrhein, N. and Bucher, M. 1998. Functional analysis and cell-specific expression of a phosphate transporter from tomato. Planta 206: 2225-233.
Daram, P., Brunner, S., Rausch, C. Steiner, C. Amrhein, N. and Bucher, M. 1999. Pht2; 1 encodes a low affinity phosphate transporter from *Arabidopsis*. Plant Cell 1: 2153-2166.
Declerck, S., de Boulois H.D., Bivort, C. and Delaux, B. 2003. Extraradical mycelium of the arbuscular mycorrhizal fungus *Glomus lamellosum* can take up, accumulate and translocate radiocaesium under root organ culture conditions. Environmental Microbiology 5: 510-516.
Delhaize, E. and Randall, P.J. 1995. Characterization of a phosphate accumulator mutant of *Arabidopsis thaliana*. Plant physiology 107: 207-213.
Dexheimer, J., Gianinazzi-Pearson, V., Gianinazzi, S. and Marx, C. 1982. Role of possible de la vacuole du chaminon symbiotique des mycorrhizas VA dans la nutrition du plantes. Actualites Botaniques 79: 45-51.
Dexheimer, J., Gianinazzi, S. and Gianinazzi-Pearson, V. 1979. Ultrastructural cytochemistry of the host fungus interface in endomycorrhizal association *Glomus mosseae-Allium cepa*. Zeitschrift fur Pflanzenphysiologie 92: 191-206.
Ezawa, T., Cavagnaro, T.R., Smith, S.E., Smith, F.A., and Ohtomo, R. 2004. Rapid accumulation of polyphosphate in extraradical hyphae of an arbuscular mycorrhizal fungus as revealed by histochemistry and a polyphosphate kinase/luciferase system. New Phytologist 161: 387-392.
Ezawa, T., Kuwahara, S., Sakamoto, K., Yoshida, T, Saito, M. 1999. Specific inhibitor and substrate specificity of alkaline phosphatase expressed in the symbiotic phase of the arbuscular mycorrhizal fungus, *Glomus etunicatum*. Mycologia 91: 636-641.
Ezawa, T., Smith, S.E. and Smith, F.A. 2001. Differentiation of polyphosphatase metabolism between the extra- and intracellular hyphae of arbuscular mycorrhizal fungi. New Phytologist 149: 555-563.
Ezawa, T., Smith, S.E. and Smith, F.A. 2002. Phosphorus metabolism and transport in AM fungi. Plant and Soil 244: 221-230.
Feng, G., Su, Y.B., Li, X.L. Wang, H., Zhang, F.S., Tang, C. X. and Rengel, Z. 2002. Histochemical visualization of phosphatase released by arbuscular mycorrhizal fungi in soil. Journal of Plant Nutrition 25: 969-980.
Feng, G., Song, Y.C., Li, X.L. and Christie, P. 2003. Contribution of arbuscular mycorrhizal fungi to utilization of organic sources of phosphorus by red clover in a calcareous soil. Applied soil Ecology 22: 139-148.
France, R.C. and Reid, C.P.P. 1984. Interaction of nitrogen and carbon in the physiology of ectomycorrhiza. Candian Journal of Botany 61: 964-984.
Ferrol, N., Barea, J.M. and Azcon-Aguilar, C. 2001. Mechanisms of nutrient transport across interfaces in arbuscular mycorrhizas. Third International Conference on Mycorrhizas, Adelaide, Australia, http://www.mycorrhizas.ag.utk.edu/icoms/icom3.html

Ferrol, N., Pozo, M.J., Antelo, M. and Azcon-Aguilar, C. 2002. Arbuscular mycorrhizal symbiosis regulates plasma membrane H^+ATPases gene expression in tomato plants. Journal of Experimental Botany 53: 1683-1687.

Franco-Zorilla, J.M., Gonzalez, E., Bustos, R., Linhares, F., Leyva, A., Paz-Ares, J. 2004. The transcriptional control of plant responses to phosphate limitation. Journal of Experimental Botany 55: 285-293.

Frommer, W.B., Mariam, R., Rikirsch, E., Zimmerman, S., Sentanac, H. and Wipf, D. 2003. Isolation of monoacid and peptide transporters from the ectomycorrhizal fungus *Hebeloma cylindrosporum* by functional complementation of yeast. Proceedings of American Society of Plant Biologists, Abstracts, p. 23.

Furihata, T., Suzuki, M. and Sakurai, H. 1992. Kinetic characteristics of two phosphate uptake systems with different affinities in suspension cultured *Catharanthus roseus* protoplasts. Plant Cell Physiology 33: 1151-1157.

Gerlitz, T.G.M. and Gerlitz, A. 2002. Effects of monovalent cations on phosphate accumulation and storage. Biochemistry-Moscow 67: 575-582.

Gianinazzi-Pearson, V. 1996. Plant response to arbuscular mycorrhizal fungi: Getting to the roots of symbiosis. Plant Cell 8: 1871-1883.

Gianinazzi-Pearson, V., Arnould, C., Oufattole, M., Arange, M. and Gianinazzi, S. 2000. Differential activation of H^+ATPase genes by an Arbuscular Mycorrhizal fungus in root cells of Transgenic tobacco. Planta 211: 609-613.

Gianinazzi-Pearson, V and Gianinazzi, S. 1978. Enzymatic studies on the metabolism of vesicular-arbuscular mycorrhiza. 2. Soluble alkaline phosphatase specific to mycorrhizal infection in onion roots. Physiologie Vegetale 14: 883-839.

Gianinazzi-Pearson, V. and Gianinazzi, S. 1986. The Physiology of improved phosphate nutrition in mycorrhizal plants. In: Gianinazzi-Pearson, V. and Gianinazzi, S. (Eds) Physiological and Genetical Aspects of Mycorrhizae. First European Symposium on Mycorrhizae. Paris, France, INRA, pp 101-109.

Gianinazzi, S. Gianinazzi-Pearson, V. and Dexheimer, J. 1979. Enzymatic studies on the metabolism of vesicular-arbuscular mycorrhiza 3. Ultrastructural localization of acid and alkaline phosphatase in onion roots infected by *Glomus mosseae*. New Phytologist 83: 127-132.

Guescini, M., Vallorani, L., Pierloni, R. Sacconi, S., Zeppa, F., Palma, F., Amicucci, A. and Stocchi, V. 2001. Nitrate reductase regulation in the ectomycorrhizal fungus *Tuber borchii* Vittad cultured on different nitrogen sources. Third International Conference on Mycorrhizas, Adelaide, Australia, http://www.mycorrhiza.ag.utk.edu/icoms./icom3.html

Guescini, M., Pierleoni, R., Palma, F., Zeppa, S., Vallorani, L., Potenza, L., Sacconi, C., Giomaro, G. and Stocchi, V. 2003. Characterization of the *Tuber borchii* nitrate reductase gene and its role in Ectomycorrhizae. Molecular Genetics and Genomics 269: 807-816.

Harrison, M.J. 1997. The arbuscular mycorrhizal symbiosis: An Underground Association. Trends in Plant Sciences 2: 54-56.

Harrison, M.J. 2003. The arbuscular mycorrhizal symbiosis and phosphate transport in plants. Samuel Roberts Noble Foundation, Ardmore, Oklahoma, http:// www. noble.org/ plantbio/ Harrison /2003/Harrison2.html

Harrison, M.J., Dewbre, G. R. and Liu, J. 2002. A phosphate transporter from *Medicago truncatula* involved in the acquisition of phosphate released by arbuscular mycorrhizal fungi. The Plant Cell 14: 2413-2429.

Harrison, M.J. and Van Buuren, M.L. 1995. A phosphate transporter from the mycorrhizal fungus *Glomus versiforme*. Nature 378: 626-629.

Hyde, G.J. and Ashford, A.E. 1997. Vacuole motility and tubule forming activity in *Pisolithus tinctorius* are modified by environmental conditions. Protoplasma 198: 85-92.

Itoh, S. and Barber, S.A. 1983. Phosphorus uptake by six plant species as related to root hairs. Agronomy Journal 75: 457-461.

Jargeat, P., Gay, G., Debaud, J.C., and Marmiese, R. 2002. Transcription of a Nitrate Reductase gene isolated from the symbiotic basidiomycete fungus *Hebeloma cylindrosporum* does not require induction by nitrate. Molecular and General Genetics 263: 948-956.

Jargeat, P. Rekangalt, D., Verner, M. C., Gay, G. Debaud, J. C., Marmeisse, R. and Fraissinet-Tachet, L. 2003. Characterization and expression analysis of a nitrate transporter and nitrate reductase genes, two members of gene cluster for nitrate assimilation from the symbiotic basidiomycete *Hebeloma cylindrosporum*. Current Genetics 43: 199-205.

Javelle, A. Andre, B., Marini, A.M. and Chalot, M. 2003a. High-affinity Ammonium transporters and nitrogen sensing in mycorrhizas. Trends in Microbiology 2003 11: 53-55.

Javelle, A., Morel, M., Rodriguez-Patrana, B., Botton, B., Andre, B., Marini, A., Brun, A. and Chalot, M. 2003b. Molecular characterization, function and regulation of ammonium transporters (AMT) and ammonium metabolizing enzymes (GS, NADP-GDH) in the ectomycorrhizal fungus *Hebeloma cylindrosporum*. Molecular Microbiology 47: 411-417.

Jayachandran, K., Schwab, A.P. and Hetrick, B.A.D. 1992. Mineralization of organic phosphorus by vesicular arbuscular mycorrhizal fungi. Soil Biology and Biochemistry 24: 897-903.

Joner, E.J. and Jakobsen, I. 1995. Growth and extracellular phosphatase activity of AM fungal hyphae as influenced by soil organic matter. Soil Biology and Biochemistry 27: 1153-159.

Joner, E.J. and Johansen, A. 1999. Phosphatase activity of external hyphae of two arbuscular mycorrhizal fungi. Mycological Research 104: 81-86.

Joner, E.J., Ravanskov, S. and Jakobsen, I. 2000a. Arbuscular mycorrhizal phosphate transport under monoxenic conditions using radiolabeled inorganic and organic phosphate. Biotechnology Letters 22: 1705-1708.

Joner, E. J., Van Aarle, I. M. and Vosatka, M. 2000b. Phosphatase activity of extra radical arbuscular mycorrhizal hyphae: A review. Plant and Soil 226: 199-210.

Kaldorf, M., Schmeler, E. and Bothe, H. 1998. Expression of maize and fungal nitrate reductase genes in Arbuscular Mycorrhiza. Molecular Plant-Microbe Interaction 11: 439-448.

Karunarathnam, R.S., Baker, J.H. and Barker, A.V. 1986. Phosphorus uptake of mycorrhizal and non-mycorrhizal roots of soybean. Journal of Plant Nutrition 9: 1303-1313.

Kleber, R. and Nehls, U. 2003. Nitrogen dependent gene expression of *Amanita muscaria*. http://www.gentik.uni-beilfeld.de/molmyk/treffen/kollo4/abstracts/nehls-abstract.shtml

Kojima, T., Hayatsu, M. and Saito, M. 1998. Characterization and partial purification of a mycorrhiza-specific phosphatase from *Gigaspora margarita-Allium cepa* symbiosis. Second International Conference on Mycorrhiza, Uppsala, Sweden, http://www.mycorrhiza.ag.utk.edu

Koide, R.T. and Kabir, Z. 2000. Extraradical hyphae of the mycorrhizal fungus *Glomus intraradices* can hydrolyse organic phosphate. New Phytologist 148: 511-517.

Kothe, E., Muller, D., Krause, K. 2002. Different high affinity phosphate uptake systems of ectomycorrhizal *Tricholoma* species in relation to substrate specificity. Journal of Applied Botany 76: 127-132.

Krajinski, F., Hause, B., Gianinazzi-Pearson, V. and Franken, P. 2002. Mtha1, a plasma membrane H^+ATPase gene from *Medicago truncatula*, shows arbuscule-specific induced expression in mycorrhizal tissue. Plant Biology 4: 754-761.

Krishna, K.R. 1997. Phosphorus efficiency of semi-dry land crops In: Accomplishments and future research challenges in dry land soil fertility research in the Mediterranean area. Ryan (Ed.) International Center for Agriculture in Dry Areas, Aleppo, Syria, pp. 343-363.

Krishna, K.R. 2002a. Soil mineral deficiency, nutrient acquisition and crop production. In: Soil Fertility and Crop Production. Krishna, K.R. (Ed.) Science Publishers Inc., Enfield, New Hampshire, USA, pp 65-90.

Krishna, K.R. 2002b. Crop improvement towards resistance to soil fertility constraints. In: Soil Fertility and Crop Production. Krishna, K.R. (Ed.) Science Publishers Inc., Enfield, New Hampshire, USA, pp 387-410.

Krishna, K.R. and Bagyaraj, D.J. 1985. Phosphatases in the rhizospheres of mycorrhizal and non-mycorrhizal peanut. Proceedings of the 6th North American Conference on Mycorrhiza, Oregon State University, Corvallis, Oregon, pp 372.

Lammers, P., Jun, J., Abubaker, J., Arreola, R., Gopalan, A., Bago, B., Hernandez-Sebastian, C., Allen, J.W., Douds, D.D., Pfefer, P.E. and Sachar-Hill, Y. 2001. The Glyoxylate cycle in an Arbuscular Mycorrhizal fungus: carbon flux and gene expression. Plant Physiology 127: 1287-1298.

Larson, J., Thignstrum, I., Jackobsen, I. and Rosendahl, S. 1996. Benomyl inhibits phosphorus transport but not fungal alkaline phosphate activity in a *Glomus*-cucumber symbiosis. New Phytologist 132: 127-13.

Leggewie, G. Willmitzer, L. and Reismeir, J.W. 1997. Two cDNA from potato are able to complement a phosphate uptake-deficient yeast mutant: identification of phosphate transporters from higher plants. Plant Cell 9: 381-392.

Lindahl, B., Finlay, R. and Olsson, S. 2001. Simultaneous, bi-directional translocation of ^{32}P and ^{33}P between wood blocks connected by mycelial cords of *Hypholoma fasciculare*. New Phytologist 150: 189-194.

Ling-lee, M., Chilvers, G.A. and Ashford, A. E. 1975. Polyphosphate granules in three different kinds of tree mycorrhizas. New Phytologist 75: 551-554.

Liu, H. Trieu, A.T., Baylock, L.A. and Harrison, M.J. 1998. Cloning and characterization of two phosphate transporters from *Medicago truncatula* roots: Regulation I response to phosphate and to colonization by arbuscular mycorrhizal (AM) fungi. Molecular Plant-Microbe Interaction 1: 14-22.

Maldonando-Mendoza, I.E., Dewbre, G.R. and Harrison, M.J. 2001. A phosphate transporter gene from the extraradical mycelium of a arbuscular mycorrhizal fungus *Glomus intraradices* is regulated in response to phosphate in the environment. Molecular Plant- Microbe Interaction 14: 1140-1148.

Maldonado-Mendoza, I.E. and Harrison, M.J. 1999. Regulation of the expression of a phosphate transfer from *Glomus intraradices* in response to exogenous levels of phosphate. Samuel Roberts Noble Foundation Plant Biology symposium, Taylor Publishing Co, Texas, USA, pp 130-132.

Marschner, H. 1995. Nutrient availability in soils. In: Mineral Nutrition of Higher Plants. Academic Press, London, pp 483-507.

Martins, A., Santos, M., Santos, H., and Pais, M.S. 1999. A ^{31}P nuclear magnetic resonance study of phosphate levels in roots of ectomycorrhizal and nonmycorrhizal plants of *Castanae sativa*. Trees-Structure and Function 13: 168-172.

Marx, C., Dexheimer, J., Gianinazzi-Pearson, V. and Gianinazzi, S. 1982. Enzymatic studies on the metabolism of VA mycorrhizas. 4. Ultra-cytoenzymological evidence (ATPase) for active transfer in the host arbuscule interface. New Phytologist 90: 37-43.

Mimura, T. 1995. Homeostasis and transport of inorganic phosphate in plants. Plant Cell Physiology. 36: 1-7.

Montanini, B., Betti, M., Marguez, A. J., Balestrini, R., Bonfonte, P. and Ottonella, S. 2003. Distinctive properties and expression profiles of glutamine synthetase from a plant symbiotic fungus. Biochemical Journal 373: 357-368.

Montanini, B., Moretto, N., Soragni, E., Percudani, R. and Ottanello, S. 2002. A high affinity ammonium transporter from the mycorrhizal ascomycete *Tuber borchii*. Fungal Genetics and Biology 36: 22-34.

Muchal, V.S., Pardo, J.M. and Raghothama, R.G. 1996. Phosphate transporters from the higher plant *Arabidopsis thaliana*. Proceedings of National Academy of Sciences, USA 93: 101519-110523.

Muchal, U.S. and Raghothama, K.G. 1999. Transcriptional regulation of phosphate transporters. Proceedings of National Academy of Sciences, USA 96: 5868-5872.

Negel, J., Dodd, J.C., Van Teunen, D., Gianinazzi-Pearson, V. and Gianinazzi, S. 2003. Fungal Alkaline Phosphatases: Approaches for identifying genes. COST Scientific meetings, Cologne, Abstract, p.1.

Nehls, U., Weise, A., Guttenberger, M. and Hampp, R. 1998. Carbon allocation in ectomycorrhiza: identification and expression analysis of *A. muscaria* monosaccharide transporter. Molecular Plant-Microbe Interaction 11: 167-176.

Nehls, U., Wiese, J. and Hampp, R. 2000. Cloning of *Picea abies* monosaccharide transporter gene and expression analysis in plant tissues and ectomycorrhizas. Univeristat Tubingen-Botanisches Institut-report http://www.uni-tuebingen.de/uni/bbp/lit/treesip.html

Nehls, U., Bock, A., Enke, M. and Hampp, R. 2001. Differential expression of the hexose-regulated fungal genes *AMPAL* and *AmMST1* within *Amanita-Populus* ectomycorrhiza. New Phytologist 150: 583-589.

Nuria, F., Barea, J.M. and Concepcion, A. 1998. Cloning and expression analysis of P-type H^+ATPase genes in the arbuscular mycorrhizal fungus *Glomus mosseae*. Second International Conference on Mycorrhizas, Uppsala, Sweden, http://mycorrhiza.ag.utk.edu

Nye, P.H. and Tinker, P.B. 1977. Solute movement in the soil-root system. University of California Press, Berkeley, California, USA, pp 278.

Olsson, P.A., van Arle, I.M., Allaway, W.G., Ashford, A.E. and Rouheir, H. 2002. Phosphorus effects on metabolic processes in monoxenic arbuscular mycorrhiza cultures. Plant Physiology 130: 1163-1171.

Paszkowski, U., Kroken, S., Roux, C. and Briggs, S. P. 2002. Rice phosphate transporters include an evolutionarily divergent gene specifically activated in arbuscular mycorrhizal symbiosis. Proceedings of the National Academy of Sciences of the United States of America 99: 13324-13329.

Pearson, S. and Jackobsen, I. 1993. The relative contribution of hyphae and roots to phosphorus uptake by arbuscular mycorrhizal plants, measured by dual labeling with ^{32}P and ^{33}P. New Phytologist 124: 489-494.

Pfeffer, P.E., Douds, D.D., Becard, G. and Sacher-Hill, Y. 1999. Carbon uptake and the metabolism and transport of lipids in an arbuscular mycorrhiza. Plant Physiology 587-598.

Plassard, C. and Gobert, A.M. 2001. Regulation of NO_3 uptake in the symbiotic association *Pinus pinaster-Rhizopogon roseolus*. Third International Conference on Mycorrhizas, Adelaide, Australia, http:// www.mycorrhiza.ag.utk.edu/icoms/icom3.html

Poirier, Y. Thoma, S., Somerville, C. and Schiefelbein, J, 1991. A mutant of *Arabidopsis* deficient in xylem loading of phosphate. Plant Physiology 97: 1087-1093.

Rausch, C., Daram, P., Brunner, P., Jansa, S., Leggewie, M., Amrhein, N.B. and Bucher, M. 2001. A phosphate transporter expressed in arbuscule-containing cells in potato. Nature 414: 462-466.

Requena, N., Breuninger, M., Franken, P. and Oeon, A. 2003. Symbiotic status, phosphate and sucrose regulate the expression of two plasma membrane H^+ ATPase genes from the mycorrhizal fungus *Glomus mosseae*. Plant Physiology 132: 1540-1549.

Rosewarne, G.M., Smith, S.E., Smith, F.A. and Schatchman, D.P. 2001. The role of plasma membrane H^+ATPases in arbuscular mycorrhizal roots. Third International Conference on Mycorrhizas, Adelaide, Australia, http://www.mycorrhiza.ag.utk.edu/icoms/icom3.html

Schmidt, M.E., Heim, S., Wylegalla, C., Helmbrecht, C. and Wagner, K.G. 1992. Characterization of phosphate uptake by suspension cultured *Catharanthus roseus* cells. J. Plant Physiol. 140: 179-184.

Smith, F.W., Ealing, P.M., Dong, B. and Delhaize, E. 1997. The cloning of two *Arabidopsis* genes belonging to phosphate transporter family. Plant Journal 11: 83-92.

Smith, F.W., Mudge, S.R., Rae, A.L. and Glassop, D. 2003. Phosphate transport in Plants. Plant and Soil 248: 71-83.

Solaiman, M.Z. and Saito, M. 2001. Phosphate efflux from intra radical hyphae of *Gigaspora margarita* in vitro and its implication for phosphorus translocation. New Phytologist 151: 525-533.

St Arnaud, M., Hazel, C., Vimard, B., Caron, M. and Fortin, J. A. 1996. Enhanced hyphal growth and spore production of the arbuscular mycorrhizal fungus *Glomus intraradices* in an in vitro system in the absence of host roots. Mycological Research 100: 328-332.

Straker, C.J. and Mitchell, D.T. 1985. The characterization and estimation of polyphosphates in endomycorrhizas of the Ericaceae. New Phytologist 79: 431-440.

Thomson, B.D., Clarkson, D.T. and Brain, P. 1990. Kinetics of phosphorus uptake by the germ tubes of the vesicular arbuscular mycorrhizal fungus *Glomus intraradices* in an in vitro system in the absence of host roots. New Phytologist 116: 647-653.

Tinker, P.B. and Gildon, A. 1983. Mycorrhizal fungi and ion uptake. In: Metal and Micronutrients: Uptake and Utilization by Plants. Roff, D.A. and Peirpont, W.S. (Eds) Academic Press Inc., New York, pp. 21-31.

Turnau, K. Berger, A., Loewe, A., Einig, W., Hampp, R. Chalot, M., Dizengrenul, P and Kottke, I. 2000. CO_2 concentration and nitrogen input affect the C and N storage pools in *Amanita muscaria/Picea abies* mycorrhizas. Tree Physiology 14: 429-434.

Uetake, Y., Kajima, T., Ezawa, T. and Saito, M. 2001. Tubular vacuoles observed in *Gigaspora margarita* hyphae using laser scanning confocal microscopy. Third International Conference on Mycorrhizas, Adelaide, Australia, http://mycorrhizas.ag.utk.edu

Versau, W.K. 1995. A phosphate repressible, high affinity phosphate permease is encoded by the *PHO-5* gene of *Neurospora crassa*. Gene 153: 135-139.

Weise, J. 1999. Charakteriseirung de Zucker transports in der ektomycorrhiza. Ph.D Thesis, Tubingen University, Tubingen, Germany pp 88.

Weise, J., Kleber, R., Hampp, R. and Nehls, U. 2000. Functional characterization of the *Amanita muscaria* monosaccharide transporter AmMST1. Plant Biology 15.

Wu, C. and Lefebvre, D. 1990. Characterization of PO4 uptake mutants of *Arabidopsis thaliana*. Plant Physiology 93: S100 (Abstract 583).

Yao, Q., Li, X.L., Feng, G. and Christie, P. 2001. Mobilization of sparingly soluble inorganic phosphates by the external mycelium of an arbuscular mycorrhizal fungus. Plant and Soil 230: 279-285.

5

PLANT-FUNGAL SYMBIOSIS VERSUS PATHOGENESIS: MOLECULAR VIEWPOINTS

Host reaction to fungal symbiont or pathogen may begin even before the establishment of any physical contact. Many steps leading to initial contact, formation of appresoria, production of inter- and intracellular structures, and molecular interactions that maintain these two types of biotrophic partnerships could be similar. Infection by fungal pathogen results in activation of an array of defense mechanisms that restrict its multiplication. Hypersensitive cell death, oxidative burst, accumulation of phytoalexins and induction of pathogenesis-related proteins are a few examples of host reaction to the pathogen. Many of these reactions occur even during the establishment of symbiotic associations with mycorrhiza or rhizobia (Duffy and Cassals, 2000, 2003). Defense reactions are triggered in host tissue with the invasion of mycorrhizal fungus, but only in a transient and uncoordinated manner that fades away as symbiosis progresses (Gianinazzi-Pearson and Gianinazzi, 2003). Whereas, several other host reactions and end reactions can be diametrically opposite. At least broadly, if host interaction with fungal pathogen leads to deleterious effects, a symbiotic relationship often imparts physiological benefits. The physiological and molecular basis for such interactive effects on biotrophic partners are being studied; yet there exist numerous lacunae in our knowledge. A comparative study of fungal pathogenesis and symbiosis may provide us with some helpful leads in selectively enhancing symbiosis; at the same time, repressing pathogenesis. For example, a carefully tailored breeding program that enables the development of crop genotypes with resistance to fungal pathogenesis, simultaneously preserving/enhancing mycorrhizal component will have immense applied value in practical agriculture and forestry. Also, a comparative study of molecular aspects could aid in search and easier identification of chemicals that selectively inhibit pathogenesis without any detriment to AM/ECM fungus. A review of recent advances in molecular biology of hosts reactions to fungal pathogens and symbionts—a few examples of mycorrhiza mediated host resistance to fungal pathogenesis and comparison of nutrient exchange at biotrophic interfaces—have been included in this chapter.

1. Host Response to Fungal Symbionts and Pathogens

1. A. Dual Cultures of Host and Symbiont or Pathogen

The study of molecular interactions between host and symbiont/pathogen will need appropriate dual cultures that avoid interferences and factors that may confound observations. Foremost, an accurately devized experimental setup that enables production of symbionts or host and pathogen, access to sampling procedures and molecular analysis is critical. In this regard, mycorrhiza researchers have employed a range of different experimental techniques to establish ECM/AM symbiosis and study the effects of specific factors on both the partners—plant and fungus. Of course, systems that mimic natural conditions, either in nursery and/or field, are available. Certain laboratory procedures have helped in monitoring sharp changes in the molecular and physiological aspects of symbionts. Most importantly, organ and cell cultures of a host tree challenged with compatible ECM fungi or specific pathogens can be used efficiently to detect modifications in morphological, physiological and molecular aspects of the symbionts. Now, let us consider one such experimental technique which is frequently used to study the plant-ECM symbiosis under controlled conditions (Sirrenberger et al. 1995). In this case, spruce (*Picea abies*) callus cells were grown in dual cultures with ectomycorrhizal fungi such as *Amanita muscaria*, *Lactarius determinus*, *Hebeloma crustuliniforme*, *Suillus variegatus* or the pathogen *Heterobasidion annosum*. Presence of spruce cells induced the growth of fungal partners but not the pathogen (Table 5.1). The ECM fungi listed in Table 5.1 physically grew intercellularly on the callus cells of the host and packed tightly into several layers of hyphae. This growth pattern is reminiscent of hyphal mantles encountered under natural conditions. In fact, both *A. muscaria* and *H. crustuliniforme* densely covered the surface of spruce callus cells by producing broad-lobed and highly branched hyphae. In case of *L. determinus*, the hyphae were densely woven into a plait-like pattern on the surface of spruce cells, whereas, *H. annosum* a pathogen, overgrew the callus rapidly, digested the cells and did not show any mantle-like mycorrhiza structures. Such dual culture of plant cells with either mycorrhiza or pathogens have helped in gaining insights into the causes of hypersensitive responses, effects of elicitors, differential signaling and morphogenesis between pathogens and symbionts.

Table 5.1 Influence of Spruce callus cells on the growth (colony diameter in centimeters) of natural ECM fungal partners and pathogens

Fungus	*Spruce callus cells*	
	Present	*Absent*
ECM fungal symbiont		
Amanita muscaria	1.7	1.2
Suillus variegatus	3.1	1.5
Lactarius determinus	3.0	1.5
Pathogen		
Heterobasidion annosum	9.0	9.0

Source: Sirrenberger et al. 1995.

Mutants of either the host plant or biotrophic partner(s) both have served as part of a model system to study the molecular interactions. Hahn and Mendgen (2001) state that such model systems to study the molecular biology of biotrophy, in general, have involved mutants of cultivable biotrophic fungi (e.g. *Cladosporium fulvum* and *Ustilago maydis*); hemibiotrophic fungi (e.g. *Colletotrichum* spp.), which are initially biotrophic but later switch to pathogenesis; or pathogens such as powdery mildew fungus on *Arabidopsis*; *Medicago* or *Lotus* colonized with symbiotic AM fungi and/or Rhizobium. There is no doubt that specific mutants introduced into these model biotrophic systems can be of immense utility while assessing the molecular interactions.

1. B. Chitinous Elicitors and Hypersensitive Responses

Early stages of establishment of mycorrhizal symbiosis or infection by pathogenic fungus, both involve production of elicitors. Let us consider some parallel situations that occur with symbionts and pathogens. Symbiotic fungi such as *Hebeloma crustuliniforme*, *Amanita muscaria* or *Suillus variegates* are known to release elicitors constitutively, almost like a pathogenic fungus (Salzer et al. 1997a, b). These elicitors, for example from *H. crustuliniforme*, induce wide-ranging biochemical changes in host (spruce) cells. Efflux of Cl^{-1} and K^{+}, influx of Ca^{+2}, phosphorylization of specific proteins (e.g. 63 KDa protein), dephosphorylation of a 65 KDA protein, extracellular alkalization and synthesis of H_2O_2 are the general responses seen in suspension cultured cells of Spruce (Fig. 5.1). At the same time, it is interesting to note that elicitors from several of the pathogenic fungi tested on their susceptible hosts are known to trigger similar reactions. Proteinaceous elicitors from *Phytophthora cryptogea* on tobacco cells; glycolipid elicitors from *Pseudomonas syringae* on soybean; and an oligopeptide elicitor from *Phytophthora megaspermae* in Parsley cells are examples belonging to pathogenic fungi (Salzer et al. 1997a).

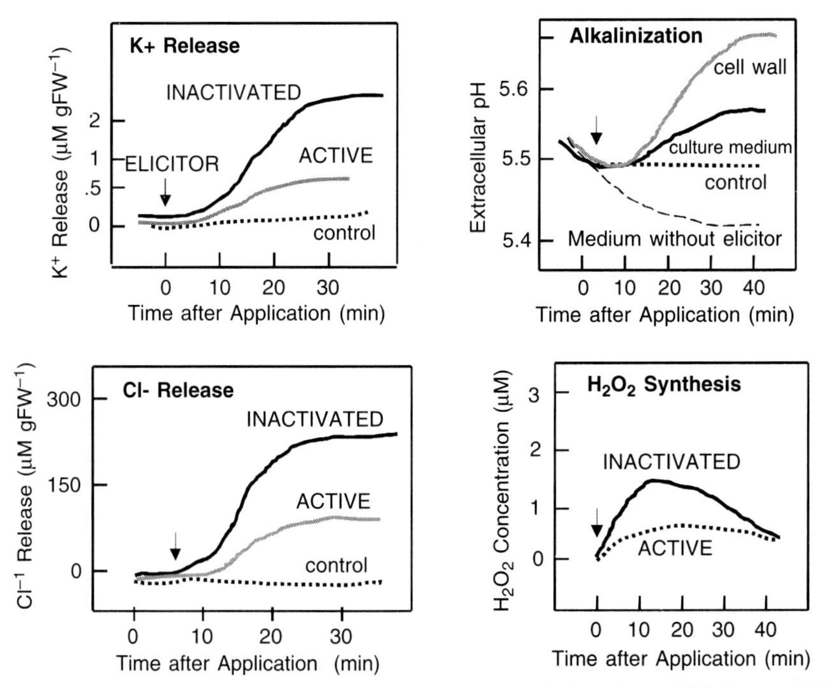

Fig. 5.1 Hypersenstive responses of Spruce cells to elicitors from ECM fungus *Hebeloma crustuliniforme*. *Note:* Arrow indicates point of application of chitinous elicitor.
Source: Salzer et al. 1997a, b.

Incidentally, release of K^+, influx of Ca^{+2} and H^+, generation of active oxygen species are common hypersensitive reactions during host defense. Several of these host defense responses also occur normally in the course of establishment of ECM symbiosis. Obviously, efficient suppression of host defense responses take place in ECM, which may be absent in a plant versus pathogenic fungus interaction. To understand the plant-ECM fungus interaction further, let us firstly consider chitin elicitors. Chitin is a common cell wall constituent of all ECM fungi and host roots do respond to such chitinous elicitors. Salzer et al. (1997a) reported that *Picea abies* cells challenged with colloidal chitin elicitors and chitotetraose react immediately with the efflux of Cl^{-1} and K^+, phosphorylation of a protein pp 63, extracellular alkalinization and H_2O_2 synthesis. These initial reactions lead to the 'hypersensitive response' of the host root cells. According to Salzer et al. (1997a) the suppression of defense reaction involved the production of apoplasmic chitinase isoforms that aid cleavage of chitin elicitors, be it artificial or that produced by ECM fungi. Cleavage of chitin elicitors generally removes the elicitor activity. Incidentally, high concentration of such chitinase common in young portions of spruce roots is known to destroy elicitor activity early during ECM formation. In some cases, lowering the concentration of chitinous elicitors below a value that goes undetected by host cell seems to be a relevant mechanism with mycorrhizas. Whatever be the mechanism, bottom line requirement is the removal of host response to chitinous elicitors for a normal establishment and function of ECM fungal symbiosis.

Auxins attenuate elicitor responses

We know that elicitors from ECM fungi *Hebeloma*, *Amanita* or the pathogenic fungi such as *Heterobasidion annosum* and others induce a defense response involving the production of enzymes, release of K^+ and Cl^{-1}, synthesis of H_2O_2 or specific proteins. Notably, in many cases, such elicitor responses were equal when spruce cells/roots were exposed to either ECM fungi or a pathogenic fungus. The symbiotic fungus and pathogen were not distinguished in terms of elicitor response. At this point, we should note that all ECM fungi have the ability to synthesize and release auxins. Such release of auxin causes production of short roots, on which mantle sheath and hartig net are commonly seen. Mensen et al. (1998) have proposed that auxins synthesized in ECM may play a vital role in attenuating elicitor-induced defense responses of the host. Such an auxin release will eventually allow the ECM fungus to overcome host rejection. In spruce cells cultured without auxins, all the elicitors triggered a transient accumulation of ionically wall-bound peroxidase. However, such early spurt in peroxidase accumulation diminished if auxin was supplied to cell cultures. We may note that in spruce-ECM symbiosis, auxins generally reduced the constitutive expression of wall-bound and symplastic peroxidases, as well as chitinases and β1,3 glucanases. In a comparable situation where tobacco cells were challenged with elicitors, the chitinase expression was downregulated as a consequence of auxin addition. It seems that such auxin-induced attenuating effect, seen with plant-pathogen interactions are specific, depending on the host species. Overall, we can surmize that hormone production, which is common in ECM/AM, may help in controlling undesired defense response by hosts. Further, it is believed that hormone production may actually regulate antagonistic reactions, by enhancing compatibility that is required to sustain the symbiotic relationship between partners (Salzer and Hager, 1993; Mensen et al. 1998). There is no doubt that modification of elicitor activity and effective control of defense reactions constitute a pre-condition for mycorrhiza formation (Salzer et al. 1997a, b).

1. C. ECM Fungi Overcome Plant's Defense Strategies

The establishment of a symbiotic or pathogenic association will essentially involve the suppression of host defense reactions, if any. In the natural course, symbiont or a pathogen overcomes such host defense responses in order to infect and establish itself in the host plant. Staples and Mayer (2004) suggest that similar to ECM fungi, AM and orchid infecting mycorrhizal fungi too have a marked ability to suppress defense-related response of the host resistance mechanism. They share this characteristic with plant pathogens. In nature, this trait is immensely responsible for successful establishment of symbiosis. Salzer et al. (1997a, b) have stated that while establishing symbiotic association, ECM fungi cope with plant defense mechanisms. The strategies used by these fungi include:

(a) reduction of pre-formed defense mechanisms,
(b) insensitivity of the fungus to host plant's defense-related proteins,
(c) enzymatic destruction of fungal elicitors by plant enzymes, and
(d) hormonal regulation of elicitor-induced defense reaction.

Phenolics are a major source of the plants' preformed defense during establishment stages of the ECM symbiosis. Plants accumulate a variety of different phenolic compounds in response to a fungal invasion, be it symbiotic or pathogenic. Reduction in specific phenols that may retard ECM/AM fungal infection/spread is, hence, a requirement under natural conditions. But, we should note that in fine roots of spruce and pines, considerable amounts of soluble and cell wall-bound phenolics accumulate (Munzenberger et al. 1990; 1995). Some of these phenolics are constitutive and possess strong anti-fungal activity that hinders ECM fungal infection and establishment. For example, ferulic acid, p-coumeric acid, p-hydorxy-acetophenone, piecin, isorhapontin, catechin and several others are known to suppress mycorrhizal activity. Agreed, that in the initial stages of ECM fungal invasion, elicitors are induced within specific tissues of the plant, but as the symbiosis progresses with mature hartig net and extensive extraradical mycelium, phenolic compounds decrease in concentration. For example, Salzer (1997a) states that such a decrease in phenolics is marked in the roots of *Larix decidua* rather than in *Picea abies*. Decrease in phenolic strength in *Picea* roots was attributable to ECM symbiosis, which establishes and functions comparatively rapidly with *Laccaria amethestia*. Yet another reason suggested for a decrease in preformed phenolics is polymerization of low molecular weight phenolics. In certain plant AM-fungal combinations, the conversion of active phenols to polyphenolic cell deposits may reduce host defense (Sylvia and Sinclair,1983).

Rendering itself insensitive to host defense-related proteins such as chitinase and 1,3 β-glucanases is yet another mechanism by which AM/ECM avoid detrimental influences of the host that are normally perceived by invading symbiotic/pathogenic fungi. For example, a combination of chitinases and glucanases from fungi such as *Trichoderma longibrachiatum* induces swelling and lysis of cells whereas, these enzymes collected from non-mycorrhizal pine had no effect on ECM fungus *Hebeloma crustuliniforme*. Obviously, insensitivity of mycorrhizal fungi to such defense enzymes of the host seems to be due to specificity. Some reports suggest that an increase in chitinases/glucanases occurs as a consequence of ECM/AM formation, but it has no negative influence on the symbiotic fungi. It seems that chitinase produced due to induction by AM or ECM was localized and held within vacuoles. In case of AM, fungal cell walls are supposedly resistant to certain types of chitinase (e.g. bean extracellular chitinase). The AM colonization levels were unaffected even by constitutive expression of chitinase by the host (*Nicotiana* sp.).

A different mechanism involves the production of extracellular enzymes by the host plant that can effectively destroy or suppress the fungal elicitors. In a simplified trial, Salzer et al. (1997a, b) have shown that treatment of fungal elicitors with extracellular extracts from spruce cell cultures decreased the effectivity of elicitors. The chitin-based elicitors are rendered ineffective because of fragmentation. Suppressor molecules that may inactivate the fungal elicitors selectively at the active site is another possibility. It needs to be proved in case of mycorrhizas, although it is quite frequently reported among plant-pathogenic fungal interaction. Hormones released by ECM/AM fungi may alter the effectivity of elicitors. Downregulation of plant's defense reaction as a consequence of auxin and cytokinin production has also been suspected in AM symbiosis.

With regard to host plant, Provorov et al. (2002) explains that it exerts strict control on AM fungal development through an 'immune-like' system. The defense reactions induced in cortex include cell wall changes, callose accumulation, synthesis of phytoalexins and induction of several symbiosis regulatory genes. Certain molecular responses may be similar to both fungal pathogenesis and AM symbiosis; still, plants' defense responses upon attack by fungal pathogens differ markedly from reactions observed in AM symbiosis at several points (Table 5.2). For example, Provorov et al. (2002) report that immediately after infection by root pathogens, key enzymes of phenyl propanoid pathway are induced. Whereas, if *Glomus* penetrates root cortex, the activity of phenylalanine ammonia lyase and related enzymes are lower and localized. Lytic enzymes are induced only during penetration of epidermis by *Glomus*, but later, as AM fungus spreads into cortex, the activity of lytic enzymes become feeble (Kapulnik et al. 1996; Parniske, 2000).

Table 5.2 Comparison of defense reactions induced during development of symbiotic and Pathogenic fungi

Plant reactions	*AM fungi (e.g.* Glomus)	*Root pathogen* (Rhizoctonia)
Modification of the cell walls	Weak thickening (without a pronounced structural modification); phenolics are not accumulated	Strong thickening with the papillae formation; phenolics are accumulated
Activity of Phenyl Propanoid Pathway	Low; PAL, SHI and SHS are active during root colonization, PAL and CHS are active during arbuscule development No activation of IFR	High synchronous induction of PAL, CHS, CHI and IFR
Synthesis and accumulation of callose	Absent at early stages, low level during the formation of arbuscules	Intensive at the early and late stages of infection
Synthesis of peroxidases, Chitinases, gluconases	At early stages only	At all stages of infection
Synthesis of pathogen-regulated	Low level in the vicinity of arbuscule only	High level without a specific subcellular localization

Source: Provorov et al. 2002.

PAL—Phenylalanine Ammonia Lyase; CHI—Chalcone isomerase, CHS—Chalcone synthase, IFR-Isoflavine reductase.

Dantan-Ganzalez et al. (2001) studied the effect of pathogens, and symbionts such as nodule bacteria or mycorrhizal infection on actin reorganization in host cells. They noticed that mono-ubiquitylation in *Phaseolus/Rhizobium* or *Phaseolus/ Phytophthora* or even *Phaseolus/*mycorrhiza interaction was a common phenomenon. The plasticity found in actin is partly attributable to post-translational modification steps such as phosphorylation and ubiquitylation. Viral infections and stress caused due to heat shock, wound or osmotic imbalance did not induce such modifications of actin in host cells. However, addition of H_2O_2 or yeast elicitor to *Phaseolus* suspension cultures did produce actin modification reactions. Such a modification of actin due to a symbiont such as mycorrhiza or nodule or a pathogen seems to infuse stability into microfilaments in the cells.

1. D. Chitinases in Symbiosis and Pathogenic Interactions

Chitinases are useful biochemical indicators during plant versus pathogen interactions. Their role in the induction of plant defense has been well documented (Collinge et al. 1993; Meins et al. 1994; Vierhelig, 1993). Chitinases are grouped into six classes (I to VI), each of which is categorized by a primary structure of protein. Chitinases hydrolyse and cleave β1,4 glycosidic bond between N-acetyl glucosamine residues of chitin. Chitin is a component of fungal wall, be it AM or ECM or pathogenic fungus. Hence, chitinases (or their isoforms) could be playing an important role in the regulation of establishment of a symbiosis or pathogenic interaction. However, we have to decipher specific roles and mechanisms of action of different isoforms of chitinases during symbiosis and pathogenesis. Recently, Salzer et al. (2000) summarized the role of chitinases in symbiosis and in plant defense mechanisms against pathogens. The chitinases of class I to V are known to possess anti-fungal activity. The class-I chitinases too enhance resistance to pathogens, say *Rhizoctonia solani*, etc. In the recent past, induction of chitinase activity and genes encoding various isozymes has been investigated. To a certain extent, variations in relevant gene expression patterns too have been deciphered. In case of *Medicago truncatula*, the induction pattern of Mtchitinase I, II, III-1 and IV is similar whenever the host is challenged with *Fusarium solani*, or *Aschochyta pisi* or *Phytophthora infestans* (Table 5.3). Whereas, chitinase isoform III-4 genes were expressed differently, depending on whether the pathogenic fungus was virulent or avirulent.

Table 5.3 A chart depicting the release of different isoforms of Mtchitinase in *Medicago truncatula* roots as a consequence of treatment with elicitors from pathogens and symbionts, namely, Rhizobium or Arbuscular Mycorrhiza

Mtchitinase	*I*	*II*	*III-1*	*III-2*	*III-3*	*III-4*	*IV*	*V*
SYMBIONTS								
Arbuscular Mycorrhiza	+++	++	++	++	+++	++	++	
Rhizobium	++	++	++			+	++++	
PATHOGENS								
Ascochyta pisi	+++	+++	+++			+	++	
Fusarium solani f.sp *phaseoli*	++++	++++	++++		+	+++		
Fusarium solani f.sp. *pisi*	+++	+	++		+	+++		

Source: Based on a summarized depiction by Salzer et al. 2000.

Note: Each plus mark denotes a specific chitinase band detected in the gel.

Our knowledge regarding the role of chitinases during symbiotic interaction and their molecular regulatory aspects is meager. Chitinases are induced in response to nodule bacteria during the process of infection. They are known to cleave Nod factors and modulate key morphogenetic signals during nodule formation. Similar to plant versus pathogen interactions, chitinase I, II and IV are induced during nodulation (Table 5.4). If compared with pathogenic or rhizobial interactions, the pattern of chitinase gene expression in mycorrhizal roots is different. Expression of class I, II and IV chitinase genes is not induced in mycorrhiza. Instead, class III chitinase genes are strongly induced Salzer et al. (2000) point out that Mtchitinase III-2 and Mtchitinase III-3 were not at all induced if infected by pathogens or rhizobia. Therefore, expression of chitinase class III-2 and III-3 seems specific to plant-AM fungus interaction. It is imperative that chitinase and their isoforms play a role in the establishment and regulation of plant microbe interactions, be it pathogenic or symbiotic. Further, the role is specific, depending on the isoform and type of interaction in question.

Generally, plant versus pathogen interactions lead to higher chitinase activity than that encountered in plant-AM symbiosis (Bananomi et al. 2001; Mohr et al. 1998). Some isozymes of chitinase are known to inhibit the growth of fungal pathogen, particularly in combination with β 1,3 glucanases (Mauch et al. 1988). Such chitinases, produced constitutively and in high levels are known to impart resistance to host plant against fungal invasion. However, it may not affect AM fungus. Within ECM symbiosis, plant chitinases are known to inactivate chitin elicitors released by the cell walls of ECM fungus. Therefore, it prevents induction of defense responses.

Other than chitinases, expression of catalases and peroxidases are often studied in relation to external stimuli such as infection by a fungal pathogen. They may also have a role in AM /ECM symbiosis. A transient increase in the activities of catalase and ascorbate peroxidase was noticed in tobacco roots infected by *Glomus mosseae*. The spurt in enzyme activity coincided with both the appressorial stage of the fungal invasion and salicylic acid accumulation. Blilou et al. (2000) argue that such biochemical manifestations indicate that the first reaction of host root cells to AM fungus is a defense response.

Polyamines occur free as cations or frequently in a conjugated form with phenolic acids, DNA or RNA. They are implicated in sustaining proper cell replication, morphogenesis and regulation of senescence. Primarily, spermines, spermidines and putrescine are ubiquitous in plants. The polyamine metabolism undergoes changes as a consequence of a fungal invasion, be it pathogen or AM fungi. Walters (2000) suggests that the pattern and extent of changes in polyamine metabolism is dependent on whether the invading fungus is pathogenic, symbiotic or necrotrophic. Compatible and incompatible interactions may show different effects on polyamine metabolism. Metabolism of both free and conjugated polymamines may differ, depending on the interaction and fungal species.

2. Nutrient Transport at Biotrophic Interfaces

Biotrophic interactions include a wide range of mutualistic situations such as symbiosis, commensalism and parasitism. In all these conditions, a biotrophic interface that develops and the transfer (or exchange) of nutrients that occurs across it is a prominent aspect that sustains the partnership (Harrison, 1999). In most situations, regulation of

Table 5.4 The chitinase activity in different plant-microbe interactions—symbiotic and pathogenic

Plant-microbe Interaction	*Plant host*	*Microbial partner*	*Chitinase induction*
SYMBIOSIS			
Arbuscular Mycorrhizas	*Allium porrum*	*Glomus mosseae*	Activity induced transiently
	Glycine max	*G. mosseae*	Activity induced in roots
	Nicotiana tabacum	*G. versiforme*	Acidic chitinase in gels and roots
		G. intraradices	Acidic chitinase induced
		G fasciculatum	Acidic chitinase induced
	Pisum sativum	*G. mosseae*	Acidic chitinase activity in roots
	Medicago truncatula	*G intraradices*	Chitinase III-2, III-3, III-4, in roots
Ectomycorrhizas	*Eucalyptus globulus*	*Pisolithus tinctorius*	Activity in roots
	Picea abies	*Amanita muscaria*	Activity in roots
		Hebeloma crustuliniforme	Activity in roots
Rhizobium	Glycine max	Bradirhizobium *japonicum*	Activity in roots
		B. japonicum mutant	Activity in nodules, roots
	Medicago truncatula	*Rhizobium meliloti*	class-I, II, mRNA in roots
		R. meliloti	class IV, III-4, mRNA in nodules
	Vicia faba	*R. leguminosarum*	class-III chitinase, mRNA, roots
PATHOGENESIS			
Fungal Pathogens	*Eucalyptus globulus*	*Phytophthora cinnamomi*	Activity in gels, roots
	Hordeum vulgare	*Erysiphe graminis*	class-II chitinase
	Medicago truncatula	*Aschochyta pisi*	class-I, II, III-I, III-4, mRNA, roots
		Fusarium solani	class-I, III-4, mRNA, roots
		Phytophthora megaspermae	class-I, II, III-I, IV, mRNA, roots
	Nicotiana tabacum	*Chalara elegans*	Acidic chitinase in gels, roots
		Phytophthora infestans	class-I basic, class-II acidic, mRNA
		Pseudomonas tabaci	class-I basic, mRNA, leaves
	Phaseolus vulgaris	*Fusarium solani*	class-I, II, IV, mRNA
		Collectotrichum lindumuthtianum	mRNA, hypocotyls
	Triticum aestivum	*Puccinia graminis tritici*	class-I, mRNA, leaves

Source: excerpted from Salzer et al. 2000.
Note: Last column depicts presence/induction of different chitinase isoforms in the tissues.

nutrient transport at the interfaces between biotrophic partners is a crucial step that determines whether a biotrophic relationship remains symbiotic or pathogenic. The extent and direction of nutrient transport at interfaces, no doubt, decides pathogenesis and symbiotic effects. With regard to the role of interfaces, Hahn and Mendgen (2001) explain that despite diametrically opposite effects experienced between plant-AM/ECM fungus (symbiosis) and plant-pathogen fungus interactions, many of the structural and functional features as well as molecular regulatory aspects exhibited can look similar, if judged appropriately. Firstly, they argue that regulation of nutrient translocation at interfaces in different biotrophic relations is equally crucial for the sustenance of fungus, be it symbiosis or pathogenesis. We may realize that nutrient flow is bi-directional in mycorrhizas, meaning that soil minerals flow into plant root tissue via fungal mycelia, and carbohydrates get drained in the opposite direction into fungus from roots cells (Fig. 5.2). The schematic diagram by Hahn and Mendgen (2001) is based on two assumptions. First, the plasma membrane H^+ATPase activity is indicative of active metabolic uptake of nutrients. Second, the site-specific expression of transporters is indicative of the nutrient uptake at interface membranes on which the transporters reside. At this juncture, it is useful to reconsider certain pertinent aspects about sugar and amino transporters, as well as the 'symbiotic regulatory genes' (see chapters 3 and 4). They too actively participate in nutrient exchange at biotrophic interfaces. In an ECM fungus *Amanita muscaria*, a hexose transporter mediates sugar uptake at the interface. The genes for hexose transport proteins are generally upregulated during symbiosis or when external sugar concentration is above 5 mM in a free-living state. The genes for high affinity P transporter such as LePT1 in tomato, MtPT1 in *Medicago*, GvPt1 in *Glomus versiforme* all mediate P transport at interfaces between the symbionts.

In parasitic fungi (e.g. rusts), both intercellular hyphae and haustoria within host cells are important locations for nutrient translocation at interfaces. Haustoria are organs of fungal pathogens that develop intracellularly in root/leaf tissue. They are comparable to arbuscule in a symbiotic AM fungal interaction, especially with regard to bi-directional nutrient translocation. For example, haustoria of powdery mildews mediate nutrient transfer between the biotrophs. Hahn and Mendgen (2001) have pointed out that high H^+ATPase activity at haustoria compared with spores or mycelium of the powdery mildew fungus is a clear indication of nutrient transport activity at haustorial interfaces. Induction and expression patterns of hexose and/or amino acid transporters also support the view that plant interfaces between plant cell with haustoria mediate nutrient transport (e.g. *Uromyces fabae* on bean). The transcripts for amino acid and/or sugar transporter proteins are easily traced in the haustoria, although intercellular hyphae may also contribute to nutrient transfer from host cell. Clearly, the haustorial interfaces with host cells are major locations of nutrient transport in plant versus pathogenic fungal interactions (e.g. rust fungus).

The bi-directional flow of nutrients in mycorrhizas stated above relates two different nutrients. One of them is the movement of carbon from plant to fungus and phosphorus vice versa. However, we should note that bi-directional movement, influx and efflux occur with respect to a single nutrient, be it carbon, nitrogen or phosphorus. At this point, it is pertinent to consider the fact that in soil-plant relations, the nutrient transport (e.g. NO_3^-, K^+, or PO_4^-) measured is actually the difference between influx and efflux of mineral ions at root cell plasma membrane—i.e. root/soil interface. Like influx of nutrients, efflux—be it feeble or conspicuous—does occur at all times and could be mediated via transporters at the plasma membrane or through passive flow. Comparing

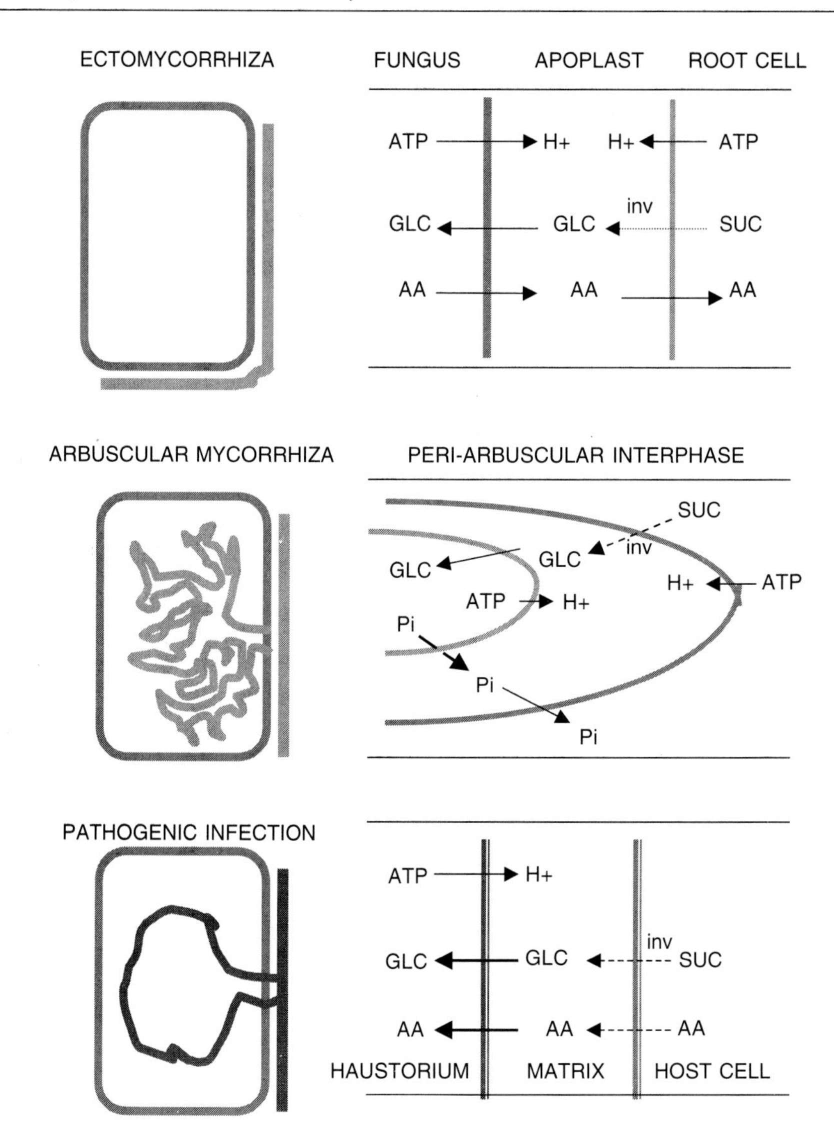

Fig. 5.2 Diagrammatic representation of nutrient transfer at the biotropic interfaces of symbiotic association (AM and ECM) and a pathogenic association with plants. Note: SUC = Sucrose; GLC = Glucose; AA = Amino acid. (Modified and redrawn from Hahn and Mendgen, 2001)

it with nutrient transport in biotrophic relationships, we must think of the bi-directional transport of the same nutrient at the interfaces. In case of symbiosis, for example in plant-AM/ECM fungus, if P transport from fungus into the root cell is considered as influx, then a minor quantity of efflux of P from the host to fungus cannot be ruled out. In this case, therefore, the measured P transport is the difference of influx and efflux of P in root cells with arbuscules. Since P drain away from the host is comparatively feeble the interactions remain symbiotic. If not, it could shift to pathogenesis. Arguments with a similar basis, but anti-parallel to symbiotic relationships could be envisaged for pathogenesis. In case of pathogenesis, at interfaces, efflux of nutrients/carbohydrates from host to fungus is predominant. Hence, the host plant suffers nutritional drain and, eventually, deterioration. However, the invading pathogen may lose out, if nutrient movement at interfaces is predominantly towards the host root/leaf cells (i.e. influx). Within an extended period of symbiosis or pathogenesis, both influx/efflux of nutrients mediated via transporters situated at the interfaces may operate simultaneously. It is the equilibrium point of this transport phenomenon at interfaces that is crucial to the biotrophic interaction. The equilibrium point determines the type of interaction between biotrophic partners.

3. Mycorrhizas Resist Pathogenic Fungal Infections: Molecular and Cell Physiology

Experimentation to understand the interactive effects of mycorrhizal symbiosis on fungal pathogen began a few decades ago. A large number of reports are available on the aspect of AM/ECM fungi and their interaction with different pathogens that commonly afflict their host species. Many a times, mycorrhizas impart bioprotection against the pathogen; especially against its physical spread inside the host and its detrimental effects. A number of tests have also shown that mycorrhizal colonization may have no effect on pathogen. In this context, Borowicz (2001) has made some interesting meta-analysis that encompasses all the reports between 1970 and 1998 regarding the interaction of mycorrhizas and pathogens. Firstly, the size of 'bioprotective effect' or 'resistance' differs due to various causes, including the AM/ECM fungal strain, pathogen and its biotype, host as well as environmental factors. Classification of interactive effects into average plant growth, pathogen growth and AM/ECM growth was useful in assessing the trend that occurred among the several reports. In most cases, the observed effect was related to specific pathogen, its species and biotype used during experimentation. Overall, inoculation with AM fungi had an immense negative effect on the growth of pathogen and its harmful effects. Reduction in AM or ECM fungal growth noticed on certain hosts was easily attributable to competition for energy source and host surface. Alternatively, reduced proliferation of pathogen meant that host resistance and/or bioprotective effect imparted by mycorrhizas were significant. Several of the trials—both those reported and perhaps many more that went unreported—relate to a situation wherein AM or ECM fungi do not influence the pathogen and its growth, nor the detrimental effects on the host plant. Certain reports have clearly indicated the suppression of AM/ECM fungal spread due to the presence of pathogen. It is interesting to note that interaction of pathogen and AM/ECM fungi have also led to a decrease in the growth of both symbiont and pathogen inoculated on a host plant. Such reciprocal suppression was frequently observed if the pathogen was also a fungus. Perhaps, both physical exclusion and molecular interactions between the pathogenic and mycorrhizal fungus leads to this situation. Clearly, Borowicz's (2001) analysis highlights the different patterns and trends in interactions between pathogen versus symbiont. In the context of this book, perhaps, it is most important for us to understand a molecular basis for such variations in symbiont versus pathogen interaction. The resistance mechanism as well as molecular regulation of bioprotective effect will be specific for each pair of pathogen versus symbiont interaction and that needs to be deciphered meticulously. Perhaps, generalization on cellular and biochemical mechanisms underlying bioprotection against specific group of pathogens and AM or say a set ECM fungi should be possible. For example, we can make an educated guess that molecular mechanisms relevant to several root infecting fungi or, say, bacterial pathogens and their interaction with AM fungi could follow a similar pattern. However, there should be no doubt that, still, each pair of pathogen and symbiont will exhibit peculiarities with regard to molecular interactions. Albeit, pre-existing, low and localized resistance may contribute to bioprotective action of mycorrhizas. In most cases, a fully established symbiosis seems necessary for bioprotective action to take effect (Gianinazzi-Pearson and Gianinazzi, 2003).

3. A. Arbuscular Mycorrhizas versus Pathogens—Examples

There are several reviews dealing with interactions between mycorrhizas and plant pathogens, including fungi. In the following paragraphs, a few examples pertaining to AM symbiosis have been discussed.

AM versus Phytophthora

With regard to biocontrol of *Phytophthora* by AM fungi, Norman and Hooker (2000) reported that interactive phenomenon begins earnestly, impeding the pathogen right at the sporulation stage. Microorganisms and extracts from mycorrhizal (*G. intraradices*) roots suppressed sporulation of *Phytophthora frageriae*. Similarly, exudates from strawberry roots colonized with *G. etunicatum* or *G. monosporum* reduced sporulation of *P. fragaria* by 67% and 64%, respectively, compared to root exudates from control (non-mycorrhizal) roots. These sporulation suppression reactions occurred within 48 to 72 h after treatment. The biochemical basis for such a biocontrol mechanism has not been understood as yet. Obviously, chemical moieties in root exudates are involved in suppression of the pathogen. Biocontrol mechanisms elucidated by AM fungi and their host against the pathogen are varied. They may be operative either singly or in different combinations.

The AM-related bioprotection to soil-borne pathogen might actually involve a range of different mechanisms. Dassi et al. (1998) argue that AM symbiosis elicits early and transient defense reactions in plants by involving several biochemical compounds. This process may sensitize the plants to respond quickly and intensely, thus thwarting pathogenic attack that may occur later. In tomato, roots pre-colonized with AM fungus (*Glomus mosseae*) provided a bioprotective effect against pathogenic fungus *Phytophthora parasitica*. It was attributed to pathogenesis-related (PR) proteins in tomato that was induced by mycorrhizal symbiosis. In their evaluations, Dassi et al. (1998) found that PR protein expression was enhanced due to *P. parasitica* infection, while only a weak increase was recorded in *G. mosseae* colonized roots. In AM-infected plants, PR protein induction was directly related to the intensity of pathogen attack. In a different study involving tobacco/*G. intraradices* system, a steady state level of the pathogen-related protein-la (PR-la) was noticeably reduced. Such suppression coincided with an increase in the cytokinin, zeatin riboside due to mycorrhizal colonization. A related test indicated that application of kinetin, 6-benzyl aminopurine also resulted in the repression of PR-la. Hence, an explanation is that enhanced cytokinin levels in mycorrhizal plants influences the gene expression pattern, resulting in reduced PR-la. Lingua et al. (2001) noticed that in tomato infected with *Phytophthora nicotianae*, the pathogen induced nuclear changes. It caused pyknosis, chromatolysis and affected duplication process in host root cells. Flow cytometry and fluorescence measurements indicated that DNA content of nuclei was lowered due to *Phytophthora* infection. Mycorrhization of seedlings may reduce such effects on the cellular function of the host roots. Tomato is a multiploid (2n and 4n) species. Infection by *P. nicotianae* increases 2n and reduces 4n nuclei in root cells. AM colonization is said to restore this situation (Lingua et al. 2003).

While assessing the resistance mechanisms of mycorrhizal plants to infection by pathogens, it is customary to test the reaction between the symbiotic fungus and cultivable pathogen, be it in vivo or in vitro. However, for interactions involving AM fungi or even a cultivable ECM fungus, it is useful to assess both symbionts and pathogen, while they exist in association with their common host (Christelle et al. 1998). Let us consider an example involving a tripartite interaction between tomato roots, symbiotic AM fungus *Glomus mosseae* and pathogen *Phytophthora nicotianae* var. *parasitica*. Immuno-cytochemical evidences prove that the presence of pathogen results in necrosis of cortical tissue along with accumulation of phenolics. The reactions do not impede a pathogen's progression into further locations in root tissue as well as the root cylinder. However, the pathogen is suppressed in mycorrhizal roots, as well as in non-mycorrhizal segments of the mycorrhizal roots. The pathogen was totally absent from arbuscular cells. In fact, arbuscule-containing cells did not necrose and phenolic accumulation was least. In this case, mycorrhiza-mediated resistance seems to involve both direct and distance effect on the pathogen. Details on molecular regulation are yet to be accumulated. Cordier et al. (1998) proposed that bioprotection conferred by *G. mosseae* is due to localized and induced systemic resistance. Phenolics and other cell-mediated defense responses reduced the pathogenic effect. They suspect that mycorrhizal roots remained immune to pathogen (*Phytophthora parasitica*). Such immunity was attributable to cell wall depositions reinforced by callose layers adjacent to the hyphae. The systemically induced resistance could be due to host wall thickenings containing non-esterified pectins and PR-la protein. None of such cell reactions was noticed in non-mycorrhizal roots. The polypeptide profiles that develop in response to inoculation with AM fungus *Glomus mosseae* and pathogen *Phytophthora parasitica* were studied by Dassi et al. (1999). It revealed that quite a few additional polypeptides were induced specifically due to interaction of AM fungus with the pathogen. Detailed comparison with polypeptide patterns from AM fungal spores suggested that at least some additional polypeptides synthesized during the interactions were coded by plant genes. A later study on interaction between AM fungus and *Phytophthora parasitica* has shown that selective expression of isozymes of β 1,3, glucanases in tomato roots could be related to resistance or bioprotection provided by the symbiont. Electrophoretic analysis conducted by Pozo et al. (1999) showed that two additional acidic isoforms are expressed by tomato roots, in response to colonization by *G. mosseae*. Tomato roots challenged with the pathogen *P. parasitica* also showed induction of glucanase, but these were unrelated to *G. mosseae* induced isoforms. In tomato roots pre-infected with *G. mosseae* and post infected with *P. parasitica* two additional basic isoforms of glucanases were expressed. The expression of extra basic isoforms of glucanases was suspected to be relevant for bioprotective action of AM fungus.

In case of tomato root, necrosis caused by *Phytophthora parasitica*, Vigo et al. (2000) observed that AM fungal-mediated bioprotection could be due to reduction in infection loci of the pathogen. Cytomorphological studies showed no variations in the root due to interaction. Influence of root surface, receptor/infection loci and biochemical aspects of infection by pathogen seemed to be influenced by the presence of AM fungus, thus reducing pathogenic infection. Sometimes, in natural soils, it is possible that AM-mediated resistance gets modified because of fluctuations in the microbial populations in rhizosphere (Cordier et al. 2003).

AM versus *Fusarium*

Chinese researchers inoculated watermelons with both AM fungus *Glomus versiforme* and pathogen *Fusarium oxysporum* var. *niveum*. They monitored membrane permeability, auto-oxidation rate and melanaldehyde (MDA) contents in roots. The presence of AM fungus reduced MDA, auto-oxidation rates and membrane permeability. The magnitude of such reduction was greater in Zhang-5, which is a variety susceptible to fusarial attack (Min et al. 2003). They summarized that a change in membrane permeability was responsible for AM-mediated resistance to fusarial infection in watermelon. In case of *Asparagus officinalis*, pre-infection with AM fungus such as *Gi. margarita* or *G. fasciculatum* prevented fusarial entry into short cells. The AM fungi preferentially elongated into short cells in the exodermis of feeder roots. This phenomenon suppressed fusarial invasion and its spread into feeder roots. Hence, nutrient acquisition by host plant was proportionately restored or unaffected (Matsubara et al. 2001). Obviously, AM fungal-related resistance to *Fusraium oxysporum* involves cellular interaction and physical exclusion. Molecular signaling that avoids the entry of pathogen into cells pre-colonized with mycorrhiza, if any, should be investigated. Identification of specific biomolecules may have applied value in disease suppression.

In certain conditions, AM fungal colonization may modify the biotic and abiotic components of rhizosphere, so that deleterious effect of pathogen is reduced. For example, Filion et al. (2003) quantified both AM fungus (*G. intraradices*) and pathogen (*Fusarium oxysporum*) in the rhizosphere and mycorrhizosphere of bean plants using real time PCR assays. They traced a reduction of pathogen in the rhizosphere of mycorrhizal plants. It was attributed to mycorrhiza-mediated modifications in microbiota and soil physico-chemical conditions in the rhizosphere. A similar investigation in Norwich, East Anglia had shown that mycorrhizal status of a host plant is crucial in suppressing the parasitic fungus *Fusarium oxysporum* (Scott et al. 1998). In the field, rhizospheres of mycorrhizal *Vulpia ciliata* var. *ambigua* supported significantly low fusarial population compared to bulk soil. Further, it was argued that sometimes, benefits to host plant due to suppression of a fungal pathogen—such as fusarium in the present case—might be of greater consequence to plant growth than just the improvement in P nutrition. Clearly, there is need to decipher the molecular basis for such interactions between AM fungus and fusarial pathogen in rhizosphere soil. If possible, we should also learn to regulate such molecular interactions to our benefit.

Some of the above discussions have dealt with monitoring AM fungi and parasitic organisms using nucleotide sequences and protein profiles. We may note that lipid profiles of AM fungi and pathogens/saprophytes can also be utilized to study their interactive effects in the soil/plant host. As we know, lipid signatures are indicative of specific saprophytes and AM fungi. More specifically, Larsen et al. (1998) reported that fatty acid 16:1 omega 5 contents differed among AM fungal species, and it was absent in saprophytes. Instead, 18:2 omega 6,9 was the dominant fatty acid in saprophytic fungi. Using such fatty acid profiles, they studied saprophytes/parasites such as *Fusarium culmorum*, *Penicillium hordei*, and *Rhizoctonia solani*. Such fatty acid profiles also helped in quantifying the mycelial growth and sporulation AM fungus *Glomus intraradices* and saprophyte *Fusarium culmorum*.

AM versus other pathogenic fungi

Orna et al. (1998) studied AM fungus *Glomus intraradices* and its interactions with pathogens affecting tobacco such as *Botrytis cinerea* or mosaic virus (TMV). They observed lesions on both mycorrhizal and non-mycorrhizal plants, but it was a bit earlier in the leaves of mycorrhizal plants. In response to application of plant defense chemical activator, 2,6-dichloroisonicotinic acid (INA) or salicylic acid (SA), the expression of tobacco pathogen-related proteins (PR1, PR2 and PR3) was lower in mycorrhizal plants. The reduction in PR proteins was confirmed by a decrease in mRNA levels of corresponding genes. Application of KH_2PO_4 did not mimic mycorrhizal effects on PR proteins. Hence, they inferred that mycorrhiza modulates expression of plant defense-related genes in tobacco, which needs to be understood in greater detail. Shaul et al. (1999) arrived at a similar conclusion that a regulatory process gets initiated in mycorrhizal roots that modify the gene expression and symptom development in tobacco leaves.

Rhizoctonia solani is known to induce several defense-related molecules in Alfalfa plants. Guenoune et al. (2002) found that *R. solani* infection induced a 5- to 10-fold increase in chalcone isomerase, isoflavone reductase and peroxidase activity, along with a marked increase in autofluorescence. These changes were suppressed in the presence of *G. intraradices*. Some of the isoflavonoids tested by them inhibited *G. intraradices* spore germination and hyphal elongation. In view of these observations, it was inferred that during early stages of AM fungal establishment, a certain degree of suppression of defense-related activities of the pathogen-infested host takes place. Molecular aspects of interaction between arbuscular mycorrhiza and a soil-borne pathogen such as *Rhizoctonia solani* are as yet unclear. Recently, Yao et al. (2003) have reported that mycorrhizal (*G. intraradices*) potato plants accumulated higher levels of phytoalexins such as rishitin and solavetivone, provided they were challenged with pathogen *R. solani*. These phytoalexins inhibited mycelial growth of *R. solani*. Guillon et al. (2001) made a beginning in this direction by monitoring the disease development and expression of mRNA for defense-related genes. Mainly, the expression of genes relevant to defense-related enzymes such as phenylalanine ammonia lyase, chalcone synthase, chalcone isomerase and hydroxyproline rich glycoprotein were examined in mycorrhizal *Phaseolus vulgaris* seedlings post-infected with *Rhizoctonia*. Pre-colonization with AM did not reduce the severity of rot disease caused by *Rhizoctonia*. Colonization by *G. intraradices* did not elicit any change in phenylalanine ammonia lyase, chalcone synthase and chalcone isomerase transcripts. However, they observed systematic alteration in the expression of these genes in response to interaction of mycorrhizal plants to pathogen. The extent of change in transcript levels—either stimulation or suppression—were actually altered, depending on the time elapsed after *Rhizoctonia* infection.

Phytoplasmas cause diseases in wild and cultivated crop species. Recently, Berta et al. (2001) reported that prior mycorrhization of tomato plants was able to resist the ill effects of phytoplasma disease. Analysis of host roots using light and electron microscopy, morphometry and flow cytometry showed that phytoplasmas were often degenerated in plants already colonized with AM fungi. Infection by phytoplasma results in decreased DNA content, but in plants treated with both AM fungus and phytoplasma, the percentage of diploid nuclei and DNA content remained at intermediate levels. It indicates that AM fungus imparts resistance to phytoplasma-mediated ill effects on the cellular contents in tomato.

AM inoculation suppresses causation and severity of canker due to *Dothierella gregaria* poplar seedlings. Mycorrhizal colonization enhanced turgidity, available P content, total phenolics, activities of peroxidase and polyphenoloxidase. Increased tolerance to canker disease was attributable to phenolic accumulation, as well as better water and nutrient status of mycorrhizal plants (Tang and Chen, 1995).

Mycorrhiza versus Mycorrhiza—A systemic suppression (or auto-immunity)

An interesting phenomenon known as 'systemic suppression of mycorrhizal infection' has been noticed in barley roots. Vierheilig et al. (2000) reported that prior colonization of one half of a split root system with a AM fungus, say *G. mosseae* or *Gi. rosea* clearly suppresses colonization by an AM fungus inoculated on to the other half at a later stage. In other words, establishment of AM fungus in one portion of root avoids new entry and infection by a later inoculum. It is a situation explainable as 'mycorrhiza against mycorrhiza' or a kind of auto-immunity. The systemic suppression was not attributable to competition for carbohydrates and P application did not mimic the condition. Further, a systemic suppression of mycorrhization in mycorrhizal roots was specific only to AM fungal genera or its species. Obviously, we ought to know the molecular basis for such specific systemic responses. We are not sure if such mechanisms involving systemic suppression by AM fungi are operative against other parasitic fungi. Resistance of mycorrhizal plants to attack by pathogenic fungi could be mediated via such a systemic suppression phenomenon. Such a mycorrhiza-induced immunity to super infection by pathogenic fungi will definitely find positive application in agricultural situations. The phenomenon of systemic suppression of mycorrhizal infection, if operative in natural soils, clearly affects the ecology of AM fungus. Particularly, the extent of root colonization, spatial distribution, succession and persistence of AM fungi in soil will be affected, rather regulated to a certain extent by this phenomenon.

3. B. Ectomycorrhizas versus Plant Pathogens

This topic has been reviewed in the past at regular intervals (Marx, 1972; Sinclair et al. 1982; Zak, 1964; Zengpu et al. 1995). The current trend is to emphasize the molecular aspects of disease tolerance bestowed by AM/ECM fungi and identify the strains of pathogen/mycorrhizal fungus using rRNA sequence differences. Let us consider a few examples that involve ECM and pathogens.

In case of *Pinus massononiana*, Li et al. (1995) found that inoculation with ECM fungi such as *Suilus grevelia* and *Boletus* sp. are known to inhibit pathogenic effect of *Fusarium solani* and *Rhizoctonia solani.* Zengpu et al. (1994) examined the antagonism between ECM fungi such as Suillus and Boletus on pathogens such as *Fusarium oxysporum, F. solani, Rhizoctonia solani, Pythium aphanidermatum* and *P. ultimum.* Secretion of non-volatile and volatile chemical compounds, and cellular lysis in pathogenic fungi were deemed to be the mechanisms of antagonism. In case of biotrophic interactions between ECM fungus *Pisolithus tinctorius* and the pathogen

Phytophthora cinnamomi, Martins et al. (2001) have shown that the timing of inoculation of symbiont and its stage of development could be crucial factors. A well-developed ECM symbiosis seems necessary to initiate molecular interactions that suppress the pathogen. A similar situation was encountered while analyzing the interaction between AM fungus and a fungal pathogen on pea, ·*Aphanomyces euteiches*. A fully developed mycorrhizal association was necessary to induce chitinolytic activity and cause suppression of pathogen (Slezeck et al. 1999).

Host genetics influences both symbionts and pathogens

Sometimes, the genetics and molecular regulation underlying the resistance/ susceptibility to pathogen as well as mycorrhiza may overlap, especially if the pathogen in question is a fungus. For example, mutation at a single point may impart resistance/ susceptibility to both mycorrhizal and pathogenic fungus. This situation is comparable to the overlapping effects of a single gene mutation on the symbionts, *Rhizobium* and AM fungus (see chapter 3). On the other hand, genes for resistance/susceptibility to pathogen and AM fungus may be different. Ruiz-Lozano et al. (1999) tested various barley mutants as well as their progenitors that were resistant or susceptible to powdery mildew caused by *Erisyphae graminis*. They examined the reaction of these barley genotypes to AM fungal (*G. mosseae*) colonization. Two powdery mildew-resistant lines 'Ingrid' and 'Sultan' also proved to be resistant to AM fungal infection. Reduction in the spread of pathogen and AM fungus was systemic and caused by the mutation of wild type, whereas, Ror and Rar mutant barley lines, in which susceptibility to powdery mildew is restored, showed an increased resistance to AM fungal infection. These mutations must have affected genes that differentially modulate symbiotic and pathogenic interactions.

Classical breeding approaches designed to gain resistance to fungal pathogens, especially root-infecting pathogens should carefully consider the consequences on AM/ECM symbiosis of the host crop species. Infection sites, early signal exchange, formation of appresoria, entry into root cortex and establishment may overlap for fungal symbiont and pathogen. As a corollary, genetic advance in host crop species for AM/ECM could involve greater susceptibility to symbiotic fungus. Concomitantly, it may induce higher susceptibility to pathogenic fungus that invades, adopting a mechanism similar to AM/ECM fungus. Hence, genetic selection procedures may have to run simultaneously in anti-parallel directions. The procedure should aim at accumulating resistance to pathogen, but susceptibility to AM/ECM fungus. In that case, success will depend on the extent of non-overlap of loci and genetic mechanisms between symbiotic and pathogenic interactions.

Mycorrhiza in transgenic plants resistant to fungal pathogens

In recent times, transgenics with broad-spectral resistance to plant pathogens are in vogue in several parts of the world. They involve transformation of plants with gene coding for antimicrobial proteins. Obviously, it might affect even the non-target organisms, say for example, symbiotic organisms like mycorrhizas, rhizobium, etc. In particular, a transgenic plant with genes introduced for resistance to fungal pathogens may also retard AM/ECM fungal spread within its root tissue. For example, Bianciotto et al. (1998) tested a number of transgenic tobacco lines with anti-fungal 'plant defensin' gene for their reaction to mycorrhizal colonization. These plant defensins are small cisteine-rich peptides with 45 amino acids, and plants with defensin genes resist infection by the pathogenic fungus *Alternaria longipes*. However, they reported that these tobacco lines did not resist AM colonization, and no undue changes in fungal hyphae, arbuscules, or vesicles occurred. They inferred that constitutive resistance brought about via transgenics could be specific, without having any deleterious effects on beneficial fungal associations. Perhaps, a case-by-case assessment is required to ascertain whether transgenic plants with resistance to fungal pathogens are deleterious to the establishment of mycorrhizal colonization.

Concluding Remarks

Hypersensitive host responses, including cellular and biochemical changes, seem to be similar for pathogenic or symbiotic interaction. In mycorrhizas, these response are milder and transient. Mycorrhizal fungi adopt mechanisms to repress host defense reactions. This involves a reduction in pre-formed defense in the host, becoming insensitive to host defense proteins, destruction of host enzymes and attenuation of elicitor-induced reactions. Several other strategies could be operative in mycorrhizas. Details on induction/repression of specific genes in response to elicitors are required. Role of chitinase isozymes in antifungal activity has been understood to a certain extent. We know that genes for chitinase isozymes are differentially expressed. Their regulation involves selective shifts in the expression of a different set of enzymes, depending on symbiosis and pathogenesis. Biotrophic interactions involve subtle changes in nutrient exchange process in addition defense reactions. Accumulating further details on nutrient transfer at biotrophic interfaces—be it haustoria in pathogenic or arbuscules in symbiotic associations—will be useful. Similarities in molecular mechanisms of nutrient transfer at biotrophic interfaces, if any, need careful comparison.

Bioprotection imparted by mycorrhizal symbiosis is an outcome of complex molecular interaction between the biotrophs—plants, pathogens and mycorrhizal fungi (Harrier and Watson, 2004). Mycorrhizal fungi adopt an array of physiological mechanisms to suppress pathogenesis. Localized defense responses are observable in arbuscules. Pathogens are entirely avoided in arbuscular cells. Induced systemic response (ISR) can provide bioprotection by preventing intercellular development of pathogens even in non-mycorrhizal portions of the host root. Both, localized and induced response involve the development of callose, pathogenesis-related proteins, phenolics and membrane changes. Subtle changes that may occur in gene expression within AM fungi, host and pathogen need careful investigation. Host genetics, especially its resistance/susceptibility traits, may also influence molecular changes in the mycorrhizal fungus and pathogen.

Mycorrhiza-mediated resistance to pathogenic fungi in plants involves a tripartite interaction. *In a well-nodulated leguminous plant, colonized with AM fungus and parasitized by a fungal pathogen, a four-way biotrophic relationship occurs.* No doubt, it complicates the construction of experimental setup, biochemical analysis and interpretation of data. Interactions among the four biotrophs occur in all permutations and combinations. Its analysis and interpretation in time and space will be that much cumbersome and complicated. Perhaps, confining our attention to few physiological aspects and genes of relevance to *tetrapartite* relations will be wiser.

References

Berta, G., Bingua, G., Massa, N., Antosiano, M. and D'Agostino, G.D. 2001. Evidence for increased tolerance to phytoplasma disease in arbuscular tomato plants. Third International Conference on Mycorrhizas, Adelaide, Australia, http://www.mycorrhizasag.utk. edu/icoms /icom3. html

Bianciotto, Valeria, X . Martini, I. and Bonfonte, P. 1998. Arbuscular mycorrhizal interactions with transgenic plants expressing anti-fungal proteins. Second International Conference on Mycorrhizas, Uppsala, Sweden. http://www.mycorrhiza.ag.utk.edu.html

Bonanomi, A., Wiemken, A., Boller, T. and Salzer, P. 2001. Local induction of mycorrhiza specific class III chitinase gene in cortical root cells of *Medicago truncatula* containing developing or mature arbuscule. Plant Biology 3: 194-200.

Borowicz, V. A. 2001. Do arbuscular mycorrhizal fungi alter plant-pathogen relations? Ecology 82: 3057-3068.

Blilou, I., Bueno, P., Ocampo, J. A. and Garcia-Garrido, J. 2000. Induction of catalase and ascorbate peroxidase activities in tobacco roots inoculated with the arbuscular mycorrhizal *Glomus mosseae*. Mycological Research 104: 722-725.

Christelle, C., Gianinazzi-Pearson, V. and Gianinazzi, S. 1998. Resistance mechanism to *Phtyophthora nicotianae* var. *parasitica* in mycorrhizal tomat: pathogen development within root tissue and host cell responses. Second International Conference on Mycorrhizas, Uppsala, Sweden, http://www.mycorrhiza.ag.utk.edu

Collinge, D.B., Kragh, K.M., Mikkelson, J.D., Nielsen, K. K., Rasmussen, U., and Vad, K. 1993. Plant chitinases. Plant Journal 3: 31-40.

Cordier, C., Pozo, M. J., Barea, J. M., Gianinazzi, S. and Gianinazzi-Pearson, V. 1998. Cell-defense responses associated with localized and systemic resistance to *Phytophthora parasitica* induced in tomato by an arbuscular mycorrhizal fungus. Molecular Plant-Microbe Interaction 11: 1017-1028.

Cordier,C., Lemoine, M.C., Jaubertie, J.P., Alabouvette, C. and Gianinazzi, S. 2003. Influence of soil antagonistic microbes on the expression of resistance induced by arbuscular mycorrhizal fungi in raspberry. UMR INRA, Universite de Bourgougne BCE-IPM, CMSE-INRA, Dijon Cedex, France, pp A:\Faro Abstracts.html, pp. 1-21.

Dantan-Gonzalez, E., Rosenstein, Y., Quinto, C. and Sachez, F. 2001. Actin monoubiquitylation is induced in plants in response to pathogens and symbionts. Molecular Plant-Microbe Interaction 14: 1267-1273.

Dasi, B., Dumas Gaudot, E., Gianinazzi, S. 1998. Do pathogenesis-related (PR) proteins play a role in bioprotection of mycorrhizal tomato roots towards *Phytophthora parasitica*? Physiological Molecular Plant Pathology 52: 167-183.

Dassi, D. Samra, A. Dumas-Gaudot, E. Gianinazzi, S. 1999. Different polypeptide profiles from tomato roots following interactions with arbuscular mycorrhizal (*Glomus mosseae*) or pathogenic fungi. Symbiosis 26: 65-77.

Duffy, E.M. and Cassals, A.C. 2000. The effect of inoculation of potato microplants with arbuscular mycorrhizal fungi on tuber yield and size. Applied Soil Ecology 15: 137-144.

Duffy, E.M. and Cassals,A.C. 2003. An investigation on the effect of arbuscular mycorrhizal fungi inoculation on the expression of pathogenesis-related (PR) proteins in micropropagated potato. UMR INRA, Universite de Bourgougne BCE-IPM, CMSE-INRA, Dijon Cedex, France, pp A:\Faro Abstracts.html, pp. 1-21.

Filion, M., St-Arnaud, M. and Jabaji-Hare S.H. 2003. Quantification of *Fusarium solani* in mycorrhizal bean plants and surrounding mycorrhizosphere soil using real time polymerase chain reaction and direct isolations on selective medium. Phytopathology 93: 229-235.

Gianinazzi, V. and Gianinazzi, S. 2003. Modulation of defense responses and induced resistance by mycorrhizal fungi. UMR INRA, Universite de Bourgougne BCE-IPM, CMSE-INRA, Dijon Cedex, France, pp A:\Faro Abstracts.html, pp. 1-21.

Guenoune, D., Galili, S., Phillips, D.A. Volpin, H., Chet, I., Okon, Y. and Kapulnik, Y. 2002. The defense response elicited by the pathogen *Rhizoctonia solani* is suppressed by colonization of the AM fungus *Glomus intraradices*. Plant Science 160: 925-932.

Guillon, C., St-Arnaud, M., Hamel, C., and Jabaji-Hare, S.H. 2002. Differential and systemic alteration of defence-related gene transcript levels in mycorrhizal bean plants infected with *Rhizoctonia solani*. Canadian Journal of Botany 80: 305-315.

Hahn, M. and Mendgen, K. 2001. Signal and nutrient exchange at biotrophic plant-fungus interfaces. Current Opinion in Plant Biology 4: 322-327.

Harrier, L.A. and Watson, C.A. 2004. The potential role of arbuscular mycorrhizal fungi in the protection of plants against soil borne pathogens in organic and /or sustainable farming. Pest Management Science 60: 149-157.

Harrison, M. J. 1999. Biotrophic interfaces and nutrient exchange in plant fungal symbiosis. Journal of Experimental Botany 50: 1013-1022.

Kapulnik, Y., Volpin, H., ithzaki, H., Ganon, D., Galili, S., David, R., Shaul, D., Elad, Y., Chet, I. and Okon, Y. 1996. Suppression of defense responses in mycorrhizal alfalfa and tobacco roots. New Phytologist 133: 59-64.

Li, X., Fu, B. and Yu, J. 1995. Inoculation of *Pinus massanoniana* with Ectomycorrhizal fungi; growth responses and suppression of pathogenic fungi. In: Mycorrhizas for Plantation Forestry in Asia. Brundrett,B., Dell, B., Malajczuk, N. and Mongquin, G. (Eds). Australian Center for International Agricultural Research (ACIAR), Canberra, Australia, pp. 72-76.

Lingua, G., D'Agostino, G., Fusconi, A. and Berta, G. 2001. Nuclear changes in pathogen-infected tomato roots. European Journal of Histochemistry 45: 21-30.

Lingua, G., D'Agustino, G. Masssa, N. and Berta, G. 2003. Mycorrhizal plant development and relationship with some pathogens. UMR INRA, Universite de Bourgougne BCE-IPM, CMSE-INRA, Dijon Cedex, France, pp A:\Faro Abstracts.html, pp. 1-21.

Larsen, J., Olsson, P. A. and Jakobsen, I. 1998. The use of fatty acid signatures to study mycelial interactions between the arbuscular mycorrhizal fungus *Glomus intraradices* and the saprophytic fungus *Fusarium culmorum* in root-free soil. Mycological Research 102: 1491-1496.

Martins, A., Gouveia, E., Coehlo, V., Estevinho, I. and Pais, S. 2001. Effect of mycorrhizal inoculation of tolerance of *Castanae sativa* plants to *Phytophthora cinnamomi* infection. Third International Conference on Mycorrhizas, http://www.mycorrhizas.ag.utk.edu/icoms/icom3.html

Marx, D.H. 1972. Ectomycorrhiza as biological deterrents to pathogenic root infections. Annual Review of Phytopathology 10: 429-434.

Matsubara, Y., Ohba, N. and Fukui, H. 2001. Effect of arbuscular mycorrhizal fungus infection on the incidence of fusarium root in asparagus seedlings. Journal of the Japanese Society for Horticultural Science 70: 2002-206.

Mauch, F., Mauch-mani, B. Boller, T. 1988. Antifungal hydrolyases in Pea tissue. II. Inhibition of fungal growth by combinations of chitinase and β 1,3, glucanases. Plant Physiology 88: 936-942.

Meins, F. Jr., Fritig, B., Linthorst, H.J.M., Mikkelson, J.D., Neuhaus, J.M. and Ryals, J. 1994. Plant chitinase genes. Plant Molecular Biology Reporter 12: 22-28.

Mensen, R., Hager, A. and Salzer, P. 1998. Elicitor-induced changes of wall-bound and secreted peroxidase activities in suspension cultured spruce (*Picea abies*) cells are attenuated by auxins. Physiologia Plantarum 102: 539-546.

Min, L., Wu-haun, W. and Run-jin, L. 2003. Influence of arbuscular mycorrhizal fungi and *Fusarium oxysporum* var. *niveum* on lipid peroxidation and membrane permeability in watermelon roots. Acta Phytopathologica Sinica 33: 3 (abstract).

Mohr, U., Lange, J., Boller, T., Wiemken, A. and Vogli-lange, R. 1998. Plant defense genes are induced in the pathogenic interaction between bean roots and *Fusarium solani*, but not in the symbiotic interaction with mycorrhizal fungus *Glomus mosseae*. New Phytologist 138: 587-598.

Munzenberger, B. Heilemann, J., Strack, D. Kottke, I. and Oberwinkler, F. 1990. Phenolics of mycorrhizal and non-mycorrhizal roots of Norway Spruce. Planta 182: 142-148.

Munzenberger, B., Kottke, I. and Oberwinkler, F. 1995. Reduction of phenolics in mycorrhizas of *Larix decidua*. Tree Physiology 15: 191-196.

Norman, J.R. and Hooker, J.E. 2000. Sporulation of *Phytophthora frageriae* shows greater stimulation by exudates of non-mycorrhizal than by mycorrhizal strawberry roots. Mycological Research 104: 1069-1073.

Orna, S., Elad, Y., Chet, I. and Kapulnik, Y. 1998. *Glomus intraradices*-induced gene expression changes in tobacco leaves. Second International Conference on Mycorrhizas, Uppsala, Sweden, http://www.mycorrhiza.ag.utk.edu

Parniske, M. 2000. Intracellular accommodation of microbes by plants: a common developmental program for symbiosis and disease. Current Opinions in Plant Biology 3: 20-328.

Pozo, M. J., Azcon-Aguilar, C., Dumas-Gaudot, E. and Barea, E. 1999. Beta 1,3 glucanse activities in tomato roots inoculated with arbuscular mycorrhizal fungi and/or *Phytophthora parasitica* and their possible involvement in bioprotection. Plant Science 141: 149-157.

Provorov, N.A., Borisov, A.Y. and Tikhanovich, I.A. 2002. Developmental genetics and evolution of symbiotic structures in Nitrogen-fixing nodules and arbuscular mycorrhiza. Journal of Theoretical Biology 214: 215-232.

Ruiz-Lozano, J.M., Gianinazzi, S. and Gianinazzi-Pearson, V. 1999. Genes involved in resistance to powdery mildew in barley differentially modulate root colonization by the mycorrhizal fungus *Glomus mosseae*. Mycorrhiza 9: 237-240.

Salzer, P., Banonomi, A., Beyer, K., Vogeli-Lange, R., Aeschbacher, A., Lange, J., Wiemken, A. Kim, S., Cook, D.R. and Boller, T. 2000. Differential expression of eight chitinase genes in *Medicago truncatula* roots during mycorrhiza formation, nodulation and pathogen infection. Molecular Plant-Microbe Interaction 13: 763-777.

Salzer, P. and Hager, A. 1993. Effect of auxins and ectomycorrhizal elicitors on wall-bound proteins and enzymes of Spruce (*Picea abies*) cells. Trees 8: 49-55.

Salzer, P., Hubner, B., Sirrenberg, A. and Hager, A. 1997a. Differential effect of purified spruce chitinases and B-1,3 glucanases on the activity of elicitors from Ectomycorrhizal fungi. Plant Physiology 114: 957-968.

Salzer, P. Munzenberger, B., Schwacke, R., Kottke, I. and Hager, A. 1997b. Signalling in ectomycorrhizal fungus-root interactions. In: Trees—Contributions to Modern Tree Physiology. Rennenberg, H., Eschrich, W and Ziegler, H. (Eds) Backhuys Publishers, Leiden, The Netherlands, pp. 339-356.

Scott, C., Lewis, B. and Watkinson, A. 1998. Mycorrhizal status affects populations of *Fusarium oxysporum* in the roots and rhizosphere of *Vulpia ciliata* ssp. *ambigua*. Second International Conference on Mycorrhizas, Uppsala, Sweden. http://www.mycorrhiza.ag.utk.edu

Shaul, O., Galili, S., Volpin, H. Ginzberg, I, Elad, Y. Chet, I. and Kapulnik, Y. 1999. Mycorrhiza-induced changes in disease severity and PR protein expression in tobacco leaves. Molecular Plant-Microbe interaction 12: 1000-1007.

Sinclair, W.A., Sylvia, D.M. and Larsen, A.O. 1982. Disease suppression and growth promotion in Douglas fir from root rot. Plant and Soil 71: 299-302.

Sirrenberger, A., Salzer, P. and Hager, A. 1995. Induction of mycorrhiza-like structures and defense reactions in dual cultures of Spruce callus and Ectomycorrhizal fungi. New Phytologist 130: 149-156.

Slezack, S., Dumas-Gaudot, E., Paynot, M. and Gianinazzi, S. 1999. Is a fully established arbusucalar mycorrhizal symbiosis required for bioprotection of *Pisum sativum* roots against *Aphanomyces euteiches*? Molecular Plant-Microbe Interaction 13: 238-241.

Staples, R.C. and Mayer, A.M. 2004. Suppression of host resistance by fungal plant pathogens. Israel Journal of Plant Sciences 51: 173-184.

Sylvia, D.M. and Sinclair, W.A. 1983. Phenolic compounds and resistance to fungal pathogens induced in primary roots of Douglas Fir seedlings by the ectomycorrhizal fungus *Laccaria laccata*. Phytopathology 73: 390-397.

Tang, M. and Chen, H. 1995. The effect of VAM on resistance of Poplar to a canker fungus. In: Mycorrhizas for Plantation Forestry in Asia. Brundrett, B., Dell, B., Malajczuk, N. and Mongquin, G. (Eds). Australian Center for International Agricultural Research (ACIAR), Canberra, Australia, pp. 67-71.

Vierhelig, H., Alt, M., Neuhaus, J.M., Boller, T. and Weimken, A. 1993. Colonization of transgenic *Nicotiana sylvestrus* plants, expressing different forms of *Nicotiana tabacum* chitinase by the root pathogen *Rhizoctonia solani* and by the mycorrhizal symbiont *Glomus mosseae.* Molecular Plant-Microbe Interaction 6: 261-264.

Vierheilig, H., Garcia-Garrido, J.M., Wyss, U. and Piche, Y. 2000. Systemic suppression of mycorrhizal colonization of barley roots already colonized by AM fungus. Soil Biology and Biochemistry 32: 589-595.

Vigo, C., Norman, J.R. and Hooker, J.E. 2000. Biocontrol of the pathogen *Phytophthora parasitica* by arbuscular mycorrhizal fungi is a consequence of effects of infection loci. Plant Pathology 49: 509-514.

Walters, D.R. 2000. Polyamines in plant-microbe interactions. Physiological and Molecular Plant Pathology 57: 137-146.

Yao, M.K., Desilets, H., Charles, M.T., Boulanger, R. and Twedell, R.J. 2003. Effect of mycorrhization on the accumulation of rishitin and solavetivone in potato plantlets challenged with *Rhizoctonia solani.* Mycorrhiza13: 333-336.

Zak, B. 1964. Role of mycorrhizae in root disease. Annual Review of Phytopathology 2:377-392.

Zengpu, I., Junran, J. and Changwen, W. 1995. Antagonism between Ectomycorrhizal fungi and plant pathogens. In: Mycorrhizas for plantation forestry in Asia. Brundrett,B., Dell, B., Malajczuk, N. and Mongquin, G. (Eds). Australian Center for International Agricultural Research (ACIAR), Canberra, Australia, pp. 77-81.

6

ECOLOGY OF MYCORRHIZAL FUNGI: MOLECULAR APPROACHES

During the last two decades, progress on ecological aspects of mycorrhizal fungi has been steady and vivid. Among different approaches, morphotyping has been most frequently utilized to analyze the dynamics of mycorrhizal fungi. Morphotyping relies excessively on sporulation behavior and morphological features. Morphotyping may not always provide accurate judgment regarding the distribution and activity of mycorrhizal fungi. For example, above ground spore patterns discerned might not tally with population, hyphal spread or intensity of activity below ground. Inferences drawn based exclusively on fungal morphotyping could be inaccurate. During the past 5 to 8 years, molecular identification techniques are being increasingly adopted to study the ecological aspects and ascertain inferences put forth previously using morphotyping of mycorrhizal fungi. Advances on molecular ecology of mycorrhizas have been rapid. Molecular analysis of these fungi has also led us to new concepts in plant-microbe interactions. Accordingly, prime focus in this chapter is on molecular approaches relevant to the study of mycorrhizal ecology. A summary of molecular techniques is followed by brief discussions on genetic variation and population genetics. With regard to AM ecology, recent improvements in our knowledge on genetic diversity, spatial and temporal variations, influence on plant diversity above ground and their role in agricultural ecosystems have been discussed in detail. Similarly, for ECM fungi, discussions are confined to their persistence in soil ecosystem, recolonization patterns that develop in disturbed forest plantations, ECM fungal succession in forest stands, nurse functions, mycoheterotrophy, influence of nutrients and microbial interactions.

1. Genetic Markers and Molecular Methods to Study Ecology of Mycorrhizas

During ecological studies on AM fungi, accurate identification of individual species and its isolates, if any, are important. Methods for in situ identification and characterization of AM/ECM fungal species composition in roots and rhizosphere are critical for effective ecological assessment (Clapp et al. 1999). For the most part, currently, taxonomy is based on morphological markers such as specific spore features and related structure. Quite often, morphologically similar isolates may still be genetically and physiologically different, resulting in widely varying effects on host and adjoining microflora in the soil (Streitwolf-Engel et al. 1997; Van der Heijden et al. 1998a,b). In fact, Sanders et al. (1996) opine that using morphological markers which are heavily influenced by environmental parameters may lead us to erroneous conclusions about the extent of mycorrhizal diversity in the soil. According to Klironomos et al. (1993) and Roldan-Fojard (1994), ecological analysis based on morphological traits may help us with valuable circumstantial evidences that can augment inferences from molecular tests. To quote an example, Clapp et al. (1999) caution that in a particular case, PCR-RFLP analysis indicated extensive colonization of roots by *Glomus* species. However, spore survey indicated none. No doubt, in such cases, survey based on molecular markers will be accurate. Currently, nucleic acid-based methods are in vogue to study the AM/ECM fungal ecology in the soil, as well as their interaction with the plant community. Potentially, such techniques allow quantification of relative abundance (colonization) of different isolates that co-exist on a single root system, in the rhizosphere or soil (Simon et al. 1992).

Ribosomal RNA genes and spacer sequences have been useful during identification and phylogenic analysis of many organisms, including AM/ECM fungi. Most commonly used molecular markers are DNA sequences of a small subunit of rRNA (SSU), internal transcribed spacers (ITS) and a partial large subunit of rRNA (LSU). The SSU are high copy number genes, which exist in tandem arrays and exhibit very little sequence variation between copies. The SSU and LSU sequences have been used to design several primers for the identification of AM fungi at the level of order (VANS1) and family (VAGLO-*Glomus*; VAACAU-*Acaulospora*; VAGIGA-*Gigaspora, Scutellospora*) (Clapp et al. 1999; Simon et al. 1992, 1993; Van Tuinen et al. 1998).

Molecular markers are useful in studying ECM ecology. In general, considerable variations are observed among ECM species with regard to distribution of their fruit bodies in different forest ecosystems (Gardes and Bruns, 1996; Jonsson et al.1999). Some mycorrhizas rarely produce fruit bodies, while others may do so frequently and profusely. As such, spread of fruit bodies above the soil surface is not entirely indicative of below ground distribution and activity of the fungus in question. Therefore, inferences pertaining to ecological patterns of ECM may not be accurate, if only morphological markers such as fruit bodies are considered. A thorough ECM analysis should include a study of the relevant genetic markers. According to Buscot et al. (2000) mycorrhizal ecologists have increasingly resorted to the use of molecular markers. To a certain extent, our understandings about environmental influences and ecological manifestations of both AM and ECM fungi have been improved due to this factor. Molecular detection via specific markers on selected ECM (or AM) is useful in autoecological studies. Such molecular analysis essentially involves design of a fungal taxon-specific primer based on length, G+C content, number, position and quality of mismatches in amplified DNA. However, a good level of sequence variation in ITS region is required in order to select species-specific primers. Let us consider an example. Ammucucci et al. (1998) amplified the DNA of fruit bodies of truffles (*Tuber* species), their external mycelium and symbiotic roots using a single-step PCR. They analyzed the truffles intensely for variation in ITS region. The analysis included several white truffles and five closely related species. Ammucucci et al. (1998) found adequate differences in the ITS region of *Tuber* species and closely related ones. A rapid identification of the fungus was also possible through single-step PCR and sequencing. To quote another example, recently, Gomes et al. (2002) studied polymorphism in ITS of rDNA to identify 26 different isolates belonging to 8 genera of ECM fungi collected from various locations within Brazil. Their assessments suggest that all 8 genera possessed specific RFLP patterns. Hence, this technique was feasible for monitoring ECM fungi involved in inoculation programs. For ecological studies of ECM, ITS region of the rDNA is a very useful marker, compared with analysis of other parts of the genome. Firstly, it varies enough to allow separation of species. Secondly, variation within species is low, so that species can be recognized even if the samples are from geographically different locations. Yokoyama et al. (2002) reported the identification of a pair of PCR primers amplifying a sequence of 235 nucleotides from the strain of *Gigaspora margarita* used in commercial cultures. An oligo-nucleotide probe was also developed. Using a combination of PCR and probe, DNA preparations from single spore and colonized roots were successfully analyzed for presence of *Gigaspora margarita*. It also helped in distinguishing the isolate among several others that occur in agricultural soils of Japan.

Molecular Methods to Study Mycorrhizal Ecology

Presently, a few different molecular techniques are being effectively used to study the ecological aspects of mycorrhizas. One of the important advances that aided identification of ECM fungal taxa and provided better insights into their ecology is the polymerase chain reaction (PCR). Most frequently, ribosomal genes and spacers have been amplified using PCR. Generally, high copy number and conserved sequences that serve for primer design, and variable regions are said to be major advantages. Horton and Bruns (2001) state that currently, use of PCR and ribosomal genes is a familiar theme in molecular ecology. However, study of the ECM community via molecular techniques differs from other aspects of mycorrhizal ecology. Here, these techniques are utilized mainly to identify ECM fungi, but less frequently, for quantification. Quantification of ECM fungi using molecular approaches may not be mandatory. Actually, ECM fungi in root tips are encountered in the form of small discrete macroscopic packages. They can be counted and weighed easily.

Analysis of restriction fragments of the internal transcribed spacer (ITS) is a common molecular technique utilized to understand the ecology of AM/ECM fungi. As stated earlier (see chapter 1), this nuclear region, commonly investigated by fungal taxonomists, lies between the small subunit (SSU) and large subunit (LSU) ribosomal RNA (rRNA) gene. It contains two non-coding regions and 5.8S rRNA gene. Typically, it is 650 to 900 bp in size. Universal primer pair (e.g. ITS1 and ITS4) or fungal specific primers (ITS1 or ITS4 and ITS4b) are used to amplify. Taxon-specific ITS primers are also available for use during ecological analysis of ECM fungi. A point to note is that misinterpretations are a possibility, if appropriate ITS and target fungal groups do not match, because primers intended for specific groups/families of fungi do not amplify sequences in every species within the intended group. For mycorrhizal ecological purposes, ITS sequence-based procedures, if utilized, need greater prior taxon sampling and thorough phylogenetic analysis. Therefore, ITS-based approaches should be limited to generic level identification. The 5.8S nuclear rRNA gene sequence is useful for broad level characterization, primarily to identify the source of unknown ITS-PCR amplification as plant, fungal or animal origin. Such a molecular procedure essentially avoids the problem of analyzing non-target DNA from extracts (Redecker et al. 1999).

Restriction fragment length polymorphism (RFLP) analysis of ITS region separates many species. This procedure is rapid, cost effective with the least technical requirements. Typically, only a set of 2 or 3 restriction enzymes are used to digest and analyze the RFLP patterns. A few digests suffice because sequence differences between taxa are due to insertions or deletions of nucleotide that cause fragment length variations. Hence, simple side-by-side comparisons provide ECM fungal identity at the species level. The number of unidentified RFLP types may be high if this technique is used alone (Horton and Bruns, 2001). The unidentified types occur even in samples drawn from locations or fungi for which ITS-RFLP data bases from sporocarp samples are available extensively. Quite often, database relates to ECM fungi that produce large sporocarps and those not frequently colonizing the root. Also, RFLP patterns may not match exactly, because size estimates for fragments differ and intraspecific variation is common. Sometimes the RFLP databases may not be standardized with regard to primers, enzymes and the way information is stored or retrieved.

Nuclear large subunit (LSU) analysis helps in the identification of plant species from the mycorrhizal root extracts. Cullings (1992) reported a method wherein a plant-specific primer, in combination with universal primer 28C, will amplify a portion of 28S rRNA gene of plants, even when fungal DNA is present. An RFLP is then conducted on the PCR products and unknown patterns are compared with databases. Horton and Bruns (2001) state that using this procedure; separation of conifers (*Pinus, Psuedotsuga*) at the genus level is possible. Although it is difficult to achieve amplification, angiosperms *Quercus*/Fagaceae and *Adenostoma*/Rosaceae could also be differentiated. Hence, developing primers for ITS regions that allow amplification will be useful. Molecular details, especially sequence data for relevant mycorrhizal fungi will help in drawing more accurate inferences and in understanding the ecological aspects better. Detailed analysis of at least the dominant RFLP types, or those of particular interest to ecologists should be possible, although not attempted frequently. Horton and Bruns (2001) point out that presently, there are several molecules of choice to conduct detailed sequence analysis, but SSU rRNA genes are preferred. This gene has not been sequenced in many ECM fungal taxa. Perhaps, it is tedious and the resolution achieved matches neither labor/time nor cost effectivity. On the other hand, 16S sequences have been deciphered in greater detail for several ECM fungi deemed ecologically prominent in certain locations. For example, detailed 18S sequence data are available for members of Pezizales, Gomphaceae and Cantherellaceae (Hibbet et al. 1997; Norman and Egger, 1999; Pine et al. 1999). The SSU or 18S sequence data can be matched with established ribosomal gene databases. Among molecular approaches employed to decipher ecological aspects of ECM fungi, the 5' end of LSU rRNA gene is yet another useful target for sequence identification. Horton and Bruns (2001) point out that since it is a more variable region, therefore, it could be a more informative target than SSU or 18S. It allows placement of ECM fungi at the generic level. Availability of sequence data for many Agaricales and Boletales has allowed the use of this technique to study their ecology (Gardes and Bruns, 1993; Moncalvo et al. 2000).

The mitochondrial large subunit gene (mt LSU) has also been effectively utilized in the identification of ECM fungi (Horton and Bruns, 2001; Taylor and Bruns, 1999). Assessment of mtLSU rRNA gene sequence provides unambiguous placements in families (e.g. Russulaceae, Amanitaceae and Cantherellaceae). It is not easily applicable on certain ECM groups such as Cortinariaceae and Tricholomataceae. Overall, a combination of rapid morphological sorting, RFLP matching and sequence analysis seems to be a successful approach for identifying ECM fungi during ecological studies (Horton and Bruns, 2001; Fig. 6.1).

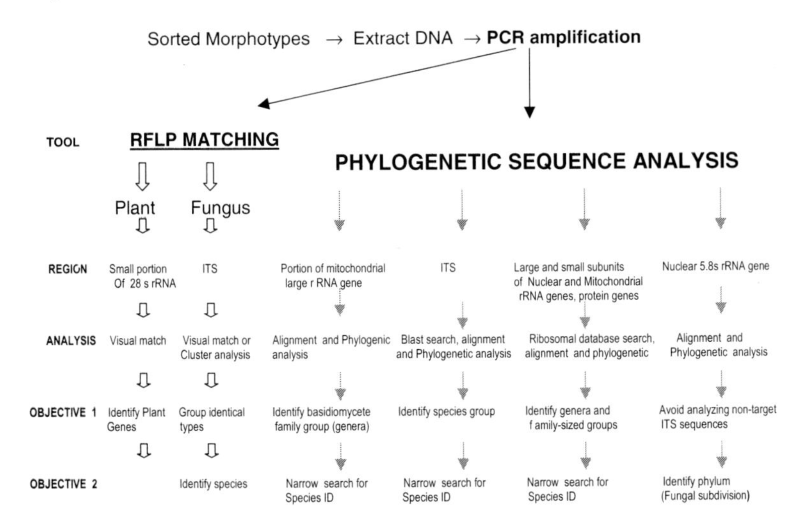

Fig. 6.1 Flow Chart of a Combined Morphotyping and Molecular analysis-based approach to identify ECM fungi in root tips collected from the field. *Note:* Rapid sorting and morphotyping is performed on the sample, followed by RFLP-matching and Phylogenetic analysis. (*Source:* Horton and Bruns, 2001)

A slightly rapid and advanced version, known as multiplex PCR, was utilized by Eberhart et al. (1999) to identify ECM fungus *Tylospora* occurring on a spruce. In a multiplex PCR, several fungal species can be detected using a mixture of species-specific primers. For example, the persistence of *Tricholoma scalpturutum* and *T. populina* has been simultaneously studied using multiplex PCR (see Buscot et al. 2000). There are also specific fungal primers such as ITS-F, ITS4-B, or NC6Bmun that hybridize ribosomal sequences across a broad range of fungal taxa (Gardes and Bruns, 1993). A second step involving RFLP or species-specific oligonucleotide probing in a southern hybridization will then help in the identification of actual fungal species. RFLP markers are cost effective and fairly easy to perform among molecular techniques. The PCR-amplified products are digested with 4 to 5 selected restriction enzymes; the digests (fragments) are then separated and traced on gel electrophoresis. Through this method, we can rapidly identify and establish species-specific restriction length patterns (ITS-RFLPs). Once the database on restriction enzyme patterns for a fungal species are standardized and confirmed, they can be useful in screening samples to study the ecology of ECM fungi. Actually, a high degree of interspecific sequence variation in ITS means that just a few restriction enzymes may be sufficient to separate related species within a genus (Pritsch et al. 1997). In case of a mismatch between ITS-RFLP patterns of mycorrhizas and sampled fruit bodies, we may compare the unmatched sequences with the available sequence database. There is no doubt that identification of sample mycorrhizal fungi to the generic level will help in narrowing down the search for successful ITS-RFLP pattern matches. The database for mitochondrial, large subunit ribosomal 5.8S rRNA and nuclear 28S ribosomal gene are available and can help in the identification of broad generic groups and species which will narrow down further search using RFLP marker.

Microsatellite Markers: The traditional approach in population genetics is to use isozyme analysis. However, considering the difficulty in obtaining sufficiently large-sized spore samples of AM or sometimes ECM, and time required for tedious analysis, their use to decipher gene flow and population genetics of AM fungi is limited. The alternative available at present is to use micro-satellite DNA or short tandem repeats. These are stretches of tandem mono-, di-, tri- or tetra nucleotide repeats of various lengths. The loci that possess micro-satellites are heritable, but selectively neutral markers because these are non-coding regions. The utility of micro-satellite DNA to study the gene flow and genetic distance between individuals has been previously reviewed (Bachmann, 1994). These genetic markers have been frequently utilized to trace soil microbes. The population structure of plant pathogenic fungi has also been deciphered using these markers. However, we are yet to adopt this technique frequently for study of ECM. Let us consider a report by Kretzer et al. (2000) that utilized micro-satellite markers to trace *Rhizopogon vinicolor.* It is a basidiomycetous ECM fungus that specifically colonizes *Psuedotsuga menzesii,* commonly found in Pacific Northwest of America. *Rhizopogon* produces fruit bodies below ground that are difficult to sample, but the tuberculate mycorrhiza are morphologically distinct and provide ample material for DNA extraction and amplification. Kretzer et al. (2000, 2004) have reported procedures for the development for micro-satellite markers, wherein under the conditions, primer pairs they tested did not amplify DNA from *Pseudotsuga menzesii* (Table 6.1), which is the only host for *R vinicolor.* Consequently, they sampled and assessed species identity using RFLP of the ITS region. Such micro-satellite markers can be employed to study the distribution of genes and the population structure of specific ECM fungal isolates. A major advantage of such a system is that primers do not amplify DNA of the host plant, *P. menzesii* in the example stated above. Therefore, root samples can be easily assessed for ECM fungal markers. Douhan and Rizzo (2003) have suggested that amplified fragment length micro-satellite (AFLM) could be used to investigate organisms like AM fungi which cannot be cultured easily and obtaining DNA samples is

Table 6.1 Examples of micro-satellite loci to trace and assess *Rhizopogon* distribution on roots and in soil

Locus	*Repeat**	*Gene bank Acc no.*	*Primer sequence (5'–3')*	*No. of alleles*	*Called sizes (bp)*	H_O	H_E
Rv02	(CAC)7	AF154076	'F'GTGAAACTGTTCAACGCACG GTAACTCGTGCTCTTCATCG	6	296.6	0.47	0.76
Rv15	(CAC)11	AF154077	'F'TCCACCTATCCACCTATTCG AACGTAGACCCAACATAAGAC	4	274.5	0.37	0.43
Rv25	(CGA)6	AF221518	AGCACCCGGAATGTTGAGG 'H'GGCTATTACAGTGGTCAATGTC	2	245.4	0.05	0.05
Rv35	(CAC)6	AF221519	CGTGCAGCCCGAATCTC 'H'GATCCATTCTGAGATGCTCC	4	285.8	0.61	0.56

Source: Kretzer et al. 2000.
Note: *Sequence motif found in *R. vinicolor* T20787; 'H' = HEX dye; 'F' = Fluorascein; @= fragment sizes as determined by the ABI GENESCAN software are highly reproducible due to internal size standard. They commonly include fractions of a base pair; $H_E = 1 - \Sigma P^2$ I, with Pi = frequency of allele 'i'.

difficult. Micro-satellite markers are useful in analyzing the genetic structure and gene flow. One such system was developed for *Pisolithus* species by Kanchanaprayudh et al. (2002). Using these markers, ECM associations between *Eucalyptus camaldunensis* and *Pisolithus* species commonly found in forests of Thailand were assessed. More recently, Hitchcock et al. (2003) utilized inter-simple sequence repeat PCR (ISSR-PCR) to develop markers for simple sequence repeat-rich (SSR) regions. It was intended to identify and investigate genetic relatedness among *Pisolithus* samples. Micro-satellite markers have also been reported for *Hebeloma cylindrosporum*, which is an important ECM fungus. Jany et al. (2003) screened EST databases and arrived at six different simple sequence repeat (SSR) markers specific to *Hebeloma*. Using such SSR markers, the occurrence of *Hebeloma cylindrosporum* genotypes could be identified unambiguously.

2. Genetic Variation and Population Genetics

Recent developments about the genetic variation and population dynamics of AM, ECM and Ectendomycorrhiza have been discussed under this section. With regards to ECM, reports are available for a large number of species, but only a couple of examples dealing with the genus *Laccaria* have been included here.

Mechanisms of Genetic Variation in Arbuscular Mycorrhizas

The population ecology of multi-genomic organism such as AM fungus—that exists in symbiotic association simultaneously in connection with multiple hosts—needs to be understood in greater detail. A clearer perspective may be possible with the application of molecular techniques (Sanders, 2002). A number of key features of AM fungal population biology have been deciphered recently, and we need to keep in mind certain phenomena such as genetic exchange and drift while assessing AM fungi.

Molecular analysis of ribosomal genes and their sequence within the order Glomales clearly reveals a wide genetic variation. This does not seem commensurate with our present knowledge, if any, regarding the sexual phase of these fungi. The general belief is that AM fungi display an asexual stage throughout the life cycles. That means AM fungi exhibit differences in ribosomal genes despite being asexual. Heterokaryotism, that is multiple nuclei within single spores/hyphae, or heterogeneity due to each single nucleus or both the mechanisms operating at different proportions might be the cause for such a variation in the ribosomal genes. As yet, there is no molecular technique that distinguishes between the two mechanisms by which such a variation in genes is caused. Heterokaryotism, which leads to the occurrence of different ribosomal genes, in different nuclei of a single spore seems to be the most plausible and pertinent. Genetically, such heterokaryotic spores/cells are conglomerates of individuals. Hence, they should be deemed genetic populations, although morphologically single entities. At present, we lack detailed knowledge about the mechanisms involved in the transfer of nucleus that generate heterokaryotism in AM fungal spores (Pawlowska and Taylor, 2004). Whereas, in higher fungi (i.e. Ascomycetes, Basidiomycetes), heterokaryotism is frequent. In these cases, exchange of genetic material is facilitated through hyphal fusion, anastomoses and other sexual reproduction processes. The broad host range, adaptation to widely different environments and general plasticity in growth/perpetuation, all of these could be attributed to broad genetic base created via heterokaryotism. However, analyses of single spores from pot cultures using PCR-RFLP reveals remarkable genetic consistency, despite being genetically diverse. Generally, multiple copies of ribosomal genes are homogenous within single genomes. In fact, it is this homogeneity which allows ribosomal genes and ITS sequences to be utilized extensively during phylogenetic investigations of AM/ECM fungi or other organisms.

AM fungi being asexual, opportunity for unequal chromosomal crossing and recombination are minimal. However, frequent mutations that stay uncorrected could generate the genetic variability observed on ribosomal gene sequences within single spores of glomales (Pringle et al. 2000). There are no authentic reports about sexual reproduction in AM fungi, but Sanders et al. (1996) argue that such a situation in no way rules out sexual reproduction. In addition, variation in chromosomal genes, within or among the nuclei of AM fungi, could also occur due to mechanisms such as gene turnover and molecular drive. There is no doubt that advances in phylogeny, evolutionary and even ecological aspects of AM fungi are dependent on our knowledge regarding the cellular/molecular mechanisms that generate and maintain genetic diversity. We should take note that anastomoses occur frequently whenever hyphae of the same isolate of AM fungus come in physical contact. It leads to the development of large mycelial networks in pre-symbiotic and symbiotic phases. Recently, Giovannetti et al. (2003) reported that anasotomses occur widely between hyphae belonging to the same and different germlings of an isolate in *Glomus mosseae*, *G. caledonium* and *G. intraradices*, whereas, hyphae from different AM fungal species never fused, suggesting fungal ability to recognize species level differences. Also, vegetative compatibility/incompatibility groups occur within an AM fungal species such as *G. mosseae*. A high frequency of anastomoses between hyphae from genetically different spores, no doubt, allows for genetic exchange and enhances diversity. We should note that spores of AM fungi are multi-nucleate and possess a high degree of genetic diversity in ITS region of rRNA genes (see Pringle, 2001; Pringle et al. 2000). AM fungi are clonal organisms, anastomoses between different germlings induces and maintains genetic diversity, especially in the absence of sexual recombination. The anastomoses ability of AM fungi is suspected to be a survival strategy of AM fungi and it might also impart genetic/physiological fitness in an environment.

Ribosomal RNA Sequence Variation and Phylogeny: Variation in rRNA sequences occurs even within a single spore of AM fungi, which lessens the validity of their use in phylogenetic analysis. Sanders et al. (1996) opine that variation witnessed in ITS region in single spores could be attributed to mutations that are unchecked and perhaps not repaired in natural course. Further, such mutations in non-coding regions are more prone to persist. This is because, deleterious mutations are selected against in a surviving population, whereas, those in non-coding regions may stay neutral, therefore go undetected in nature.

With reference to variation in the VANS1 region in *Scutellospora*, it is suggested that 18S ribosomal gene sequence analysis relates to the phylogeny of rRNA gene, rather than the AM fungi. Similar situations have been encountered with phylogenetic analysis of bacteria. The next question relates to the extent of variation in rRNA sequence that occurs within single spores. Sanders et al. (1996) reported that differences for only a small number of bases were discerned within single spores of

G. mosseae. In case of *G. geosporum*, slightly more number of bases differed in the ITS region. Generally, differences in base sequences are far greater when spores of different morphology are evaluated compared with the single spore types. Overall, molecular phylogeny established based on analyses of ribosomal gene has been utilized in different laboratories to assess AM fungi derived from ecologically different conditions.

We already know that AM fungal spores contain between hundred and several thousand nuclei. In their recent effort using fluorescent in situ hybridization, Kuhn et al. (2003) found that two variant ITS sequences segregate in different frequencies among the nuclei of one spore. It means that nuclei found in spores are genetically different. They actually identified as many as 16 different variant sequences within *G. intraradices* spore. According to them, none of the 16 variant ITS appeared to be psuedogenes. They have interpreted that variant gene sequences should be appropriately considered. Different variants of a gene could be expressed at different times or in different environmental conditions. Obviously, studying only one variant of ITS sequence provides a partial view. Inferences drawn after investigating the population of ITS sequences will differ from those based on a single variant sequence. Perhaps, multigenic nature is the cause for genetic drift in AM fungi. It may also confer functional diversity to the fungus and better ability to adapt to fluctuations in environment. Powlowska and Taylor (2004) believe that polyploid organization of genomes in spores/hypae could acccomodate intranuclear rDNA polymorphism and buffer these apparently asexual organisms against ill effects of accumulating mutations.

Populations of a Basidiomycetous ECM: *Laccaria bicolor*

The basidomycetous ECM fungi are an important component of temperate forests. Their unique ability to form symbiosis and maintain physiological link with host tree makes them an interesting aspect of soil biology, particularly with reference to their population ecology and genetics. Most of these basidiomycetous ECM species do not form conidia and only few produce chlamydospores. Hence, their reproduction and dispersal are dependent on production of basidia, mycelial fragments, or sometimes, sclerotia. There are several studies on the population structure, genetic variability, reproduction strategies and gene flow in natural populations of *Laccaria* and *Pisolithus* (Bastide et al. 1995). Largely, the Holobasidiomycetes are heterothallic and possess multi-allelic bipolar and tetra polar mating systems. *Laccaria bicolor*, for example, is also heterothallic with bifactorial mating system and includes several intersterility groups. Mueller and Gardes (1991) identified three intersterility groups within *L. bicolor*, which actually constitutes three distinct North American species, including *L. bicolor senso stricto*. Douderick and Anderson (1989) have already underlined the importance of sexual reproduction in *L. bicolor*. But a parallel intersterility compatibility system also operates both intra- and interspecifically. Bastide et al. (1995) suggest that a large number of mating type factors encountered confirms the role of basidospore dispersal in establishing new genotypes of *L. bicolor*. Basidiospore dispersal will maintain the genetic exchange among allopatric populations whenever intersterility barriers do not restrict the gene flow. Crosses between dikaryotes and monokaryotes also induce a genetic variation in *L. bicolor*. Presence of dikaryotic mycelium often induces basidiospore germination. In axenic cultures, co-inoculated dikaryons, di-mon crosses dikaryotize the monokaryons. According to Bastide et al. (1995) formation of such di-mon crosses results in nuclear/cytoplasmic genome combinations.

Laccaria bicolor propagation is effected both via sexual basidiospores and vegetative mycelium. Basidiospore production, no doubt, induces genetic exchange among compatibility strains, whereas vegetative mycelium facilitates the expansion of successful genotypes into soil and plant host. Obviously, the reproduction strategy of *L. bicolor* or any other basidomycetous ECM fungus may vary, depending on environmental conditions. In *L. bicolor,* mycelial growth rate is a polygenic trait, and a useful indicator/determinant of competitive ability of a fungal genotype in nature. According to Deacon and Flemming (1992), ecological interactions during root colonizations may involve competitive exclusion, rapid mycelial growth and inoculum potential of mycobionts.

Let us consider a long-term analysis of the pattern of colonization and spread of yet another basidiomycete, namely *Hebeloma cylindrosporum*. Molecular analysis was conducted on samples drawn at a forest site in France from 1993 to 1997. Gryta et al. (2000) mapped the fruit bodies of this tetra polar heterothallic fungus and studied the mating types and molecular markers (i.e. rDNA polymorphism and RAPD). They inferred the presence of two non-overlapping genets based on information from haploid nucleus and dikaryotization of isolates. In fact, most of the isolates collected belonged one of the genets, and their territories spread at a rate of 0.45 to 0.60 m per year. Further, it was concluded that in case of *H. cylindrosporum,* the origin of additional genotypes observed during 1993 to 1997 was attributable to selfing within the genets, rather than due to out crossing. There were indications that only a fraction of the area sampled was favorable for colonization. Native genets could inhibit genetic diversification through somatic incompatibility system.

Burgandy Truffle (*Tuber uncinatum*) was studied with a view to understand the existing genetic diversity among populations. RAPD techniques were used to type the fruiting bodies collected from various plantations and woodlands (Luis et al. 2001). Relatively high diversity was noticed among populations. The pair-wise distance among genotypes ranged from 3 to 50%. Individuals of the largest clone were spread over 300 km^2 area. However, most genets were limited to single or a few sites. *Tuber uncinatum* persists vegetatively, but the genetic structure noticed within the population studied is commensurate with high frequency of sexual reproduction.

Ectendomycorrhizas

Let us consider an interesting study on the distribution and fungal community composition in a location from Australia. A 10 m^2 plot within a dry sclerophyll forest was sampled for long root hair pieces from *Woollsia pungens*. Molecular analysis of the ITS region revealed the co-existence of at least six endophyte taxa. However, quite a number of plants possessed a single taxon belonging to Heliotales (Chambers et al. 2000; Liu et al. 1998), suggesting that single ericoid fungal species may dominate. Study of RFLPs indicates that most fungal isolates drawn from a single plant were actually derived of a single genotype. This implies that single mycelium may infect and spread extensively into eparcid root systems. Like ECM/AM symbiosis, fungal interconnections between proximate ericoid plants occur. In nature, it is possible that ectendomycorrhizae mediate solute (nutrient) transfer between plants. In another study, Midgley et al. (2001) mapped the spatial distribution of ectendomycorrhizal fungi colonizing *W. pungens* seedlings using a digital image of the root system, and ITS region analysis. In all, 5 taxa of endophytes could be identified, but approximately 75% of fungal assemblages were dominated by a single taxa. They also reported that of the five ITS-RFLP types, type-1 comprised six genotypes. Correspondingly, the root system was colonized by the dominant taxa, and most isolates belonged to the single genotype. Inter-simple sequence repeat (ISSR-PCR) analysis revealed that the other four taxa occurred only in patches and in small zones on the root. Overall, most frequently isolated endophyte taxon identified using ITS sequence belonged to *Hymenoscyphus* clade. This finding contrasted with the observation that unspecified Heliotales dominate these soils (Liu et al. 1998). Certain endophytes such as *H. ericae* are known to possess considerable interspecific physiological heterogeneity, with reference to N utilization. Cairney and Ashford (2002) believe that in nature, occurrence of a composite of genotypes may help maximize or sustain N acquisition by the endophytes from the soil. In fact, a low degree of functional diversity is common in natural population of soil fungi, including ectendomycorrhizas.

Ericoid mycorrhizas may have been traced in the roots of all Ericoid plants, but not necessarily all year round. Cairney and Ashford (2002) believe that ericoid mycorrhizas, particularly those colonizing the eparcids in Australia, are strongly seasonal. Loss of hair roots, along with the mycorrhizal fungal component, seems common in dry season or drought-affected plants. Hair roots reappear in autumn and their length increases with soil moisture. These roots become mycorrhizal, and the highest colonization rates are around spring time. Studies utilizing ITS sequence analysis indicate that in the Northern Hemisphere, ericoid mycorrhizas persist throughout the year (Read and Kearley, 1995), except in dry years. Within limits, the Ericoid mycorrhizal distribution may not be affected much by the latitude, altitude, edaphic or climatic parameters (Read, 1996). Location-specific effects are equally important. For example, in *W. pungens* grown in Australia, the mycorrhizal infection may not be correlated with minor changes in moisture, but was negatively correlated with temperature.

3. Ecology of Mycorrhizal Fungi with Emphasis on Molecular Analysis

It is interesting to note that 90 to 95% of plants on the globe are associated with at least one type of mycorrhizas. Among them, arbuscular mycorrhizas are the most widely distributed. In view of this observation, Smith and Read (1997) have argued that it is 'mycorrhizas', and not roots, which transport the greatest amounts of nutrient to support vegetation. Bago et al. (2000) in fact, commented that root biology could be a 'mycorrhizal biology' in certain situations. Yet we know very little about their ecology. Efforts to understand their ecology using molecular methods are only very recent, perhaps at best 5 to 8 years old.

3. A. Molecular Analysis of AM Fungal Diversity and its Relation to Ecosystematic Functions

Arbuscular Mycorrhizal fungi are highly versatile. They are well distributed throughout different biomes that include natural vegetation, agroecosystems and zones with extreme environments. Individual species are highly plastic in terms of their adaptability to different environments. For example, AM fungi reported from semi-arid sites in Australia also occurred in meadows in European temperate zones. Broad host range along with the ability to proliferate in different environments is indicative of a species' plasticity (Sanders et al. 1996).

To a great extent, information about the diversity of AM fungi and species composition in natural ecosystems are based on morphological traits of spores. Typically, 5 to 20 different species of AM fungi have been traced in a soil community. However, Sanders et al. (1996) suggest that AM fungal diversity could be much greater, if investigations are based on fungal genetic components using the PCR - RFLP methods. For example, Sanders et al. (1996) recorded high heterogeneity in the ITS region in natural populations of AM fungi. The ITS region of ten morphologically identical *Glomus* was different. It is important to note that analyses of ribosomal gene sequences gave us the first accurate indications regarding genetically diverse AM fungi colonizing and co-existing in the same root segment. Such molecular methods are also helpful in characterizing the distribution of AM fungi on roots in time and space. For example, occurrence of *Acaulospora* and *Scutellospora* in blue bell roots was confirmed using spore morphology and *Glomus* species were reported as infrequent or absent. However, genetic analysis on the same sample revealed the presence of *Glomus*-specific sequences and, indeed, a good spread of this species within the roots. The general trend indicates that molecular analyses of roots/soil from the natural ecosystems show up the presence of many more AM fungal species, when compared with morphology-based taxonomic analysis. Within this chapter, every effort has been made to highlight inferences drawn using molecular analysis. However, to obtain a complete story, trials based on morphotyping of mycorrhizal fungi have also been discussed.

Arbuscular mycorrhizal fungi contribute significantly to soil/crop ecosystem through their involvement in nutrient dynamics, particularly carbon and phosphorus. It would be interesting to ascertain the effects of their genetic diversity and fluctuations in individual species (isolates) on ecosystematic functions. There are clear suggestions that AM fungal diversity below ground can have profound effects even on the composition annual and perennial plant species above ground (Grime et al. 1987; Sanders and Koide, 1994). Sanders et al. (1996) state that for quite some time, inferences on the ecological aspects were based on the assumption that AM fungi possess low host specificity. In fact, very low attention was paid to genetic diversity within species and isolates. At present, we know for sure that individual isolates of AM fungi can differ in their ability to colonize and induce P uptake. With certain genetically distinct (PCR-RFLP typed) isolates of AM fungi, their effects can be far reaching, affecting the plant community above ground. Albeit, there is a general tendency to estimate the potential effects of AM fungi occurring in a pool, rather than assessing the net genetic diversity and its impact on the soil ecosystem. Parameters such as extent of root colonized by a particular AM fungal genotype or different genotypes infecting a root system, and seasonality among genetically different isolates need greater attention while assessing their ecological effects. Molecular methods could be used advantageously to arrive at accurate inferences. For example, hyphal morphology may vary, depending on the host, its physiological status and environment, leading to ambiguity in identification. In the field, nucleic acid-based methods can overcome such inaccuracies and allow accurate identification despite a fluctuation in host physiology and environment (Bruns and Gardes, 1993; Gardes et al. 1991).

3. B. Spatial and Temporal Changes in AM Fungal Communities

As we try to improve accuracy in estimating AM-mediated effects on crop plants and develop suitable inoculation strategies, it is important that we consider spatial and temporal changes to both native and introduced AM fungal community. Castelli and Casper (2003) hypothesized that a kind of feedback exists between AM flora, especially individual species or its isolates and host plants. This feedback phenomenon may regulate the spatial and temporal distribution of AM fungi. In nature, such feedback is dependent on specificity between host plant and fungal species. These AM fungi that initially exhibit specificity for a host can be guided by negative (depressive) or positive (improving) feedback. In order to test this assumption, they examined a range of plant species and AM flora in a serpentine grassland dominated by *Andropogon gerardi, Schizachyrium nutans* and *Sporobolus heterolepis.* Spore abundance was utilized in order to ascertain specificity. Their study was based on morphotyping; however, molecular identification using ITS sequence and PCR-RFLP would have enhanced accuracy of such studies. Seven different AM fungal species were traced as unevenly distributed in the root zone of grasses. Generally, *Gigaspora gigantea* exhibited specificity to *Sporobolus* compared to *Andropogon* or *Sorghastrum. Glomus microcarpus* preferred to colonize *Schizachyrium* more than *Andropogon* or *Sporobolus. Glomus etunicatum* exhibited specificity to *Andropogon.* These are examples for a positive feedback between plant hosts and AM fungal species. Preferences and positive feedback was confirmed using trap cultures in pots. In some cases, a good indication of positive feedback was also perceivable through enhanced nutrient and growth simulation of the host. *Glomus etunicatum* that was specific to *Andropogon* exhibited a negative feedback, since it depressed the growth slightly, unlike other AM fungi. Similarly, one particular strain of *G. etunicatum* that was specific to *Schizachyrium* reduced the biomass and nutrient status of host. This clearly indicates a negative feedback. Intraspecific variation in negative/positive feedback between host and AM fungal species and its influence on the spatial and temporal distribution needs to be studied using both morphotyping and molecular methods. The intensity of AM fungal sporulation and senescence might itself be affected by a feedback phenomenon involving physiological signals between symbionts. Bever (2002) opines that among mutualistic associations such as AM symbiosis, negative feedback plays an important part in regulating abundance and co-existence of fungal species in a given soil environment. In support of this suggestion, he reported that in experiments with *Plantago*, negative feedback caused asymmetry in fungal abundance and growth benefits of host. Growth of *Plantago* was actually induced better by AM fungus that is preferred by a second host species *Panicum sphaerocarpum.* The resulting AM fungal dynamics might reduce nutritional benefits to the host, but contributes to the co-existence of different AM fungal species and the two plant host species. Relevance of fungal isolates to negative feedback-mediated regulation of population, host growth improvement, if any, needs to be ascertained.

Molecular techniques based on SSU rRNA sequences and RFLP variation were applied to study the alterations in spatial and temporal distribution of AM fungi under trees as well as under story shrubs in Panamanian tropical forests (Husband et al. 2002a,b). Under the root zones of nearly 48 plant species examined, diversity of AM fungal species was much greater than that noticed on trees from temperate climatic zones. Overall, diversity was high at 30 different AM fungal species in tropical soils. These tropical AM fungal species showed considerable spatial heterogeneity and non-random associations with different hosts. Strong shifts in mycorrhizal communities were frequently observed. AM fungal types dominant with seedlings were almost entirely replaced by the next sampling done a year later. We may, therefore, infer that high diversity in AM fungal populations detected across time points, sites and hosts causes distinct ecological effects on symbionts. Sometimes, monoculture of the host may support the diverse AM fungi. Such a situation, noticed under *Plantago lanceolata*-based grassland was attributed to modes of colonization and AM fungal preferences to a particular type of host (Johnson et al. 2004).

A study of the ecological models for AM fungi and their hosts suggests that diversity can be maintained, if functionally different communities are kept spatially separated. To substantiate this argument, Lovelock et al. (2003a) examined the spatial and temporal variations in AM fungal communities in a wet tropical forest of Costa Rica. Correlations between AM fungi and host tree species, especially their life histories, relative importance of soil type, seasonality and rainfall pattern were estimated. Host tree species differ in their associated fungal communities, which can sometimes be generalized based on life histories. They also concluded that changes in relative abundance of a few AM fungal species could affect the host distribution pattern. Soil P fertility, ranging between 5 and 9 Mg ha^{-1}., did not affect the composition of fungal communities, but sporulation pattern was affected. Sporulation was generally more abundant in soils with low fertility. In terms of seasonality, sporulation was abundant in dry season. Interestingly, *Glomus* species were abundant in sites with a higher seasonal rainfall. It is important to note that any change in one or few species of AM fungi, may at times, immensely influence the fungal community pattern on a wider scale.

Abiotic factors influence the symbiosis and emergent variations in AM fungal patterns. Among them, soil moisture and temperature interactions are important parameters that decide the fungal population levels and patterns. To quote an example, the study of temporal and spatial distribution AM fungi in the Negev Desert of Israel has shown that increasing soil moisture and/or organic matter content positively influences the AM population (He et al. 2002). In terms of environment, spore density was positively correlated with total soluble N in soil. The percentage colonization and spore density was greatest during September to November. They suggested that whatever be the technique utilized, AM colonization pattern and spore density changes during a year and at different depths in soil can be fairly good indicators of AM fungal dynamics. Mohammad et al. (2003) utilized morphotyping to ascertain the influence of abiotic and biotic factors that may influence the AM fungal diversity and population in a semi-arid agroecosystem prevalent in northern Jordon. Firstly, AM fungal population varied, depending on the location (soil type) and host crop. *Glomus mosseae* was the most frequent. It occurred in 85% of the soil samples, *Glomus geosporum* was recorded in 20% and *G. caledonium* in 8% of the soil samples. Their analysis suggested that physical properties of the soil such as pH, electrical conductivity, moisture, silt and clay contents strongly influenced AM fungal density and their activity. They suspected that abiotic factors and cropping pattern influences the AM fungal diversity and population levels to a greater extent than other factors examined. Application of molecular techniques could reveal details on fungal isolates that tolerate extremes and dominate a given soil environment.

In a different study, salt marshes and bush land were examined for AM fungal diversity using geo-statistical techniques. Spatial distribution of spores, hyphae, and soil properties were analyzed using nested core sampling methods. Semi-variograms have shown that spore density is strongly correlated to patchiness of AM fungal distribution and the dominant genera. The patch sizes differed between the genera. In salt marshes, distance from the host plant was crucial. In maquis (bushland) spore density, distribution and patchiness were related to organic matter content of the soil region (Carvalho et al. 2003). No doubt, soil physical properties played a crucial part in determining the spatial distribution of AM fungi. Disturbance to plant cover and /or soil has its impact on AM fungal spatial and temporal distribution. Let us consider an example that evaluates the effect of disturbance to vegetation on AM fungi. Brundrett (2000) traced between 5 to 13 species in undisturbed tropical habitat near Nedlands, Western Australia, but the disturbed sites in the area had much less diversity. In highly disturbed mine sites, AM fungal species were very few and localized only around the vegetation as patches. AM fungal abundance and spore levels in disturbed sites reached those found in undisturbed sites as the plant cover recovered. Morphotyping and bioassays indicated that in undisturbed sites, all major genera such as *Scutellospora, Gigaspora, Glomus* and *Acaulospora* were represented. AM fungi seem to adopt different propagule and life history strategies, depending on the availability of host. Isolate preferences judged using molecular methods could provide better accuracy.

Succession in AM fungal communities is an important ecological manifestation. It influences host nutrition and growth pattern. Such AM fungal patterns may vary in time and space, depending on a range of biotic and abiotic factors and their intensities. For example, in soils at Slowinski, Poland, over 21 AM fungal species were traced during eight successional phases. Early colonizers were associated with hosts belonging to Graminae and Juncaceae. At the middle successional stage, ericaceous hosts replaced them, eventually followed by trees in advanced stages. The AM fungal colonization increased from early to middle stages, but declined gradually (Tadych and Blaszkowski, 2000). Host-dependent selection and induction of fungal population was conspicuous. Among the five most frequent AM fungal species, Glomus-107 was an early colonizer that diminished in population in later stages. *Acaulospora koski* dominated the middle stages, followed by *G. aggregatum.* Understanding AM fungal succession in tropical soils helps us to arrive at better judgment regarding inoculation strategies.

Temporal variations in AM fungal population can be subtle, depending on the frequency of sampling and age of the host. Using AM fungal-specific primers to amplify SSU rRNA gene sequence, Husband et al. (2002a,b) could classify over 550 clones occurring in a Panamanian forest location into 18 different AM fungal species. Their assessments revealed that as seedlings matured, fungal diversity decreased. Previously, rare species or isolates replaced the dominant fungal types. Seedlings of different age, if sampled at one point type, showed markedly different fungal populations. Seasonal dynamics of AM fungal populations has its share of influence on the soil microbial flora and host vegetation. To quote an example, Allen et al. (1998) noticed that most tree species and all herbaceous hosts showed the highest colonization during dry/summer season. However, spore production was not higher during the same season.

3. C. Mycorrhizal Fungal Diversity Influences Plant Biodiversity

Terrestrial ecosystems, be it agricultural or natural, support diverse plant species with varying intensities. The ecological mechanisms that impart dynamism and sustain plant diversity may differ on the basis of a variety of reasons. These are being debated and revised as we accumulate newer evidences. Physical competition, resource partitioning, rapid multiplication and spread, soil-plant interactions, photosynthetic ability and a range of other factors regulate plant diversity in a given terrestrial ecosystem. Reynolds et al. (2003) has opined that plant-microbe interactions in soil contributes to the diversity and ecosystematic functions within a plant community. Two key aspects of soil microbial component, namely structure and dynamics, may influence the niche differentiation, feedback and finally, plant community above ground. In this regard, Van der Heijden et al. (1998a,b) state that plant-microbe interactions, especially AM fungi and their influence on host diversity and ecoystematic functions, have not been studied in great detail. Ecologically, AM fungi might play a crucial role in determining the above-ground plant diversity. The presence of AM fungi is known to increase plant diversity (Grime et al. 1987). At the same time, we may note that in natural soils, not single or few, but a community of AM fungi comprising diverse species/isolates exist. Some soil ecosystems may support high AM fungal diversity but others play host to a comparatively less number of species. Since different AM fungal species induce differential response in terms of growth, nutrition and clonal propagation in any vegetation, they will surely impose a certain degree of influence on the plant diversity. Let us consider a few examples that support this contention. First, Van der Heijden et al. (1998b) conducted greenhouse trials and complementary ecological field experiments to estimate the below ground diversity of AM fungi and its relation to plant biodiversity in a European grassland and American natural field setting. Simulations indicated that plant diversity in European grassland setting decreases if AM fungal diversity is altered, whereas, if AM fungal species richness is increased, then nutrient capture, productivity and net plant biodiversity are enhanced. Hence, they concluded that microbial interactions in the soil—in this context the interaction and biodiversity of AM fungi—could enormously affect above ground plant diversity, its productivity and variability. Let us consider a different example. Moise et al. (1998) studied two endomycorrhizal tropical tree species from French Guiana. One of them, *Dicorynia guianensis,* exhibits an aggregative habit with seedlings growing better under parent trees, whereas, *Eperua falcate* is dispersed and its seedlings survive better at a distance from the mother plant. They reported that *D. guianensis* is more growth responsive to AM fungi than *E. falcate.* The parent tree is a good source of AM fungal inoculum for emerging seedlings under *D. guianensis,* but it is not so with *E. falcate* seedlings that are placed apart. These aspects may also accentuate differences in nutrient capture modes and growth pattern, finally affecting the diversity of plants aboveground. Koide and Dickie (2002) have argued that specific interactions between AM fungi and plant host contribute variations in plant population and biodiversity. Firstly, AM fungi can induce inequality in growth and reproductive capability in a given plant population. In some cases, AM fungi increase reproduction, leading to higher population size. Established mycorrhizal plants in an area can serve as a source of AM fungal inoculum to conspecific seedlings, aiding their better regeneration and patches of dominance by a plant species. Korb et al. (2003) have narrated an interesting application of AM fungal effects on the host plant community in a forest dominated by Ponderosa pine. They found that a relative amount of fungal propagules was positively correlated with graminoid cover and herbaceous understory species, but negatively related to overstorey canopy-forming tree species. They suggest that such site-specific results could be utilized during forest restoration programs to selectively enhance the under- or over-story plant species.

We can argue that resource heterogeneity—be it abiotic or biotic—can influence plants, their species diversity and general vegetation to a certain extent. The AM fungi, as part of the biotic component, could also affect composition and growth of plant species. In a trial using several taxa of AM fungi, Van Der Heijden et al. (2003) found that depending on the AM flora, P and N distribution in the plant community can alter. In a way, composition of AM communities could influence plant species, their co-existence and nutrient allocation. During recent times, molecular methods have been utilized advantageously to decipher such changes in the AM fungal community in soil and relate its effect on composition of plant community and growth of vegetation. There are numerous reports indicating that AM fungi alter the plant diversity and competitive ability of host plants (Gange et al. 1990; Hart et al. 2003; Klironomos et al. 1993, 2000; O'Connor et al. 2002; Marler et al. 1999; Van Der Heijden et al. 1998b). In this regard, Hart et al. (2003) have commented that AM fungal effects on plant diversity and their co-existence can be judged at two levels; namely coarse and fine scale effects. Coarse scale effects relate to the influence of the presence or absence of AM fungi on plant diversity and co-existence. Generally, coarse-scale effects of AM fungi on host diversity have been well understood. Fine scale changes that pertain to interactions between AM fungal communities, their species or genotype composition and physiological activities on host diversity have not been discerned in any great detail. Especially, factors that influence plant-AM fungal feedback effects, shared mycelial networks, and nutrient transfer due to interconnections need greater attention. In natural soils, AM fungi occur as communities with species and isolate diversity. For example, Bever et al. (2001) found upto 37 AM fungal species in one soil location. Actually, a multi-dimensional niche determined by host plant community and soil environment decides the composition of fungal community. We also know that AM fungal taxa differ in their ability to translocate nutrients and induce growth. As a consequence, AM fungal community too influences the manner in which plants co-occur in a plant community. A study by Van Der Heijden et al. (1998b) showed that composition of certain grassland plant species varies if AM fungal communities were altered. However, AM fungal-mediated effects on plant host were not conspicuous. The dominant grass species such as *Bromus* remained unaffected by changes in AM flora. However, to a certain extent, AM fungal flora could affect nutrient allocation among co-existing plant species.

3. D. Agricultural Practices and AM Fungi

Arbuscular mycorrihzal fungi are most abundant in agricultural soils. They form a symbiotic association with over 80% of the crop plants and play a key role in nutrient transfer from the soil to the roots and influence the below ground nutrient cycling in agricultural ecosystems. Hence, Dodd (2000) suggested a need to understand the below ground microbial ecology with a major emphasis on AM fungi. The genetic and functional diversity of AM fungi in different agroecosystems has to be understood with greater clarity. Their relevance to soil-plant interactions, especially crop nutrition, nutrient transfer between plants, drought stress and soil pathogens have to be deciphered. There are indeed innumerable reports on AM fungi and their influence on various aspects of crops. However, at this point, emphasis is on major agricultural operations that affect soil ecosystems, especially the dynamics of AM fungi.

Tillage is an important soil-related activity that influences the persistence and ecosystematic functions of AM fungi. Firstly, tillage intensity is a key factor that affects various physico-chemical properties of soil, which in turn, modify the activity of various microbes, including AM fungi. Reduced or no-tillage, chisel plow and conventional tillage are common soil preparation methods that can have variable effects on AM symbiosis (Yocum et al. 1985). To understand this better, both molecular and morphological assessments of AM fungal communities have been conducted under long-term tillage studies. Jansa et al. (2001) have reported that AM fungi belonging to three genera, namely, *Gigaspora, Scutellospora, Entrophospora* were more prevalent in reduced tillage treatments. The ITS rDNA sequencing confirmed the occurrence of at least 17 different AM fungal species in the fields (Jansa et al. 2002). Later, specific DNA marker targeted to 25s rDNA was utilized to confirm that *Scutellospora* and *Gigaspora* were predominant in maize fields exposed to long-term tillage treatments. *Glomus* species were generally, more frequent in fields provided with intensive soil tillage and management. Undisturbed mycelium and survival strategy of individual AM fungal species were suspected to cause this variation. Tillage had significant influence on sporulation of different AM fungal species in the soil. From a later study conducted at Tanikon in Switzerland, Jansa et al. (2003) have concluded that changes in AM fungal community surrounding maize roots might be due to at least four different reasons. They are:

(a) Differences in tolerance to tillage-induced disruption of the AM fungal hyphae; among the different AM fungal species;
(b) Changes in availability of nutrients in soil depending on tillage intensity;
(c) Changes in physico-chemical conditions of soil and microbial activity; and
(d) Changes in weed population in response to tillage and intercultural activity.

A combination of crop rotation that includes wheat or maize and a non-host (Canola) provided with three different tillage treatments were assessed in two sites in Switzerland. Generally, maize roots from no-tillage treatments were colonized intensely as compared with those from conventional tillage plots. Wheat roots were comparatively weakly colonized (35%) in all the treatments. Canola, as expected, was uninfected by mycorrhizal fungi in all the treatments (Mozafar et al. 2000). It was also inferred that a combination of mycorrhizal intensity and root density could affect nutrient acquisition by the crops. In Portugal, a wheat-sunflower-triticale rotation, given different tillage treatments, was examined for AM symbiosis. Again, no-tillage proved to be beneficial and there was enhanced AM colonization of wheat and triticale by 18 to 20% over conventional tillage (Britto et al. 2001). Ridged-tillage is yet another soil management practice that causes the least disturbance to soil structure and favors soil organic matter accumulation. It should be more congenial to AM fungal population build-up and activity. A study by Landry et al. (2001) has shown that ridged tillage influences both AM fungi and soil P dynamics favorably. Therefore, nutrient recommendations, especially P application under the ridged-till system, may have to consider the role of AM fungal on soil P dynamics and crop growth. Galvez et al. (2001) believe that low-input systems along with no-till practices may be more congenial to AM fungal population in agricultural soils. Observations on maize grown under low input suggested AM fungal spore populations under no-till were higher than in mold board-treated plots. AM fungal activity and diversity was also dependent on soil depth. Rillig and Field (2003) have reported that coarse endophytes were comparatively more frequent in subsoil layers (15 to 45 cm). They also inferred that interactions between CO_2 and soil depth significantly influenced the AM fungal distribution.

Soil aggregate stability is an important parameter affected by tillage intensity as well as microbial load, including AM fungi. Plant roots also contribute to soil aggregation and structural stability. Thomas et al. (1993) state that it is possible to separate the relative effects of plant roots and mycorrhizal mycelia in the soil. Mycorrhizae are known to modify suboptimal growth environment to their advantage by secreting a glycoprotein glomalin (Rilling and Steinberg, 2002). Concentrations of glomalin may range from 2 to 15 mg g^{-1} of soil in temperate climates. It may reach 60 mg g^{-1} in tropical agricultural fields (Lovelock et al. 2003b). Rillig et al. (2003) have reported that soils underneath the canopy of plants contained higher glomalin pools than open microsites. AM fungal hyphal length also differed significantly between these sites. To understand this phenomenon, AM fungal traits such as length and density of extraradical mycelium, glomalin production and physico-chemical properties of soil at different depths were analyzed by Borie et al. (2000). Glomalin production increased proportionately with length of time, in fields on which no-till practices were adopted. Total and easily extractable glomalin content reached 0.36% of organic carbon due to no-till treatment. Similarly, pH, available P, organic carbon and water stable aggregates were also affected by no-till practices. Glomalin contents were highly correlated with organic-C and mycorrhizal hyphal density. A test with glass beads of different sizes has revealed that glomalin yield increased if the beads were small sized (<106 μm). However, soil aggregate stability was governed by several other interacting factors in soil in addition to aggregate size. Bearden and Petersen (1999) analyzed the soil aggregate stability under a sorghum crop grown on an Indian Vertisol. Turbidimetric measurements indicated that aggregate stability was highest in treatments provided with AM fungus, especially those having higher levels of fungal hyphae. The AM fungal effect on soil aggregates was clearly discernible after one growing season. It was related to both the time required to fungal proliferation and root growth. Assessment of vegetation within forest plantations in Ohio has revealed definite linkages between AM fungal ability to produce glomalin, density/ diversity of herbaceous plants and soil nutrient availability (Knorr et al. 2003). Lovelock et al. (2003b) have stated that accumulation of total glomalin and easily extractable glomalin was actually related to soil fertility. Low fertility situations aided production of higher levels of glomalin. It was argued that higher levels of glomalin in low fertility soils are mainly a result of slow microbial degradation. Differences in glomalin concentration across fertility gradients could also be due to composition of AM fungal community, their biomass soil type interactions. Application of molecular techniques may help us in understanding the differences, if any, in the ability of AM fungal isolates to secrete glomalin and affect soil aggregation.

The host crop is a major factor that can influence AM symbiosis and related nutritional advantages. Normally, crop genotypes sown by farmers may differ depending on a wide range of factors related to its physiology, soil conditions, season and even economic consideration. Such a dynamic shift in host species will definitely affect the AM flora and nutritional traits related to plant-AM symbiosis. Sanginga et al. (2000) have shown that dynamics of major nutrients such as C, N, and P are affected by the host genetic constitution. The dominant cowpea variety sown in savanna zones of northern Nigeria varies, depending on several factors. Accordingly, they examined about 43 cowpea-breeding lines on savanna soils. Depending on their ability to garner P, cowpea genotypes could be grouped into poor performers, P-responders and non-responders to P. The P intake and use efficiency were highly correlated with AM fungal colonization in many genotypes grouped as P-responders. Some of these genotypes were also graded as efficient for N fixation. They fixed upto 22 kg N ha^{-1}. in one growing season. Hence, it was recommended that in semi-arid soils with poor inherent fertility, P and N efficient genotypes that utilize AM fungi/Rhizobium symbiosis effectively should be preferred. Detection and identification of efficient AM fungal and Rhizobial strains will be possible by using appropriate molecular techniques. Changes in AM fungal community in response to shifts in host genotype and cropping pattern may also be studied using ITS sequence and PCR-RFLP based analysis.

Crop rotation induces changes in a soil's physico-chemical conditions and microflora, including AM fungi. A large number of reports substantiate this inference, many of them providing details on changes in AM root colonization levels, diversity of symbionts and extent of benefits, if any, to host crop and fungus. Let us consider a few examples from Sahelian, West Africa. Bagayoko et al. (1999) examined rotations involving pearl millet/cowpea and sorghum/groundnut in Niger and Burkina Faso. The sandy soils support a wide range of AM fungal species. In general, rotation with legumes showed higher levels of AM population compared with sole cereal sequences. They have also suggested that in these Sudano-Sahelian sandy soils, better crop nutrition due to rotation is easily attributable to better microbial load, favorable nitrogen mineralization rates and AM fungal mediated nutrient recovery from the soil. Preferences or specificity, if any, for AM fungal isolates could be known through the use of molecular markers, ITS region and PCR-RFLP analysis of AM flora. However, through morhpotyping, Sanginga et al. (1999) were able to record such variation in AM flora caused due to crop rotation. In a maize/soybean rotation practiced in northern Nigeria, they identified four AM fungal genera comprising 29 species. *Glomus* was the dominant genus (56%) followed by *Gigaspora* (26%) and *Acaulospora* (14%). *Sclerocystis* was least represented at 4%. We should expect host preferences to certain isolates of individual species that may flourish better with higher population build up and nutrient translocation traits. Again, the use of molecular techniques will help us substantiate specificities and dominance by a particular isolate/ genotype of fungus. The AM flora will also vary if genotypes of host crop(s) are changed by farmers.

We may take note that in addition to crop sequences and rotations, dual cropping will also influence AM flora and its effect on the nutrient transfer. There are clear reports that nutrients are shared, translocated and retranslocated between plants situated closely in field, using mycorrhizal fungal bridges. All major nutrients, C, N, P, and K are known to move between plant roots connected via AM fungal hyphae. One of the examples, a study of perennial grass and white clover showed that ^{15}N was transferred between these two plant species (Rogers et al. 2000). Direction of nutrient transfer between plants is an important aspect to note, because it has relevance in practical agriculture. Such a phenomenon will influence nutrient distribution and utilization in a cropping system. We should be able to trace AM fungal preferences, if any, for a certain pairs of crop species commonly cultivated by farmers situated in different cropping belts.

Nutrient and water interactions are perhaps the most studied items in agricultural soils, especially with reference to soil fertility, including organic matter and microbial components. Actually, a wide range of soil chemical properties influences AM fungal flora. Seasonal droughts that are frequent in many agricultural zones do affect the AM fungal biology, especially its spread and sporulation in the soil. Egerton-Warburton et al. (2003) have argued that plant habit and its preference to exploit water/nutrients from different horizons in soil will affect AM fungal diversity and population levels. They examined several species of AM and ECM fungi from 0 to 200 cm depth of soil, including regions with bedrock and stored water zones exploited by host plant species. Potentially strong correlations were discerned between abundance of AM fungi (spore and hyphae), ascomycetous ECM fungi and soil features such as bulk density at various depths. Nutrient (N, P, K, Ca and Mg) availability in different horizons, as well as bedrock moisture content also affect fungal abundance. Their evaluations indicated that in addition to spores, mycorrhizal hyphae could be the key to survival of AM fungi in various strata of an agricultural soil.

Nitrogen inputs into agricultural fields can immensely influence AM fungal population and diversity. The level of N enrichment and extractable soil N in soil seems to influence AM fungal population differently. Egerton-Warburton and Allen (2000) found that increasing N inputs induced replacements of large-spore-sized *Scutellospora* and *Gigaspora* with small-spore sized *Glomus* species. Nitrogen enrichment is known to reduce spore abundance. It actually affected the timing of sporulation in AM fungi and reduced root colonization levels of many plant species found in the area. Wolf et al. (2003) examined the impact of plant species and elevated CO_2 levels on AM population for three consecutive years from 1998 to 2000 at Cedar Creek in Minnesota, USA. Among the 11 AM fungal species, *Glomus clarum* was induced by elevated CO_2. Nitrogen inputs reduced the AM population slightly. Overall, plant species diversity, CO_2 and nitrogen altered the AM fungal activity and species diversity. Johnson et al. (2002) measured the extent of C transfer from host plant to AM fungi in a grassland. The amount of ^{14}C allocated to mycorrhizal mycelium in 70 h was nearly 3 to 5% of the ^{14}C fixed initially by the plant. These observations confirm the rapidity of photosynthate allocation to AM mycelium and demonstrate the importance of short-term dynamics of C fluxes to AM symbiosis. Organic amendments are known to increase soil microbial activity, including AM fungi. For example, application of compost to fields transplanted with Pistacea seedlings that were pre-inoculated with AM fungi proved best in terms of AM colonization, P nutrition and growth. Soil enzyme activity, labile C fractions and bulk density were favorably modified as a result of the above treatment (Caravaca et al. 2002).

Rutto et al. (2002) studied the effect of orchard management practices such as inter culture, fertilization, fungicide application, pruning, and harvesting on AM fungal colonization, sporulation, and persistence. High AM fungal colonization could be sustained in orchards that were unfertilized and not applied with fungicides. Root colonization levels were also influenced by season. It decreased during July to August. In orchards without understory, AM colonization in roots and spore density decreased immediately after fungicide sprays. Orchards with sod supported high levels of AM fungal population. They concluded that excessive fertilization; frequent fungicide spray and clean culture were detrimental to AM fungal population and diversity.

Contamination of soils with heavy metal ions impairs soil microbial activity in general, including AM fungi. The tolerance or susceptibility exhibited by AM fungi towards heavy metal accumulation can vary, depending on the species and isolate (Table 6.2). Examination of AM fungal species on agricultural soils that were frequently applied with sewage sludge containing heavy metals has shown that tolerance to heavy metals varies among fungal species. Among the six AM fungal species tested, tolerance to heavy metal ranged from sensitive to relatively tolerant. Total spore counts and colonization reduced with increasing toxicity due to heavy metals, but the

Table 6.2 A molecular analysis of arbusuclar mycorrhizal colonization of *Medicago truncatula* seedlings as influenced by different sewage treatments

Sewage treatment	*Mycelium in soil (%)*		
	G. mosseae	*G. intraradices*	*Gi. rosea*
Sand	100	100	33
Sand plus sewage sludge = A	100	100	0
Sand plus sewage sludge = B	66	0	0
Sand plus sewage sludge = C	33	100	33

Source: Jacquot et al. 2000
Note: Values indicate percentage of samples containing AM fungal mycelium when detected using molecular identification methods e.g. nested PCR. Sewage sludge-A was enriched with organic pollutants such as pyrene; Sewage sludge-B contained phenanthrene and fluoranthene; and Sewage sludge-C contained heavy metals Cu, Pb, Ni, Cd.

extent of suppression in AM fungal population was species specific. This clearly affected the net genetic diversity measured at the end of the treatment period. The dominant AM fungal species varied with the kind of heavy metal contamination applied. For example, *Glomus claroideum* was easily isolated from soils without sludge, but *Glomus mosseae* isolates were frequent in soils with medium/least contamination. In other words, the AM fungal diversity, dominant species and its population level will be guided by their inherent ability to adapt to contamination and tolerate toxicity. Acid rain is a severe problem in many agricultural locations. This phenomenon affects the ecosystematic functions of the vegetation, including soil microbial component. Simulations of acid rain conditions, accompanied with high aluminum concentrations, have shown that ecophysiology of AM fungi beginning with germination, then symbiotic functions and life cycle are all affected (Vosatka et al. 1999). A strain of *Acaulospora tuberculata* from the soil with low pH exhibited a higher tolerance to acid rain effects when compared to *Glomus mossea* or *Glomus fistulosum*. Higer tolerance was attributable to previous acclimatization. It was also reported that histochemical tests and estimation of alkaline phosphatase and NADH-diaphorase activities in the extraradical mycelium (ERM) of the AMF are a sensitive indication of acid rain plus aluminum effects. These enzymes were more sensitive to changes in soil pH and Al toxicity than commonly measured parameters such as hyphal growth and colonization in roots.

3. E. Spatial and Temporal Distribution of ECM Fungi

Regulation of spatial patterns and genetic diversity is an important aspect of both ECM and AM symbiosis. We ought to understand the specific factors that influence fungal/plant host patterns under natural conditions. For instance, coniferous forests of northern latitudes may have more than 1000 species of ECM fungi, but all over the tropical deciduous forests, we may still encounter only 25 species of AM fungi. According to Allen et al. (1995) distribution of AM/ECM on a broad scale is influenced by biomes. The AM fungi are predominant in tropical, semi-arid and arid biomes, while ECM flourish in the temperate forest vegetation. They believe that AM fungi prefer soils with low organic matter; hence, are common on tropical soils that are generally low in organic matter. On the contrary, ECM fungi are predominant in temperate forest soils that are rich in organic matter. ECM fungi exhibit greater host specificity; hence their distribution, to a good extent, will be dependent on specific host species. Whereas, AM fungi are generalists with comparatively low host specificity among tropical plant species. Further, they suggest that physiological and genetic diversity, along with interaction with the plant, may actually decide the distribution pattern and diversity. The ECM fungi may respond to environment directly or via their host. Introduction/loss of a mycorrhizal fungus too has its influence on flora.

During the past 8 years, since adopting molecular techniques to investigate mycorrhizal symbiosis, our understanding of not just taxonomy and genetic nature, but spatial structure, distribution and succession of mycorrhizal fungi in natural settings has also become more detailed (Bruns et al. 2001). Across-the-site comparisons of a large number of pine forest sites in North America showed that members of Russulaceae, Thelophoraceae and various non-thelophoroid non-resupinate taxa are dominant. In fire-disturbed sites, Ascomycetes and Suilloid taxa are typically dominant. Within individual sites, species-level community composition varies at all spatial scales, and correlates with distance only at scales below two meters. Mapping of species from dominant ECM families, namely *Russula*, *Amanita* and *Thelephora* revealed that individuals are spatially restricted. Individual spore patterns may often have subtle and direct effects on fungal community composition.

In addition to genetic diversity and population density, aspects of ECM fungal distribution pattern need appropriate attention. At present, molecular identification procedures such as ITS-RFLP have been widely adopted to know the below ground distribution of ECM fungi. Such knowledge about the distribution directly affects inferences drawn regarding species richness in a given location. Horton and Bruns (2001) have pointed out that data on species richness is often based on inadequate number of samples, leading to inappropriate judgments about distribution and ecology of ECM. Patchiness of ECM fungal distribution can occur at a very fine scale. For example, around 50 fungal species were reported in 25 samples from 0.25 ha. of Ponderoza Pine, but over 200 species were observed in 0.5 ha. of Douglas Fir stand. Hence, sampling intensity and related procedures do affect inferences about the true distribution pattern of ECM fungal distribution and ecological functions. The distribution of ECM fungi can be clustered for many species. Some occur in only 10% of the soil core samples; others are traced in 25% of the samples. Individual soil or root tip samples may contain multiple fungal species. Sometimes, the most abundant species may get concentrated into just one or two core samples. Thus, a shift in sampling location or intensity can alter inferences about species diversity and distribution pattern (Horton and Bruns, 2001; Standell et al. 1999). The ECM fungal distribution can also be in larger patches. On isolated trees within a pine stand, a single ECM fungal species may dominate the entire tree root system and its rhizosphere. It is believed that the structure of pine stand with loosely scattered trees and dry climate may cause such a clustering of a species around a tree. Molecular analysis has also revealed a relatively uniform distribution of ECM fungi such as *Tomentella sublilacina* in Californian pine stands or *Tylospora fibrilosa* in Northern European spruce forests (Dahlberg et al. 1997; Standell et al. 1999). Similarly, *Cenococcum geophilum* and *Piloderma* are known to occur uniformly in all the soil cores sampled in several European forest locations. As stated earlier, members of Russulaceae, Thelephoraceae and non-thelephoroid resupinates are among the most dominant taxa of ECM fungi on roots of conifers in North America and Europe. These same taxa may also dominate angiosperms in other temperate locations. Among resupinate sporocarp types,

Tomentella spp. dominate. In a study conducted at Gard in France, *Rhizopogon rubescens* dominated the pine seedlings. The ITS-RFLP analysis of 233 root tips showed that *R. rubesens* occurred in 55% of them. Horton and Bruns (2001) opine that several physiological and ecological aspects, as well as good competitive ability of these fungi, may be the reason for their dominance in these environments. They may also be playing crucial functional roles in the ECM communities. Overall, major inferences drawn from molecular analysis are: (1) the ECM communities are diverse, and their distribution is patchy; and (2) there is poor correlation between fungi that are dominant, based on sporocarp studies versus those occurring in the root tips.

Knowledge about spatial and temporal distribution, as well as persistence of ECM fungi is useful while devizing seedling inoculation programs. It may provide an understanding about ecology of both the natural flora and introduced ECM fungus. Mycorrhizal succession studies in forest plantations of Northern Hemisphere have shown that *Laccaria* is most commonly distributed as an early-stage ECM with young trees in disturbed sites. Its sporophore distribution is temporally variable (Fig. 6.2) and has a broad host range. The arc of sporphore distribution around host trees is said to range between 0.07 and 0.1 m y^{-1}. In the Canadian locations, Bastide et al. (1994) mapped the sporophore distribution of *Laccaria bicolor* on *Picea abies* for four consecutive seasons. They superimposed it with results from RAPD analysis. RAPD tests have provided evidence that a single individual of *Laccaria bicolor* can persist in the host root system for over 3 years and continuously vary in its spatial distribution. Often a single or several genetically identical mycelia can be mycorrhizal both in space and time. Persistence of single/or dominant fungal genotypes has implications during designing inoculum and its delivery. The ECM fungal genetic diversity around

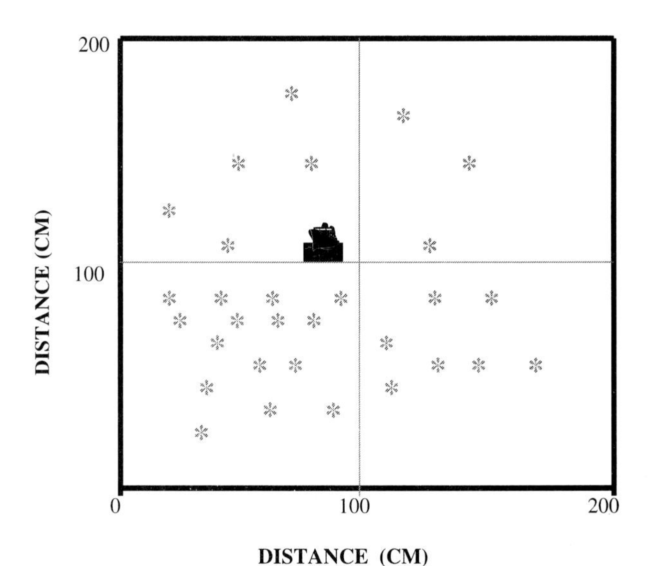

Fig. 6.2 A representation of spatial distribution of a basidiomycetous ECM fungus *Laccaria bicolor* around a single Norway spruce tree in a forest stand in North American location.
(*Note:* indicates location of the tree; * shows approximate position of fungal sporophores.)
Source: Redrawn from Bastide et al. (1994).

the host root may also be dependent on it. Clearly, we ought to know the factors that lead to the dominance of rhizosphere by a certain single/or group of ECM fungal genotypes, and reasons for exclusion of several other isolates from the zone. The intra- and interspecific competitive ability of a fungal isolate may be a crucial factor (Bastide et al. 1994).

Let us consider yet another example. Guidot et al. (2002b) examined the colonization process and population dynamics of *Hebeloma cylindrosporum* in disturbed dune and forest habitat at a location on the Atlantic Coast. The dunes supported *Pinus pinaster* trees that occurred in a scattered pattern at the fringes of dune, whereas, forest habitat possessed an evenly distributed thick plantation of *Pinus pinaster* trees. Data collected for several years on different locations within the experimental zone indicated that fruit bodies occurred within discrete patches of ground. The type of habitat had a significant effect on the size of patches and its internal density. Compared to patches in forest locations, dunes consistently supported larger but less denser patches of *Hebeloma* fruit bodies. Patches rarely occurred at the same spot year after year. The fruit body size indicates its genet size. The genets in forest locations were annual and averaged 26mm. In the dunes, *Hebeloma* produced perennial genets with sizes around 40 mm. Clearly, characteristics of habitat influences genet size, spatial patterns of patches of fruit bodies and seasonality. Guidot et al. (2002a) focused on understanding the spatial and temporal distribution of underground mycelia of *Hebeloma* in the soil. They have argued that ECM fungal mycelia and below ground distribution could be of greater ecological and functional significance to its symbiotic relationship with pine. Competitive PCR methods that allow quantification of *Hebeloma* DNA in soil samples were devized. The target sequence chosen for the PCR analysis was a 533 bp fragment of nuclear ribosomal ITS region amplified using two specific primers. They collected a series of soil samples surrounding each fruiting body and around a host tree (*Pinus pinaster*). Their assessments suggested that the below ground biomass was concentrated directly underneath the fruit bodies, or at best, close to them. Mycelia or DNA of *Hebeloma* was not detected beyond 50 cm from a fruit body. In the vicinity of a fruit body, the DNA content varied 10 ng at the nearest point to 0.07 ng at the farthest spot. Guidot et al. (2001, 2004) have pointed out that many population studies on ECM fungi have relied on fruit body sampling, assuming that it reflects the spatial organization of organisms below ground. It may not be so. Molecular analysis of genets and their distribution may provide a different a picture. Now, using molecular tests, they found that all the genotypes of *Hebeloma* present above ground were also traceable below ground. However, ITS sequence analysis has proved that in patches occupied by several different fruiting genotypes, additional non-fruiting ones could be present on the roots. In most cases, mycorrhizas of one genotype were found to be no more than 20 cm away from the fruit body. Disappearance of the fruit body was also related with the disappearance of fungus, although strict coincidence may not be possible in all the cases. Guidot et al. (2001, 2004) noticed a rapid turnover of ECM fungal species on the root system, provided no disturbances occurred. Patches of basidiospores were created either by the local elimination of competitors or by local nutrient enrichments. Sometimes, an ECM species could be completely eliminated in one year and need not contribute to the next generation. Let us consider a different example. A recent study of basidiomes of *Amanita alboverrucosa*, *A. ochraphylla* and *A. pyramidifera* from Australian sclerophyll forests has shown that genotypes of Amanita spread across an area 10 to 60 m in diameter from the tree. Molecular analysis using inter-simple repeat sequence (ISSR) revealed that for wider range distribution of genotypes of this fungus, vegetative spread via large belowground mycelial genets might be important (Sawyer et al. 2003).

Landeweert et al. (2003) reported a molecular method for the identification of ECM fungi in different horizons of soil at different depths. This method differs from the classical root tip approach, but provides better judgment on isolates and ECM fugal communities in the soil. They have used a basidiomycete-specific primer pair to amplify fungal ITS sequences from total DNA extracts of the soil horizons. The amplified DNA was then cloned, sequenced and identity established through homology with sequences from gene banks. These techniques are useful in studying the distribution of ECM fungi in soil. An interesting study relates to a hypothesis that niche differentiation based on soil substrates and horizons regulates ECM fungal diversity in addition to spatial and temporal distribution of different isolates. Dickie et al. (2002b) have devized a technique based on DNA extraction, PCR and RFLP analysis to judge the vertical distribution of ECM hyphae in a soil column below forest vegetation. Using it, they identified fungal communities underneath a *Pinus resinosa* forest stand in northern United States of America. Fungal communities differed markedly among the four soil layers. Cluster analysis had revealed six different patterns of resource utilization; namely, litter-layer specialists, litter-layer generalists, F-layer and H-layer species, B-horizon species and multi-layer generalists. The spatial partitioning and distribution were mostly dependent on substrate/resource gradient. Hence, niche differentiation and substrate availability could be major determinants of ECM diversity.

Interconnections with multiple hosts, especially those from canopy and understory seem quite frequent. This phenomenon will definitely influence the distribution of ECM fungal species and genets. A recent study of Coastal Californian forests comprising a canopy (*Psuedotsuga menzieseii*) and understory (*Lithocarpus densiflora*) by Kennedy et al. (2003) has shown a greater abundence of multiple-host ECM taxa. Use of molecular techniques revealed that 13 of the 17 multiple host ECM taxa identified in this study were present on both hosts. In addition, distribution and abundance of shared ECM taxa was often unequal between hosts. It has been proposed that common mycorrhizal networks between canopy and understory may play a crucial role in the distribution of ECM fungi, both in temperate and mixed tropical evergreen forests.

3. F. Persistence of an Introduced ECM Fungus (e.g. *Laccaria bicolor* isolates)

Knowledge about the persistence of ECM fungi in natural soils is important. Reports on persistence of several ECM fungal species are available. However, in the present context, let us consider a single ECM fungus *Laccaria bicolor* and understand the underlying principles relevant to the persistence of an ECM fungus. *Laccaria* species are worldwide in distribution, frequently encountered in nursery and mature forest plantations of Douglas fir (*Pseudotsuga menzesii*) and Norway spruce (*Picea abies*). Accurate knowledge regarding the persistence of *Laccaria* on different genotypes of the host trees, particularly on their roots, within the rhizosphere, in field soil, and patterns of propagule distribution are mandatory while developing an appropriate ECM inoculation strategy. In fact, it is suggested that patterns of survival and distribution of both native and introduced strains of *L. bicolor* should be known in as much detail as possible. At Nancy in France, molecular methods such as PCR/RFLP of the internal transcribed spacer (ITS) and the intrageneric spacer (IGS) of rRNA and/or large mitochondrial rDNA (LrDNA), high performance capillary electrophoresis (HPCE) of IGS, RAPD and micro-satellite primed PCR were utilized to investigate the persistence of *L. bicolor.* In their nursery, persistence of the introduced (inoculant) genotype of *L. bicolor* as well as the competing native strains was studied. Obviously, short roots and ECM fungus present were counted. The survey conducted using the variation in ITS and IGS1 RFLP typing aimed at deciphering the persistence and spread of an inoculant strain (*L. bicolor* S238N) imported from the USA versus the native isolates that are frequent in French forest plantations. Nearly 80 to 90% of the short roots of 2-year-old Douglas fir and Norway spruce contained the inoculant American strain *L. bicolor* S238N. In most parts of the study area, the ITS and IGS types belonging to indigenous isolates of *L. bicolor, L. laccata, Thelophora terrestris* and *Rhizopogon subaereolatus* were feeble (<20%) or absent. Molecular tests have also proved that in un-inoculated plants grown in nursery, indigenous ECM fungi colonized nearly all the short roots upto 99%. The above trial with the American isolate of *L. bicolor* (S238N) shows that a good inoculum will have high potential to colonize host roots, better competitive ability and rapid distribution. Such an ECM fugal inoculum may lead to better plant response and performance in the outfield.

French researchers conducted a molecular analysis to judge the performance of American isolate of *L. bicolor* on established plantations of Douglas fir and Norway spruce. Populations of ECM fungi, mainly that of the introduced strain (*L. bicolor* S238N) and indigenous *L. laccata* were studied on 8-year-old plantations at St Brisson, in Central France. They utilized RAPD-based analysis. Genotyping using heteroduplex analysis of nuclear rDNA IGS1 and 51 RAPD markers revealed 4 genets of *L. laccata* and 5 genets of *L. bicolor.* Nearly 37% of the *L bicolor* genets were identical with inoculant strainS238N (Selosse et al. 1998a, b; 1999). Another study at the same location (St. Brisson) assessed the persistence of *L bicolor* S238N using cytoplasmic and nuclear markers. It proved that over a period of 7 years after transplantation, the inoculant strain predominated in only 37% of the 429 samples judged for ECM fungi. Therefore, it was construed that the American isolate may have limited potential in the outfield situation, although it may be robust and competitive in nursery-inoculation programs. The gene flow between *L. bicolor* S238N (inoculant) and native strains seems to be restricted. Restricted sexual compatibility due to geographically different origins, along with lack of acclimatization to different hosts might have impeded gene flow into native strains. However, such restricted gene flow might have actually allowed genetically stable persistence of the inoculant strain of *L. bicolor* for over 7 years in plantation. Such a long-term persistence of inoculant strain (*L. bicolor* S238N) is in contrast to the survival pattern noticed with Norway spruce examined in the same location. Molecular evidences accrued for *L. bicolor* S238N persistence on Norway spruce indicate that only 3% seedlings possessed it after a 4-year period in the plantation. Most Norway spruce plantlets also had unidentified ECM fungi. The inoculant carried from the nursery literally vanished with only 15% of the short roots showing the S238N strain of *L. bicolor.* It was concluded that indigenous *L. bicolor* and other ECM fungal isolates on birch, oak or spruce found in that forest plantation outcompeted the inoculant strain present on Norway spruce roots. In summary, the American isolate *L. bicolor* S238N was more likely to perform better on Douglas fir than on Norway spruce. Such pilot tests using molecular approaches can provide us with more accurate information regarding the persistence, infection potential and competitiveness of ECM fugal isolates. More importantly, it can help us judge the gene flow between inoculant and native strains, as also forecast genetic consequences ensuing out of an inoculation program.

To study the persistence of *Russula brevipes*, sporocarps were mapped and genets identified using species-specific hyper-variable micro-satellite markers (Bergemann and Miller, 2002). Most sporocarps of a single genet localized within 3 m. Largest distance measured between sporocarps was 18 m. Depending on host, populations of *R. brevipes* on Sitka spruce were comprised largely of related individuals, whereas in lodge pole pine, most genets were unrelated. High genetic variation of the mycosymbiont under Sitka spruce suggests frequent mating and recombination between the local inhabitants from primary establishment of basidiospores. The presence of only few unrelated genets under lodgepole pine suggests infrequent mating and long distance dispersal mechanisms.

Molecular Analysis to Assess Ectomycorrhizal Recolonization—A Case Study from California

Discussions here relate to a study reported about the ectomycorrrhizal survival and recolonization after a stand replacing wildfire that occurs periodically on Bishop's Pine (*Pinus muricata*) in the coastal zones of southern California. In this trial, Bruns et al. (2002b) aimed at understanding the genetic composition of ECM fungi that re-establish after a fire. They argued that for a re-establishment to take place after a wildfire, the three possibilities are: (a) re-establishment may be through resistant propagules that were held in soil; (b) dispersal of new propagules from adjacent unburned forest plantations; and (c) ECM fungi may survive as mycelia. Previous investigations in this location had revealed that Russulaceae and several *Amanita* species were dominant ECM fungi in the pre-fire plantations. During two years after the fire event *Rhizopogon, Wilcoxinia* and *Tuber* species were frequent. Certain *Hebeloma* species appeared to re-establish through new propagules, whereas *Wilcoxinia* species persisted as resistant propagules. In all of these fungi listed above, mycelial survival as a mode for re-establishment was not a great possibility. There was no evidence to support re-infections via mycelia. Generally, *Russula* species that is dominant during a pre-fire period was found to re-colonize immediately after a fire event. It survived, regenerated on the roots and spread into greater depths in the forest soil through propagules. However, Taylor and Bruns (1999) suspect that rapid recolonization events may be through mycelia-mediated survival and regeneration.

In their study, Bruns et al. (2002b) specifically analysed two ECM fungi, namely *Suillus pungens* and *Amanita franchetti* for their ability to recolonize a fire-ravaged forest plantation. They sampled 9-month-old post-fire seedlings for mycelial survival, archived spore samples, and mapped and genotyped the mushrooms. Molecular tests conducted were either single-stranded conformational polymorphism (SSCP) of isolated randomly amplified polymorphic DNA (RAPD) fragments (Bonello et al. 1998) or amplified fragment length polymorphisms (AFLP) (Bruns and Gardes, 1993; Redecker et al. 2001). Analyses of post-fire forest proved that of the two ECM fungi, *Amanita francheti* was not a common colonizer. It was not detected below ground, although it was seen as dominant fungus in the pre-fire survey. Fruiting never occurred during the first five years after the fire event. Thus, *Amanita francheti* did not survive, colonize or propagate within the post-fire forest plantation. Both the modes of recolonization, namely via mycelia and resistant spores, were ruled out. We may note that such lack of recolonization was not characteristic of other *Amanita* species traced in the forest. For example, *Amanita gemmata* recorded within the pre-fire forest plantation was observed to recolonize during the post-fire period comparatively quickly.

Now let us consider the activity of the other ECM fungus *Suillus pungens,* whose survival is largely dependent on Bishop's Pine, because it is *exclusive* to this host. Bruns et al. (2002b) therefore, commented that along with Bishop's pine, *S. pungens* too might have undergone upheavals in its population and functioning in response to each periodic wildfire event. Molecular tests revealed that *S. pungens*, in particular, recolonized rapidly after a fire. It leads to a relatively better abundance of this ECM fungus in the soil. Overall, of the two ECM fungi in spotlight, *Amanita francheti* was not a common survivor after a plantation fire event. Molecular tests also indicated that with *S. pungens*, spore-based survival and reestablishment on post-fire plantations is a clear possibility, whereas, mycelial survival through a forest fire and its spread later, leading to regeneration of ECM fungi seems to be remote (Bruns et al. 2002b).

3. G. Successions in Ectomycorrhizal Fungi

Ectomycorrhizal fungal genets (clones) and their pattern of succession may profoundly influence the ecology of soil fungi. We ought to know a great deal about ECM fungal succession in temperate forests, especially the sequential changes in genotype and their genets that occur under a tree. Specifically, it might have relevance to ECM inoculation programs. While dealing with aspects of ECM succession in northern Californian forests, Redecker et al. (2001) have made certain interesting remarks about fungal ecology and need for detailed molecular analysis of ECM fungal succession patterns, clone (genet) sizes, mode of spread, sexual/asexual propagation, gene flow, etc. Firstly, size of the fungi is measured in terms of its hyphal network, extending inside and above the soil. Individual hyphae do not show up easily identifiable traits; therefore, species recognition using its morphology is difficult or rather not feasible. As a consequence, basic biological processes such as breeding strategies, recombination and segregation are not easily studied using traits recognizable on fungal hyphae. However, using molecular techniques, the fine structure of fungal population can be deciphered. Sensitive DNA fingerprinting allows identification of genotypes and fungal individuals. Individuals within a given genotype are termed 'clones' or 'genets'. Within the realm of ECM ecology, knowledge about the persistence of genets, their sizes, propagation modes and genetic stability are crucial aspects that affect fungal dynamics. It may have significant effects on ECM functioning and tree growth. At this juncture, we should note that our knowledge about ECM fungal genets and their succession is itself dependent on accurate analyses and identification of ECM fungi.

Normally, ECM fungal species encountered early in succession are said to be efficient colonizers mediated mainly through spores. Early colonizers may often possess small and non-persistent genets (Gherbi et al. 1999). These early fungi are ruderal, and follow rapid establishment and reproduction strategy. The size of the genets of early colonizers seems important. Pioneer colonizers like *Hebeloma cylindrosporum* or *Laccaria bicolor* possess small and non-persistent genets. Maximum genet size for *H. cylindrosporum* ranged upto 3.5 m diameter, and that for *L. bicolor* was 12.5 m^2. Another early ECM species, *L. amethystine* occurring on hardwood plantations produced numerous but small (<1.5 m dia) genets (Gherbi et al. 1999). These early and ruderal species are propagated solely by spores. Therefore, they tend to be over represented in the soil bioassays. These pioneer ECM species may not persist for long. Stronger and competitive ECM fungi usually replace them.

Suillus species follow a mixed strategy and persist into late stages of ECM fungal succession. Genets of *Suillus* species are comparatively larger. For species like *S. bovinus, S. pungens* and *S. variegatus*, the genet sizes ranged from 20 to 40 m diameter. Certain *Suillus* species may put forth genets as large as 300 m^2 (Bonello et al. 1998). An early colonizer but persisting *Pisolithus* sp also produced genets of 20 to 40 m dia. *Cortinarius* is usually the first late-stage ECM fungus to be traced. Its genet's size was around 30 m diameter (Sawyer, 1999). *Suillus* and other examples cited above are known to adopt a mixed strategy in their natural conditions. They combine ruderal, rapid growth qualities with combative and stress tolerant traits. Donelly et al. (2004) have reported significant intra- and inter-ECM fungal species differences in extraradical mycelial growth. Generally, older mycelial systems persisted as mixture of mycelial cords interlinking mycorrhizal tips. In case of *Suillus bovinus*, mycelial cords and diffuse fans spread quickly into explorable soil. These variations affect foraging activities, spread and persistence of ECM fungi in soil microcosms.

Generally, ECM fungal species such as *Russula, Amanita* and *Cortanarius* are considered to be typical protagonists of later stages of ECM fungal succession in temperate forests (Redecker et al. 2001). Their spores germinate slowly and with difficulty. Hence, bioassay of nursery soils may not show the presence of these late-stage ECM fungi. Obviously, they do not colonize readily via spores, but their dominance in fields is attained through hyphal spread. Some late-stage ECM fungi may initially propagate via spores, but then switch to slow hyphal growth. Most late-stage ECM fungi are combative, stronger, tolerate stress and compete better, all of which suit their survival strategy in forest soils. Redecker et al. (2001) analyzed late-stage fungi occurring in northern California, using both morphological and molecular techniques. Morphological tests using pileus color suggested the occurrence of two different species of *Russula*, *R. cremoricolor* and *R. silvicola*. However, molecular identification showed that it was only single species with variable pileus color. Molecular analysis of *Lactarius xanthogalactus, Russula cremoricolor* and *Amanita francheti* which are generally traced in late stage of ECM succession, showed that none of the expectations about genet number or size, preferred methods of propagation or persistence seemed to coincide. None of the three late-succession ECM fungi produced larger genets. Hyphal growth was not the predominant factor determining the genet size. In fact, Redecker et al. (2001) have concluded that proliferation through sexual spores plays a greater role than previously understood. Molecular tests have clarified that the mere appearance of an ECM fungal species as an early or late stage successor is a poor indicator of actual colonization and persistence strategy.

On a site near Mount Fuji in Japan, Nara et al. (2003) recorded a total of 11450 sporocarps belonging to 23 different species of ECM fungi. They were all exclusively associated with the alpine dwarf willow (*Salix reinii*). *Laccaria* and *Inocybe* were pioneer colonizers; subsequent fungal species were added up as the plantation grew. There were no exclusions of ECM fungi due to succession. The biomass accumulation by sporocarps was unusually high, equating to 19% of the leaf biomass in productive associations. Host growth, especially its photosynthetic rate, had a strong influence on annual sporocarp production.

3. H. Forest Regeneration Practices and ECM Fungi

In the northern latitudes of Europe, Scots pine (*Pinus sylvestris*), Norway spruce (*Picea abies*), birch (*Betula pendula*) and aspen (*Populus tremula*) may support as many as 500 different ECM species. Karen (1999) opines that the number of ECM species stated above is uncertain and an underestimate because it is mainly based on the fruiting behavior of the fungal species. Many ECM fungal species form inconspicuous sporocarps and some are asexual. Typically, 50 to 150 species of ECM fungi can be traced in oligotrophic forest stands of northern Europe. For example, a 4-year sporocarp study of Scots pine revealed 115 species of Agaricales and Boletales. The number of ECM fungi increased if Asocymetes were included. Allen et al. (1995) argue that since the number of host plant species is few compared to the large number of ECM fungi, diversity of ECM fungi noticed is perhaps influenced marginally by the host plant species.

Modern tree management and land use methods have spread vastly, covering most parts of temperate forest zones. Without doubt, forestry practices do affect ECM diversity, species richness and functional aspects in several ways. Species richness may increase with the age of plantation (Keizer and Arnolds, 1994). The ECM fungal community structure itself changes with tree age, which is caused by succession. The slow-growing ECM fungi will need older plantations in order to flourish and maintain a viable population. Obviously, in forest plantations that experience shorter rotation periods, slow-growing ECM fungal species will be avoided. Forest-regeneration methods too affect ECM fungal composition. In the normal course, as trees are harvested, ECM fungi get severed from carbon, energy and nutrient source to a good extent. Fresh infection and fungal colonies generated immediately after are mainly through ECM mycelia, whereas, after a lapse of time, dispersal via spores or other propagules (e.g. sclerotia of *Cenococcum geophilum*) are crucial to the regeneration of ECM flora in the regenerated stands. But then, in established plantations, mycelial spread seems important. Molecular analysis indicates that as forest plantations become older, the number of genets of different ECM fungi decrease but their size may increase. Recently, many more of these aspects have been analyzed using molecular methods and then compared with previous inferences derived from morphotyping.

Let us consider a case study from Swedish forests. Karen (1999) examined the influence of regeneration methods on ECM community composition in Scots pine. She examined the ECM fungal species composition of forest plantations maintained as secondary stands with continuous tree layer- shelter wood (S), and compared it with clear-cut replanted stands (P). The change in ECM flora as a result of fungal succession was also studied. The inventory of mycorrhizal fungi on short roots indicated that only one of the many species- 'RFLP type-1' was immensely affected by regeneration techniques. This species 'RFLP type-1' occurred more frequently (40%) in the short roots from the trees of replanted forests (P) compared with 20% in shelter woods (S). *Russula decolorans* and some *Cortinarious* spp. tended to be common in shelter woods, compared with replanted stands (P). In the replants, ITS-RFLP patterns from ECM fungi did not match with *R. decolorans*. Karen (1999) explains that partial cuts in a shelter wood might have induced rapid re-growth of ECM fungi that easily adapt to soil disturbance. The presence of a living host tree might have induced higher population. Generally, as forest plantations age, mycorrhizal communities undergo succession. Such a secondary succession of ECM fungi might have induced species richness and abundance in the shelter woods. In a parallel bioassay of soils from Siljasfors forests, nearly 22 different RFLP types were identified from 62 mycorrhizal short root tips examined. Over 50% of RFLP types matched with those found in the outfield plantation. However, none of the RFLP-types from this bioassay matched *Piloderma croceum* or *Tylospora fibrillose* that were the most common among mycorrhizas sampled in the outfield. Karen (1999) explains that tree age, differences in soils and a few other factors could have caused the differences noticed in bioassay and out-filed plantation. Overall, for ECM fungal species tested using ITS-RFLP procedures, their ability to persist during regeneration phase and spread to adjacent tree stands was highly affected by regeneration methods and nutrient replenishment schedules. In addition, large array of survival and dispersal strategies found among ECM fungi might also explain the effects of regeneration methods on ECM community in the above study.

3. I. Nurse Functions: Source Plants and ECM Fungi

The concept termed wood-wide web involves interconnections (bridges) among the forest plants, developed predominantly through extraradical mycelium of ECM fungi. Such interconnected web may aid in transport and obtaining a sort of equitable distribution of mineral nutrients and carbohydrates within a plant community. Mycorrhizal fungi can extend from one plant's roots to another plant's roots to form a 'common mycorrhizal network' (CMNS) (He et al. 2003). Yet another term to note is the 'nurse function', wherein trees grown in a dark are supplied with photo-assimilates through ECM fungal bridges with other plants (Fig. 6.3). Wiemken and Boller.

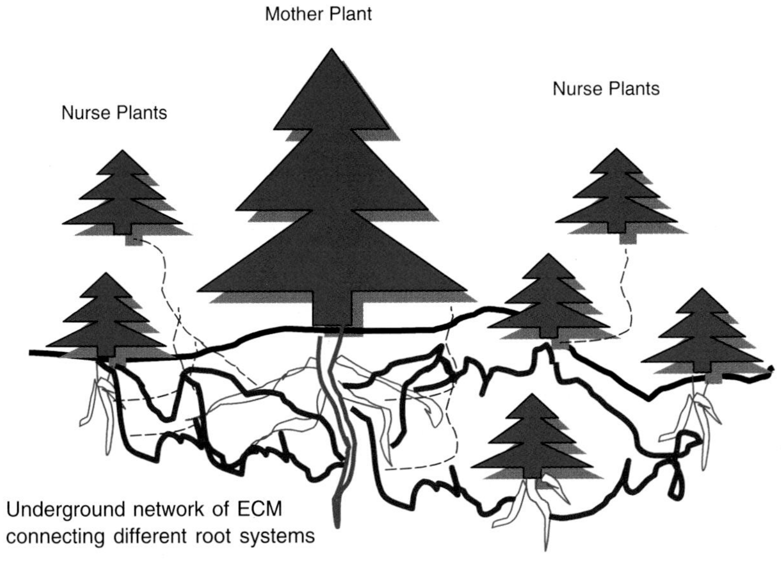

Fig. 6.3 Nurse functions of a mother plant in providing mycorrhizal inoculum to nurse plants and small seedlings in temperate forest location. *Note:* Blue lines are indicative of the spread of ECM fungal mycelia and Inoculum to nurse plants from mother plant, so that a network of ECM connections are established in a forest stand.

(2002) suggest that in wood-wide web, the most abundant and dominant ectomycorrhizal fungi play a key role in allowing the dominant trees to serve as nurses to the seedlings surrounding them. These trees may supply carbohydrates and nutrients to seedlings. They quote a study by Johnson et al. (2002) wherein ITS-RFLP analysis revealed that almost 50% of the fungal species found on pine tree species could also be traced in the nursed seedlings. At this juncture, we may note that molecular techniques such as ITS and RFLP-based methods, if used in conjunction with ^{14}C radio isotope labeling, can help us in ascertaining the exact fungal species/genotypes supporting 'nurse function' in the forest plants. In nature, we may encounter pine seedlings/heterotrophic plants that are nursed throughout with carbohydrates and nutrients via ECM web connections. Such plants have been termed 'mycoheterotrophic'. To quote examples, $^{14}CO_2$ label was traced in mycohetrotrophic plant *Corallorhiza trifida* that was interconnected to *Betula pendula* or *Salix repens*. In their studies, McKendrick et al. (2000) found that *Salix* was a good source (nurse) plant, and the fungus that formed the web was *Thelophora-Tomentella* complex. A few other mycohetrotropic plants identified using the ITS-RFLP and ^{14}C label techniques are *Sarcodes sangina*, *Pterospora andromeda* and web fungus involved in these cases was *Rhizopogon* sp. (Bruns and Read, 2000). Dickie et al. (2002a) opine that established (nurse) trees could be effective right at the seedling stage by providing inoculum to establish mutualistic associations. They examined AM and ECM infected trees for nurse functions on seedlings in the surroundings. *Quercus rubra* planted close to the nurse trees were able to reach ECM inoculum, establish mutualistic association and extract N and P better if trees in surroundings nurse them. An experiment in tropical Cameroon has revealed that an ectomycorrhizal network develops on adult trees and seedlings planted in concentric circle. Seedling survival and ectomycorrhiza formation via nurse effects were again dependent on tree species (Onguene and Kuyper, 2002). Such nurse functions may be equally frequent among AM fungal hosts.

3. J. Plant Symbionts as Epiparasites via ECM Fungi—A Case Study of Monotropidae

Typically, ECM symbiosis formed by large woody plants such as pines, oaks, birches, eucalyptus and many other temperate and tropical forest species are known to form "diffuse symbiosis". Under such a diffuse symbiotic state, both plant host and the symbiotic fungus have large parts of their thalli not associated directly with their partners. Hence, they are free to be simultaneously associated with several unrelated tree hosts/ or fungus. It means that in nature, individual ECM fungi can be simultaneously associated with several unrelated tree hosts. Similarly, individual tree hosts may be associated with multiple ECM fungi (Bruns et al. 2002a). Commonly, in a localized patch of forest, tens of ECM fungi occur on tree roots. We should realize that at the regional/continental level, innumerable ECM fungi might associate with a single plant species. *Alnus* species that associates with only a few ECM fungi and *Pisonia grandis* are examples of exceptions to the above generalists, because of their specificity.

Now let us consider "epiparasites" which are non-photosynthetic angiosperms that obtain all of their fixed carbon from ECM fungi linked to neighboring plants. These plants are called 'mycorrhizal cheaters' because they indirectly obtain fixed carbon from the surrounding photosynthetic plants through the ECM fungi that are associated with both plants. Bruns et al. (2002a) state that such epiparasites are also the host plants that exhibit a high level of host specificity as a rule. Such specificity was actually confirmed using several molecular identification methods, but mainly based on fungal mitochondrial large subunit rRNA gene sequence. The analysis indicated that *Pterospora andromaeda* (Monotropidae) is associated with only a single species of *Rhizopogon*. Similarly, *Monotropa uniflora* was restricted to members of Russulaceae. Molecular phylogenetic analysis have confirmed that Monotropidae form specific ECM associations. Different monotrope species may target different fungal lineages (Bidartondo et al. 2000; Bidartondo and Bruns, 2001). For example, distribution of *Rhizopogon ellenae* is greatest under the roots of *Sarcodes sanguina*. It is almost restricted to this plant species. Recently, Taylor and Bruns (2004) analyzed certain aspects of specificity further among the genotypes or races of individual fungal species. Bruns et al. (2002a) opine that perhaps there are no exceptions to the rule that such epiparasites are specialists. Molecular analyses have also indicated that plants from closely related genera of Monotropidae (Ericaceae) targeted only narrow clade of related ectomycorrhizal fungi. This pattern of specificity extends to many non-photosynthetic orchids associated with ECM fungi. For example, *Cephalenthera austinae* and *Corallorhiza maculata* are restricted to Thelephoraceae and Russulaceae, respectively. Bruns et al. (2002a) have remarked that such high specificity is comparable with typical host-fungal parasite relations that occur in nature. They believe that specialization among such epiparasites allows the plants to adopt better to fungal partner. Such specialization could be the result of development of resistance for other fungi via gene for gene action. According to Molina et al. (1992) plants' preference to specialized fungi will indirectly reduce the chances of competing plant species. This situation, called the "Facultative Epiparasitism", occurs when one plant indirectly parasitizes a second plant host through ECM fungal interconnections. Obviously, one plant reaps the benefits (mineral, carbon or water) from another plants' symbiotic relationship with ECM fungus (see Bruns et al. 2002a).

3. K. Nutrients, Mycorrhizal Fungal Ecology and Related Aspects

Nitrogen availability affects the abundance, diversity and function of ectomycorrhizal fungi (Lilleskov and Bruns, 2001; Lilleskov and Fahey, 2003). We know that variability in natural N and anthropogenic N deposition, both can influence ECM activity. Recent evidences in European locations indicate that atmospheric N deposition, along with multiple air pollutants, has indeed reduced production of fruit bodies of several different EC fungi. In their study, Lilleskov and Fahey (2003) aimed at understanding the effect of N deposition on ECM fungi colonizing white spruce (*Picea glauca*) in North America. In the location selected for study, N deposition was caused by a nearby NH_3 production facility. The background levels in the locations not far away were quite low, at < 2 kg ha.$^{-1}$ yr^{-1}. Lower level of ECM population was expected near the NH_3 production facility. High precipitation-N near the NH_3 production facility also led to higher N availability, foliar N: P, N: cation ratios and low pH. The ECM fungal species diversity was generally lower in the vicinity of NH_3 production facility and in locations receiving higher N deposition. Abundance of fruit bodies of four common fungi; namely *Laccaria laccata, L. bicolor, Lactarius theiogalus* and *Paxillus involutus* showed no change or increase at high N sites. Fruit body production by the ECM fungal genera *Cortinarius*, *Russula*, *Tricholoma* and *Hebeloma* were negatively correlated with N in the organic horizon of soil. The decline in the population and diversity of *Cortinarius* and *Russula* was marked. These observations support the general view that high N is detrimental to ECM fungal survival and proliferation. Peter et al. (2001) examined both above and below ground ECM species in a subalpine spruce given nitrogen for 2 years. The ECM diversity was assessed using PCR-RFLP analysis of ITS region of rDNA extracted from mycorrhizal root tips, sporocarps and mycelia. At least 44% of all ECM were formed by species belonging to Thelephoraceae and Corticiaceae that produced inconspicuous sporocarps. Sporocarp surveys indicated a reduction in ECM diversity due to N addition. However, below ground ECM diversity seemed to be unaffected by N inputs. A change in ECM species could be observed only 2 years after N addition. A shift in ECM abundance from large sporocarpic species to those with no or resupinate was easily discernable 2 years after N inputs.

The biochemical composition of soil N pool influences symbionts. Preference to different biochemical forms can also affect the mycorrhizal community, and indirectly, the composition of plant species. In this regard, Wiemken and Boller (2002) state that one of the advantages of ECM symbiosis with trees is its ability to short-circuit the nutrient cycling steps, by overcoming requirements of certain chemical transformations of N. ECM fungus can directly harness nutrients from organic matter in soil and lead it to plant roots. Plants without mycorrhiza primarily exploit N sources that are released by saprotrophs, but ECM directly degrades proteins to absorb amino acids. There are several examples depicting such short-circuiting of nutrient cycles. Biochemical and molecular assessment has indicated that ECM fungus *Amanita muscaria* secreted aspartic proteases to acquire nitrogen from protein contained in organic material. Similar studies by Nordin et al. (2001) and Persson and Nashohlm (2001) using *Pinus pinaster-Hebeloma cylindrosporum* indicated that molecular form of N available in soil affected the symbionts. Generally, glutamates as N source were preferred over NO_3-N. It is said that at the ecosystem level, amino acids, particularly glycine, are crucial in the temperate and boreal climates where pine-ECM symbionts are most frequent.

The competitive ability of ECM fungi is influenced by their to ability utilize different carbon and nitrogen sources, particularly from complex organic substrates. In nature, saprophytic fungi utilize carbon drawn by them to support N uptake into their thalli. Whereas, ECM fungus in symbiotic state draws up ready photosynthate supply from its host to support the energy required for N uptake and utilization (Miller et al. 2003). Two basidiomycetous saprophytes—*Agaricus augustus* and *Coprinus quadrifidus* and two ECM fungi *Suillus tomentosus* and *Hebeloma birrum* were selected to test this hypothesis. They studied the ability of these fungi to utilize cellulose and protein provided as C and N sources. Three C:N ratios, namely low (1:25), medium (1:75) and high (1:500) were tested. At high C:N ratios, the ECM outgrew saprophytes, but the reverse was true if C:N ratios were very low. Both *Agaricus* and *Coprinus* were able to utilize cellulose at all C:N ratios and absorb N, whereas *Suillus* was unable to utilize cellulose if glucose was absent, and *Hebeloma* used up very little of it that was provided in the medium. Based on their observations, Miller et al. (2003) believe that in nature, ECM fungi are at an advantage while utilizing C and N present in different complex organic sources, because they are attached to a ready supply of glucose from the host, so that energy requiring N uptake and metabolism can be accomplished without paucity.

Atmospheric nitrogen deposition is a serious problem in many temperate forest locations. The excessive N enrichment it causes in the soil may have its own effect, firstly on the host tree species and secondly on ECM fungus existing in symbiotic and/or saprophytic phase. As a consequence, it affects ECM fungal diversity. Nitrogen input alters the carbohydrate allocation to roots and fungal symbiont. Eaton and Ayers (2002) found that ECM fungi are sensitive to changes in carbohydrate availability. It may also affect protein mineralization capacity. In fact, 7 of the 10 ECM species tested showed reduced growth if C supply was hindered. The ECM fungi with slow growth rate and better protein mineralization ability were all late success ional species. The dependence on N source and quantity seemed to vary with the ECM fungal species. For example, *Cenococcum geophilum* was strongly dependent on N supply for its growth. Nitrogen enrichment achieved through natural atmospheric deposition or via fertilizer inputs, both will have their own influence on the diversity of ECM fungi in soil and in tree roots. In a field trial at Norleiden, Sweden, Karen (1999) found that ammonium nitrate application significantly reduced *Suillus variegatus, Piloderma croceoum* and several *Cortinarius* species, whereas, frequencies of *Lactarius reufus* and a few other ECM fungal species in short roots increased due to N inputs. Reduction in overall sporocarp-based ECM inventory in that location was mainly due to a decrease in the number of *Ordinaries* species. Reduction in colonization and mycelial growth was cited as yet another cause of changes in ECM fungal flora.

The ratio of NH_4^+ to NO_3 is known to change in forest soils. Since preferences for N-form differs depending on the tree species, fluctuations in dominant N form in soil can influence tree succession and vegetation. Generally, at high NH_4^+ concentrations, early successional species such as Douglas fir or aspen may accumulate higher levels of N and thrive better. Whereas, a low N accumulator such as white spruce may thrive better as late successional species. Kronzucker et al. (2003) argue that tree diversity and growth in forest stands, to a certain extent, is influenced by tolerance/susceptibility of ECM to high/low NH_4^+/NO_3 levels in soils. Discrimination between NO_3 and NH_4 forms can become a driving force in selecting tree species and their ECM partners.

Mycorrhizal symbiosis involves an intricate transfer of carbon between the host plant and fungus. It may also significantly influence carbon cycling within natural or agroecosystem. We know that parameters for carbon cycling vary with plant traits and type of vegetation. Cornelissen et al. (2001) tested an assumption that AM or ECM fungi may significantly affect carbon cycling as well as feedback between vegetation productivity and litter turnover. They examined crop/forest stands that were contributed by a range of plant species for major productivity indicators such as relative growth rate (RGR), foliar nutrients and litter decomposition rates. In general, they traced important variations between mycorrhizal types and carbon turnover as well as vegetation productivity. Plant species with ericoid mycorrhizas were associated with low inherent growth rates, low foliar N and P as well as poor litter decomposability. Ectomycorrhizal plant species were linked to intermediate levels of RGR, foliar N and P and litter turnover. Whereas, arbuscular mycorrhizal fungi that are common on fast-growing annuals/biennials and tropical shrubs or trees, frequently had high RGR, foliar N and P as well as better litter turnover rates. Overall, temperate flora having ericoid and ECM associations adapt to low carbon turnover strategies, but AM fungal species prefer to flourish in ecosystems with rapid/high C turnover.

In a forest plantation, elevated CO_2 (e.g. 700 ppm) can induce significant changes in the nutritional status of roots and belowground community composition of ECM fungi (Fransson et al. 2001). The magnitude of changes in ECM community seems to accentuate if treatment with elevated CO_2 was continued for a long term, say from 3 to 15 years. Therefore, it is useful to possess prior knowledge regarding C distribution into different parts of plant, its influence on ECM fungus and other aspects of ecosystem. There is no doubt that alterations in ECM community can potentially affect carbon allocation and nutrients within a forest ecosystem. Using a novel autoradiographic technique, Leake et al. (2001) were able to visualize, quantify and chart out the spatial and temporal changes in carbon allocation in the roots of forest plants, ECM fungal mycelia and litter patches. The ^{14}C pulse labeling showed that nearly 60% of the C transferred from the host plant to external mycorrhizal mycelium was actually allocated in patches that occupied only 12% soil area available for mycelial spread. Within the litter patches, recently formed mycelia garnered most ^{14}C, amounting to 27 to 50% of total C flux into external mycorrhizal mycelium. The amount of C transferred to *Suillus bovinus* was around 167 nmol, reaching a maximum by 24 to 36 h after labeling. Ultimately, these mycorrhizal systems tend to deploy C in a dynamic manner as they forage for other nutrients in soil. Definitely, ECM fungi will influence below ground dynamics of C and other nutrients accordingly.

Langely et al. (2002) exposed a regenerating scrub oak system to elevated carbon levels. They used ^{13}C label to judge the extent of C integration into various components of the ecosystem. The ^{13}C integrated into the shoot system in the first season, but below ground portion displayed extremely slow integration of the new C inputs. Roots sampled from ingrowths showed 33% of C in newly formed roots originated from source other than recent photosynthesis. Resprouting of rhizome is a common phenomenon in the regenerating oak ecosystem. As such, remobilization of below ground C is an important aspect because it can support roots and mycorrhizas for several seasons. The below ground storage of C—be it in tree roots, soil organic matter or ECM fungi—is an important aspect of C dynamics in a forest plantation. According to Langely et al. (2002) extensive and higher levels of C storage can be helpful in regenerating oak forests and ECM partners. Remobilization of C from belowground portions is known to support roots and ECM fungi for several years during the re-establishment of a forest stand. Under such conditions, regenerated portions obviously contain very little new photosynthate. It is said that long-term storage and cycling of C is generally helpful in forests prone to frequent disturbances.

Amendments employed within a forest plantation affect the ECM ecology. Forest plantations receive different amendments in addition to inputs of major nutrients. Such amendments, generally improve soil condition, remove adverse reactions, and reduce the effects of pathogens and other detrimental biotic factors. Many of such amendments affect the distribution of ECM fungi. Dolomite inputs into spruce stands are common in Sweden. Jonsson et al. (1999) used ITS typing to assess the effect of dolomite addition on ECM fungi. They found 16 different ITS-RFLP patterns and most of them were easily matched to an ECM genus that could be identified. High dolomite did influence ECM fungal community structure. Three species, namely, *Russula ochroleuca*, *Lactarius nector* and *Boletus chrysenteron* occurred both as mycorrhizas and fruit bodies. There were 5 taxa, which occurred on over 5% of the screened roots. These were *Thelephora terrestris* (21%), *Tylopilus fellus* (13%), *Tylospora fibrillose* (13%) and two unidentified taxa 910 and 6%, respectively. Together, these five taxa colonized over 60% of the mycorrhizal roots assessed. They reported that in addition to changes deciphered using ITS-RFLP patterns, fruit body data also revealed shifts in ECM fungal community in response to dolomite addition into spruce stands.

Microbial Interactions

Ectomycorrhizal fungi interact with a wide variety of microbes in the soil. Such interactions may be antagonistic, synergistic, commensalistic, parasitic, etc. Barr and Stanton (2000) selected a wide range of saprophytes common to pine forests and examined their ability to interact with ECM fungi such as *Cenococcum geophilum, Laccaria bicolor* and *Rhizopogon luteolus.* Interactions were species-specific. Overgrowth was the most common consequence in 43% of the samples, inhibition due to diffusive chemical was encountered in 26% samples and contact inhibition occurred in 16% of the cases. Application of a nitrogen source such as ammonia affected interactions significantly in many cases. In a particular situation, competitive strength of *L. bicolor* was enhanced enormously and only ECM hyphae were seen, since it completely suppressed saprophytes such as *Clitocybe* species. Interestingly, the antagonistic effect of *Laccaria laccata* on co-cultured *Trichoderma virens* was attributable to direct physical deformation of trichoderma conidia, rupture of wall and partial degradation leading low spore germination (Werner et al. 2002).

There are many reports dealing with changes in soil microflora in response to AM fungal colonization (Allen, 1992; Ames 1987; Azcon-Aguilar and Barea, 1996; Bagyaraj, 1984; Krishna et al. 1983; Linderman, 1988). The rhizospheres of symbiotic roots harbor and interact with various types of nitrogen-fixing bacteria (Diouf et al. 2003; Lata et al. 2002; Provorov, 2002), phosphate-solubilizing bacteria (Andre et al. 2003; Lata et al. 2002; Paulitz and Linderman, 1989; Schreiner et al. 1997), plant pathogens and their antagonists (Barea et al. 1998; Citernesi et al. 1996). The specific induction or suppression of a particular microbe may be dependent on several factors that influence the symbiosis. In terms of environment, soil microbial load tends to be higher in tropical agricultural fields; therefore interactions with AM fungal component in soil may be accentuated. Let us consider an example from temperate soil cultivated with cucumber. Mansfeld-Giese et al. (2002) found that under such soils, inoculation with *G. intraradices* immensely influenced aerobic heterotrophic bacterial communities. Nearly 1400 colonies could be isolated and identified using HPLC patterns. It comprised 87 species belonging to 43 genera of bacteria. *Arthrobacter*, *Psuedomonas* and *Burkholderia* were most frequent in the rhizosphere of mycorrhizal plants. Most bacterial genera tended to proliferate and stay either in close proximity or on the AM fungal hyphae.

We may note that interaction between AM fungi and soil microorganisms is constantly influenced by host plant on which AM fungi thrive. A change in the host species or even its genotype can eventually affect ensuing microbial interactions. Let us consider a recent study involving genotypes of wheat and their influence on interaction between *Azotobacter chroococcum* and AM fungus *Glomus fasciculatum.* Behl et al. (2003) reported that induction of higher population of *A. chroococcum* and AM fungus was related to root biomass and rhizosphere environment under a given wheat genotype. AM fungal infection often enhanced *A. chroococcum* population and survival rates, but it varied with host genotype. Sood (2003) suggests that specific interactions may be mediated through root exudates and chemotactic responses. In their study, *A. chroococcum* showed stronger chemotactic response towards mycorrhizal roots that exuded specific amino acids, sugars and organic acids in a higher quantity. Interaction between soil yeast, namely *Rhodotorula mucilaginosa* and AM fungi such as *Glomus mosseae* and *Gigaspora rosea* was studied by Fracchia et al. (2003). Exudates from *R. mucilaginosa* stimulated hyphal growth rates of *Glomus mosseae*. Percentage colonization and spore density of both the AM fungal species were higher if soil yeast *R. mucilaginosa* was inoculated ahead of the fungi. In a tripartite relationship involving plant, rhizobium and AM fungi, Andre et al. (2003) have noticed that relationships between symbionts could be specific at the physiological and molecular level. The stimulatory effect of mycorrhizosphere and changes in population dynamics of individual rhizobial inoculant was specific; hence, developing a proper understanding about specificity and interactions between AM fungi and inoculant rhizobial strains at molecular level may be important.

Microbial interaction with mycorrhizal fungi may also be traced internally. In case of AM fungi, transmission electron microscopy reveals that hyphae and sporocarps/ spore walls of *Glomus mosseae* harbor different microorganisms, especially bacteria, actinomycetes and fungi. Such microbial cells or fragments are also encountered within arbutoid and ectomycorrhizas (Buscot, 1994; Fillipi et al. 1995; Fillipi et al. 1998; Garbaye et al. 1996; Varrese et al. 1996). These microbes are intricately linked to the host sporocarp and may not get cleansed despite using a concentrated (10%) sterilizing agent for over 129 min (Filippi et al. 1998). It is possible that interhyphal spaces, spore walls and sporocarps provide a rich environment for the establishment and metabolic maintenance of these microbes. Cytoplasmic bacterium-like organisms (BLOs) were detected ultrastructurally in the spores of *Gi. margarita* (Bianciotto et al. 1996). It is suspected that upto 2.5×10^5 bacteria may survive inside each spore of *Gi. margarita*. Comparison of small sub-unit rRNA gene proved that endosymbiont belongs to group 2 pseudomonads, genus *Burkholderia*. Advantages of endosymbiosis between mycorrhizal fungus and bacteria are unknown. However, we have to note that in natural settings, mycorrhizal systems actually include plants, fungal community and endosymbiotic bacteria. Details about molecular level interactions between the host, mycorrhizal fungus and endosymbiotic bacteria are lacking.

Mycorrhiza in Saline Environments

Plant species from saline and sodic soils harbor AM fungal isolates that are adapted to soils with high ionic strength and electrical conductivity. Roots of salt marsh species such as *Artemisia*, *Aster* and *Plantago* were all strongly mycorrhizal, with well-developed arbuscules and vesicles (Landwehr et al. 2002). Several other plant species such as those of Chenopodaceae, as expected, did not show AM fungi. The RFLP patterns of ITS regions showed that many of AM fungi were ecotypes of *Glomus geosporum*. They suspect that specific ecotypes of AM fungi might be involved in symbiosis under saline and sodic conditions. A different study involving screening for salinity-tolerant AM fungal species that colonize neem (*Azadirachta indica*) also showed that specific ecotypes of *G. fasciculatum* and *Gi. margarita* were more common and well adapted to saline soils of northwest India (Pande and Tarafdar, 2002). A European project on Mycorrhizas (MYCOREM) aimed at studying the molecular mechanisms relevant to salt tolerance in mycorrhizal plants growing in the marshes of Netherlands and Germany. The molecular identification of spores available in the seaside saline soils revealed that 80% of them belonged to *Glomus geosporum*/ *G. caledonium* clade. Saline marshes in the southern Germany showed *G. geosporum* to be the dominant clade. In view of this ecological preference for this clade of AM fungi, Bothe et al. (2003) concluded that *G. geosporum* imparts salt tolerance to the host plant. In salt-stressed tomato plants inoculated with *G. geosporum*, the expression levels of a tonoplast aquaporine and Na^+/H^+ transporter are reduced compared to non-mycorrhizal ones. In a different study with tripartite symbiosis involving *Trifolium alexandrium*, nodule bacteria and AM fungus, the mechanism of salt tolerance was related to improved P nutrition and growth due to mycorrhizal colonization. Proline accumulation was notably higher in plants provided with dual inoculation (Ben Khaled et al. 2003). In case of maize seedlings exposed to salinity stress, Feng et al. (2002) found that osmoregulation was associated with the ability of mycorrhizal plants to accumulate higher levels of soluble sugars and electrolytes. This effect seemed to be independent of P nutritional status. The extent to which plant host or the symbiotic AM fungus can withstand saline or sodic conditions seems to differ. Investigations in the Canadian boreal zones affected by saltpan have shown that AM fungal colonization is common on plants found in mildly saline soil conditions. The AM fungi were in roots of plants found in extremely saline soils. Experiments involving transplantation of seedlings from saline to extremely saline conditions and vice versa has clearly shown that mycorrhizal fungi are limited by edaphic factors, not by the absence of host (Johnson-Green et al. 2001). Mycorrhizal fungi are less tolerant to extreme salinity compared the host *P. nuttuliana*, a salt-tolerant halophyte. The phylogenetic information on AM fungi that colonize such boreal soils affected by saltpan and mechanisms that restrict their ability to tolerate salinity needs to be understood.

Mycorrhiza in Boreal Locations

Boreal and cool temperate climates support vegetation that forms mycorrhizal association. There are many reports describing their diversity and ecosystematic functions in such environments (Dalpe, 2003; Eriksen et al. 2002). During recent years, molecular techniques have been extensively utilized to identify, enumerate and study the ecology of ECM fungi colonizing the Boreal plants, for example, those found in Norway (Vralstad, 2001). Molecular analysis done on ECM-root tips involved PCR and ITS-based approaches. It aimed at identifying and ascertaining the phylogenetic position, molecular diversity and studying their life cycle. Most characteristics of the ECM-morphotypes encountered was *Picierhiza bicolorata* (pb ECM) that occurred abundantly on conifer roots. Molecular analyses based on ITS1 rDNA sequence placed it closely with ericoid mycorrhizal fungus *Hymenoschyphus ericae* or clade-II of ericoid fungi (Carney and Ashford, 2002). Tests on the ability to form ECM revealed that a single isolate lacked the ability/potential to form both ecto and ectendomycorrhiza; however, both ECM and ectendofungus co-exist on a single host root system. In this particular study, ITS analysis indicated that pbECM shared 99% ITS sequence homology with the ectendofungus *H. ericae*. The ITS-based phylogeny strongly supported the fact that *H. ericae* is a monophyletic group. But it was also inferred that families of Helotiacea, Dermatiacea and Hyaloschyphacea identified in these boreal locations might be polyphyletic. Kernaghan et al. (2003) studied 12 different mixed wood stands of similar age grown on similar soil deposits in boreal locations of the northwestern Quebec. Diversity of canopy trees, understory flora and ECM fungal diversity were measured. They traced a positive correlation between canopy tree diversity and ECM fungal community composition. Such a correlation was suspected to be due to ECM fungal host specificity.

Heavy metal accumulation is known to influence soil microbiota. To understand the effect on ECM fungi, Markkola et al. (2002) exposed Scots pine seedlings to wet deposits of nickel (Ni). The lichen cover was removed. Nickel deposits affected colonization ECM fungi in different ways. Highest frequencies of tubercle morphotypes were traced in quadrats exposed to 100 mg m^{-2} Ni in lichen-covered conditions. Removal of lichen increased diversity of ECM fungal community. It improved the condition of the short roots, because it reduced the frequencies of poor and senescent short root. Detrimental effects of acid rains and aluminum toxicity in the temperate forest zones and boreal locations are well known. A variety of alleviation methods are employed to overcome Al-mediated ill effects in soils. Ahonen-Jonnarth et al. (2003) evaluated the effects of elevated levels of Al on ECM symbiosis. They noticed that Al did not influence the growth of *Pinus sylvestris* seedlings if they were well colonized by *Laccaria bicolor*. The Al-mediated suppression of P uptake was also not perceivable, mainly because mycorrhizal seedlings generally possessed higher concentrations of P. Mycorrhizal association enhanced pH of the soil solution. Therefore, it avoided the ill effects of Al-mediated decrease in rhizosphere pH. Also, the presence of ECM mycelia seems to decrease the loss of cations. There are suggestions that careful combination of strains of heavy metal detoxifying bacteria such as *Brevibacillus* along with mycorrhizal fungal inoculum can be efficient in reducing the ill effects (Vivas et al. 2003).

Methyl Halides and Mycorrhizal Fungi

Treseder (2000) argues that a significant portion of ozone depletion caused by methyl halides is unaccounted for. It is possible that ECM fungi partly contribute the missing fraction of methyl halide. The host vegetation for ECM flora is densely distributed in temperate latitudes. Most trees posses ECM, and they allocate nearly 20% of their photosynthates to the fungal partner. Consequently, it is believed that ECM constitutes 15% of the soil organic matter. On a wider scale, the total biomass of ECM is difficult to know exactly, but it is guessed at 1×10^{15} g. Since ECM fungi produce methyl halides, it may have consequences on the atmospheric ozone. Elevated CO_2 induces ECM proliferation and alters the community composition. It is possible that indirectly, the production of methyl halides is also affected. Most ECM fungi produce methyl halides, but their extent may vary with species, its density and the prevailing environmental parameters. *Cenococcum geophilum*, a common ascomycetous ECM fungus with worldwide distribution emits 0.008 to 61 μg g^{-1} fungus d^{-1} methyl bromide and 0.46 to 230 μg g^{-1} fungus d^{-1} methyl iodide. Assuming global fungal biomass at 1×10 15 g y^{-1}, the computed methyl halide emission rates equate approximately to 5×10 10 g y^{-1} methyl bromide and 10 11 g y^{-1} methyl iodide, respectively. Molecular methods including ITS-RFLP and monoclonal antibodies are being utilized to estimate the quantitative/qualitative distribution of different species or their isolates and relate them to methyl halide production. Such techniques could help in pinpointing the ECM fungal isolates that emit larger amounts of methyl halides. For example, certain strains of *Laccaria laccata* tested in California produced 61 μg g^{-1} d^{-1} methyl bromide and 230 μg g^{-1} d^{-1} methyl iodide. Whereas, an isolate of *Hebeloma crustuliniforme* in the same location produced only 0.008 μg g^{-1} d^{-1} methyl bromide and 2.3 μg g^{-1} d^{-1} methyl iodide.

Concluding Remarks

During the past 8 years, molecular procedures based on SSU rRNA sequence, PCR-RFLP and micro-satellite markers have been frequently utilized to understand the ecology of mycorrhizas. The impact of molecular assessment has been remarkable on inferences. In most cases, data obtained through molecular analysis has corroborated, removed incorrectness, if any, and added accuracy compared to those derived using morphotyping. Morphotyping can be error prone, especially when spore traits and their distribution do not tally with ecological manifestations of AM/ECM fungus. This anomaly is effectively removed by molecular analyses of different parts of AM/ECM fungus. Overall, molecular procedures have imparted greater accuracy to our knowledge on AM and ECM fungal succession, spatial and temporal changes of fungal genets and community structure. Mycorrhizal fungal diversity below ground influences the plant diversity obtained above ground. With regard to agricultural settings, tillage, interculture, rotations and crop genotypes adopted during field crop production are known to influence the ecological aspects of AM fungi enormously, especially succession, community structure and its ability to transfer nutrients. AM fungi affect soil structure by producing glomalin, which is a protein known to enhance soil aggregation. We also know that crop genotype can influence below ground ecology of AM fungi.

Molecular techniques have enabled exemplary progress in our knowledge on the ecology of ECM fungi, especially their spatial and temporal changes, dominant species, size and distribution of the fungal genets, mechanisms of their spread and survival as well as other relevant aspects. Excellent examples are available from the University of Berkeley, California, Nancy in France and certain locations in Sweden. Prior knowledge about the spread of inoculant strain and its ability intermate with natural ECM flora, as well as consequent gene flow is a necessity. There are excellent examples on the succession, distribution and gene flow of ECM fungi in temperate forests of North America. Such studies should be replicated on tropical forest plantations and agricultural settings. Detailed molecular analysis on the fate of inoculant strains of AM fungus is lacking. Knowledge about gene flow will be useful while deciding on inoculant strains of AM fungi.

Forest fires and regeneration practices affect the ecology of ECM fungi below ground. There are several examples depicting successional changes in ECM fungi, dominant symbiont and the rate of spread as a consequence of different regeneration practices. Many of them are location-specific. Researchers at Berkeley have reported significant advances on 'epiparasitism' or 'mycoheterotrophy'. 'Nurse function' is yet another concept that may find greater interest in due course of time, especially its relevance to practical agriculture and forestry. Application of molecular techniques has improvized data accrual and inferences on almost every aspect of ecology of mycorrhizal fungi. It includes ecological topics such as the influence of abiotic and biotic factors, microbial interactions in the mycorrhizosphere, and mycorrhizal activity in extreme environments like boreal locations.

References

Ahonen-Jonnarth, U., Goransson, A. and Finlay, R.D. 2003. Growth and nutrient uptake of ectomycorrhizal *Pinus sylvestris* seedlings in a natural substrate treated with elevated Al concentrations. Tree Physiology 23: 157-167.

Allen, M.F. 1992. Mycorrhizal Functioning. Routledge Publishers, London, U.K. pp 332.

Allen, E.B., Allen, M.F., Helm, D.J., Trappe, J.M., Molina, R. and Rincon, E. 1995. Patterns and regulation of mycorrhizal plant and fungal diversity. Plant and Soil 170: 47-62.

Allen, E.B., Rincon, E., Allen, M.F., Perezlimenez, A. and Huante, P. 1998. Disturbance and seasonal dynamics of mycorrhizae in a tropical deciduous forest in Mexico. Biotropica 30: 261-274.

Ames, R.N. 1987. Mycorrhizosphere; morphology and microbiology. In: Mycorrhizas in the Next Decade. Sylvia, D.M., Hung, L.L. and Graham, J.H. (Eds) University of Florida, Gainesville, Florida, USA, pp. 181-183.

Ammucucci, A., Zambonella, A., Giomora, G., Potenza, L. and Stocchi, V. 1998. Identification of ectomycorrhizal fungi of the genus *Tuber* species-specific ITS primers. Molecular Ecology 7: 273-277.

Andre, S., Neyra, M. and Duonnois, R. 2003. Arbuscular mycorrhizal symbiosis changes the colonization pattern of *Acacia tortalis* spp. Radianna rhizosphere by two strains of rhizobia. Microbial Ecology 45: 137-144.

Azcon-Aguilar, C. and Barea, J.M. 1996. Arbuscular Mycorrhizas and biological control of soil borne pathogens. An overview of the mechanisms involved. Mycorrhiza 6: 457-464.

Bachmann, K. 1994. Molecular markers in Plant Ecology. New Phytologist 126: 403 - 418.

Bagyaraj, D.J. 1984. Biological interactions with VA mycorrhizal fungi. In: VA Mycorrhiza. Powell, C.L. and Bagyaraj, D.J. (Eds) CRC Press, Boca Raton, Florida, USA, pp 131-153.

Bagayoko, M., Buekert, A., Lung, G., Bationo, A. and Romheld, V. 1999. Cereal/legume rotation effects on cereal growth in Sudao-Sahelian West Africa: Soil mineral nitrogen, mycorrhizae and nematodes. Plant and Soil 218: 103-116.

Bago, B., Pfeffer, P.E. and Sachar-Hill Y. 2000. Carbon Metabolism and Transport in Arbuscular Mycorrhizas. Plant Physiology 124: 948-958.

Barea, J.M., Andrade, G., Bianciotto, D., Dowling, S., Lohrke, P., Bonfonte, F., O'Gara, F. and Azcon-Aguilar, C. 1998. Impact on arbuscular mycorrhiza formation of *Psuedomonas* strains used as inoculants for biocontrol of soil-borne funal plant pathogens. Applied and Environmental Microbiology 64: 2304-2307.

Barr, J. and Stanton, N.L. 2000. Ectomycorrhizal fungi challenged by saprophytic basidiomycetes and soil micro-fungi under different ammonium regimes in vitro. Mycological Research 104: 691-697.

Bastide, P.Y., Kropp, B.R. and Piche, Y. 1994. Spatial distribution and temporal persistence of discrete genotypes of ectomycorrhizal fungus *Laccaria bicolor*. New Phytologist 127: 547-556.

Bastide, P.Y., Kropp, B.R. and Piche, Y. 1995. Population structure and mycelial phenotype variability of the ectomycorrhizal basidiomycete, *Laccaria bicolor*. Mycorrhiza 5: 18.

Bearden, B.N. and Petersen, R.L. 1999. Influence of arbuscular mycorrhizal fungi on soil structure and aggregate stability of a vertisol. Plant and Soil 218: 173-183.

Behl, R.K., Sharma, H., Kumar, V. and Narula, N. 2003. Interactions amongst mycorrhiza, *Azotobacter chroococcum* and root characteristics of wheat varieties. Journal of Agronomy and Crop Science 189: 151-155.

Bergemann, S.E. and Miller, S.L. 2002. Size, distribution and persistence of genets in local populations of the late stage ectomycorrhizal basidiomycete, *Russula brevipes*. New Phytologist 156: 313-320.

Bever, J.D. 2002. Negative feedback within a mutualism: host-specific growth of mycorrhizal fungi reduces plant benefit. Proceedings of the Royal Society of London, Series B, Biological Sciences 269: 2595-2601.

Bever, J.D., Schultz, P.A., Pringle, A. and Morton, J .B. 2001. Arbuscular mycorrhizal fungi: more diverse than meets the eye, and the ecological tale of why. BioScience 51: 923-931.

Ben Khaled, L., Gomez, A.M., Ouarraqi, E.M. and Oihabi, A. 2003. Physiological and biochemical responses to salt tolerance of mycorrhizal and/or nodulated clover seedlings of *Trifolium alexandrium*. Agronomie 23: 571-580.

Bianciotto, V., Bandi, C., Minerdi, D., Sironi, M., Vokertichi, H. and Bonfonte, P. 1996. An obligately endosymbiotic mycorrhizal fungus itself harbors obligately intracellular bacteria. Applied and Environmental Microbiology 62: 3005-3010.

Bidartando, M.I. and Bruns, T.D. 2001. Extreme specificity in epiparasitic Monotropidaceae (Ericaceae): Widespread phylogenetic and geographic structure. Molecular Ecology 10: 2285-2295.

Bidartonodo, M., Kretzer, A.M. and Bruns, T.D. 2000. High root concentration and uneven ectomycorrhizal diversity near *Sarcodes sanguinea* (Ericaceae): A cheater that stimulates its victims. American Journal of Botany 87: 1783-1788.

Bonello, P., Bruns, T.D. and Gardes, M. 1998. Genetic structure of a natural population of ectomycorrhizal fungus *Suillus pungens*. New Phytologist 138: 533-542.

Borie, F.R., Rubio, R., Morales, A. and Castillo, C. 2000. Relationships between arbuscular mycorrhizal hyphal density and glomalin production with physical and chemical characteristics of soils under no-tillage. Revista Chilena De Historia Natural 73: 749-756.

Bothe, H., Hildebrandt, U., Wilde, P., Ouziad, F. and Janetta K. 2003. The EU project MYCORREM and arbuscular mycorrhiza in saline habitats. COST scientific meetings, Cologne, Germany, Abstract, p 5.

Britto, I., Antunes, P. and Carvalho, M. 2001. Effect of conventional tillage versus no-tillage on indigenous arbuscular mycorrhizal fungi of triticale and wheat under field conditions. Third International Conference on Mycorrhizas, Adelaide, Australia http://www.mycorrhizas.ag.utk.edu/icoms/icom3.html

Bundrett, M. 2000. Understanding the diversity of Glomalean fungi in Tropical Australian habitats. http://www.munchkinsoftware.com/mycology/2000-abstracts.html

Bruns, T.D. and Gardes, M. 1993. Molecular tools for the identification of ectomycorrhizal fungi-taxon-specific oligonucleotide probes for Suilloid fungi. Molecular Ecology 2: 233-242.

Burns, T.D., Bidartondo, M.I. and Taylor, D.L. 2002a. Host specificity in Ectomycorrhizal communities: what do the exceptions tell us? Integrative and Comparative Biology 42: 352-359.

Bruns, T.D., Lilleskov, E.A., Bidartondo, M.I. and Horton, T.R. 2001. Patterns in ectomycorrhizal community structure in pinaceous ecosystems. Phytopathology 91: S 163-164.

Bruns, T.D. and Read, D.J. 2000. In vitro germination of non-photosynthetic, myco-heterotrophic plants stimulated by the fungus isolated from adult plants. New Phytologist 148: 335-342.

Bruns, T., Tan, J., Bidartondo, M., Szaro, T. and Redecker, D. 2002b. Survival of *Suillus pungens* and *Amanita francheti* ectomycorrhizal genets was rare or absent after a stand-replacing fire. New Phytologist 155: 517-523.

Buscot, F. 1994. Ectomycorrhizal types and endobacteria associated with ectomycorrhizas of *Morchella elata* with *Picea abies*. Mycorrhiza 4: 223-232.

Buscot, F., Minch, J.C., Charcosset, J.Y., Gardes, M., Nehls, U. and Hampp, R. 2000. Recent advances in exploring physiology and biodiversity of ectomycorrhizas highlight the functioning of these symbiosis in ecosystems. FEMS Microbiology Reviews 24: 601-614.

Cairney, J.W.G. and Ashford, A.E. 2002. Biology of Mycorrhizal Association of Epacrids (Ericaceae). New Phytologist 154: 305-326.

Caravaca, F., Barea, J. M. and Roldan, A. 2002. Synergistic influence of an arbuscular mycorrhizal fungus and organic amendment on *Pistacea lentiscus* seedlings afforested in a degraded semi-arid soil. Soil Biology and Biochemistry 34: 1139-1145.

Carvalho, L.M., Correia, P.M., Ryel, R.J. and Martins-Loucao, M.A. 2003. Spatial variability of arbuscular mycorrhizal fungal spores in two natural plant communities. Plant and Soil 251: 227-236.

Castelli, J.P. and Casper, B. B. 2003. Intraspecific AM fungal variation contributes to Plant-Fungal feedback in a serpentine grassland. Ecology 84: 323-336.

Citernesi, A.S., Fortuna, P., Filippi, G., Bagnoli, G. and Giovannetti, M. 1996. The occurrence of antagonistic bacteria *Glomus mosseae* pot cultures. Agronomie 16: 671-677.

Clapp, J.P., Fitter, A.H. and Young, J.P.W. 1999. Ribosomal small subunit sequence variation within spores of an arbuscular mycorrhizal fungus, *Scutellospora* sp. Molecular Ecology 8: 915-921.

Cornelissen, J.H.C., Aerts, R., Cerebolini, B., Werger, M.J.A and Van Der Heijden, M.G.A. 2001. Carbon cycling traits of plant species are linked with mycorrhizal strategy. Oecologia 129: 611-619.

Chambers, S.M., Liu, G. and Cairney, J.W.G. 2000. ITS rDNA sequence comparison if ericoid mycorrhizal endphytes from *Woollsia pungens*. Mycological Research 104: 168-174.

Cullings, K.W. 1992. Design and testing of a plant specific PCR primer for ecological and evolutionary studies. Molecular Ecology 1: 233-240.

Dahlberg, D.S., Jonsson, L. and Nylund, J.E. 1997. Species diversity and distribution of biomass above and below ground among ectomycorrhizal fungi in an old-growth Norway Spruce Forest in South Sweden. Canadian Journal of Botany 75: 1327-1335.

Dalpe, Y. 2003. Mycorrhizal fungi biodiversity in Canadian soils. Canadian Journal of Soil Science 83: 321-330.

Deacon, J.W. and Flemming, L.V. 1992. Interactions of ectomycorrhizal fungi In: Mycorrhizal Functioning. Allen, M.F. (Ed.) Routledge, Chapman and Hall Publishers, New York, pp 249-300.

Dickie, I.A., Koide, R.T. and Steiner, K.C. 2002a. Influence of established trees on mycorrhizas, nutrition and growth of *Quercus rubra* seedlings. Ecological Monographs 72: 505-521.

Dickie, I.A., Xu, B. and Koide, R.T. 2002b. Vertical niche differentiation of ectomycorrhizal hyphae in soil as shown by T-RFLP analysis. New Phytologist 156: 527-535.

Diouf, D., Diop, T.A. and Ndoye, I. 2003. Actinorhizal, mycorrhizal and rhizobial symbiosis: How much do we know? African Journal of Biotechnology 2: 1-7.

Dodd, J.C. 2000. The role of Arbuscular Mycorrhizal fungi in agro and natural ecosystems. Outlook on Agriculture 29: 55-62.

Donelly, D.P., Boddy, L. and Jonathan, L.R. 2004. Ectomycorrhizal mycelial systems in soil microcosms. Mycorrhiza 14: 37-45.

Douderick, R.L. and Anderson, N.A. 1989. Incompatibility factors and mating competence of *Laccaria* species associated with Black Spruce in Northern Minnesota. Phytopathology 79: 694-700.

Douhan, G.W. and Rizzo, D.M. 2003. Amplified fragment length micro-satellites (AFLM) might be used to develop micro-satellite markers in organisms with limited amounts of DNA applied to Arbuscular Mycorrhizal (AM) fungi. Mycologia 95: 368-373.

Eaton, G.K. and Ayers, M.P. 2002. Plasticity and constraint in growth and protein mineralization of ectomycorrhizal fungi under simulated N deposition. Mycologia 94: 921-932.

Eberhart, W., Walter, L. and Kottke, I. 1999. Molecular and morphological discrimination between *Tylospora fibrillosa* and *Tylospora asteophora* mycorrhizae. Canadian Journal of Botany 77: 11-21.

Eggerton-Warburton, L.M. and Allen, E.B. 2000. Shifts in arbuscular mycorrhizal communities along an anthropogenic nitrogen deposition gradient. Ecological Application 10: 484-496.

Eggerton-Warburton, L.M., Graham, R.C. and Hubbert K.R. 2003. Spatial variability in mycorrhizal hyphae and nutrient and water availability in a soil-weathered bedrock profile. Plant and Soil 249: 331-342.

El-Karkouri, K., Martin, F. and Mousin, D. 2002. Dominance of the mycorrhizal fungus *Rhizopogon rubescens* in a plantation of *Pinus pinea* seedlings inoculated with *Suillus collinutus*. Annals of Forest Science 59: 197-204.

Eriksen, M., Bjureke, K.E. and Dhillon, S.S. 2002. Mycorrhizal plants of traditionally managed boreal grasslands in Norway. Mycorrhiza 12: 117-123.

Feng, G., Zhang, F., Li, C., Tian, C. and Rengel, Z. 2002. Improved tolerance of maize plants to salt stress by arbuscular mycorrhiza is related to higher accumulation of soluble sugars. Mycorrhiza 12: 185-190.

Filippi, C., Bagnoli, G. and Giovannetti, M. 1995. Bacteria associated with Arbutoid mycorrhiza in *Arbutus umedo*. Symbiosis 13: 57-68.

Filippi, C., Bagnoli, G., Citernesi, A.S. and Giovannetti, M. 1998. Ultrastructural spatial distribution of bacteria associated with sporocarps of *Glomus mosseae*. Symbiosis 24: 1-12.

Fracchia, S., Godeas, A., Scervino, J.M., Sampedro, I., Ocampo, J.A. and Garcia-Romera, L. 2003. Interaction between the soil yeast *Rhodotorula mucilaginosa* and the arbuscular mycorrhizal fungi *Glomus mosseae* and *Gigaspora rosea*. Soil Biology and Biochemistry 35: 701-707.

Fransson, P.M.A., Taylor, A.F.S. and Finlay, R.D. 2001. Elevated atmospheric CO_2 alters root symbiont community structure in forest trees. New Phytologist 152: 431-442.

Galvez, I., Douds, D. D., Drinkwater, L.E. and Wagoner, P. 2001. Effect of tillage and farming system upon VAM fungus populations and mycorrhizas and nutrient uptake of maize. Plant and Soil 228: 299-308.

Gange, A.C., Brown, V.K. and Farmer, I.M. 1990. A test of mycorrhizal benefit in an early successional plant community. New Phytologist 115: 85-91.

Garbaye, J., Duponnois, R. and Wahl, J.L. 1996. The bacteria associated with *Laccaria laccata* ectomycorrhizas and sporocarps: The effect on symbiosis establishment on Douglas fir. Symbiosis 9: 267-273.

Gardes, M. and Bruns, J.D. 1993. ITS primers with enhanced specificity for basidomycetes-application to the identification of mycorrhiza and rusts. Molecular Ecology 2: 1113-118.

Gardes, M. and Bruns, T.D. 1996. ITS-RFLP community structure of ectomycorrhizal fungi in *Pinus muricata* forest: above and below ground view. Canadian Journal of Botany 74: 1572- 1583.

Gardes, M., White, M.J., Fortin, J.A., Bruns, T.D. and Taylor, J.W. 1991. Identification of indigenous and introduced symbiotic fungal ectomycorrhizae by amplification of nuclear and mitochondrial ribosomal DNA. Canadian Journal of Botany 69: 180-190.

Gherbi, H., Delareuella, C., Sellosa, M.A. and Martin, F. 1999. High genetic diversity in a population of the ectomycorrhizal basidiomycete *Laccaria amethystina* in a 150-year-old beech forest. Molecular Ecology 6: 353-364.

Giovannetti, M., Sbrana, C., Strani, P., Agnolucci, M., Rinaudo, V. and Avio, L. 2003. Genetic diversity of isolates of *Glomus mosseae* from different geographic areas detected by vegetative compatibility testing and biochemical and molecular analysis. Applied and Environmental Microbiology 69: 616-624.

Gomes, E.A., Kasuya, M.C.M., DeBarrows, E.G., Borges, A.C. and Arajo, E.F. 2002. Polymorphism in the internal transcribed space (ITS) of the ribosomal DNA of 26 isolates of ectomycorrhizal fungi. Genetic and Molecular Biology 25: 477-483.

Grime, J.P., Mackay, J.M., Hillier, S.M. and Read, D.J. 1987. Floristic diversity in a model system using experimental microcosms. Nature 328: 420-422.

Gryta, H., Debaud, J.C. and Marmiesse, R. 2000. Population dynamics of the symbiotic mushroom *Hebeloma cylindrosporum*: mycelial persistence and in breeding. Heredity 84: 294-302.

Guidot, A., Debaud, J.C., Effose, A., Marmeisse, R. 2004. Below ground distribution and persistence of an ectomycorrhizal fungus. New Phytologist 161: 539-547.

Guidot, A., Debaud, J.C. and Marmeisse, R. 2002a. Spatial distribution of the below ground mycelia of an ectomycorrhizal fungus inferred from specific quantification of its DNA in soil samples. FEMS Microbiology Ecology 42: 477-486.

Guidot, A., Debaud, J.C. and Marmiese, R. 2001. Correspondence between genet diversity and spatial distribution of above- and below- ground populations of the ectomycorrhizal fungus *Hebeloma cylindrosporum*. Molecular Ecology 10: 1121-1125.

Guidot, A., Gryta, H., Gourbierre, R., Debaud, J.C. and Marmiesse, R. 2002b. Forest habitat characteristics affect balance between sexual reproduction and clonal propagation of the ectomycorrhizal mushroom *Hebeloma cylindrosporum*. Oikos 99: 25-36.

Hart, M.M., Reader, R.J. and Klironomos, J.N. 2003. Plant co-existence mediated by Arbuscular Mycorrhizal fungi. Trends in Ecology and Evolution 18: 418-423.

He, X.H., Critchley, C. and Bledsoe, C. 2003. Nitrogen transfer within and between plants through common mycorrhizal networks (CMNs). Critical Reviews in Plant Sciences 22: 531-567.

He, X.L., Mourtov, S. and Steinberger, Y. 2002. Temporal and spatial dynamics of vesicular-arbuscular mycorrhizal fungi under canopy of *Zygophyllum dumosum* in Negev desert. Journal of Arid Environments 52: 379-387.

Hibbet, D.S., Pine, E.M. and Langer, E. 1997. Evolution of grilled mushrooms and puffballs inferred from ribosomal DNA sequence. Proceedings of the National Academy of Sciences of the USA 94: 12002-12006.

Hitchcock, C.J., Chambers, S.M., Anderson, I.C. and Cairney, J.W.G. 2003. Development of markers for simple sequence repeat regions that discriminate between *Pisolithus albus* and *P. microcarpus*. Mycological Research 107: 699-706.

Horton, T.R. and Bruns, T.D. 2001. The molecular revolution in ectomycorrhizal ecology: peeking into the black box. Molecular Ecology 10: 1855-1871.

Husband, R., Herre, E.A., Turner, S.L., Gallery, R. and Young, J.P.W. 2002a. Molecular diversity of arbuscular mycorrhizal fungi and patterns of host association over and space in a tropical forest. Molecular Ecology 11: 2669-2678.

Husband, R., Herre, E.A. and Young, J.P.W. 2002b. Temporal variation in the arbuscular mycorrhizal communities colonizing seedlings in a tropical forest. FEMS Microbiology Ecology 42: 131-136.

Jacquot, E., VanTuinen, D., Gianinazzi, S. and Gianinazzi-Pearson, V. 2000. Monitoring species of Arbuscular Mycorrhizal fungi *in planta* and in soil by nested PCR: Application to the study of the impact of sewage sludge. Plant and Soil, pp. 170-188.

Jansa, J., Mozafar, A., Anken, T., Ruh, R., Sanders, I. R. and Frossard, E. 2002. Diversity and structure of AMF communities as affected by tillage in a temperate soil. Mycorrhiza 12: 225-234.

Jansa, J., Mozafar, A., Kuhn, G., Anken, T., Ruh, R., Sanders, I.R. and Frossard, E. 2003. Soil tillage affects the community structure of mycorrhizal fungi in maize roots. Ecological Applications 13: 1164-1176.

Jansa, J., Mozafar, A., Ruh, R., Anken, T., Kuhn, G., Sanders, I. and Frossard, E. 2001. Changes in community structure of AM fungi due to reduced tillage. Third International Conference on Mycorrhizas, Adelaide, Australia, http//:www.mycorrhizas.utk.ag.edu/icoms/icom3.html

Jany, J.L., Bousquet, J. and Khasa, D.P. 2003. Microsatellite markers for *Hebeloma* species developed from expressed sequences tags in the ectomycorrhizal fungus *Hebeloma cylindrosporum*. Molecular Ecology Notes 3: 659-662.

Johnson, D., Leake, J.R. and Read, D.J. 2002. Transfer of recent photosynthate into mycorrhizal mycelium of an upland grassland: Short Term respiratory losses and accumulation of ^{14}C. Soil Biology and Biochemistry 34: 1521-1524.

Johnson, D., Vandenkoornhuyse, P.J., Leake, J.R., Gilbert, L., Booth, R.E., Grime, J.P., Young, J.P.W. and Read, D.J. 2004. Plant communities affect arbuscular mycorrhizal fungal diversity and community composition in grassland microcosms. New Phytologist 161: 503-515.

Johnson-Green, P., Kenkel, N.C. and Booth, T. 2001. Soil salinity and arbuscular mycorrhizal colonization of *Pucciniella nuttaliana*. Mycological Research 105: 1094-1100.

Jonson, L., Dahlberg, A., Nielson, M.C., Karen, O. and Zakrisson, O. 1999. Continuity of ectomycorrhizal fungi in self-regenerating boreal *Pinus sylvestrus* forests studied by comparing mycobiont diversity on seedlings and mature trees. New Phytologist 142: 151-162.

Jonsson, T., Kokaji, S., Finlay, R. and Erland, S. 1999. Ectomycorrhizal community structure in a limed spruce forest. Mycological Research 103: 501-508.

Kanchanaprayudh, J., Lian, C., Zhou, Z., Hogetsu, T. and Sihanout, P. 2002. Polymorphic microsatellite markers of a *Pisolithus* species from a *Eucalyptus* plantation. Molecular Ecology Notes 2: 263-264.

Karen, O. 1999. Effect of air pollution and forest regeneration on the community structure of ectomycorrhizal fungi. PhD. Thesis, Swedish University of Agriculture, SLU, Uppsala, Sweden, pp. 1-43.

Keizer, P.J. and Arnolds, E. 1994. Succession of ectomycorrhizal fungi in roadside verges planted with common Oak (*Quercus robus*) in Dreuthe, The Netherlands. Mycorrhiza 4: 147-159.

Kennedy, P.G., Izzo, A.D. and Bruns, T.D. 2003. There is high potential for the formation of common mycorrhizal networks between understory and canopy trees in a mixed evergreen forest. Journal of Ecology 91: 1071-1080.

Kernaghan, G., Widden, P., Bergeron, Y., Legare, S. and Pare, D. 2003. Biotic and abiotic factors affecting ectomycorrhizal diversity in boreal mixed-woods. Oikos 102: 497-504.

Klironomos, J.N., McCune, J., Hart, M. and Neville, J. 2000. The influence of arbuscular mycorrhizae on the relationship between plant diversity and productivity. Ecology Letters 3: 137-141.

Klironomas, J.W., Montoglis, P., Kendrick, B. and Wilden, P. 1993. A comparison of spatial heterogeneity of vesicular arbuscular fungi in two maple forest soils. Canadian Journal of Botany 11: 1472- 1480.

Knorr, M. A., Boerner, R.E.J. and Rillig, M.C. 2003. Glomalin content of forest soils in relation to fire frequency and landscape position. Mycorrhiza 13: 205-210.

Koide, R.T. and Dickie, I.A. 2002. Effects of mycorrhizal fungi on plant populations. Plant and Soil 244: 307-317.

Korb. J.E. Johnson, N.C., and Covington, W.W. 2003. Arbuscular mycorrhizal propagule densities respond rapidly to ponderosa pine restoration treatments. Journal of Applied Ecology 40: 101-110.

Kretzer, A.M., Molina R. and Spatofora, J.W. 2000. Micro-satellite markers for the ectomycorrhizal basidiomycete *Rhizopogon vincolor*. Molecular Ecology 9: 1171-1193.

Kretzer, A.M., Dunham, S., Molina, R. and Spatofora, J.W. 2004. Microsatellite markers to read the below ground distribution of genets in two species of *Rhizopogon* forming tuberculate ectomycorrhizas on Douglas fir. New Phytologist 161: 313-320.

Krishna, K.R. Balakrishna, A.N. and Bagyaraj, D.J. 1983. Interactions between AM fungus and *Streptomyces cinnamomeous* and their influence growth and nutrient uptake of *Eluesine coracana*. New Phytologist 92: 401-404.

Kronzucker, H.J., Siddiqui, M.Y., Glass, A.D.M. and Britto, D.T. 2003. Root ammonium transport efficiency as a determinant in forest colonization patterns: a hypothesis. Physiologia Plantarum 117: 164-170.

Kuhn, G., Koch, A., Hire, M. and Sanders, I.R. 2003. Implications and consequences of multiple genomes in AM fungi for studies of functional genes and ecological role. WWW.dijon.inra.fr/cost-838/scientific-metings/Cologne. Abstract. html

Landry, C.P., Simard, R.R., Hamel, C. and Vanasse, A. 2001. Agronomic significance of arbuscular mycorrhizal fungi activity in soil P dynamics in ridged-tillage corn production. Third International Conference on Mycorrhizas, Adelaide, Australia, http://www.mycorrhizas.ag.utk.edu/icoms/icom.html

Landeweert, R., Leeflang, P., Kuyper, T.W., Hoffland, E., Rosling, A., Wernars, K. and Smith, E. 2003. Molecular identification of ectomycorrhizal mycelium in soil horizons. Applied and Environmental Microbiology 69: 327-333.

Landwehr, M. Hildbrandt, U., Wilde, P., Nawrath, K., Toth, T., Biro, B. and Bothe, H. 2002. The arbuscular mycorrhizal fungus *Glomus geosporum* in European saline sodic and gypsum soils. Mycorrhiza 12: 199-211.

Langley, J.A. Drake, B.G. and Hungate, B.A. 2002. Extensive belowground carbon storage supports roots and mycorrhizae in regenerating scrub oaks. Oecologia 131: 542-548.

Lata, Saxena, A.K. and Tilak, KV. B.R. 2002. Biofertilizers to augment Soil fertility and Crop Production. In: Soil Fertility and Crop Production. Krishna. K.R. (Ed.) Science Publishers Inc. Enfield, New Hampshire, USA, pp 279-312.

Leake, J.R., Donelly, D.P., Sauders, E.M., Boddy, L., and Read, D.J. 2001. Rates and quantities of carbon flux to ectomycorrhizal mycelium following ^{14}C pulse labeling of *Pinus sylvestris* seedlings: effects of litter patches and interaction with a wood-decomposer fungus. Tree Physiology 21: 71-82.

Linderman, R.G. 1988. Mycorrhizal interactions with the rhizosphere microflora: the mycorhizosphere effect. Phytopathology 78: 366-371.

Lilleskov, E.A. and Bruns, T.D. 2001. Nitrogen and Ectomycorrhizal fungal communities: What we know, what we need to know. New Phytologist 149: 156-158.

Lilleskov, E.A and Fahey, T.J. 2003. Patterns of ectomycorrhizal diversity over an atmospheric nitrogen deposition gradient near Kenai, Alaska, http://www.mycorrhiza.ag.utk.edu

Liu, G., Chambers, S.M. and Cairney, J.G.W. 1998. Molecular diversity of ericoid mycorrhizal endophytes isolated from *Woollsia pungens*. New Phytologist 140: 145-153.

Lovelock, C.E., Andersen, K. and Morton, J.H. 2003a. Arbuscular mycorrhizal communities in tropical forests are affected by the host tree species and environment. Oecologia 135: 268-279.

Lovelock, C.E., Wright, S.F., Clark, D.E. and Ruess, R.W. 2003b. Soil stocks of Glomalin produced by arbuscular mycorrhizal fungi across tropical rain forest landscape. Smithsonian Environmental Research Center, Edgewater, Maryland, USA, pp 1-31.

Luis, P., Delaruelle, C., Dupre, G., Chevalier, G. and Martin, F. 2001. Genetic diversity in *Tuber uncinatum* in plantations and woodlands of Northeastern France. Third International Conference on Mycorrhizas, Adelaide, Australia, http://www.mycorrhiza.ag.utk.edu/icoms/icom3.html

Mansfeld-Giese, K., Larsen, J. and Bodker, L. 2002. Bacterial populations associated with mycelium of the arbuscular mycorrhizal fungus *Glomus intraradices*. FEMS Microbiology Ecology 41: 133-140.

Markkola, A.M., Ahonen-Jonnarth, U., Roitto, M., Strommer, R. and Hyvarinen, M. 2002. Shift in ectomycorrhizal community composition in Scots pine (*Pinus sylvestris*) seedling roots as a response to nickel deposition and removal of lichen cover. Environmental Pollution 120: 797-803.

Marler, M.J., Zabinski, C.A. and Callaway, R.M. 1999. Mycorrhizae indirectly enhance competition effects of an invasive fork on a native bunchgrass. Ecology 80: 1180-1186.

McKendrick, S.L., leake, J.R. and Read, D.J. 2000. Symbiotic germination and development of myco-hetrotrophic plants in nature: Transfer of carbon from ectomycorrhizal *Salix repens* and *Betula pendula* to the orchid *Corallorhiza trifida* through shared hyphal connection. New Phytologist 145: 523-537.

Midgley, D.J., Chambers, S.M. and Cairney, J.W.G. 2001. Diversity and distribution of fungal endophytes in different genotypes in the root system of *Woollsia pungens* (Ericaceae). Third International Conference on Mycorrhizas, Adelaide, Australia, http://www.mycorrhiza.ag.utk.edu

Miller, S., Terry, M. and Guthrie, L. 2003. Differential utilization of cellulose and protein reveals competitive advantage of ectomycorrhizal fungi over saprotrophs for complex organic substrates. http://www.mycorrhiza.ag.utk.edu

Mohammad, M.J., Hamad, S.B. and Malkawi, H.I. 2003. Population of arbuscular mycorrhizal fungi in semi-arid environment of Jordon as influenced by biotic and abiotic factors. Journal of Arid Environments 53: 409-417.

Moise, B., Louisanna, E., Barigah, T. and Garbaye, J. 1998. Endomycorrhizas in the tropical rainforest: how do they influence the regeneration patterns of two tree species in French Guiana. Second International Conference on Mycorrhizas, Upssala, Sweden, http://www.mycorrhizas. ag.utk.edu/latest.html

Molina, R., Smith, E.J., Mckay, D. and Meluille, L.H. 1992. Biology of the Ectomycorrhizal fungi. New Phytologist 137: 519-528.

Moncalvo, J.M., Lutzoni, F.M. and Rehner, S.A. 2000. Phylogenetic relationships of agaric fungi based on nuclear large subunit ribosomal DNA sequences. Systematic Biology 49: 278-305.

Mozafar, A., Anken, T., Ruh, R. and Frossard, E. 2000. Tillage intensity, mycorrhizal and non-mycorrhizal fungi and nutrient concentrations in maize, wheat and canola. Agronomy Journal 92: 1117-1124.

Mueller, G.M. and Gardes, M. 1991. Intra- and interspecific relations within *Laccaria bicolor* sensu lato. Mycological Research 95: 592-601.

Nara, K., Nakaya, H. and Hogetsu, T. 2003. Ectomycorrhizal sporocarp succession and production during early primary succession on Mount Fuji. New Phytologist 158: 193-199.

Nordin, A., Hogberg, P. Nasholm, T. 2001. Soil nitrogen form and plant nitrogen uptake along a boreal forest productivity gradient. Oecologia 129: 125-132.

Norman, J.E. and Egger, K.N. 1999. Molecular phylogenetic analysis of Peziza and related genera. Mycologia 91: 820-829.

O'Connor, P.J., Smith, S.E. and Smith, F.A. 2002. Arbuscular mycorrhizas influence plant diversity and community structure in a semi-arid herbland. New Phytologist 154: 209-218.

Onguene, N. and Kuyper, T. 2002. Importance of the ectomycorrhizal network for seedling survival and ectomycorrhiza formation in rain forests of south Cameroon. Mycorrhiza 12: 13-17.

Pande, M. and Tarafdar, J.C. 2002. Effect of phosphorus, salinity and moisture on AM fungal association in neem. Symbiosis 32: 195-209.

Paulitz, T.C. and Linderman, R.G. 1989. Interactions between fluorescent Pseudomonas and VA mycorrhizal fungi. New Phytologist 113: 37-45.

Persson, J. and Nasholm, T. 2001. Amino acid uptake: a widespread ability among boreal forest plants. Ecology Letters 4: 434-438.

Peter, M., Ayer, F. and Egli, S. 2001. Nitrogen addition in a Norway spruce stand altered macromycete sporocarp production and below-ground ectomycorrhizal species composition. New Phtyologist 149: 311-325.

Pine, E., Hibbet, D.S. and Donoghue, M.J. 1999. Phylogenetic relationships of Cantherellaceae and claveroid homobasidiomycetes based on mitochondrial and nuclear rDNA sequences. Mycologia 91: 944-963.

Powlowska, T.E. and Taylor, J.W. 2004. Organization of genetic variation in individuals of arbuscular mycorrhizal fungi. Nature 427: 733-737.

Pringle, A. 2001. Ecology and genetics of arbuscular mycorrhizal fungi. PhD Dissertation. Duke University, Durham, North Carolina, USA, pp 132.

Pringle, A., Moncalvo, J.M. and Vilgalys, R. 2000. High levels of variation in ribosomal DNA sequences within and among spores of a natural population of the arbuscular mycorrhizal fungus *Acaulospora colossica*. Mycologia 92: 259-268.

Pritsch, K., Munch, J.C. and Buscot, F. 1997. Morphological and anatomical characterization of black alder mycorrhizas by PCR/RFLP analysis of rDNA internal transcribed spaces (ITS). New Phytologist 137: 357-359.

Provorov, N.A., Borisov, A.Y. and Tikanovich, I.A. 2002. Developmental genetics and evolution of symbiotic structures in nitrogen-fixing nodules and Arbuscular Mycorrhiza. Journal of Theoretical Biology 214: 215-232.

Redecker, D., Hijri, M., Duleiu, H. and Sanders, I.R. 1999. Phylogenetic analysis of a dataset of fungal 5.8S rDNA sequences shows that highly divergent copies of internal transcribed spacers reported from *Scutellospora castanae* are of Ascomycete origin. Fungal Genetics and Biology 28: 238-244.

Redecker, D., Szaro, T.M., Bowman, J.R. and Bruns, T.D. 2001. Small genets of *Lactarius xanthogalactus, Russula cremoricolor*, and *Amanita francheti* in late stage mycorrhizal successions. Molecular Ecology 10: 1025-1034.

Read, D.J. 1996. The structure and function of mycorrhizal roots. Annals of Botany 77: 365-374.

Read, D.J. and Kearley, S. 1995. The status and function of ericoid mycorrhizal systems. In: Mycorrhiza: Structure, Function, Molecular Biology and Biotechnology. Verma, A. and Hock, B. (Eds) Springer Verlag, Berlin, pp 499-520.

Reynolds, H.I., Packer, A., Bever, J.D. and Clay, K. 2003. Grassroots ecology: Plant-Microbe soil interactions as drivers of plant community structure and dynamics. Ecology 84: 2381-2291.

Rillig, M.C. and Field, C.B. 2003. Arbuscular Mycorrhizae respond to plants exposed to elevated atmosphere CO_2 as a function of soil depth. Plant and Soil 254: 383-391.

Rilling, M.C. and Steinberg, P.D. 2002. Glomalin production by an arbuscular mycorrhizal fungus: A mechanism of habitat modifications. Soil Biology and Biochemistry 34: 1371-1374.

Rillig, M.C., Maestre, F.T. and Lamit, L.J. 2003. Microsite differences in fungal hyphal length, glomalin and aggregate stability in semiarid Mediterranean steppes. Soil Biology and Biochemistry 35: 1257-1260.

Rogers, J.B., Laidlaw, A.S. and Christie, P. 2000. The role of arbuscular mycorrhizal fungi in the transfer of nutrients between white clover and perennial rye grass. Chemosphere 42: 153-159.

Roldan-Fajardo, B.E. 1994. Effect of indigenous Arbuscular Mycorrhizal endophytes on the development of six wild plants colonizing semi-arid area in the Southeast Spain. New Phytologist 127: 115-122.

Rutto, K.L., Mizutani, F., Moon, D.G. and Kadoya, K. 2002. The relationship between cultural practices and Arbuscular Mycorrhizal activity in orchards under different management systems. Journal of the Japanese Society of Horticultural Science 71: 601-609.

Sanders, I.R. 2002. Ecology and evolution of multigenomic Arbuscular Mycorrhizal fungi. American Naturalist 160: S128-S141.

Sanders, I.R., Clapp, J.P. and Weimken, A. 1996. The genetic diversity of arbuscular mycorrhizal fungi in natural ecosystem—a key to understanding the ecology and functioning of the mycorrhizal symbiosis. New Phytologist 133: 123-134.

Sanders, I.R. and Koide, R.T. 1994. Nutrient acquisition and community structure in co-occurring mycotrophic and non-mycotrophic old-field annuals. Functional Ecology 8: 77-84.

Sanginga, N., Carsky, R.J. and Dashell, K. 1999. Arbuscular mycorrhizal fungi respond to rhizobial inoculation and cropping systems in farmer's fields in the Guinea Savanna. Biology and Fertility of Soils 30: 179-186.

Sanginga, N., Lyasse, O. and Singh, B.B. 2000. Phosphorus use efficiency and nitrogen balance of cowpea breeding lines in a low P soil of the derived savanna zone in West Africa. Plant and Soil 220: 119-128.

Sawyer, N.A., Chambers, S.M. and Cairney, J.W.G. 1999. Molecular investigation of genet distribution and genetic variation of *Cortanarius rotundisporus* in eastern Australian sclerophyl forests. New Phytologist 142: 561-568.

Sawyer, N.A., chambers, S.M. and Carney, J.W.G. 2003. Distribution of *Amanita* spp. genotypes under eastern Australian sclerophyll vegetation. Mycological Research 107: 1157-1162.

Schreiner, R.P., Mihara, K.L., McDaniel, H. and Bethlenfalvay, G.J. 1997. Mycorrhizal fungi influence plant and soil functions and interactions. Plant and Soil 188: 199-209.

Selosse, M.A. Jacquot, D., Bouchard, D., Martin, F. and LeTacon, F. 1998a. Temporal persistence and spatial distribution of an American inoculant strain of the ectomycorrhizal basidiomycete *Laccaria bicolor* in a French forest plantation. Molecular Ecology 7: 561-573.

Selosse, M.A., Martin, F., Bouchard, D. and LeTacon, F. 1999. Structure and dynamics of experimentally introduced and naturally occurring *Laccaria* sp. Discrete genotypes in a Douglas fir plantation. Applied and Environmental Microbiology 65: 2006-2014.

Selosse, M.A., Martin, F. and LeTacon, F. 1998b. Survival of an introduced ectomycorrhizal *Laccaria bicolor* strain in a European forest plantation monitored by mitochondrial ribosomal DNA analysis. New Phytologist 140: 753-761.

Simon, L., Leversque, R.C. and Lalonde, M. 1992. Rapid quantification by PCR of endomycorrhizal fungi colonizing roots. PCR Methods and Applications 2: 76-80.

Simon, L., Lalonde, M. and Bruns, T.D. 1993. Specific amplification of 18s fungal ribosomal genes from VA endomycorrhizal fungi colonizing roots. Applied and Environmental Microbiology 58: 291-295.

Smith, S.E. and Read, D.J. 1997. Mycorrhizal Symbiosis. Academic Press, San Diego, USA, pp 605.

Sood, S.G. 2003. Chemotactic response of plant growth promoting bacteria towards roots of vesicular arbuscular mycorrhizal tomato plants. FEMS Microbiology Ecology 45: 219-227.

Standell, E.R., Horton, T.R. and Bruns, T.D. 1999. Early effects of prescribed fire on the structure of the ectomycorrhizal fungus community in a Sierra Nevada Panderoza Pine forest. Mycorrhizal Research 103: 1353-1359.

Streitwolf-Engel, R., Boller, T., Weimken, A. and Sanders I.R. 1997. Clonal growth traits of two *Prunella*species are determined by co-occurring arbuscular mycorrhizal fungi from calcareous grassland. Journal of Ecology 85: 181-191.

Tadych, M. and Blaszkowski, J. 2000. Succession of Arbuscular Mycorrhizal fungi in a deflation hollow of the Slowinski National Park. Acta Societatis Botanicorum Poloniae 69:223-236.

Taylor, J. and Bruns, T. 1999. Community structure of ectomycorrhizal fungi in a *Pinus muricata* forest: minimal overlap between the mature forest and resistant propagule communities. Molecular Ecology 8: 1837-1850.

Taylor, D.L. and Bruns, T.D. 2004. Evidences for mycorrhizal races in a cheating orchid. Scientia Horticultura 271: 35-43.

Thomas, R.S. Franson, R.L. and Bethlenfalvay, G.J. 1993. Separation of vesicular arbuscular mycorrhizal fungus and root effects on soil aggregation. Soil Science Society of America Journal 57: 77-81.

Treseder, K.K. 2000. Effect of anthropogenic nitrogen of methyl halide production by ectomycorrhizal fungi. University of Pennsylvania, Unpublished project report, pp 1-8.

Van Der Heijden, M.G.A., Boller, T., Weimken, A. and Sanders, I.R. 1998a. Different Arbuscular Mycorrhizal fungal species are potential determinants of plant community structure. Ecology 79: 2082-2092.

Van Der Heijden, M.G.A., Klironomos, J.N., Ursic, M., Moutogolis, P., Streitwolf-Engel, R., Boller, J., Weimken, A. and Sanders, I.R. 1998b. Mycorrhizal fungal diversity determines plant diversity, ecosystem variabilities and productivity. Nature 396: 69-72.

Van Der Heijden, M.G.A., Weimken, A. and Sanders, I.R. 2003. Different arbuscular mycorrhizal fungi alter co-existence and resource distribution between co-occurring plants. New Phytologist 157: 569-578.

Van Tuinen, D., Jacquot, E., Zhao, B., Gollotte, A. and Gianinazzi-Pearson, V. 1998. Characterization of root colonization profiles by a microcosm community of arbuscular fungi using 25S rDNA-targeted nested PCR. Molecular Ecology 7: 879-887.

Varese, G.C., Portinara, S., Trotta, A., Scannerini, S., Luppi-Morca, A.M. and Martinotti, M.G. 1996. Bacteria associated with *Suillus grevillei* sporocarps and ectomycorrhizae and their effects on in vitro growth of the mycosymbiont. Symbiosis 21: 129-147.

Vivas, A., Azcon, R., Biro, B., Barea, J.M. and Ruiz-Lazano, J.M. 2003. Influence of bacterial strains isolated from lead-polluted soil and their interactions with arbuscular mycorrhizae on the growth of *Trifolium pratens* under lead toxicity. Canadian Journal of Microbiology 49: 577-588.

Vosatka, M., Batkhuugyin, E. and Albrechtova, J. 1999. Response of three arbuscular mycorrhizal fungi to simulated acid rain and aluminum stress. Biologia Plantarum 42: 289-296.

Vralstaad, T. 2001. Molecular ecology of root-associated mycorrhizal and non-mycorrhizal Ascomycetes. PhD. Thesis. University of Oslo, Oslo, Norway, pp 183.

Werner, A., Zadworny, M. and Idzikowksi, K. 2002. Interaction between *Laccaria laccata* and *Trichoderma virens* in co-culture and in the rhizosphere of *Pinus sylvestris* grown in vitro. Mycorrhiza 12: 139-145.

Wiemken, V. and Boller, T. 2002. Ectomycorrhiza: Gene expression, metabolism and the worldwide web. Current Opinion in Plant Biology 5: 3555-361.

Wolf, J., Johnson, N.C., Rowland, D.L. and Reich, P.B. 2003. Elevated CO_2 and plant species richness impact arbuscular mycorrhizal fungal spore communities. New Phytologist 157: 579-588.

Yocum, D.H., Larsen, H.H. and Boosalis, M.G. 1985. The effects of tillage treatments and a fallow season on VA mycorrhiza of winter wheat. In: Proceedings of 6th North American Conference on Mycorrhiza. Randy, M. (Ed.) Oregon State University, Corvallis, Oregon, USA. pp 297.

Yokoyama, K., Tateishi, T., Marumoto, T. and Saito, M. 2002. A molecular marker diagnostic of a specific isolate of an arbuscular mycorrhizal fungus *Gigaspora margarita*. FEMS Microbiology Letter 212: 171-175.

7

TRANSFORMATION, TRANSGENICS AND GENETIC ENGINEERING OF MYCORRHIZAS

1. A. Genetic Transformation of Mycorrhizal Fungi: Introduction

Many filamentous fungi are endowed with natural ability to generate genetic variation via chromosomal crossing over and other sexual cycle events. In the laboratory, a large number of filamentous have been transformed using molecular genetic methods. The majority of such transformation procedures used with filamentous fungi are dependent on cell membrane permeability. Procedures utilizing both polyethylene glycol and electroporation need formation of protoplasts or osmotically sensitive cells of the recipient in order to achieve transformation. Methods based on *Agrobacterium tumefaciens*-mediated DNA transfer are also used with filamentous fungi. Several fungal species have been transformed using biolistic approaches, especially microprojectile bombardment techniques. The list includes a large number of saprophytes and pathogens (Bailey et al. 1993; Chaure et al. 2000; Fincham, 1989; Talbot et al. 2002). Some of the ECM fungi such as *Laccaria bicolor, Paxillus involutus* (Bills et al. 1995, 1998; Martin et al. 1994) and AM fungi such as *Gigaspora rosea* (Forbes et al. 1998; Harrier and Millam, 2001) have also been transformed using biolistic approaches.

Specific markers to select and identify transformants should be carefully chosen. Commonly, resistance to hygromycin D, phleomycin, benomyl, bialophos and sulfonylurea are used as easily selectable markers on the vectors. The design and construction of reporters, vectors and efficient methods of selection/screening system for transformants are crucial. The AM or ECM fungal enhanced reporters should include efficient homologous promoters, genome location targeting mechanism and reliable reporter genes. Targeting of the reporter genes to a specific area of the genome for integration is important. Obviously, integration of reporter at a suitable genomic location will increase both efficiency of integration and expression. At this stage, we ought to realize that in AM fungi, perhaps also in many ECM fungi, functional organization of genes have not been dissected and understood properly. Some of the most widely used genetic reporter system are chloramphenicol acetyl-transferase (CAT), β-galactosidase, β-glucuronidase and bioluminescence-based markers. In the case of AM/ECM symbiosis, reporter genes should be detectible in both fungal and plant tissue. In addition, it might be difficult to trace a fluorescent-based marker, in cases where background or auto-fluorescence is bright. Luciferase genes are the commonly used bioluminescence-based markers. These reporters are preferred because of the tight coupling of protein synthesis and enzyme activity, as well as lack of introns in the gene (Wood, 1995). Dominant selectable markers such as resistance to antibiotics (e.g. Oligomycin, Hygromycin) or fungicide (Benomyl) have been used as selectable markers for transformants of AM/ECM fungi (Harrrier et al. 2002). Screening transformants under non-selective conditions may decrease transformation efficiency. According to Harrier and Millam (2001), vectors containing transposable elements may greatly increase the efficiency of integration into the genome following a biolistic transformation event. Obviously, suitable transposable elements should be identified in AM/ECM fungi that need to be transformed.

Applications of Genetic Transformation of Mycorrhizal Fungi

In case of AM fungi, development of genetic transformation procedures will be useful in studying life cycles, regulation of specific genes during asymbiotic and symbiotic phases. Application of genetic engineering techniques may lead us to AM/ECM fungal isolates with greater symbiotic efficiency. Transforming AM/ECM fungal inoculant strains with genes relevant to higher potential growth and nutritional advantages to host, increased survival and adaptability in the soil, better sporulation and spread in soil/root are a few possibilities.

Recombination events that lead to genetic variation in AM fungi are yet to be understood in detail. Since AM fungi produce clonally and lack a sexual structure, definite genetic crossing-over events have not been confirmed. Inability to culture AM fungi without host support has made it difficult to study the sexuality and recombination events. However, it is believed that development of transformation methods could be helpful. Increased opportunities to introduce recombination vectors into AM fungi will allow in situ detection of recombination events and products of unequal sister chromatid exchange. Exchange of genetic information via anastomoses or zygospore differentiation, if any, could also be detected using suitable markers (e.g. luciferase) and biolistic transformation methods (Harrier et al. 2002).

Gene function within ECM or AM fungi can be detected using specific transformants (Harrier et al. 2002). The strategy is to introduce a strong promoter with the coding region of the gene under analysis, but in the opposite direction that will lead to production of antisense RNA. When it binds to the original transcript in the cytoplasm, the newly formed double-stranded RNA is recognized and digested by cellular machinery. Through this process, expression of corresponding proteins can be interrupted and newly appearing phenotype or genetic lesion will explain the function of the gene in the normal course (Kumria et al. 1998). Vectors with reporter genes can also be tailored to study several cellular processes and transcriptional regulation, such as RNA processing, protein secretion pathway and signal transduction.

In situ detection of AM/ECM fungi via transgenics is a clear possibility. Harrier et al. (2002) suggest that the ability to insert genetic markers and tagging of AM or ECM fungi via biolistic transformation or other methods will be advantageous during ecological studies. Generally, tagged fungi will allow easier assessment of their growth, activity in soil/root, survival patterns and interaction with environment. Gene flow patterns within AM/ECM fungal community can also be studied using transformed AM/ECM fungi.

Genetic features available in other organisms but lacking in mycorrhizal fungi could be added by adopting suitable transformation procedure. For example, the ability to detoxify contaminated soil is available in several other microbes. It could be introduced into mycorrhizal fungi via genetic transformation. We can aim at transferring the genetic traits relevant to nutrient uptake, tolerance to drought, heavy metal, resistance to specific pathogens, etc. Ability to utilize recalcitrant metabolites could also be transformed from soil bacteria.

Molecular procedures to transform mycorrhizal fungi and development of transgenic AM/ECM fungi are discussed in detail in following sections of this chapter. Aspects such as the influence of transgenic plants on soil microflora have been highlighted. Possibility of transforming AM fungi with genes that impart ability to grow in vitro without host support has also been explored.

1. B. Methods for Genetic Transformation of AM and ECM Fungi

Natural Genetic Crossing

Gene transfer and recombination arising through natural crossing in ECM as well as other fungi, including their essential features of sexual reproduction, such as nuclear fusion and meiosis have been well studied and reported (Elliot, 1994; see also chapter 3). If the fusing nuclei differ with respect to several traits, then the fusion products (transformants) will be heterozygous for those relevant genes. Meiotic products, i.e recombinants should then be subjected to progeny analysis so that new combinations could be identified. In nature, inducing genetic variation and formation of new genotypes is said to be the main purpose of sexual reproduction in fungi as in other organisms. Heterothalism and homothalism are two main breeding systems among ECM fungi. Some of these aspects are not well understood for AM fungi, because we are yet to discover the sexual structures and stages. So far, only few incompatibility groups are known in AM fungi (Giovanneti et al. 2003). An ECM fungus—be it ascoymcete or basidiomycete—could be termed heterothallic, if two different strains or individuals are required for sexual interaction and to complete the life cycle. Strains can be self-sterile or self-incompatible but cross compatible, whereas, homothallic fungal species complete a sexual cycle on their own. Details about sexual cycle, genetic nature of mating types, gene transfer rates, selection and identification of recombinants as well as progeny analysis have been well reported for several ECM fungi such as *Rhizopogon*, *Pisolithus*, *Amanita*, *Laccaria, Cenococcum* (Kropp, 1997; Lamahamadi and Fortin, 1991; Lobuglio et al. 1996; Lobuglio and Taylor, 2002; Rosado et al. 1994). A few aspects of genetic of ECM fungi have been dealt with in chapter 3. Within this chapter, major emphasis is on recent approaches, especially molecular genetic methods that induce genetic transformation in ECM and AM fungi.

In case of zygomycetes, asexual reproduction is characterized by production of sporangia. Sexual reproduction involves different mating types known as (+) and (–) strains. Specialized branches called zygophores grow towards each other, establish contact and fuse at the tips. Swellings that develop at the tips are called progametangia. Formation of a cell wall delimits the gametangia that fuse later in order to facilitate somatogamy, then karyogamy and formation of zygospore. Sexual cycle in zygomycetes generates recombinants and variants in the progeny. Transfer of genes from one strain to the other and selection using appropriate media are all well-known phenomena. However, such studies have not been possible with AM fungi. Although they belong to zygomycetes, features such as zygophores or zygospores and various other stages of sexual reproduction have not been discovered yet. In this regard, Pringle et al. (2000) have aptly remarked that we lack information on the genetic system of AM fungi, ploidy levels, sexual and/or asexual processes, if any, that lead to recombination. Instead, it is generally assumed that AM fungi are asexual and individual mycelia reproduce by forming ostensibly clonal spores.

Anastomoses and Gene Transfer in AM Fungi

At the outset, we should note that anastomoses as a physiological manifestation is well known and reported for many fungi, including most ECM species. Certain ECM fungi such as *Hebeloma crustuliniforme* exhibit anastomoses frequently, whereas, *Lactarius rubrilacteus* or *Tuber* species and a few others rarely show up anastomoses (Agerer, 1991; BCERN, 2003).

Anastomoses is a well-known aspect of the life cycle that allows protoplasmic and genetic continuity in many fungi. Our knowledge about this phenomenon in AM fungi is very recent. Anastomoses or hyphal fusion has been observed regularly in mycelia originating from germinated spores. Giovannetti et al. (2003) conducted vegetative compatibility tests that showed that six geographically different *G. mosseae* isolates were capable of self-anastomizing. Frequency of anastomoses ranged from 65 to 85%, which is comparable to those reported for other fungi such as *Rhizoctonia solani.* The protoplasmic connection that ensues after the establishment of hyphal bridges provides for genetic transfer and continuity. Nuclei have been detected in hyphal bridges in many *G. mosseae* isolates. Since genetically different nuclei occur within spores (Pringle, 2001), nuclear movement via anastomoses can facilitate genetic exchange. Nuclear transfer through anastomoses, in fact, represents a natural phenomenon that maintains genetic diversity, especially in the absence of sexual mating and recombination among AM fungi.

Vegetative compatibility tests have been performed for many fungi, but primarily to study their population genetics. For ECM fungi, such vegetative compatibility tests have generally tallied with genetic and biochemical analysis. Giovannetti et al. (2003) have recently stated that vegetative compatibility groups occur even among isolates of *G. mosseae.* Attraction, hyphal tropism and fusion between hyphae occur frequently and is easily observable under microscope. Tropism or recognition responses were absent between hyphae of different species or genera of AM fungi that were incompatible. In fact, hyphal tips change the direction of growth and branching pattern to facilitate physical contact, if they come across conspecific/compatible hyphae. Clearly, specific recognition signals are involved during hyphal fusion and nuclear transfer via anastomoses. Generally, incompatible interactions lead to formation of septa and empty hyphal tips devoid of nuclei, either prior to or following hyphal fusion if it occurs. Therefore, to begin with, incompatibility prevents hyphal fusion; as a consequence cytoplasmic and nuclear exchange is obstructed. In nature, such strong incompatibility/barriers to hyphal fusion between isolates from widely different geographic locations hinders heterokaryon formation. We may also note that incompatibility suppresses rampant nuclear exchange and spread of harmful genetic elements between isolates of AM fungi. So far, a molecular basis for self and non-self recognition between germlings/isolates of AM fungi has not been deciphered. Cultural methods and chemicals that overcome physiological incompatibility and induce hyphal fusion need to be searched. It may find important application in understanding genetics of AM fungi. Consequently, wide genetic crosses between AM fungal species could become feasible. In other words, hyphal fusion could then be used as a routine method for gene transfer between AM fungi. Albeit, so far we know that anastomoses allows nuclear movement between hyphae of germlings of an AM fungal isolate. Cytological evidences for nuclear flow through hyphal bridges are available. However, evidence for transfer of a particular genetic trait as a result of nuclear transfer via hyphal fusion and its expression in a transformed (recipient) AM fungus is yet to be accrued. Genetic markers that are easily amenable for transfer via anastomoses and recognizable in the recipient fungal isolate need to be searched so as to prove this point. It may find plenty of uses as we investigate genetics of AM fungi and develop as many transgenic strains of AM fungi.

Let us consider examples of anastomoses in ECM fungi. *Rhizoctonia solani* is morphologically homogeneous, but it is complex and heterogeneous with reference to functionality. Isolates vary widely in function. They may be saprophytic, pathogenic or symbiotic. Pope et al. (1998) point out that the range of variation in functionality really makes it an interesting candidate for studying molecular/genetic basis of pathogenesis and symbiosis. They reported that in Australia, anastomosis groups (e.g. AG-6) of *Rhizoctonia* are involved in ectendomycorrhizal association with a native orchid *Pterostylis acuminata*. Genetic crossing between pathogenic and endophytic/symbiotic strains may lead us to useful information on the molecular regulation of pathogenesis/symbiotic conditions. We may be able to mask pathogenesis by anastomizing pathogenic strains of *Rhizoctonia* with strongly symbiotic ones.

Protoplast and Cell Fusion Techniques

Interspore hyphal fusion was examined in certain AM fungal species; namely *Gi. margarita* and *Gi gigantean* by Delmas et al. (1998). These fungi were able to reconnect injured hyphae leading to hyphal fusion. Events such as cell-cell chemical attraction, wall hydrolysis, membrane fusion and new organization of cytoskeletal components are some likely steps involved during hyphal fusion (Delmas et al. 1998). Experiment with germ tubes of spores belonging to the same ('monosporal') progeny showed that nearly 80% of the fusions were successful. Hyphal fusions were drastically low, at only 10%, if germ tubes not belonging to monosporal progeny were tested for fusion. We already know that compatibility groups exist among AM fungi, especially with regard to hyphal fusions and anastomoses (Giovannetti et al. 2003). Perhaps, vegetative compatibility genes govern intraspecific hyphal fusions. Attempts to obtain hyphal fusions between AM fungal species have been made by several other research groups (TERI, 2002). Hyphal fusions may aid exchange of genetic material via nuclear exchange in anastomozed individuals. Hence, by selecting appropriate strains and maker genes, we can attempt genetic crossing between AM fungi via hyphal fusions. We are yet to accumulate details on plasmogamy and karyogamy in fused AM fungal hyphae and their consequences to phenotypic expressions.

Production and Transformation of Protoplasts of Mycorrhizal Fungi

The fungal source material is a key component. Generally, young hyphae or spores of ECM fungi are utilized, since the extent of protoplast released is dependent on mycelial age. Talbot (2002) states that ideally, overnight cultures of most ascomycetes or basidomycetes are congenial for protoplast production. Firstly, fungal wall made of complex polysaccharide and chitin is digested, resulting in blebbing of cytoplasm and release of spherical protoplasts. Digestion of fungal wall is generally achieved using commercially available enzyme mixtures, such as Novozyme 234 and Glucanex, which are derived from *Trichoderma* sp. In cases of insufficient digestion, small additions of helicase, driselase or other lytic enzymes may improve the production of protoplasts. Protoplasts released may be anucleate, multinucleate or even contain a single nucleus. Protoplasts are generally released into an osmotically stabilized medium. The osmotic stabilizers are made of fungal propagation medium supplemented with 0.6 to 0.8 KCl, 1-1.2 M $MgSO_4$, 0.8 to 1.2 M NaCl and 1 M sucrose (Table 7.1; see Barret et al. 1989).

Table 7.1 Genetic transformation of protoplasts of filamentous fungi

Reagents

- CzV8 Czapeck Dox liquid medium
- 1 M $MgSO_4$
- Novozyme 234
- Sorbitol
- PEG solution: 66% PEG_{4000} pH 7.4, 100 mM $CaCl_2$
- regeneration medium: 1 M sucrose, 0.5% myco-peptone, 0.1% yeast extract

Method

1. Transfer 100 ml CzV8 Czepeck dox liquid medium supplement 10% V* juice (Campbells), aseptically to 2 × 50 ml Falcon tubes. Spin to remove V8 debris. Pour into 250 ml Ehrlenmyer's flask.
2. Inoculate with 5 × 2-3 mm^2 mycelial plugs from a fresh plate culture of *S. nodorum*, prepared on potato dextrose agar (Difco).
3. Shake for three days at 25°C to give a dispersed mycelial culture of young hyphae.
4. Harvest mycelium by centrifugation at 3000 g for 5 min. Resuspend in 50 ml of 1 M $MgSO_4$.
5. Centrifuge again and resuspend mycelium in 10 ml of filter sterilized Novozyme 234 (5 mg ml^{-1}) in 1 $MgSO_4$ in a falcon tube. Place on rocking platform for 60 to 90 min. at room temperature.
6. Protoplasts appear after an hour and microscopic examination should show them budding (blebbing) away from hyphae.
7. Add 0.6 m sorbitol to protoplasts and spin at 4000 g for 30 to 45 min. Carefully transfer protoplasts at the interface into a new tube.
8. Wash protoplasts with 10 ml of 1 M sorbitol and centrifuge at 2000 g for 5 min. resuspend in 10 ml of 1 M sorbitol. Repeat and resuspend protoplasts at a final concentration of 5×10^8 ml^{-1} in 1 M sorbitol, 100 mM $CaCl_2$.
9. Gently mix 100 ml protoplasts (5×10^7) with 25 μl PEG solution and 5 μg DNA. For restriction enzyme mediated insertion (REMI) carry out treatment with 20 to 40 U restriction enzyme (BamHI or EcoR1) for 30 to 60 min. at room temperature before addition of PEG.
10. Incubate protoplasts on ice for 20 min. Add 1 ml PEG drop-wise over 5 min. with gentle mixing. Immediately add 10 ml of 1 M sorbitol, 100 mM $CaCl_2$. Centrifuge at 200 g for 5 min. Resuspend in 10 ml regeneration medium.
11. Optionally, the protoplasts can be incubated overnight at room temperature with gentle rocking at this stage.
12. Mix aliquots of protoplasts with 50 ml molten regeneration medium and pour five petri dishes. Overlay with 10 ml top agar containing 200 μl hygromycin B. Incubate at 17°C under black light, until colonies appear (usually about eight to ten days).

Source: Talbot, 2002.

Quite a large number of filamentous fungi are amenable for transformation using protoplast-based techniques (Talbot et al. 2002). In the present context, let us briefly consider examples pertaining to ecto- and ectendomycorrhizas. In case of basidomycetous ECM *Hebeloma cylindrosporum*, PAN7.1 plasmid containing the *E. coli* hygromycin B phosphotransferase gene was utilized to transform its protoplasts. Hygromycin-resistant transformants were selected at a frequency of one to five per ng of transforming DNA (Marmiesse et al. 1992). Southern Blot analysis has clearly shown the successful integration of multiple copies of transforming DNA in the recipient genome. This method of selection could be used to introduce several other genes. For example, two plasmids—one of them containing tryptophan biosynthesis genes and other NADP glutamate dehydrogenase gene from the saprophytic basidiomycete *Coprinus cinerius*—were successfully transformed. The cotransformation efficiency for these genes was high at 70%. The hygromycin-resistance phenotype could be stably maintained during the cultivation of transformants. Also, all transformants maintained their ability to form ECM symbiosis with suitable host such as *Pinus pinaster*, thus, making it suitable for inoculation into seedlings meant for forestry.

Lyophyllum shimeji is yet another basidiomycete example, wherein protoplasts have been transformed using a transformation vector. The vector (pIS-hph) was constructed using promoter and terminator of glyceraldehydes-3-phosphate dehydrogenase (GPD) gene and introduced into fungal protoplasts (Saito et al. 2001). Protoplasts of ECM fungi have also been utilized in order to study the physiological aspects of the fungus. For example, Hampp et al. (1995) produced protoplasts of *Amanita muscaria* and *Cenococcum geophilum* to investigate the different aspects of sugar transport, especially sites for invertase production and sucrose hydrolysis.

Let us consider an ascomycetous ECM fungus. Stulten et al. (1995) have successfully standardized isolation and maintenance of protoplasts from *Cenococcum geophilum*. Under optimum cultural conditions (pH, temperature, osmotic buffer), preincubation with lytic enzymes (*Trichoderma* extracts), yielded 1 to 3 $\times$ 10^8 per gram fresh weight of fungus. Such protoplasts exhibited total plasma membrane integrity, which was confirmed using fluorascine diacetate staining. Acridine orange tests revealed that nearly 50% of protoplasts tested possessed nucleus. Most importantly, they succeeded in obtaining regeneration of the protoplasts, which reached upto 13%, depending on the culture medium. Such *Cenococcum geophilum* cultures obtained using protoplast regeneration methods formed ECM symbiosis effectively with spruce (*Picea abies*) seedlings. We should note that successful isolation and regeneration of protoplasts is a prerequisite for most of the gene transfer studies on ECM fungus.

Next, let us consider an example from Ericoid mycorrhizas. Bardi et al. (1998) successfully isolated protoplasts of an ericoid mycorrhizal fungus *Oidiodendron maius* and regenerated them. As a first step, the protocol involved cell wall digestion with lytic enzyme. Protoplasts were then separated by centrifugation. Regeneration of protoplasts was achieved using a hypertonic medium containing sorbitol as osmotic stabilizer. Protoplasts were viable and amenable for transformation with commonly used gene markers such as Benomyl and Hygromycin B. In addition to obtaining protoplasts, Bardi et al. (1998, 1999) aimed at understanding the mechanism of tolerance to heavy metals and soil contaminants. Tests on Zn++ and Cd++ sensitivity has shown that protoplasts of *C. geophilum* are susceptible to these heavy metals, but at the same concentration, mycelial stages were tolerant and grew normally. It shows that tolerance to certain heavy metals by this fungus is resident in the cell wall. A metal exclusion mechanism was suspected to be operative on the cell wall, thus providing tolerance to mycelia against heavy metals. Methods such as cell fusion, hyphal fusion, electroporation (see Table 7.2) and cell microinjection have been applied to transform different mycorrhizal fungi.

Table 7.2 Transformation of filamentous fungi by electroporation

Reagents

- Czapeck dox agar (Oxoid)
- Sorbitol
- Novozyme 234
- Miracloth
- STC: 1.2 sorbitol, 10 mM Tris-HCl pH 7.4, 10 mM $CaCl_2$
- Regeneration medium: 1 M sucrose, 0.5% yeast extract, 0.1% bactopeptone.

Method

1. Harvest 10^6 conidia/spores or equivalent hyphae into 10 ml of Czepeck Dox medium. Allow the conidia to germinate overnight at 18°C in a shaking incubator (100 rpm).
2. Collect the germinated conidia by centrifugation at 2500 g for 5 min. Wash each pellet once with 5 ml of ice-cold sterile water and once with 5 ml of ice-cold 1 M sorbitol. Centrifuge at 2500 g for 10 min. to collect the cells between the washes.
3. Resuspend the 4 germling pellets in a total of 10 ml of 1 M sorbitol, containing 5 mg ml^{-1} Novozyme 234. Incubate in a shaking 25°C incubator at 50 rpm for 2 h.
4. Filter the resulting protoplast suspension through four layers of sterile Miracloth.
5. Transfer the protoplasts into a sterile 30 ml corex tubes and collect the cells by centrifugation at 100 g for 5 min. Wash the protoplast pellets three times by gently resuspending the cells in 25 ml of 1 M sorbitol.
6. After the third wash, resuspend the protoplasts in 10 ml ice-cold sterile STC. Count the protoplasts using a haemocytometer then centrifuge at 100 g for 5 min and remove the STC with an aspirator.
7. Gently resuspend the protoplast pellet to a final concentration of 10^8 ml^{-1}. Add 50 μl of protoplasts to 3 to 6 μl of DNA (at 1 μg/μl). Flick the tube to mix, then transfer to a pre-chilled electroporation cuvette with a 0.2 cm gap (Bio-Rad cuvette No 165-2086).
8. Electroporate at 200 Ohms, 25 μF, and 0.7 kV (filed strength of 3.5 kV/cm, time constant of 0.4 msec).
9. Following the electroporation, resuspend protoplasts immediately in 1 ml of sterile ice-cold regeneration medium. Allow the protoplasts to regenerate their cell wall by incubation at room temperature overnight and then plate the cells on selective medium.

Source: Talbot, 2002.

Agrobacterium-mediated Genetic Transformation of Mycorrhizal Fungi

Soil bacterium, *Agrobacterium tumefaciens* has the ability to mediate the transfer of small sections of plasmid borne DNA in plants, yeast and filamentous fungi. It transfers a part of its Ti-plasmid, the T-DNA to plant cells during tumerogenesis. Hence, it behaves as a vector for DNA transfer. So far, a wide range of filamentous fungi have been transformed by adopting *Agrobacterium*-mediated DNA transfer, notably, several soil fungi, plant pathogens and AM /ECM fungi (Table 7.3). This method of gene transfer is comparatively simpler than other methods of transformation. DeGroot et al. (1998) reported that T-DNA mediated gene transfer into protoplasts improved the

Table 7.3 Examples of *Agrobaceterium*-mediated DNA transfer in filamentous fungi, including ecto- and arbuscular mycorrhizal fungi

Filamentous fungus	*Genes transferred*	*Reference*
Soil inhabitants *Aspergillus awamori* *Trichoderma reesi* *Neurospora crassa*	Phleomycin resistance (*Shble* gene); Random insertional mutagenesis	Pardo et al. 2002; deGroot et al. 2002
Plant pathogens *Fusarium* sp. *Botrytis* sp. *Phytophthora infestans* *Colletotrichum* sp.	Random insertional mutagenesis	Pardo et al. 2002; deGroot et al. 2002
Ectomycorrhizal fungi		
Hebeloma cylindrosporum	Tryptophan synthesis genes and NADP-glutamate dehydorgenase gene from *Coprinus cinerius.*	Marmiesse et al. 1992
	Hygromycin B resistance	Benjdia, 2003
	Plasmid inserts detected using Southern Blot	Combier et al. 2002; Combier et al. 2003
Agaricus bisporus	T-DNA at random site on host genome.	Mikosch et al. 2001
	Hygromycin B resistance	Van de Rhee, 1996
Suillus bovinus	Hygormycin B resistance	Hanif et al. 2002
Paxillus involutus	Hygromycin B phosphotransferase gene (*hph*)	
Arbuscular Mycorrhiza		
Glomus intraradices	Resistance to fungicides such as Benomyl, Hygromycin; Herbicide-Phosphinothricin (BASTA), Antibiotic-Neomycin.	Requena, 2003; Requena and Levya, 2003

frequency of transformants by 600 folds over conventional techniques of protoplast transformation. Firstly, the DNA segments to be transferred, including a selectable marker and genes of interest, have to be cloned between left and right T-DNA borders in a binary transformation vector such as pBIN19. Such a vector is then transformed into *A. tumefaciens*. Bacteria containing the cloned DNA are inoculated (co-cultivated) on to the fungus to facilitate gene transfer. Bacteria are then removed using antibiotic treatment (Talbot, 2002). A generalized protocol for *Agrobacterium*-mediated DNA transfer is shown in Table 7.4. Now let us consider a few examples that deal with *Agrobacterium*-mediated transformation of mycorrhizal fungi. In case of ECM fungus, *Hebeloma cylindrosporum*, Benjdia (2003) utilized a hypervirulent strain of *A. tumefaciens* AGL-1. It contained a plasmid *pBGghg*, which had a disarmed T-DNA attached with hygromycin B phosphortransferase gene (*hph*) under the control *of Agaricus bisporus* (*gpd*) promoter. The integration of T-DNA in the genome was confirmed using PCR and Southern Blot techniques. Resistance of transformants to hygromycin B was tested using yeast mannitol agar containing 200 μgml^{-1} hygromycin B. They reported 11% transformants by adopted the above procedure *Agrobacterium*-mediated insertional mutagenesis has also been reported for *Hebeloma cylindrosporum* (Combier et al. 2003). Southern Blot analysis was used to confirm the random

Table 7.4 Protocol for *Agrobacterium tumefaciens*-mediated genetic transformation of filamentous fungi in general-examples.

Step 1 Preparation of competent cells of *Agrobacterium*

Reagents

- 2YT media
- TE pH 7.5
- Liquid nitrogen

Method

1. Inoculate 10 ml of 2YT with a single colony of *Agrobacterium* on a fresh plate and grow with shaking at 28 to 30°C overnight.
2. Add the overnight culture to 200 ml 2YT and grow with shaking for 3 to 4 h.
3. Harvest cells by centrifugation at 3000 g for 10 min. Decant the supernatant and wash the surface of the pellet with 10 ml ice cold TE pH 7.5.
4. Resuspend the pellet in 20 ml ice-cold 2YT and immediately flash-freeze 0.5 ml aliquots in liquid nitrogen.
5. Competent *Agrobacterium* cells can be stored at –70°C for 6 to 12 months.

Step 2 Transformation of competent cells of *Agrobacterium*

Reagents

- As shown above for preparation of competent cells, plus
- Competent cells of *Agrobacterium*
- Plasmid DNA

Method

1. Thaw the frozen competent *Agrobacterium* cells on ice and add 1 μg plasmid DNA with gentle mixing.
2. Incubate on ice for 5 min. and then quickly freeze in liquid nitrogen for 5 min.
3. Incubate cells at 37°C for 5 min. and add 1 ml of 2YT and incubate at 28 to 30°C for 2 to 4 h with shaking.
4. Plate 1 to 200 μl aliquots on a selective medium (LB + 50 g/ml kanamycin for pBIN19) and incubate at 28°C for two days until colonies appear.

Step 3 DNA-mediated transformation of fungi using *Agrobacterium*

Reagents

- Agrobacterium minimal medium: 3 g K_2HPO_4,1 gNaH_2PO_4, 1 g NH4Cl, 0.3 g $MgSO_4 \cdot 7H_2O$ 0.15 g KCl, 0.05 g $CaCl_2$ 0.0025 g $FeSO_4 \cdot 7H_2O$, 5 g sucrose, make upto 1 liter with d H_2O, pH 7.0 with HCl and autoclave.
- Induction medium (IM) for 100 ml use: 1 g NH_4Cl, 0.3 g $MgSO_4 \cdot 7H_2O$, 0.15 g KCl, 0.01 g CaCl2, 0.0025 g $FeSO_4 \cdot 7H_2O$, 40 mM MES pH 5.6, 10 mM glucose, 5 ml glycerol, 4 mM acetosyringone.

Method

1. Inoculate a single *Agrobacterium* colony into 2 ml minimal medium. Grow for 48 h with shaking at 28°C.
2. Dilute the culture to OD_{660} of 0.15 in induction medium and grow for 6 h at 28°C with agitation at 200 rpm.
3. Mix 200 μl fungal spores at 10^3, 10^4 and 10^5 spores/ml. Pipette the mixtures onto nitrocellulose filters placed on IM plates. Incubate at 20°C for 48 h.
4. Transfer the filters to fungal minimal medium agar plates containing the selective antibiotic (e.g. Hygromycin B) and 500 μg/ml Claferin (Cefataxime). Incubate at 20°C for 48 h.
5. Remove filters and place in 10 ml fungal minimal medium containing selective antibiotic and Claferin in a 50 ml Nalgene tube. Vortex briefly to resuspend the fungal spores. Plate out the suspension on selective medium—0.5 per 9 cm Petri dish is usually adequate for development of well-separated colonies.

Source: Talbot, 2002.

plasmid insertions in as many as 83 of the transformants. Basidiomycetous ECM fungi such as *Suillus bovinus* and *Paxillus involutus* are also amenable for *Agrobacterium*-mediated DNA transfer (Pardo et al. 2002). The transformants selected using a marker, *Shble* gene were normal in their ability to establish ECM symbiosis with the host *Pinus sylvestris. Agaricus bisporus* is yet another ECM fungus, wherein *Agrobacterium*-mediated DNA transfer has been successfully achieved (Table 7.3). The above results clearly substantiate possibility of selective gene transfer in ECM/AM fungi using *A. tumefaciens*. Transforming mycorrhizal fungi with genes that enable greater symbiotic benefits will be useful. So far, mycorrhizal fungi have been transformed with few selective markers and genes coding for enzymes using *Agrobacterium*-based procedures.

Transformation of AM Fungi Using Microprojectile Bombardment

Biolistic transformation is a useful technique to transfer genes into different organisms. The donor DNA or constructs are coated with gold or other suitable microcarrier and bombarded with forces to achieve entry into recipient tissue/cells. Harrier et al. (2002) have stated that such a biolistic method of gene transfer is particularly useful to transform AM fungi. AM fungi are not amenable for in vitro culture without host support. Also, they posses coenocytic mycelium that are supposedly recalcitrant and do not easily yield protoplasts competent to receive transforming DNA. There are actually several parameters that need careful consideration and standardization in order to achieve efficient transformation. Let us consider those major aspects. Firstly, characteristics of the biological material should be amenable for biolistic transformation. The target tissue should be competent and receptive. It should allow high rates of entry of DNA-coated microprojectile particles, maintain good levels of cell survival and growth after bombardment. In their study with AM fungi, Harrier and Millam (2001) used spores for biolistic transformation. They have argued that AM spores are easy to handle during surface sterilization. Damage, if any, after the bombardment can be easily detected. Choice of appropriate fungal species is also crucial. Harrier et al. (2002) selected *Gi. rosea* because it produces large-sized spores (230-305 μM) with a thin spore wall (2.4 to 7.5 μM). These characteristics allowed better penetration of micro-projectiles. In addition, *Gi. rosea* did not possess bacteria-like organisms in spores that may complicate detection of expression of molecular markers inside transformed samples. The multinucleate status of AM fungal spores could improve chances of formation of transformants, but it leads to formation of heterokaryons. It is not known as to how many individual nuclei need to be transformed before we can detect expression of a reporter gene. Also, the transformed gene may get diluted in due course (Harrier et al. 2002). Table 7.5 depicts a generalized protocol that could be utilized while attempting transformation using the particle-bombardment method.

Vector that enables gene transfer and integration into recipient genome is definitely an important aspect during biolistic transformation. Generally, plasmid constructs must have appropriate reporter or selective genes and promoters. Size of the transforming DNA should be conducive. Genes may be introduced as DNA or RNA, in circular, linear, single-stranded and /or double stranded form. Table 7.6 lists the protocols for extraction of nucleic acid that could be eventually utilized to construct vectors meant for micro-projectile bombardment and a few other methods of

Table 7.5 Transformation of filamentous fungi through particle bombardment

Reagents

- M10 particles (Bio-Rad Laboratories, Hercules, CA, USA) were prepared and coated with plasmid DNA according to manufacturer's instructions.

Method

1. To a 1.5 ml Microcentrifuge tube add 25 μl tungsten particle suspension (1.5 mg in 50% glycerol), 5 μl plasmid DNA (0.5μg/μl), 25 μl of 2.5 m $CaCl_2$, and 5 μl of 1 M spermidine free base. After each addition, the suspension is vigorously vortexed and incubated on ice for 10 min.
2. Spin down tungsten particles for 5 to 10 sec at full speed in a microcentrifuge. Decant the supernatant and wash particles twice in 100 ml of 95% ethanol.
3. Resuspend the particles in 50 μl ethanol, and pipette 10 μl particles onto a macrocarrier disc (Bio-Rad).
4. Spread conidia/hyphae of the fungus to be transformed into 10 ml of selective medium in a 9-cm Petri dish, briefly air-dried under sterile conditions. Use for microprojectile bombardment within 1 to 3 h.
5. In general, a helium pressure of 1000 psi, and a target distance of 6 cm can be used for bombardment. Transformants are visible three to six days after bombardment.

Source: Talbot, 2002.

DNA-mediated transformation. For example, Harrier and Millam (2001) have utilized a construct pNOM102, which is actually GUS gene developed using coding sequence of *E. coli gus A* that was attached to glyceraldehyde-3 phosphate dehydrogenase (*gpd*) promoter from *Aspergillus nidulans.* A plasmid vector (pSPC01) carrying the gene for hygromycin resistance was preferred to transform *Gi decipiens* and *Gi. margarita* by Parsad-Chinnery (1997).

Microprojectile particle can be accelerated in different ways. For example, Harrier et al. (2002) have used a PDS-1000/helium device (Fig. 7.1). Such a device is known to deliver microprojectiles and associated debris uniformly in a wider zone on the target tissue/cells. There are several parameters that affect the delivery of microprojectile, e.g. gas pressure, vacuum levels, macro-carriers, instrument settings. Standardization of the microprojectile bombardment device is required in order to obtain optimum particle penetration. Microprojectile particles could be tungsten, platinum, iridium or gold. Gold particles are frequently selected because they are non-toxic to cellular activity and do not catalytically attack donor DNA. Harrier et al. (2002) state that the size of the gold particle is a key determinant of biolistic transformation efficiency. Table 7.7 lists various parameters that need standardization while attempting a biolistic transformation. Forbes et al. (1998) had earlier reported that 60% of spore surface area was covered by the bombarded particles, leading to transformation frequencies of 40 to 50%. The transformed *Gi. rosea* spores obtained by Harrier et al. (2002) could initiate colonization of *Allium porrum* roots. Expression of transferred GUS gene could be easily detected through PCR using GUS-specific oligonucleotide primers in the colonized roots and even in the second generation spores. It, therefore, confirms a stable transformation of *Gi. rosea* using the microprojectile bombardment technique.

Table 7.6 Methods and protocols for nucleic acid extraction useful during transformation of fungi in general—examples

1. Extraction and purification of DNA from fungi: CTAB method

Reagents

- 500 ml CTAB Buffer: 2% hexadecyltrimethyl ammonium bromide, 10 g (SigmaH-5882); 100 mM Tris 6.06 g; 10 mM EDTA 1.46 g; 0.7 M NaCl make upto 500 ml with double distilled H_2O and store at room temperature.
- CIA: Chloroform/Isoamyl alchohol (24:1 v/v)

Method

1. Grow a fungal culture (e.g. *Hebeloma cylindrosporum*) in 50 to 100 ml of rich growth medium for two to five days, depending on species. Young mycelium gives better results.
2. Extract 5 to 10 g of mycelia by filtering through fine cloth or collect the fungus as pellets. Dry the mycelium by blotting and store –70°C for future use.
3. Grind the mycelial using liquid nitrogen in a pestle and mortar.
4. Transfer powdered mycelium to Nalgene tubes and add 4 ml CTAB lysis buffer, which has been heated previously to 65°C and disperse the fungal material evenly.
5. Add 40 ml mercaptoethanol and incubate for 30 min. at 65°C, ensure adequate mixing of the contents.
6. Add 3 to 4 ml of CIA very carefully to the tubes and let them stand uncapped in a fume cupboard for 1 min.
7. Incubate tubes on a shaking platform for 20 min. at 60 rpm.
8. Extract by centrifugation at 9650 g for 20 min. at 4°C (in a fixed angle rotor).
9. Remove the aqueous (top) phase, being careful not to remove denatured proteins and debris at the interface.
10. Add an equal volume of CIA, shake the tube gently and repeat centrifugation.
11. Remove the aqueous phase and add to 0.5 vol of Isopropanol in a fresh Nalgene tube to precipitate DNA.
12. Leave the tubes on ice for 5 min. and then centrifuge at 9650 g for 20 min. at 4°C in a swinging bucket rotor.
13. Wash pellet in 10 ml of 70% ethanol and repeat centrifugation.
14. Dry the pellet on the bench by inverting the tubes on paper towels. Leave them for 30 min. to 1 h. Resuspend pellet in 100 ml TE buffer pH 8.0 and store 4°C.

Re-extraction and reprecipitation is suggested if DNA does not digest perfectly with restriction enzymes. Generally, cloudiness in the DNA sample extracted by this procedure does not interfere with action by restriction enyzymes.

2. Isolation of nuclear DNA from filamentous fungi

Equipments and reagents

- Dounce Homogeniser
- Waring blender-chill blender cups at –70°C
- Bead beater
- Washed Mira cloth
- Acid washed glass beads (size 300 μm)
- 10 × SSE (recipe for 500 ml): 10 mM spermine-HCl, 10 mM spermidine-HCl, 1.0 M KCl, 100 mM EDTA, 100 MM Tris-HCl pH 7.0, store at –20°C.
- PMSF: 100 mM phenylmethylsulfonyl fluoride, 0.871 g/10 ml 95% EtOH; store at –20°C
- 0.5 × SSE (1 litrer): 100 ml of 10 × SSE salts, 1 ml of 2-mercaptoethanol, 10 ml of 100 mM PMSF (0.871 g in 50 ml 95% EtOH),171.2 g sucrose to be prepared fresh.
- 4 liters ddH_2O chilled (4°C).
- TNE: 10 mM Tris-HCl pH 7.6, 15 mM NaCl, 1 mM EDTA
- 20% (w/v) N-lauryl sarcosine (sodium salt): 20 g/100 ml H_2O

(*Contd.*)

(*Contd.*)

Method

1. Grow fungal culture in 500-ml rich medium and incubate on a shaker for at last 24 h. Young hyphae are best for DNA extraction.
2. Harvest mycelium by filteration through Mira cloth and wash extensively with cold ddH_2O, blot the mycelium and preserve dry at –70°C until use or place the mycelium in a beaker and add 0.5 × SSE buffer to prepare a mycelial paste.
3. Precool a waring blender cup to –70°C, transfer mycelium and grind at high speed until the mycelium is well powdered.
4. Transfer the mycelial powder to a bowl and thaw it at 0 to 4°C, stir continuously.
5. Precoll a bead beaer with glass beads, in a wet ice bath. Add both thawed mycelium and precooled beads. Grind until all hyphal fragments are fully broken.
6. Filter the preparation through miracloth and rinse the beads with 0.5 × SSE. Increase the volume in steps to reach 250 ml.
7. Retain the filterate and process by centrifugation for 20 min. Transfer the pellet to Dounce homogenizer. Gently resuspend the pellet in 40 ml of fresh 0.5 × SSE.
8. Decant suspension to a Nalgene tube and process by centrifugation at 6000 g for 20 min. in a swinging bucket rotor. Resuspend the pellet by adding 30 ml of 0.5 × SSE and adding to the Dounce homogenizer once again. Wash, rinse and resuspend the pellet two more times with 0.5 × SSE. During these latter washing steps, add 0.2% (v/v) Nonidet P-40 to lyze the mitochondria.
9. Resuspend the final pellet in 10 ml of 0.5 × SSE. Add 20% N-lauryl sarcosine to a final concentration of 1%, with gentle agitation. Warm to room temperature. The resulting solution will be viscous.
10. Add proteinase K (0.5 g) to the suspension and incubate at 37°C for 1 h.
11. Dialyse at room temperature against 1 liter of 1 × TNE. Exercise great care when transferring to avoid shearing of DNA. Change the TNE dialysis buffer every 1 to 1.5 h. Cell debris settle at the bottom of the tubing.
12. Process the dialyzed DNA samples by centrifugation to remove cell debris (8000 g for 5 min. at 4°C in a swinging bucket rotor). Discard the pellet.
13. Prepare a standard Caesium chloride gradient at 49 (w/v). Do not expose to bright light to avoid DNA damage. Centrifuge overnight at 47,000 g.
14. Extract the DNA band using an 18 gage needle with indirect UV. Do not increase the volume.
15. Dialyse the DNA in three changes of 1 × TNE. Quantify of the DNA by Spectrophotometry A concentration of 300 to 500 ng/μl is normal. The yield will be 5 to 10 ml from this procedure.
16. The DNA can be concentrated by Isopropanal precipitation or using a Speed-Vac, if required.

3. Isolation of total RNA from filamentous fungi

Total RNA from filamentous fungi, including AM/ECM fungi has been isolated using different methods. Procedures involving chaopatric agents such as guanidinium thiocyanate, along with a reducing agent 2-mercaptoethanol or those utilizing CTAB or hot phenol extraction method, all aim at denaturing the proteins quickly and inhibiting the action of ribonucleases until RNA is extracted and purified. RNAases are stable enzymes and they may contaminate laboratory equipment. Detergents are to be used to protect RNA samples, say by regularly storing in SDS. The protocol stated below has been used routinely for many years to isolate and purify RNA from fungal cells derived from a wide range of species.

Reagents

- Extraction buffer: 4.8 g PAS (p-Aminosalicylic acid, sodium salt; 0.8 g TNS (tri-isopropylnaphthalene sulfonic acid, sodium salt). Dissolve each detergent separately in 40 ml of ddH_2O. Add the PAS to the TNS with constant stirring.

(*Contd.*)

(*Contd.*)

- TELS: 10 mM Tris-HCl pH 7.6, 0.1 mM EDTA, 0.2% SDS.
- 5 × RNB: 1.0 TRIS base, 1.25 M NaCl, 0.25 M EGTA, titrate to pH 8.5 with NaOH and store at –20°C. Add 20 ml 5 × RNB to the PAS-TNS mix (PAS = 4.8%, TNS = 0.8%). A light brown precipitate will form, but does not interfere with the buffer. Store the solution on ice and always use fresh.

Method

1. Harvest a fresh culture by filtration, blot the mycelium, dry and grind it in a mortar containing liquid N_2 until a fine powder is achieved.
2. Transfer the powder to a Nalgene tube containing 4 ml of extraction buffer and 4 ml phenol/chloroform. Vortex for 5 min. in 30 sec bursts, followed by 30 min. on ice.
3. Process by centrifugation at 9650 g for 10 min. at 4°C. Transfer the aqueous phase to a new tube and keep on ice.
4. Add 5 ml of extraction buffer to the phenol/chloroform interface. Mix and heat to at 68°C for 5 min. Cool on ice and centrifuge as above. Pool the aqueous phases.
5. Re-extract the pooled aqueous phase twice (or until there is no interface) with phenol/chloroform.
6. Slowly add 0.25 vol of 10 M LiCl (final concentration 2 M). Store overnight on ice. Pellet the RNA by centrifuging at 9650 g 15 min. at 4°C.
7. Resuspend the pellet by gentle pipetting in 5 ml of 0.1 M sodium acetate pH 6/70% ethanol. Centrifuge as before, but for 20 min. Repeat once.
8. Resuspend the pellet in 50 μl TELS. If the RNA is to be used for poly (A) purification and cDNA library construction, then store in ddH_2O which has been DEPC-treated.

4. Isolation of total RNA from filamentous fungi using Trizol

Reagents

- Trizol (Gibco BRL)

Method

1. Harvest a fungal culture by filteration in the process described previously. Grind mycelium in liquid nitrogen in a pestle and mortar to a fine powder.
2. Transfer the powdered mycelium to a Nalgene tube containing 20 ml of Trizol, vortex the mixture for 15°C. Incubate the mixture at room temperature for 20 min.
3. Add a 4 ml aliquot of chloroform and shake the tubes vigorously for 15 to 30 sec. Incubate them for a further 3 min. on the bench.
4. Process the RNA preparations by centrifugation at 12000 g for 15 min at 4°C in a fixed angle rotor.
5. Remove the aqueous phase to a Nalgene tube avoiding interface material.
6. Add 10 ml Isopropanol (1 vol) and leave on the bench for 10 to 15 min.
7. Precipitate RNA by centrifugation at 12000 g for 20 min. at 4°C.
8. Pour away supernatant and resuspend the pellet in 70% ethanol. Centrifuge again at 6000 g for 10 min. at 4°C.
9. Dry the pellet on the bench and resuspend in 200 ml TELS. Resuspend in ddH_2O for further manipulations other than Northern blotting.

Source: Talbot, 2002

2. Transgenics and Genetic Engineering in Mycorrhizas

2. A. Mycorrhiza in Transgenic Plant Hosts

Historically, domestication of plants and selection for genetic traits began at least 10,000 years ago. Knowledge regarding genetics provided better rationale during the

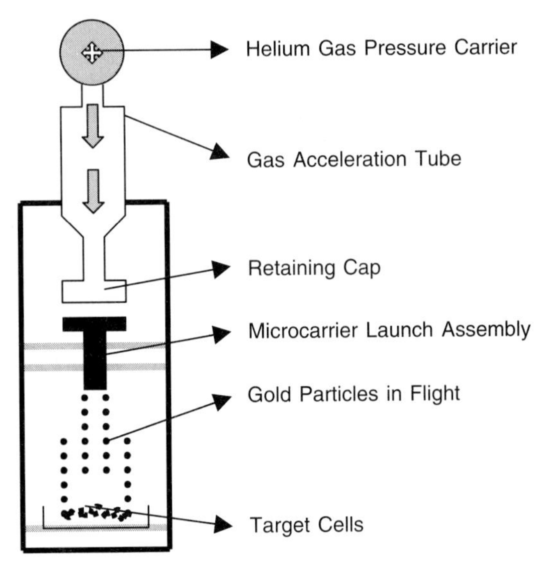

Fig. 7.1 Diagrammatic representation of helium particle bombardment device.

Table 7.7 Standardized Parameters for particle bombardment of *Gigaspora rosea*—an example

Biolistic parameter	*Optimized values*
Amount of gold particles	1.2 mg
Macrocarrier gold load per bombardment	150 μg
Microprojectiles (size)	0.6 μM gold
Plasmid concentration	100 μg ml^{-1}
Amount of plasmid utilized per shot	5 μg
DNA load per bombardment	625 μg
Platform setting	2
Stage setting	5
Acceleration pressure	1350 psi
Chamber vacuum pressure	27 in. Hg
Gap distance	7 cm
Macrocarrier travel distance	12 cm
Stopping plate aperture	0.8 cm

Source: Bills, 1998; Harrier and Millam, 2001; Harrier et al., 2000.

selection of plant types. Plant breeders have consistently employed different strategies based on natural crossing, compatibility and combining the ability to transfer specific genes into new varieties and hybrids. We know that host genetic component influences mycorrhizal ability (Krishna et al. 1984). However, we lack accurate estimates regarding the influence of such plant improvement on mycorrhizal ability. With the advent of molecular methods such as recombinant DNA techniques, gene cloning, transformation, development of transgenic crop genotypes have become possible. The release of transgenic crop genotypes has aroused debates regarding the environmental aspects, especially about the flow of cloned genes in the plant community. According to Azevedo and Araujo (2003), although environmental risks arising through introduction of transgenic crop genotypes are being carefully assessed, their effects on associative and beneficial soil microbes such as nitrogen fixers, mycorrhiza and plant growth-stimulating bacteria are not being evaluated earnestly. There is also need to assess gene-flow that may occur horizontally between transgenic plants, associated microbes and the disturbance it might cause to symbionts that interact very closely with transgenic hosts. Evidently, a transgenic crop may influence the symbionts in many different ways.

During plant improvement, it is common to include genes for resistance to diseases, including those that impart resistance to fungal diseases. This may or may not interfere with our ability to enhance susceptibility of these same genotypes to symbiotic fungi. A similar situation occurs while dealing with transgenic plants that are already endowed with genes that suppress fungal diseases. The anti-fungal proteins that transgenic host releases may affect AM or ECM fungal infection and proliferation, depending on the particular case. There is no doubt that this aspect needs very careful consideration while developing transgenics tolerant/resistant to pathogenic fungi. Valeria et al. (1998) states that molecular techniques have generally aimed at achieving durable and broad-spectral resistance by transforming plants with genes coding for anti-fungal proteins. Obvious concern is the effect of these genes and their products on non-target and beneficial symbiotic microbes. Let us consider an example. Plant defensins are small cistein-rich peptides of about 45 amino acids and transgenic tobacco plants expressing the defensin Rs-AFP2 show an enhanced resistance to foliar pathogen *Alternaria longipes*. Tests by Valeria (1998) have shown that all the seedlings were colonized by AM fungi, irrespective of the level of expression of 'defensin' genes. The morphogenetics of the AM fungus inside the host root remained unaffected. Thus, in this particular case, defense-related genes do not seem to interfere with symbiotic potential of plants.

In due course of time, transgenic trees are likely to be introduced into forestry. Some of them may be carrying genes that resist fungal diseases. Such genes may also influence mycorrhizal symbiosis. For example, mycorrhizal colonization of transgenic aspen was investigated for over 15 months at Jena, in Germany (Buscot et al. 2001; Cummins, 2002; Fladung et al. 1999; Kaldorf et al. 2002). These transgenic aspen carrying 35S-rolC genes are characterized by dwarfed phenotype, precocious bud break, small wrinkled leafs and altered cytokinin levels. Wood formation by 35S-rolC aspen trees show stable alterations such as delayed formation of cells, occurrence of thin walled and less lignified fibres and lack typical late wood. Transgenic trees with altered lignin content can influence mycorrhizal development. In the above experiment by Kaldorf et al. (2002) ectomycorrhiza was well developed in all the samples tested. Nearly 90% of the samples had the dominant ECM fungal species and diversity of ECM fungi was almost similar between transgenic and non-transgenic trees. Only one ECM fungal species represented by *Phialocephala fortini* (E2/5) was rare on transgenic trees and the effect seemed to be clone specific.

Let us consider a couple of more examples. In case of barley, transgenic lines with the ability to produce antifungal proteins and resist fungal diseases such as powdery mildew has been developed. Transgenic lines over-expressing antifungal proteins in the entire plant at all times could definitely hamper the development of mycorrhizal association (Riso, 2000). Hence, an altered strategy was to produce transgenic lines that selectively accumulated antifungal proteins in only leaf epidermis. Such modification effectively overcomes any detriment to mycorrhizal component and other non-target microbial flora. Similarly, transgenic wheat lines (e.g. GM-Wheats: KP4-Griena, KP4-Golin) exhibit antifungal activity, especially against

Ustilago maydis and *Tellitia tritici.* Its influence on rhizosphere microflora, including mycorrhiza, will have to be known (Institute of Plant Sciences, 2000). Gene flow, particularly those related to antifungal activity, needs verification.

2. B. Transgenic Mycorrhizal Fungi

Glomalin is a protein synthesized by AM fungi. It helps in soil aggregation and maintenance of soil structure. Hence, it finds applications in agricultural situations, particularly in sequestering carbon and reducing loss of soil quality (Comis, 2002). Shmaefsky (2003) has suggested that developing transgenic plants with the ability to produce glomalin through their roots is a possibility. Tissue-specific expression of glomalin could be a major hurdle while transferring this trait from AM fungus to plants. It is suspected that success depends on ensuring accurate post-translational modifications and excretion of this protein from the root to the exterior (soil). The carbon costs for production of high amounts of glomalin by roots is as yet unknown. Transgenic AM fungal strains endowed with increased copies of glomalin gene is yet another possibility that needs attention. We may develop AM fungal isolates that are superior in glomalin production.

According to Lopez-Bucio et al. (2000) the ability of plant roots to use insoluble/sparingly-soluble P can be enhanced using genetic engineering. It involves imparting the roots with genes that help increased production of organic acids, especially citrates. They have reported that citrate over-producing tobacco can utilize insoluble P sources better, leading to higher yields. This argument can be applied further to mycorrhizas if suitable genes that aid production and excretion of higher amounts of organic acids can be produced by genetic transformation. Both plant and symbiotic fungus could be transgenic and possess the ability to over-produce organic acids.

Intensive agriculture envisages the use of high doses of pesticides, fungicides and herbicides to keep the detrimental effects of insects, pathogens and weeds within threshold levels. No doubt, non-target and beneficial organisms may be affected. Population density and community structure of mycorrhizas in the soil may get affected severely when high doses of pesticides/herbicides are applied. Hence, there is every reason to transform inoculant strains of AM/ECM fungi with genes that impart resistance/tolerance to such agricultural chemicals. Some examples are transgenic AM/ECM fungi with genes for resistance to fungicide such as benomyl, antibiotics like hygromycin B, phleomycin. We may also explore the possibility of producing transgenic AM/ECM fungi that resist bacterial infections, either by releasing bacteriostatic substances or specific antibacterial antibiotics. Appropriate gene sources need to identified and transferred into inoculant strains of mycorrhizal fungi. This may at least thwart infection by bacterial pathogens via soil/roots. Transgenic AM/ECM fungi with the ability to release enzymes that enhance rates of pesticide degradation may find application in practical agriculture.

ECM fungi with the ability to fix atmospheric nitrogen, in addition to P transport benefits to host plant have been reported long ago. Actually, Nitrogen-fixing asymbiotic bacteria *Azotobacter* sp. were co-cultured with protoplasts of *Rhizopogon* and allowed to be transformed in the presence of polyethylene glycol. Transformed protoplasts of Rhizopogon were regenerated. The ECM ability of transformed *Rhizopogon* remained unaffected and N_2 fixation by engulfed *Azotobacter* continued for over 48 hours after transformation. Indeed, AM or ECM fungi with the ability to maintain N fixation and aid P transport function will be a boon if well-stabilized transgenic fungal isolates are introduced into the soil. However, we have to ascertain the environmental effects and gene flow whenever transgenic organisms are introduced into natural soil. We have to trace the perpetuation and movement of *Azotobacter* among fungal populations and their consequence, if any, on fungal/soil ecology.

Yet another use for transgenic AM /ECM fungi can be found in the production of secondary metabolites. Several of these fungi are known to produce very useful secondary metabolites, steroids, etc. Accentuation of these genes by transforming them with multiple copies of required genes will be beneficial.

Molecular Approaches to Achieve in vitro Culture of AM Fungi

In the past, there have been several attempts to culture AM fungi. Mycorrhiza researchers have adopted a wide variety of different hypotheses, scientific reasoning and methods while tackling this problem. However, none of them have proved successful so far. The AM fungi still need host plant support to grow and complete their life cycle. Initially, various media and concoctions were tested. Some combinations of nutrients resulted in minor stimulation in hyphal growth in vitro that subsided later. Such hyphal growth was dependent on nutrients and energy stored in the mother spore. There were trails to ascertain if inhibitors were involved in the suppression of independent growth. Their idea was to chelate the suspected auto-inhibitors and allow for an uninterrrupted growth of germ tubes and hyphae, if possible. Watrud et al. (1978) examined the effect of different chelators on germination and hyphal growth. They reported an extended hyphal growth that subsided later. Hepper (1983) proved that certain amino acid combinations could support AM fungal growth a little better than minimal medium, but once again, independent growth halted after a while and the life cycle was not completed. Recently, Hildbrandt et al. (2002) investigated the co-cultures of AM fungus with slime-forming bacteria *Paenibacillus validus*. They found that hyphae of *G. intraradices* grew, branched profusely and formed certain coiled structures. Such hyphal coils were comparable in structure to arbuscules, but a bit more densely coiled. Still, they did not detect the formation of the next generation of spores, wherein the daughter spores would have germinated normally and continued the life cycle. Biochemical factors that stimulate extended growth of hyphae are yet to be identified. Organ culture of host plants has been effectively used to produce axenic AM fungi. Root explants and hairy root cultures derived via transformation with *Agrobacterium* were utilized frequently (Balaji et al. 1995; Becard and Fortin, 1988; Declerck et al. 1996, 1998; Diop et al. 1992; Harrison, 1999; Mugnier and Mosse, 1987; Pawlowska et al. 1999; Pfeffer et al. 1999). In this case, either a host or its differentiated organ supports the growth of AM fungus. There are indeed several other procedures that allow production of axenic cultures of AM fungi. For example, sterilized pot cultures, hydroponic and aeroponic cultures (Jarstfer and Sylvia, 1992).

Several authors have opined that the inability to culture AM fungi in vitro on simple laboratory medium has been a major drawback in our quest to decipher their genetic and molecular aspects. On the other hand, progress in molecular aspects of fungi in general has been the phenomenal. A variety of molecular techniques have been standardized and utilized to understand physiological processes and manipulate the fungi. In many cases, genes/loci that contribute to rapid growth rate or ability to utilize a wide range of carbon and other nutrient sources have been identified. Obviously, the question here is: why not use molecular gene transfer techniques to impart AM fungi with ability to grow in vitro? Earliest of the efforts and suggestions to apply molecular approaches involving gene transfer to obtain in vitro culture of AM fungus without host support was made by Krishna (1995). It was intended to transform hyphae/protoplasts of AM fungi with genes for prolific growth on a simple minimal medium such as potato dextrose medium. This suggestion relied on arguments that several closely related fungi belonging to Zygomycetes and Oomycetes possessed genes relevant for luxuriant growth in vitro without host support. Definitely, among the various research approaches adopted, such molecular options need due consideration. Firstly, there is need to understand the location of exact genetic lesion(s) or molecular inability in AM fungi that hampers their independent growth in vitro in a laboratory medium. Donor fungi with corresponding genes and techniques that allow their transfer and integration into specific strains of AM fungi are other minimum requirements. Among different sources of genes for rapid growth, *Phycomyces blakesleeanus* was suggested since it is a closely related coenocytic fungus whose genetic nature and molecular aspects were comparatively better studied (Binding and Weber, 1974; Cerda-Olmeda 1977, 1987; Cerda-Olmeda and Lipson, 1987; Eslava et al. 1975). Standardized methods for isolation of protoplasts and their fusion are available (Suarez et al. 1987). Growth of *P. blakesleeanus* is prolific on simple growth media such as potato dextrose agar. Specific genes that aid independent growth of AM fungi may also be available in other fungi which are less closely related, such as *Neurospora* spp, *Penicilium* spp. or other organisms. At present such molecular approaches are in their initial stages and need careful attention. We know that several plant hosts support a luxuriant growth of AM fungi that exist in symbiotic state. There is every reason to search plant cells for relevant gene(s) that impart AM fungi with ability for independent growth. We need to know the exact biochemical lesion in AM fungi that is being overcome/supplemented by host roots in nature. Only then can we search and clone specific plant genes that aid in vitro growth of AM fungi.

Concluding Remarks

In summary, molecular techniques to transform AM and ECM fungi are available. In due course of time, greater refinement and better yield of stable transformants can be expected. Widely different and useful genes could be introduced into AM/ECM fungi. Indeed, innumerable opportunities are available to utilize transgenics and improve symbiotic efficiency. While releasing transgenic crops/mycorrhizal fungi, greater caution is needed, especially regarding their influence on non-target beneficial/detrimental soil microflora. If transgenic plants involve antifungal traits, then information on gene flow and influence on mycorrhizal community in soil is required. With regard to AM fungi, we may attempt to culture them in vitro by transforming them with necessary genes from other fungi or host plants. Overall, molecular approaches seem to promise us with the greatest number of opportunities to improve benefits of mycorrhizal phenomenon.

References

Agerer, R. 1991. Characterization of Ectomycorrhiza. In: Techniques for the Study of Mycorrhiza. Norris, J.R., Read, D.J. and Varma, A.K. (Eds) Academic Press, London, pp. 50-51.

Azevedo, J.L. and Araujo, W.L. 2003. Genetically Modified Crops: environmental and human health concerns. Mutation Research—Reviews in Mutation Research 54: 223-233.

Bailey, A.N., Mena, G.L. and Herrera-Estrella, L. 1993. Transformation of four pathogenic *Phytophthora* species by Microprojectile bombardment on intact mycelium. Current Genetics 23: 42-46.

Balaji, B., Poulin, M.J., Vierheilig, H., Piche, Y. and Azcon-Aguilar, C. 1995. Responses of an arbuscular mycorrhizal fungus *Gigaspora margarita*, to exudates and volatiles from Ri T-DNA transformed roots of non-mycorrhizal mutants of *Pisum sativum*. Experimental Mycology 19: 275-283.

Bardi, L., Perotto, S. and Bonfonte, P. 1998. Protoplasts from *Oidiodendron maius* strains: isolation, regeneration and sensitiveness to heavy metals. http://ww.icom2.slu.se/abstracts/bardi.html

Bardi, L., Perotto, S. and Bonfonte, P. 1999. Isolation and regeneration of protoplasts from two stains of the ericoid mycorrhizal fungus *Oidiodendron maius*: Sensitivity to chemicals and heavy metals. Microbial Research 154: 105-111.

Barret, V., Lemke, P.A. and Dixon, R.K. 1989. Protoplast formation from selected species of ectomycorrhizal fungi. Applied Microbiology Biotechnology 30: 381-387.

BCERN, 2003. British Columbia Ectomycorrhizal Research Network. BCERN-CDE: Index of Descriptions. File://A.\BCERN-CDE.html. pp. 1-3.

Becard, G. and Fortin, J.A. 1988. Early events of Vesicular-Arbuscular mycorrhiza formation on Ri T-DNA transformed roots. New Phytologist 133: 273-280.

Benjdia, M. 2003. Genetic transformation of ectomycorrhizal *Hebeloma cylindrosporum*-mediated by *Agrobacterium tumefaciens*. http:/www.uni-tuebingen.de/plantphys/hebeloma/index.html

Bills, S.N., Podilla, G.K. and Hiremath, S. 1998. Genetic engineering of the ectomycorrhizal fungus *Laccaria bicolor* for use as a biological control agent. Mycologia 91: 237-242.

Bills, S.N., Richter, D.L. and Podila, G.K. 1995. Genetic transformation of the ectomycorrhizal fungus *Paxillus involutus* by Particle Bombardment. Mycological Research 99: 237-242.

Binding, H. and Weber, H.J. 1974. The Isolation, regeneration and fusion of *Phycomyces* protoplasts. Molecular and General Genetics 135: 273-276.

Buscot, F., Kaldorf, M., Fladung, M. and Muhs, J.H. 2001. Establishment of mycorrhizas on rolC-transgenic Aspen in a field trial. Fourth International Conference on Mycorrhizas, Adelaide, Australia, http://mycorrhizas.ag.utk.edu

Chaure, P., Gurr, S.J. and Spanu, P. 2000. Stable transformation of *Erisyphe graminis*, an obligate biotrophic pathogen of barley. Nature Biotechnology 18: 205-207.

Cerda-Olmeda, E. 1977. Biochemical Genetics of *Phycomyces blakesleeanus*. Annual Review of Microbiology 31: 535-547.

Cerda-Olmeda, E. 1987. Standard growth conditions and variations. In: *Phycomyces*, Cerda-Olmeda, E. and Lipson, E.D., Cold Spring Harbor, New York, pp. 337-339.

Cerda-Olmeda, E. and Lipson, E.D. 1987. A Biography of *Phycomyces*. Cold Spring Harbor, New York, pp. 7-26.

Combier, J., Melayah, D., Marmiese, R. and Gay, G. 2002. Transformation du Champignon ectomycorhizien *Hebeloma cylindrosporum* per *Agrobacterium tumefaciens*. Phytopathologie conference, Resume poster Interaction.htm, CIRAD, Assols, Savok, France, pp. 1-7.

Combier, J.P., Melayeh, D., Reffier, C., Gay, G. and Marmeisse, R. 2003. *Agrobacterium tumefaciens*-mediated transformation as a tool for insertional mutagenesis in the symbiotic fungus *Hebeloma cylindrosporum*. FEMS Mikrobiology Letters 220: 141-148.

Comis, D. 2002. Glomalin: Hiding place for a third of the world's stored carbon. Agricultural Research 50: 4-7.

Cummins, J. 2002. Transgenic trees may cause problems to mycorrhiza. Sustainable Agricultural Network Discussion Group. University of Florida, Gainesville, Florida, USA, sanet-mg@lists.ifs.ufl.edu

Declerck, S., Strullu, D.G. and Planchette, C. 1996. In vitro mass-production of the vesicular arbuscular mycorrhizal fungus *Glomus versiforme*, associated with Ri T-DNA transformed carrot roots. Mycological Research 104: 293-300.

Declerck,S., Strullu,D.G. and Planchette, C. 1998. Monoxenic culture of the intraradical forms of *Glomus* sp. isolated from a tropical ecosystem: a proposed methodology for germplasm collection. Mycologia 90: 579-585.

DeGroot, M.J., Bundoch, P., Hooykuss, P.J., Beijerbergen, A.G. 1998. *Agrobacterium tumefaciens*-mediated transformation of filamentous fungi. National Biotechnology 16: 839-842.

Delmas, S.N., Magnier, A. and Becard, G. 1998. Hyphal fusion and intersporal crossing in arbuscular mycorrhizal fungi. Second International Conference on Mycorrhizas, Uppsala, Sweden, http:www.mycorrhiza.ag.utk.edu.html

Diop, T.A., Becard, G. and Piche, Y. 1992. Long-term in vitro culture of endomycorrhizal fungus *Gigaspora margarita* on Ri T-DNA transformed roots of carrot. Symbiosis 12: 249-259.

Elliot, C.G.1994. Reproduction in fungi: Genetical and Physiological Aspects. Chapman and Hall, London, U.K., pp. 310.

Eslava, A.P., Alvarez, M., Burke, P.V. and Delbruck, M. 1975. Genetic recombination between mutants of *Phycomyces*. Genetics 80: 445-462.

Fincham, J.R.S. 1989. Transformation in fungi. Microbiological Reviews 53: 148-170.

Fladung, M., Kaldorf, M., Muhs, J.H. and Buscot, F. 1999. Mycorrhizal status of Transgenic *Populus* in a field trial. Forest Biotech Posters. http:// www.dainet.de/bfh

Forbes, P.J., Millam, S., Hooker, J.E. and Harrier, L.A. 1998. Transformation of the arbuscular mycorrhizal fungus *Gigaspora rosea* using particle bombardment. Mycological Research 102: 497-501.

Giovannetti, M., Sbrana, C., Strani, P., Agnolucci, M., Rinaudo, V. and Avio, L. 2003. Genetic diversity of isolates of *Glomus mosseae* from different geographic areas detected by vegetative compatibility testing and biochemical and molecular analysis. Applied and Environmental Microbiology 69: 616-624.

Hampp, R., Schaeffer, C., Wallenda, T., Stulten, C., Johann, P. and Einig, W. 1995. Changes in carbon partitioning or allocation due to ectomycorrhiza formation: Biochemical evidence. Canadian Journal of Botany 73 (supplement 1): S548-S556.

Hanif, M., Pardo, A.G., Gorfer, M. and Raudoskoski, M. 2002. T-DNA transfer and interaction in the ectomycorrhizal fungus *Suillus bovinus* using Hygromycin B as a selectable marker. Current Genetics 41: 183-188.

Harrier, L. A. and Millam, S. 2001. Biolistic transformation of Arbuscular Mycorrhizal fungi: Progress and Perspectives. Molecular Biotechnology 18: 25-33.

Harrier, L. A., Millam, S. and Franken, P. 2002. Biolistic transformation of Arbuscular mycorrhizal fungi: Advances and Applications. In. Mycorrhizal Technology in Agriculture. Gianinazzi, S., Schuepp, H., Barea, J.M. and Haselwandter, K. (Eds) Birkhauser Verlag, Switzerland, pp. 59-70.

Harrison, M. J.1999. Molecular and cellular aspects of the Arbuscular Mycorrhizal symbiosis. Annual Review of Plant Pathology and Plant Molecular Biology 50: 361-389.

Hepper, 1983. Limited independent growth of Vesicular Arbuscular Mycorrhizal fungus *in vitro*. New Phytologist 93: 537-542.

Hildbrandt, U., Janetta, K. and Bothe, H. 2002. Towards growth of Arbuscular Mycorrhizal fungi independent of a host plant. Applied and Environmental Microbiology 68: 1919-1924.

Institute of Plant Sciences 2000. Field test summary of crop genetic diversity group. Institute of Plant Sciences, Zurich, Switzerland, pp. 3.

Jarstfer, A.G. and Sylvia, D.M. 1992. Inoculum production and inoculation technologies of Vesicular Arbuscular mycorrhizal fungi. In: Soil Technologies: Applications in Agriculture, Forestry and Environmental Management. Metting, B. (Ed.). Marcel Dekker Inc., New York, pp. 349-377.

Kaldorf, M., Fladung, M. and Muhs, H.J. 2002. Mycorrhizal colonization of transgenic aspen in a field trial. Planta 214: 653-660.

Krishna, K.R., Shetty, K.G., Dart, P.J. and Andrews, D.J. 1984. Genotype dependent variation in mycorrhizal colonization and response to inoculation of Pearl Millet. Plant and Soil 86: 113-125.

Krishna, K.R.1995. Towards culturing Arbuscular Mycorrhiza In vitro: A molecular gene transfer method to grow AM fungi *in vitro.* Project report of Exchange Program No P 1 1285. Soil and Water Science Department, University of Florida, Gainesville, Florida, USA, pp. 1-14.

Kropp, B.R. 1997. Iinheritance of the ability for ectomycorrhizal colonization of *Pinus strobus* by *Laccaria bicolor.* Mycologia 89: 578-585.

Kumria, R., Verma, R. and Rajaram, M.V. 1998. Potential application of antisera RNA technology in plants. Current Science 74: 35-41.

Lamahamdi, M.S. ad Fortin, J.A. 1991. Genetic variation of ectomycorrhizal fungi: Extramatrical phase of *Pisolithus* species. Canadian Journal of Botany 69: 1927-1934.

Lobublio, K.F., Berbee, M.L. and Taylor, J.W. 1996. Phylogenetic origins of the asexual mycorrhizal symbiosis *Cenococcum geophilum* and other mycorrhizal fungi among the Ascomycetes. Phylogenetics and Evolution 6: 2870294.

Lobuglio, K.F. and Taylor, J.W. 2002. Recombination and genetic differentiation in mycorrhizal fungus *Cenococcum geophillum.* Mycologia 94: 777-780.

Lopez-Bucio, J., de la Vega, O.M., Guevara-Garcia, A., Herrara-Estrella. L. 2000. Enhanced phosphorus uptake I transgenic tobacco plants that overproduce citrate. Nature Biotechnology 18: 450-453.

Marmiesse, R., Gay, G., Debaud, J.G. and Casselton, L.A. 1992. Genetic transformation of the symbiotic Basidiomycete fungus *Hebeloma cylindrosporum.* Current Genetics 22: 41-45.

Martin, F., Tommerup, I.C. and Tagu, D. 1994. Genetics of Ectomycorrhizal fungi: Progress and Prospects. Plant and Soil 159: 170-181.

Miksoch, T.S., Levrijsien, B., Sonenberg, A.S., VanGriensven, L.J. 2001. Transformation of the cultivated mushroom *Agaricus bisporus* using TDNA from *Agrobacterium tumefaciens.* Current Genetics 39: 35-39.

Mugnier, J. and Mosse, B. 1987. Vesicular Arbuscular mycorrhizal infections in transformed root inducing T-DNA roots grown axenically. Phytopathology 77: 1045-1050.

Pardo, A.G., Hanif, M., Raudososki, M., Gerfor, M. 2002. Genetic transformation of ectomycorrhizal fungi mediated by *Agrobacterium tumefaciens.* Mycological Research 106: 132-137.

Parsad-Chinnery, S. 1997. Mycorrhizal Research Group Report. University of West Indies, West Indies. http: Users.Sunbeach.net/users /lec/mycrg.html

Pawlowska, T.E., Douds, D.D. and Charvat, I. 1999. In vitro propagation and life cycle of the arbuscular mycorrhizal fungus *Glomus etunicatum.* Mycological Research103: 1549-1556.

Pfeffer, P.E., Douds, D.D., Becard, G. and Silchar-Hill, Y. 1999. Carbon uptake and the metabolism and transport of lipids in arbuscular mycorrhiza. Plant Physiology 120: 587-598.

Pope, E.J., McGee, P.A. and Carter, D.A. 1998. Evolutionary relationships between symbiotic and pathogenic *Rhizoctonia solani.* Second International Conference on Mycorrhiza, Uppsala, Sweden, ICOM website.

Pringle, A., Moncalvo, J. and Vilgalys, R. 2000. High levels of variation in ribosomal DNA sequences within and among spores of a natural population of the arbuscular mycorrhizal fungus *Acaulospora colossica.* Mycologia 92: 259-268.

Pringle, A. 2001. Ecology and Genetics of Arbuscular Mycorrhizal Fungi. Ph.D dissertation. Duke University, Durham, North Carolina, pp. 132.

Requena, N. 2003. Transformation of the arbuscular mycorrhizal fungus *Glomus intraradices* with *Agrobacterium tumefaciens.* Workshop on Transformation of Mycorrhiza. Universitat Beilefeld, Germany, p. 10.

Requena, N. and Levya, A. 2003.*Agrobacterium*-mediated transformation of arbuscular mycorrhizal fungi. http://www.Uni-tuebingen.de/uni/bp/requena

RISO, 2000. Annual Report 2000. Plant Research Department, Copenhagen, Denmark, A:/ Annual report 2000-Plant Microbe Symbiosis.html

Rosado, S.C.S., Kropp, B.R. and Piche, Y. 1994b. Genetics of Ectomycorrhizal Symbiosis. 2. Fungal variability and heritability of ectomycorrhizal traits. New Phytologist 126: 111-117.

Saito, T., Tanaka, N. and Shinozawa, T. 2001. A transformation system for an ectomycorrhizal basidiomycete, *Lyophillum shimeji.* Bioscience Biotechnology and Biochemistry 65: 1928-1931.

Stulten, C.H., Kong, F.X. and Hampp, R. 1995. Isolation and regeneration of protoplasts from the Ectomycorrhizal ascomycete *Cenococcum geophilum.* Mycorrhiza 5: 259-266.

Shmaefsky, B. 2003. Glomalin: A future area of transgenic plant research. ISB News report: Agriculture and Environment. Virginia Polytechnic University, Virginia, http://www.isb.vt.edu/brargmeeting.htm. pp. 11-14.

Suarez, T., Orejas, M., Arnau, J. and Torres-Martinez, S. 1987. Protoplast formation and fusion. In: *Phycomyces.* Cerda-Olmeda, E. and Lipson, E.D. (Eds). Cold Spring Harbor, New York, USA, pp. 351-353.

Talbot, N. 2002. Molecular and Cellular Biology of Filamentous Fungi-Practical Approach. Oxford University Press, Oxford, England, pp. 267.

TERI 2002. Tata Energy Research Institute Report. http://dbtindia.nic.in/raid/biofertilizer.html

Valeria, B., Martini, I. and Bonfonte, P. 1998. Arbuscular mycorrhizal interactions with transgenic plants expressing anti-fungal proteins. Second International Conference on Mycorrhizas, Uppsala, Sweden. http://www.mycorrhizas.ag.utk.edu

Van de Rhee, M., Graca, P.M.A., Huizig, H.J. and Mooibroek, H. 1996. Transformation of the cultivated mushroom, *Agaricus bisporus* to hygromycin B resistance. Molecular Genetics and Genomics 250: 252-258.

Watrud, L.S., Heithaus, J.J. and Jaworski, E.G. 1978. Evidence for production of inhibitor by the vesicular arbuscular mycorrhizal fungus *Gigaspora margarita.* Mycologia 70: 821-828.

Wood, K.V. 1995. Marker protein for expression. Current Opinion in Biotechnology 6: 50-58.

8

EPILOGUE

Mycorrhiza is a worldwide symbiotic phenomenon, encompassing nearly 80 to 90% of plant species and several fungi. It extends into agricultural belts, forests, natural vegetation and several extreme climates such as arid, semi-arid, saline and boreal zones. Mycorrhizal symbiosis represents a 400-million-year old co-evolution of plants and specialized fungi. Bryophytes of Devonian period such as *Aglaophyton* were the earliest of land plants to harbor vesicular arbuscular fungi-like endophytes. Ectomycorrhizas that possess a mantle and a hartig net evolved on conifers, much later, around 130 to 180 million years ago. Molecular typing suggests that ericoid mycorrhizas evolved from AM-like ancestors. Fossil records of ericoid mycorrhizas prove their existence at least 80 million years ago. Brundrett (2002) opines that the evolutionary sequence of mycorrhizas begins with AM-forming ancestral plants, progressing to ECM, then on to arbutoids and finally into Ericoid mycorrhizas. Mycoheterotrophic monotropid plants could be placed next in this evolutionary trend.

Primordial environment that aided the formation of different types of mycorrhizas is crucial, but as yet unknown. There is lack of knowledge about natural forces and their subtle variations that drive the evolution of plants and fungi towards symbiosis and not pathogenesis. For example, nutrient dearth in the environment, host plant with inefficient roots, and fungi-lacking versatility in metabolic process may have aided evolution of mycorrhizas. During the past 400 million years, mycorrhizal symbionts might have experienced both severe and milder fluctuations in the congenial environment. These environmental fluctuations might have hastened or retarded or modified the direction and rates of mycorrhizal evolution.

ECM and Ericoid mycorrhizas show specificity with regard to symbiont pairs, whereas, specificity is not a significant aspect among arbuscular mycorrhizas. In nature, there are also plant species and fungi that never display any tendency to form mycorrhizas (e.g. Cruciferae). There are no examples of plants or fungi that have shown a reversal of the evolutionary trends from mycorrhizal to non-mycorrhizal status. It will be interesting to study the factors that might hasten plant species to lose their mycorrhizal ability. Perhaps a series of mutations at regular intervals that affects early signaling, entry and establishment steps will naturally lead to evolution of a non-mycorrrhizal version. Molecular analysis of plants/fungi that might have alternated between being mycorrhizal and non-mycorrhizal—if at all they exist—will be useful. Molecular evolution of certain host species that have co-evolved with both ECM and AM fungi need to be investigated.

Classical taxonomy of mycorrhizal fungi that relies heavily on morphotyping of spores and other fungal structures has been in vogue for a few decades. Phylogenetic keys for AM, ECM and Ericoid fungi have been periodically updated. Morphotying can be helpful in broad identification and restricting the number of samples that need detailed molecular analysis. Application of molecular techniques based on SSU rRNA sequence, PCR-RFLP and micro-satellite markers has yielded useful phenograms. Molecular analysis has also led to several taxonomic revisions of fungal genera and species. For example, in Redecker's (2002) classification, two new AM fungal families Archaeosporaceae and Paraglomaceae are added. Similarly, in case of an ECM fungus *Rhizopogon*, Grubisha et al. (2001) have meticulously re-examined and revized the phylogeny of this genus using ITS sequence. Previous taxonomic versions relied much on morphotyping. This trend—to revize classification, introduce or club fungal species—may continue, depending on the discovery of new fungal species and sharper classification tools. Generally, taxonomic positions obtained using molecular techniques seem to corroborate with those derived from morphotyping, but there are several exceptions. Use of both morphotyping and molecular analysis is advizable so as to arrive at accurate taxonomic positions.

Arbuscular Mycorrhizas: Developmental Physiology and Metabolic Aspects

Physiological aspects of symbionts are vital to the development and sustenance of mycorrhizal symbiosis. During the past decade, considerable knowledge has been accrued about the cytological and molecular changes related to regulation of mycorrhizal symbiosis. Yet there are innumerable physiological processes that require detailed investigation. The developmental physiology of AM symbiosis begins with dormant spores. Dormancy period is variable, depending on the availability stimuli and congenial environment. Extension of germ tubes and pre-symbiotic mycelium depend on the energy stored in spores. It is a crucial phase lasting 24 to 48 h, until the pre-symbiotic mycelium encounters a congenial host root. Formation of appresoria, infection and establishment of fungus inside the root cortex requires another 48 to 72 h. Hectic cytological and physiological changes occur inside fungal spores, mycelium and root cells. Franken et al. (2002) have argued that pre-symbiotic mycelium should possess a full complement of active genes required for growth and extension. Obviously, metabolic processes that sustain a limited growth fungal growth include uptake and utilization of carbohydrates. Several other metabolic processes such as cell cycle events, chromatin duplication, respiratory activity and molecular signals that occur in pre-symbiotic mycelium and during its transition to symbiotic phase are being intensively investigated. Molecular physiology of host and fungus during formation of appresoria, invasion into cortical tissue and establishment of AM fungus inside the roots has been delineated excellently using a series of host mutants (Marsh and Schultz, 2001).

Carbohydrate physiology of extraradical mycelium involves tricarboxylic acid cycle, glycolysis and pentose phosphate pathway. These metabolic pathways create sufficient carbon fluxes into fungal symbiont. Sugars are major carbon transport moieties. Invertases occurring in symplastic interface regulate the glucose availability to fungus. AM fungi are known to accumulate excess C as trehalose and glycogen. Reverse flow of carbon compounds has also been reported, especially in mycoheterotrophic plants. Carbon transport from intra- to extraradical mycelium and vice versa occurs as tryacylglycerides (Bago et al. 2002). Lipids are also energy storage substances in extraradical mycelium. Molecular regulation of lipid biosynthesis and degradation in response to C status are being investigated.

Investigations on nitrogen metabolism and transport in AM are relatively few. Tests utilizing ^{15}N stable isotope have proved that AM fungal hyphae explore and translocate substantial quantities of N from soil. The fraction of N transport mediated by AM fungi is unknown. It may be negligible, moderate or significant, when compared to N translocated by roots. The energy expended on N transport through AM fungi and root needs to be compared for efficiency. The rate of N translocation is dependent on genetic aspects of AM fungi, such as hyphal spread and rate of N transport. AM fungi prefer NO_3-N, while NH_4-N suppresses hyphal growth and vesicle formation. Glutamate dehydrogenase, glutamate synthase and nitrate reductase are the key regulators of nitrogen metabolism in AM fungi. Some aspects of secondary metabolism have been investigated. Especially, biochemical nature and accumulation pattern of yellow pigment 'mycorradicin', flavonoids, sterols and auxins are being researched.

Mycorrhizal activity in soil is closely related with phosphorus exploitation and translocation to root cortex. Plant roots and most free-living soil microbes are endowed with the ability to hydrolyze organic P. In case of AM fungi, extraradical hyphae release phosphatases that hydrolyze organic-P. Regulatory aspects of genes coding extracellular phosphatase need detailed investigation. Quantity of organic-P hydrolyzed and made available through extracellular hyphae may not be significant. In fact, Joner et al. (2000) opine that the release of phosphatases is more a routine biological manifestation in many microbes, including mycorrhizal fungi. It is generally accepted that mycorrhizae exploit the available sources of P in soil.

Phosphorus acquisition occurs simultaneously via plant roots and mycorrhizal mycelium. Phosphate transporters coded by plant and fungus, both mediate P transfer. Knowledge about the location of their secretion and pattern of gene expression is important. Several P transporter genes have been characterized (Burleigh and Bechmann, 2002). For example, plant-coded P transporters such as LePT1 or PHT1 are active in roots and mediate energy-driven P transfer from soil solution to root cells. Whereas, P transporters such as StPT3 are upregulated in arbuscular cells and mediate P transfer at biotrophic symplast in arbuscular cells. Proportion of plant and fungal coded P transporters active at any time may be dependent on the P availability (see Fig. 4.4). For example, P scarcity upregulates high affinity fungal transporters, whereas, high P induces root-coded P transporters. The energy-driven P transport in mycorrhizal systems is facilitated by a set of isozymes of ATPases. Genes for several of these ATPase isozymes have been studied. Requena et al. (2003) point out that the pattern of gene expression corresponding to various ATPase isozymes in relation to stage of symbiosis seems to be important.

Cytological and molecular evidences support the view that tubular vacuoles mediate long-distance P transport along the fungal hyphae (Ashford and Allaway, 2002). Vacuoles act as a subcellular continuum of P movement. P-bearing compounds such as polyphosphate are frequently transported via bulk flow along the length of hyphae. Ashford (2002) has pointed out several aspects related to vacuole function that needs greater attention by researchers. They are: (a) Proportion of vacuolar P transport in relation to other modes, if any; (b) P movement through vacuoles could be bi-directional; (c) role of vacuolar P transport in fungal pathogenesis in comparison to mycorrhizas; and (d) influence of vacuolar P transport on symbiotic efficiency.

Genetics and Molecular Regulation of Arbuscular Mycorrhizas

Host-genetic constitution influences AM symbiosis, especially parameters like intensity of colonization, fungal spread in the soil, P uptake and growth stimulation. Genetic variation available for P nutrition can be exploited advantageously by applying appropriate breeding techniques on host plant species (Krishna, 1997; Manske, 2000; Thingstrup et al. 2000). Inheritance studies indicate that additive gene action and sometimes heterosis are operative on mycorrhizal traits. Genetic nature of fungal partner also influences symbiosis. We have to learn more about genetic crossing and transfer of traits in mycorrhizal fungi (Bever and Morton, 1999). Fungal isolates differ in their ability to colonize roots and exploit P available in the soil. Hence, selection for improved symbiotic performance seems appropriate for the present. Inability to culture AM fungi in vitro, lack of knowledge regarding their sexual phase and genetic crossing over events means that fungal-crossing techniques cannot be applied yet to achieve improvement in fungal strains.

Details on molecular regulation of AM symbiosis, especially aspects related to early signaling, root penetration, fungal establishment and nutrient exchange, are being rapidly accumulated (Franken, 1999; Franken et al. 2002; Podila, 2002; Stommel et al. 2001; Sundaram et al. 2004; VanBuuren et al. 1999). AM symbiosis involves meticulous regulation of innumerable genes, but only a few have been sequenced, cloned and studied. It is difficult to guess, but several more genes relevant to AM symbiosis need to be studied. Recently, Voiblet et al. (2002) have proposed a concept called 'Symbiosis Regulated genes'. It pertains specifically to genes that are induced/ repressed in response to symbiosis. Identification of SR gene is based on the extent of induction/repression. It helps researchers to focus on only a few genes. Induction and/or repression of SR genes occur at different stages of symbiosis such as presymbiosis, early symbiosis or senescence. They may selectively affect signaling, nutrient exchange phenomenon, primary and secondary metabolism. These SR genes could belong to either host or fungus.

In legumes, molecular regulations relate to a tripartite association. So far, several genes specific to AM and Rhizobium symbiosis have been characterized in *Medicago* and lotus (Bastel-Corre et al. 2000; Bonfonte et al. 2000; Provorov et al. 2002). Interestingly, regulation of certain genes is common to microsymbionts, AM fungus and Rhizobium.

Ectomycorrhizas: Physiology, Genetics and Molecular Biology

Rhizosphere signals initiate a series of highly programmed morphogenetic and physiological events in symbionts, leading to the development of ECM symbiosis (Martin et al. 2001). Early signaling may be complex. Flavonoids, diterpenes, rutin, harmones (zeatin) and nutrients can serve as signal substances and induce morphogenetic changes in host roots. Hyphaporine accumulates immediately after initial contact and induces cytoskeletal changes in root hairs. It may also influence ion fluxes and nutrient exchange (Theiry et al. 1998). Modifications in gene expression in response to hypaphorine accumulation need investigation. Auxins regulate root morphogenesis and ECM development. They modify the growth polarity of cortical cells and initiate short root formation. Auxin production may not induce cytoskeletal changes, nor affect the polysaccharide accumulation. Auxin effects are definitely mediated by changes in gene expression during early mycorrhization steps.

ECM fungi are capable of utilizing different organic sources supplied through a synthetic medium. Under natural conditions, root exudates, host cell wall components and rhizosphere microbes are good carbohydrate sources. It is argued that ECM fungi ordinarily consume about 30% of photosynthates. Hence, C derived from the above sources may not be significant. In plants, sucrose is the major photo-assimilate transported through phloem and unloaded into apoplast. However, most ECM fungi lack the ability to utilize sucrose. Hence, molecular regulation of invertases at the apoplast decides the carbon supply to fungal symbiont. Glucose and fructose are utilized by ECM fungi, but excess carbohydrate gets accumulated as trehalose and mannitol. Monosaccharide transport system in the ECM fungus mediates hexose transport at the apoplastic surface. Monosaccheride transporters are upregulated in the presence of hexoses. In case of *Amanita muscaria*, 18 to 24 h were needed to perceive a specific shift in gene expression (*AMST1*) from constitutive to inducible state. In nature, extraradical mycelium of the ECM fungus encounters relatively low carbohydrate concentration in soil, whereas, at apoplast, host cells are richer in carbohydrates. ECM fungus is known to adapt to such variable carbon fluxes by converting C into trehalose, mannitol or glycogen. A few sugar transporters from ECM fungi have been investigated, but several others are yet to be characterized and their regulation understood.

ECM fungi are endowed with NO_3 uptake apparatus that is inducible linearly with external supply of NO_3. Presence of NH_4 or glutamate regulates NO_3 uptake in tree roots. ECM fungi actually respond to NH_4/NO_3 ratios available in soils. Mostly, they accumulate excessive NH_4 in root cytosol, but efficiency of fluxes varies. ECM fungi discriminate N forms and isotopes. Such isotope preferences can affect the ^{15}N pattern in mycelium and accuracy of measurements about NH_4 accumulation and assimilation. Organic N sources such as peptides, proteins and amino acids are utilized by ECM fungi through the action of proteolytic enzymes utilize. Regarding NH_4 assimilation, glutamate dehydrogenase, glutamine synthase and glutamate synthase are important enzymes that regulate N metabolism in ECM. Glutamine and glutamate are the preferred N sources for fungal growth.

Molecular aspects of ammonium transporter gene families from several microbes, including ECM fungi, are well understood (Chalot et al. 2002; Javelle et al. 2003 a,b). Activity of ammonium transporters in extraradical mycelium aids N uptake from soil. These ammonium transporters mediate an energy-coupled N movement across apoplast, but their regulatory aspects are still to be deciphered in detail. Expression of ammonium transporter genes is regulated by fluctuations in N supply. In case of *Tuber borchii*, ammonium transporters (TbAMT1) sense both NH_4 and NO_3 concentration in the medium and aid their transport appropriately. Hence, TbAMT1 is said to be crucial for N uptake by hyphae proliferating in the soil. Nitrate assimilation by ECM fungi is again substrate regulated. Nitrate reductase genes ECM fungi have been cloned and characterized. The presence of inorganic NO_3 or organic-N induces nitrate gene cluster, but the presence of NH_4 represses them, depending on the availability of photo-assimilates.

Classical genetics of ECM fungi is a relatively better studied aspect. Details on the sexual cycle and genetic recombination and gene action are available for major ECM fungi (Diez et al. 2001; Kropp, 1997; Lobuglio and Taylor, 2002). Ectomycorrhizal symbionts show a vast variability that can be exploited to enhance P nutrition and tree growth. A few common ECM fungi such as *Pisolithus, Laccaria, Rhizopogon, Suillus* and *Cenococcum* have been examined specifically for variation in nutrient absorption traits. Selective genetic crossing is also possible with many ECM fungi, so that mycelial and nutrient acquisition traits could be improved. Improvement of host tree for mycorrhizal traits is also possible, despite being tedious and time consuming. Of course, cell and tissue cultures may hasten the seedling selection and advancement.

Remarkable progress has occurred on the identification and characterization of 'Symbiosis Regulate' genes in ECM (Franken, 1999; Tagu and Martin, 1995; Vioblet et al. 2001). Genes induced or repressed by at least 2.5 fold in symbiotic state, when compared either of the individual hosts, are being considered SR gene. Certain SR genes are stimulated ten folds over an individual host or fungus. SR genes are induced/repressed at all stages; namely pre-symbiosis, early establishment, maintenance and senescence. They may pertain to a variety of symbiotic functions such as signaling, cellular structure, metabolic and housekeeping functions, and nutrient exchange at the apoplastic interface. SR genes may be host or fungus coded. Sometimes, sequential or simultaneous alteration of host and fungal SR genes may be required to accomplish a metabolic step. There are SR genes that supplement or rectify genetic lesion in the other partner. In view of the above variations, firstly, SR genes should be classified appropriately (see Fig. 3.7). So far, very few SR genes have been analyzed in detail. Understanding the molecular regulation of SR genes will become proportionately intricate since expression of each SR gene varies with time and space. Martin et al. (2001) anticipates a maze of regulatory genes affecting the metabolic aspects of symbionts. In this regard, the concept of SR genes is useful because it restricts the focus and only a few selected genes could be intensively analyzed, e.g. genes for tubulin, SRAP, phenylalanine ammonia lyase, lignolysis. It is clear that SR genes are those affected conspicuously by mycorrhizal association. However, several other genes expressed routinely, but not accentuated, may also be playing a crucial role in sustaining the partnership. A mutation in one of these loci will surely enlighten us regarding its metabolic ramifications.

Molecular Ecology of Mycorrhizas

Host plant response to pathogenic and symbiotic fungi may be similar on several aspects. Resistance reactions that occur in response to mycorrhizal colonization such as influx of Ca^{+}, efflux of Cl^{-1} and K^{+}, synthesis of H_2O_2 are supposedly milder, transitory and fade away with time. Mycorrhizal fungi overcome such plants' defense reactions by adapting different strategies (Gianinazzi-Pearson and Gianinazzi, 2003; Salzer et al. 1997). For example, reducing pre-formed defense mechanism and becoming insensitive to plant' defense-related proteins, or attenuation of elicitor-induced defense reaction by hormones reduces host-rejection processes.

Chitinases are important biochemical indicators of different plant-microbe interactions. Genes for several chitinase isozymes relevant to pathogenesis as well as mycorrhizal symbiosis have been cloned and characterized (Salzer et al. 2000). Molecular regulation of genes for different classes/isozymes of chitinases induced during symbiosis is being deciphered accurately (see Table 5.4). Nutrient movement across biotrophic interface that develops during symbiosis or pathogenesis is a crucial aspect. Details on molecular regulation of the nutrient movement, involvement of H^{+}ATPase and influence of strains of pathogen/symbiont are being rapidly accumulated. Mycorrhizal fungi may also impart a 'bioprotective effect' against pathogenic fungi. The molecular basis for such specific bioprotective effect are known for many cases. Knowledge about subtle biochemical responses and their regulation, along with the ability to manipulate bioprotective reactions will find application in agricultural situations.

Molecular identification procedures and micro-satellite markers have been frequently used to understand the ecological aspects of mycorrhizas. In many cases, new inferences about the ecological manifestation of mycorrhizal fungi have corroborated those obtained via morphotyping. However, molecular analysis has provided new inferences and ideas regarding ecology of mycorrhizas. Details on population genetics, spatial and temporal variations, and succession of AM fungal communities obtained through molecular analyses are available. Molecular typing has clearly proved that major agricultural practices such as tillage, interculture, cropping pattern and genotypes immensely influence AM fungal ecology. Glomalin is a unique AM fungal product known to improve soil aggregation and carbon sequestration. Efforts to clone glomalin genes, characterize them and understand their regulatory aspects are in progress. Production of transgenic AM fungus having a greater ability to produce glomalin is also being explored.

In case of ECM fungi, knowledge about genet sizes, spatial and temporal distribution, succession, persistence and influence of forest regeneration practices have vastly improved due to application of molecular techniques (Bruns et al. 2001, 2002). We must realize that often, sporulation patterns above ground do not match with hyphal network and its spread below ground. It can severely affect the accuracy of inferences. Such errors are avoided during molecular analysis. Hence, it is believed that information gathered using molecular techniques might be useful during production of ECM inoculum and developing inoculation strategies. Nurse functions of ECM-bearing trees and development of a common mycorrhizal network underground are vital to maintain ecosystematic functions of symbionts (Weimken and Boller, 2002). Mycoheterotrophy involves non-photosynthetic epiparasitic plants that are concurrently connected to ectomycorrhizal plants via fungus. These 'epiparasitic plants' or mycorrhizal cheaters derive carbon from the surrounding photosynthetic plants through the ECM fungal network (Bruns et al. 2001). Molecular tests have shown that mycohetrotrophic plants exhibit high specificity. Mycorrhizal fungi interact with a variety of soil microorganisms. The end result could be synergistic, antagonistic, commensalistic, parasitic, etc. In nature, such interactions immensely influence ecosystematic functions of mycorrhizal fungi. Molecular techniques have also been applied to understand the ecological aspects of mycorrhizal fungi that colonize plants in saline soils, arid zones, low temperature boreal area, etc.

Mycorrhizas—The Future

Advances in taxonomy of mycorrhizal fungi will depend on inquisitiveness and enthusiasm to survey, identify and reclassify them. In future, we may overcome several lacunae in our knowledge about the physiology of symbionts and molecular regulation of symbiosis. We may discover more about SR genes and their relevance to symbiotic efficiency. Application of molecular techniques to enhance symbiotic efficiency will be a major preoccupation for some time. Attempts to culture AM fungi in vitro could be intensified using molecular approaches. Overall, it is necessary to understand molecular biology of mycorrhizas in greater detail and apply those facts to improve symbiotic performance.

References

Ashford, A.E. 2002. Tubular vacuoles in Arbuscular Mycorrhiza. New Phytologist 154: 545-547.

Ashford, A.E. and Allaway, W.G. 2002. The role of motile tubular vacuoles in mycorrhizal system. Plant and Soil 244: 177-187.

Bastel-Corre, G., Dumas Gaudet, E., Poinsot, V., Dieu, M., Dierick, J.F., Van Tueinen, D., Ramacle, J., Gianinazzi-Pearson, V. and Gianinazzi, S. 2000. Proteome analysis and identification of symbiosis-related proteins from *Medicago truncatula* by two-dimensional electrophoresis and mass spectrophotometry. Electrophoresis 23: 122-137.

Bago, B., Zipfel, W., Williams, R.M., Jun, J., Arreola, R., Lammers, P.H., Pfeffer, P.E. and Sachar-Hill, Y. 2002. Translocation and utilization of fungal storage lipids in the arbuscular mycorrhizas. Plant Physiology 128: 108-124.

Bever, J.D. and Burton, J. 1999. Heritable variation and mechanisms of inheritance of spore shape within a population of *Scutellospora pellucida*, an arbuscular mycorrhizal fungus. American Journal of Botany 86: 1209-1216.

Bonfonte, P., Genre, A., Facio, A., Martini, I., Schauser, L., Stougard, J., Webb, J. and Parniske, M. 2000. The *Lotus japonicus LjSym 4* gene is required for the successful infection of root epidermal cells. Molecular Plant-Microbe Interactions 13: 1109-1120.

Brundrett, M.C. 2002. Co-evolution of roots and mycorrhizas of land plants. New Phytologist 154: 275-304.

Bruns, T.D., Lilleskov, E.A., Bidartondo, M.I. and Horton, T.R. 2001. Patterns in ectomycorrhizal community structure in pinaceous ecosystems. Phytopathology 91: S 163-164.

Bruns, T., Tan, J., Bidartondo, M., Szaro, T. and Redecker, D. 2002. Survival of *Suillus pungens* and *Amanita francheti* ectomycorrhizal genets was rare or absent after a stand—replacing fire. New Phytologist 155: 517-523.

Burleigh, S.H. and Bechmann, I. 2002. Plant nutrient transporter regulation in arbuscular mycorrhizas. Plant and Soil 244: 247-251.

Chalot, M., Javelle, A., Blaudez, D., Lambillote, R., Cooke, R., Sentenec, H., Wipf, D. and Botton, B. 2002. An update on nutrient transport processes in ectomycorrhizas. Plant and Soil 244: 165-175.

Diez, J., Anta, B., Manjon, J.L. and Honrubia, M. 2001. Genetic variability of *Pisolithus* isolates associated with native hosts and exotic eucalyptus in the western Mediterranean region. New Phytologist 149: 577-587.

Franken, P. 1999. Trends in molecular studies of AM fungi. In: Mycorrhiza. 2nd Edition. Varma, A. and Hock, B. (Eds). Springer Verlag, Berlin, Heidelberg, pp. 37-59.

Franken, P., Kuhn, G. and Gianinazzi-Pearson, V. 2002. Development and molecular biology of arbuscular mycorrhizal fungi. In: Molecular Biology of Fungal Development. Oseiwacz, H.D. (Ed) Marcel Dekker, New York, pp. 325-348.

Gianinazzi, V. and Gianinazzi, S. 2003. Modulation of defense responses and induced resistance by mycorrhizal fungi. UMR INRA, Universite de Bourgougne BCE-IPM, CMSE-INRA, Dijon Cedex, France, pp A:\Faro Abstracts.html pp. 1-21.

Grubisha, L.C., Trappe, J.M., Molina, R. and Spatofora, J.W. 2002. Biology of the ectomycorrhizal genus *Rhizopogon*. VI Re-examination of intrageneric relationships inferred from phylogentic analysis of ITS sequences. Mycologia 94: 607-619.

Javelle, A., Andre, B., Marini, A.M. and Chalot, M. 2003a. High-affinity ammonium transporters and nitrogen sensing in mycorrhizas. Trends in Microbiology 11: 1153-1159.

Javelle, A., Morel, M., RodriguezpPatrana, b., Botton, B., Amdre, b., Marini, A., Brun, A. and Chalot, M. 2003b. Molecular characterization, function and regulation of ammonium transporters (AMT) and ammonium metabolizing enzymes in the ectomycorrhizal fungus *Hebeloma cylindrosporum*. Molecular Ecology 47: 411-417.

Joner, E.J., Van Aarle, I.M. and Vosatka, M. 2000. Phosphatase activity of extraradical arbuscular mycorrhizal hyphae: A review. Plant and Soil 226: 199-210.

Krishna, K.R. 1997. Phosphorus efficiency of semi-dry land crops. In: Accomplishments and Future Research Challenges in Dry Land Soil Fertility Research in the Mediterranean Area. Ryan. (Ed). International Center for Agriculture Research in Dry Areas, Aleppo, Syria, pp. 343-363.

Kropp, B.R. 1997. Inheritance of the ability for ectomycorrhizal colonization of *Pinus strobus* by *Laccaria bicolor.* Mycologia 89: 578-585.

Lobuglio, K.F. and Taylor, J.W. 2002. Recombination and genetic differentiation in the mycorrhizal fungus *Cenococcum geophilum.* Mycologia 94: 777-780.

Manske, G.G.B. 1990. Genetical analysis of the efficiency of VA mycorrhiza with Spring Wheat 1. Genotype differences and a reciprocal cross between efficient and non-efficient variety. In: Genetic Aspects of Plant Mineral Nutrition. El-Bassam, N. (Ed.) Kluwer Academic Publishers, Dordrecht, Netherlands, pp 397-405.

Manske, G.G.B., Ortiz-Monasterio, J.I., Van Ginkel, M., Gonzalez, R. M., Rajaram, S., Molina, E. and Vlek, P.L.G. 2000. Traits associated with improved P-uptake efficiency in CIMMYT's Semi-Dwarf Spring Bread Wheat grown on acid Andisols in Mexico. Plant and Soil 221: 189-204.

Marsh, J.M. and Schultz, M. 2001. Analysis of arbuscular mycorrhiza using symbiosis-defective plant mutants. New Phytologist150: 525-532.

Martin, F., Duplessis, S., Ditengou, F., Lagrange, H., Voiblet, C. and Lapeyrei, F. 2001. Development cross talking in the ectomycorrhizal symbiosis: Signals and Communication genes. New Phytologist 151: 145-154.

Podila, G.K. 2002. Signaling in mycorrhizal symbiosis—Elegant mutants lead the way. New Phytologist 154: 541-551.

Provorov, N.A., Borisov, A.Y. and Tikanovich, I.A. 2002. Developmental genetics and evolution of symbiotic structures in nitrogen-fixing nodules and Arbuscular Mycorrhiza. Journal of Theoretical Biology 214: 215-232.

Redecker, D. 2002. Molecular identification and phylogeny of arbuscular mycorrhizal fungi. Plant and Soil 244: 67-73.

Requena, N., Brueninger, M., Franken, P. and Oeon, A. 2003. Symbiotic status, phosphate and sucrose regulate the expression of two plasma membrane H^+ ATPase genes from the mycorrhizal fungus *Glomus mosseae.* Plant Physiology 132: 1540-1549.

Salzer, P., Hubner, B., Sirrenberg, A. and Hager, A. 1997. Differential effect of purified spruce chitinases and B-1,3 glucanases on the activity of elicitors from ectomycorrhizal fungi. Plant Physiology 114: 957-968.

Salzer, P., Banonomi, A., Beyer, K., Vogeli-Lange, R., Aeschbacher, A., Lange, J., Wiemken, A. Kim, S., Cook, D.R. and Boller, T. 2000. Differential expression of eight chitinase genes in *Medicago truncatula* roots during mycorrhiza formation, nodulation and pathogen infection. Molecular Plant-Microbe Interaction 13: 763-777.

Stommel, M., Mann, P. and Fraken, P. 2001. EST-library construction using spore RNA of the arbuscular mycorrhizal fungus *Gigaspora rosea.* Mycorrhiza 10: 281-285.

Sundaram, S., Brand, J.H., Hymes, M.J., Hiremath, S., and Podila, G.K. 2004. Isolation and analysis of a symbiosis-regulated and RAS-interacting vesicular assembly protein gene from the ectomycorrhizal fungus *Laccaria bicolor.* New Phytologist 161: 529-538.

Tagu, D., Phthom, P., Cretin, C. and Martin, F. 1993. Cloning symbiosis-related cDNAs from *Eucalyptus* ECM by PCR-assisted differential screening. New Phytologist 125: 339-343.

Thiery, S., Huang, J. and Lapeyrei, F. 1998. Host plant stimulates the tryptophan, betaine hyphaporine accumulation in *Pisolithus tinctorius* hyphae during ectomycorrhizal infection. Fungal hyphaporine controls K^+ uptake, H^+ extrusion and root hair development. Second International Conference on Mycorrhizas, Uppsala, Sweden, http://www Mycorrhizas .ag.utk.edu.html

Thingstrup, I., Kahiluoto, H. and Jakobsen, I. 2000. Phosphate transport by hyphae of field communities of arbuscular mycorrhizal fungi at two levels of P fertilization. Plant and Soil 221: 181-187.

Van Buuren, M.L., Maldonanado-Mendoza, E., Treiu, A.T., Blaylock, L.A. and Harrison, M.J. 1999. Novel genes induced during an arbuscular mycorrhizal (AM) symbiosis formed between *Medicago truncatula* and *Glomus versiforme*. Molecular Plant-Microbe Interactions 12: 171-181.

Voiblet, C., Duplessies, S., Encelot, W. and Martin, F. 2001. Identification of symbiosis regulated genes in *Eucalyptus globulus*- *Pisolithus tinctorius* ectomycorrhiza by differential hybridization of arrayed cDNAs. The Plant Journal 25: 181-191.

Wiemken, V. and Boller T. 2002. Ectomycorrhiza: gene expression, metabolism and the worldwide web. Current Opinion in Plant Biology 5: 355-361.

INDEX